D1376778

DATE DUE

SPACE COMMUNICATION AND NUCLEAR SCINTILLATION

. . . .

SPACE COMMUNICATION AND NUCLEAR SCINTILLATION

Edited by

Nirode Mohanty

HUGHES AIRCRAFT COMPANY

VNR VAN NOSTRAND REINHOLD
_____ New York

Library of Congress Catalog Number 89-70502
ISBN 0-442-23696-4

Printed in the United States of America

Typeset in Hong Kong by Asco Trade Typesetting Ltd

Van Nostrand Reinhold
115 Fifth Avenue
New York, New York 10003

Chapman and Hall
2-6 Boundary Row
London, SE1 8HN, England

Thomas Nelson Australia
102 Dodds Street
South Melbourne 3205
Victoria, Australia

Nelson Canada
1120 Birchmount Road
Scarborough, Ontario M1K 5G4, Canada

16 15 14 13 12 11 10 9 8 7 6 5 4 3 2 1

Library of Congress Cataloging-in-Publication Data

Space communication and nuclear scintillation / edited by Nirode
   Mohanty.
      p.   cm.
   Includes bibliographical references.
   ISBN 0-442-23696-4
   1. Astronautics—Communication systems. 2. Radio waves-
Scintillation. 3. Electromagnetic pulse. 4. Space-based radar.
I. Mohanty, Nirode.
TL3025.N83   1991
621.382'54—dc20
                                                              90-12042
                                                                   CIP
                                                                  Rev.

Contributors

L. William Bradford
Mission Research Corporation
26382 Carmel Rancho Lane
Carmel, CA 93923-8732

John Choma, Jr.
Department of Electrical Engineering
University of Southern California
Los Angeles, CA 90089-0271

R. Dybdal
The Aerospace Corporation
P.O. 92957
Los Angeles, CA 90009

S. Bruce Franklin
TRW
One Space Park
Redondo Beach, CA 90278

Marvin Gantsweg
R&D Associates
P.O. 9695
4640 Admiralty Way
Marina del Rey, CA 90295

S. Hovanessian
The Aerospace Corporation
P.O. 92957
Los Angeles, CA 90009

Dennis L. Knepp
Mission Research Corporation
26382 Carmel Rancho Lane
Carmel, CA 93923-8732

P. Vijay Kumar
Department of Electrical Engineering
University of Southern California
Los Angeles, CA 90089-0272

Alfonso Malaga
Signatron Inc.
Lexington, MA 02173

Laurence B. Milstein
Electrical Engineering Department
University of California—San Diego
Box 109
La Jolla, CA 92093

Nirode Mohanty
Hughes Aircraft Company
2000 East El Segundo Boulevard
El Segundo, CA 90245

Charles L. Rino
Vista Research Inc.
P.O. 998
Mountain View, CA 94042

J. Gregory Rollins
Department of Electrical Engineering
University of Southern California
Los Angeles, CA 90089-0271

K. T. Woo
TRW
One Space Park
Redondo Beach, CA 90278

Preface

Signal propagation in the presence of nuclear-detonated propagation media is an extremely critical, complex, and challenging area of study. Communications, radar, and electrooptical signals may be severely attenuated in the presence of nuclear-induced scintillation. Semiconductor devices and satellite systems may be damaged by nuclear radiation. The survivability of communications, navigation, surveillance, weather satellite systems, and other space systems depends upon the extraction of weak, distorted signals. In 1986 a symposium was held at The Aerospace Corporation, Los Angeles, California, to discuss nuclear scintillation effects on communications and radar signals with a few selected speakers and invited audience.

The purpose of this book is to address these topics, with special emphasis on areas of interest important to communications and radar engineers and scientists, who should investigate these topics in much depth and rigor in the interest of mankind. The book provides an elementary introduction to communication and radar signal propagation in the presence of a nuclear-detonated disturbance. The book can be used as a text or reference in the areas of communication, radar, and propagation. The prerequisite for the book is an elementary knowledge in linear systems, probability and electronics.

The book begins with space communications. It deals with space systems, deterministic and random signals and noise, analog and digital communications, synchronization, interference analysis, earth stations, satellites, multiple-access communications, ionospheric propagation, nuclear scintillation and effects, and communication link performance. The second chapter is on error-correcting codes. Various codes including Hamming code, Reed-Muller code, Golay code, BCH code, Reed-Solomon code, Fire code, convolutional code, and Viterbi decoding algorithm and maximum-length code are explained.

Antennas and propagation effects for satellite communications are discussed in Chapter 3. SHF/EHF antenna technology, multiple beam antenna, coverage, adaptive interference cancellation, propagation phenomena, and hydrometer attenuation and depolarization are covered in this chapter. Chapter 4 contains a brief introduction to radar systems. Pulse radars, pulse Doppler radars, clutter, ambiguity functions, adaptive filtering, range equations, search radars, tracking radars, scan antennas and detection of radar signals with various examples and design problems are included in this chapter. Spread spectrum communications systems to protect communication links from jamming and intentional interference are reviewed in Chapter 5. New developments in propagation theory and their impact on communication and radar signal survivability are given in Chapter 6. Scintillation effects on space radar are dealt with in Chapter 7. Spatial irregularities in the ionospheric electron density can produce rapid random fluctuations in the amplitude, phase, and angle of arrival of propagating electromagnetic waves. These fluctuations, known as scintillation, have been characterized in the context of radar signals. If integrated circuits are to function properly in a nuclear-detonated environment, provisions must be made to either allow for radiation induced degradation or attempt hardening measures to reduce the degradation produced by radiation. Radiation effects on semiconductors are explained in Chapter 8. Satellite-earth propagation through nuclear disturbed ionosphere and communications channel characterization, modulation and mitigation techniques are dealt with in Chapter 9. The following two chapters, 10 and 11, are devoted to channel simulations and parameter characterization. Amplitude and phase scintillation, Rayleigh fading channel, flat fading, frequency selective fading, decorrelation time, and frequency behavior of the channel are depicted in Chapter 10. The final chapter, Chapter 11, is a study of communication systems for the Strategic Defense Initiative—the so-called Star Wars systems. The major elements of the system, such as the boost surveillance and tracking system, space surveillance and tracking system, space-leased intercept system, ground-based surveillance tracking system, ground-based radar, and exoatmospheric reentry interceptor systems require communication systems to provide message connectivity between the many space- and ground-based platforms supporting sensors and command and control facilities. The design of communication links with appropriate modulation, coding, interleaving, and other mitigation techniques are outlined in this chapter.

I express my sincere appreciation to all authors for their contribution, commitment, and cooperation. I would like to thank my colleagues, staff, and management of Hughes Aircraft Company for their help, and creating an excellent professional environment to support technical advancement. I also want to thank the production staff at Van Nostrand Reinhold and Spectrum Publisher Services for their assistance. Finally, I thank Ms. Dianna Littwin, of Van Nostrand Reinhold Company for her understanding and assistance.

Nirode Mohanty

Contents

NIRODE MOHANTY

1

Fundamentals of Space Communications

In 1945 it was known that only three satellites in a circular orbit in the equatorial plane at an altitude of 36,000 km can provide communication links for the entire world. There are now hundreds of satellites in space at various orbits and altitudes for the purpose of navigation, weather, communication, surveillance, and defense. Modern warfare can be conducted by using these satellites, which can be the targets of physical damage or transmission blockade. Nuclear bomb explosions, whose radiance and heat have been compared with thousands of suns, can do such damage. In this book we address the issue of nuclear detonation effects on space communication and surveillance.

The key parameters in the design of such a communication link is the choice of frequency, bandwidth, signal duration, data rates, and modulation. The link can be degraded in the presence of atmospheric disturbances, man-made disturbances, natural ionizations, and nuclear-induced ionizations. Several mitigation techniques such as coding, interleaving, frequency hopping, adaptive equalization, adaptive antenna beam forming, diversity methods, and radiation hardening can be used.

This chapter introduces the principles of communication systems using satellites as repeater stations in the presence of severe disturbances caused by nuclear detonation. The space communication link is described in Section 1.1. Various deterministic and random signals and their spectra used in communication systems are discussed in Sections 1.2 and 1.3. Analog communication, baseband processing (such as sampling, quantization and source coding), and digital modulation are treated in Section 1.4. A brief treatment of synchronization is presented in Section 1.5. Degradation resulting from filtering, amplification, and transmission is given in Section 1.6. Sections 1.7 and 1.8 deal with

1

earth and space segments. Three multiple-access systems for satellite communications are covered in Section 1.9. Propagation effects, performance, and various parameters that characterize fading are discussed in Section 1.10. Nuclear effects on the atmosphere and communication links are outlined in Section 1.11. A few exercises are added to illustrate and extend the main text. Other chapters deal with coding, anti-jamming modulation, mitigation techniques, and antennas and nuclear hardening. There are many topics that are left out. For additional information readers may consult the brief bibliography at the end of this chapter.

1.1 SPACE COMMUNICATIONS

Space communications began in 1946 when the U.S. Army established radar contact with moon. In 1957 the Soviet Union launched the first satellite, and in 1958 SCORE, the first U.S. communications satellite by transmitted voice signals. Now, more than 30 years later, there are more than 100 satellites in space. These satellites render various types of services including fixed satellite service (FSS), broadcast satellite service (BSS) and mobile satellite services (MSS). These satellites from the Soviet Union, Canada, the United States, and Europe can be categorized as experimental, international, military, domestic, and regional. The satellite communication technology has progressed significantly with regard to frequency, frequency reuse, bandwidth, power, number of circuits, lifetime, coverage, structure, propulsion, and weight. There are satellites to track, monitor and connect other satellites. Satellites are used for navigation, surveillance, and natural resource exploitation in addition to data, voice, and video transmission.

Satellites are active relay and switching stations in space with wide area coverage, accessibility, and bandwidth. Only three satellites in a geosynchronous orbit at distance of 36,000 km from the earth can provide connections to users dispersed around the earth. Such a system can provide connectivity to fixed users and mobile users, ships, and aircraft at any time and anywhere except the polar regions. In terms of bandwidth and capacity, Intelsat VI, a satellite network of the international organization Intelsat, with 110 member nations, can provide about 30,000 telephone circuits. In satellite communications there is no need for hopping signals from one line-of-sight (LOS) microwave tower to another, usually spaced 80 km apart. A typical satellite communication system is shown in Fig. 1.1. A satellite system has earth segments, with multiple ground stations and satellite control earth stations and space segments with a network of satellites. The satellite consists of a communication payload and spacecraft bus. The communication payload has an antenna system and a *transponder* system. The spacecraft bus has various subsystems such as electrical and propulsion subsystems, as well as thermal, tracking, tebemetry, and command (TT&C) subsystems. The satellites are placed in circular, elliptical, geosynchronous, or inclined orbits depending on the nature of service. Most of the satellites are now in geosynchronous orbit with a

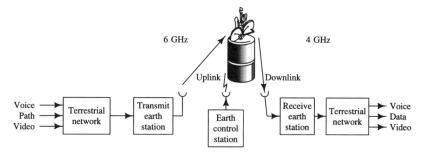

FIGURE 1.1   Satellite communication system.

24-hour service time. But there are inclined, elliptical orbits such as the Russian satellites Molnya and Tundra, to provide communication to remote places in the Soviet Union with only 12-hour service times. There are also low-earth circular orbits of 500 to 800-km altitudes for navigational and relay satellites. Early, 40-kg satellites use 6-GHz for uplink and 4 GHz for downlink frequency. By contrast, the 1670-kg. Intelsat VI has uplink and downlink frequencies of 14 and 12 GHz, respectively. The earth segment consists of antennas, feeds, high-power amplifiers, low-noise amplifiers, up- and down-converters, multiplexer, modems for terrestrial links, power supplies, and maintenance and monitoring systems.

Signals arriving at the earth stations through the terrestrial network include voice channels, analog TV channels, and data from computers and other sources. These signals, usually called baseband signals, are deterministic or random. They are combined into a composite baseband signal. The combination of diverse signals originating from different sources is called multiplexing. The multiplexed signal is converted to an appropriate form for transmission by changing the amplitude, phase, or frequency of a sinusoidal signal. This process is known as *modulation*, and it can be analog or digital. The present trend is toward digital communications. The main advantages of digital modulation or communication are its error-correcting capability, insensitivity to noise and nonlinearity, flexibility, and regeneration. However, one of the problems in digital communication is that of synchronization. A digital satellite communication system is shown in Fig. 1.2. The analog signals for the sake of digital modulation are converted to binary data in source codes. These binary data are coded for error correction in a channel coder. The coded binary bits are modulated by a number of digital modulation schemes using radio-frequency (RF) modulators. The signals are further spread-spectrum-modulated for interference rejection. These signals are up-converted and amplified and transmitted through transmit antenna systems. These signals pass through space, subject to atmospheric, ionospheric, and other ionized disturbances, and arrive at the satellite antenna. Signals from the other earth stations also arrive at the same time at the satellite antenna. The simultaneous access by several earth stations to the satellite transponder for

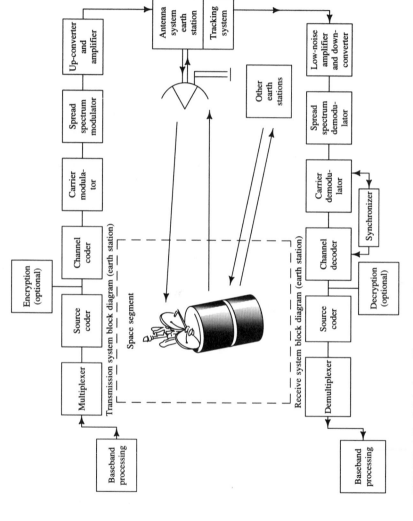

FIGURE 1.2 Digital satellite communication system.

the relay is achieved by means of a *multiple-access system*. The signals are down-converted at various transponders, amplified, and transmitted back to the earth stations by appropriate methods such as beam switching and the use of multiple beams. These signals are passed through low-noise amplifiers at the earth stations, down-converted, despread, demodulated, and decoded at channel and source to baseband signals, and then demultiplexed and sent to the appropriate destinations reading the header. The major disadvantage of satellite communication is a transmission delay of 1/4 second. For voice communications this delay is annoying, and sometimes it creates echoes. Even though various echo control schemes are common now, voice packet switching has not been used widely. In the following sections and chapters we will address the performance of space communications and radar signals in the presence of atmospheric, ionospheric, and nuclear-detonated disturbances.

1.2 DETERMINISTIC SIGNALS

A signal is a detectable physical quantity or impulse—such as electric current, voltage, or magnetic field—by which messages or information can be transmitted. It is a function of time t and message α or some parameters such as phase, frequency, amplitude, position, duration, or some combination thereof. It is denoted as $s(t, \alpha)$ or simply $s(t)$. If α is a deterministic quantity, then $s(t)$ is called a deterministic signal. If $s(t)$ takes continuous values, then $s(t)$ is called a *continuous-time signal*. If it takes discrete values (a finite set of values), it is called a *discrete-time signal* or *digital signal*. The following four equations are examples of continuous-time signals:

1. $s(t) = A \cos(2\pi f_c t + \phi) \qquad 0 \leq t \leq T$ \hfill (1.1)

   where A is the amplitude, f_c is the carrier frequency, and ϕ is the phase. This is a sinusoidal signal and is deterministic when A, f_c, and ϕ are known.

2. $s(t) = [A_1 + A_2 m(t)] \cos(2\pi f_c t + \phi) \qquad -T \leq t \leq T$

   where $m(t)$ is the message or information, A_1 and A_2 are constants, f_c is the carrier frequency, ϕ is the phase, and $2T$ is the duration of the signal. This signal is called an *amplitude-modulated* (AM) signal.

3. $s(t) = A_3 \cos\left(2\pi f_c + A_4 \int_0^t m(r)\, d\tau + \phi\right) \qquad -\infty < t < \infty$

   This is called a *frequency-modulated* (FM) signal.

4. $s(t) = A, \qquad -T/2 \leq t \leq T/2$

   $\qquad = 0, \qquad$ elsewhere \hfill (1.2)

   This signal is a rectangular pulse of height A and duration T.

The following two equations are examples of discrete-time signals:

1. $s(t) = A \cos(2\pi f_c t + \phi)$ $n \Delta t \le t \le (n + 1) \Delta t$

   $\quad = -A \cos(2\pi f_c t + \phi)$ $(n + 1) \quad \Delta t \le t \le (n + 2) \Delta t$ (1.3)

   $\qquad\qquad\qquad\qquad\qquad\qquad n = 0, 1, 2$

   This is called binary phase-shift keying (BPSK).

2. $s(t) = A \cos 2\pi f_c t$ $n \Delta t \le t \le (n + 1) \Delta t$ $n = 0, 1, 2,$

   $\quad = 0$ $(n + 1) \Delta t \le t \le (n + 2) \Delta t$

   This is called binary amplitude-shift keying (ASK).

A signal is called *periodic* of period T_0 if

$$s(t + T_0) = s(t) \tag{1.4}$$

where T_0 is the smallest positive number that satisfies Eq. (1.4); $f_0 = 1/T_0$ is called the frequency; $s(t) = A \cos 2\pi f_c t$, $-\pi \le t \le \pi$; and $f_c = \frac{1}{2}\pi s(t)$ is a periodic signal with period 2π. The time-average value or direct-current (dc) value of a signal $s(t)$, $-\infty < t < \infty$, is given by

$$\bar{s} = \lim_{T \to \infty} \frac{1}{T} \int_{-T/2}^{T/2} s(t)\, dt$$

$$\quad = \frac{1}{T_0} \int_{-T_0/2}^{T_0/2} s(t)\, dt \tag{1.5}$$

if $s(t)$ is periodic with period T_0.

If a signal $s(t)$ is zero outside $(-T/2, T/2)$, then the signal is *time-limited* and of duration T. Its time average value is

$$\bar{s} = \frac{1}{T} \int_{-T/2}^{T/2} s(t)\, dt \tag{1.6}$$

The *energy* of a signal $s(t)$ is defined as

$$\varepsilon = \lim_{T \to \infty} \int_{-T/2}^{T/2} |s(t)|^2\, dt$$

$$\quad = \int_{-T_0/2}^{T_0/2} |s(t)|^2\, dt \qquad \text{for periodic signal } s(t) \tag{1.7a}$$

where $|s(t)|^2 = s(t) \cdot (s(t))^*$; $s^*(t)$ is the complex conjugate of $s(t)$. The average *power* of a signal $s(t)$ is defined as

$$P = \lim_{T \to \infty} \frac{1}{T} \int_{T/2}^{T/2} |s(t)|^2\, dt$$

$$\quad = \frac{1}{T_0} \int_{T_0/2}^{T_0/2} |s(t)|^2\, dt \qquad \text{for periodic signal } s(t) \tag{1.7b}$$

A signal is called an *energy signal* if the energy, $\mathscr{E}$ is finite and is called a *power signal* if P is finite and nonzero. The unit for power is watts. To express the power in small units, it is expressed by $10 \log_{10} P$ dBW. Hence, 100 watts is 20 dBW.

If the signal

$$s(t) = A \cos(2\pi f_c t + \phi) \qquad 0 \leq t \leq T, f_c = 1/T$$

the time average value is

$$\bar{s} = \lim_{T \to \infty} \frac{1}{T} \int_0^T A \cos(2\pi f_c t + \phi) \, dt$$

$$= \lim_{T \to \infty} \frac{1}{T} \left[A \cos \phi \int_0^T \cos 2\pi f_c t \, dt - A \sin \phi \int_0^T \sin 2\pi f_c t \, dt \right.$$

$$= 0$$

The average power is

$$P = \lim_{T \to \infty} \frac{1}{T} \int_0^T A^2 \cos^2(2\pi f_c t + \phi) \, dt$$

$$= \lim_{T \to \infty} \frac{1}{T} \int_0^T \frac{A^2}{2} [1 + \cos(4\pi f_c t + 2\phi) \, dt]$$

$$= \frac{A^2}{2}$$

The root-mean-square (RMS) value of a signal $s(t)$ is given by

$$S_{\text{rms}} = \left[\lim_{T \to \infty} \frac{1}{T} \int_{-T/2}^{T/2} s^2(t) \, dt \right]^{1/2} \tag{1.8}$$

A signal $s(t) = A \cos[2\pi f_c t + \phi(t)]$ can be written as

$$s(t) = \text{Re}\{A \exp[i(2\pi f_c t + \phi(t)]\}$$

$$\text{Re}[A e^{i2\pi f_c t} e^{i\phi(t)}] = u(t) e^{i2\pi f_c t} \tag{1.9}$$

where $u(t)$ is called the complex *envelope* of the signal $s(t)$ where $u(t) = A e^{i\phi(t)} = A[\cos \phi(t) + i \sin \phi(t)]$, $i = \sqrt{-1}$.

Let $m(t)$ be a message or information signal with finite energy. The spectrum (Fourier transform) of the message is defined as

$$M(f) = \int_{-\infty}^{\infty} m(t) e^{-i2\pi ft} \, dt \tag{1.10a}$$

$$\triangleq |M(f)| \exp[i\psi(f)] \tag{1.10b}$$

where $|M(f)|$ is called the *message amplitude* spectrum and $\psi(f)$ is called the *phase spectrum*. If the message is real, then

$$|M(f)| = |M(-f)| \tag{1.11}$$

that is, the amplitude spectrum is an even function (see Mohanty, 1987) and

$$\psi(f) = -\psi(-f) \tag{1.12}$$

and hence the phase spectrum is an odd function.

The *bandwidth* of the message is defined as the range of values for which $|M(f)|$ is nonzero. Let a message $m(t)$ be given by

$$m(t) = \frac{B \sin \pi t B}{\pi t B} \qquad -\infty < t < \infty$$

then

$$|M(f)| = 1 \qquad \frac{-B}{2} \leq f \leq \frac{B}{2}$$

$$= 0 \qquad \text{elsewhere}$$

The bandwidth of the message is B.

Note that the message is not time-limited, whereas the message bandwidth is band-limited; that is, $B < \infty$. If the signal is the time-limited, then it is not band-limited. Let $m(t)$ be a non–time-limited signal and define

$$m_T(t) = m(t) \qquad -T/2 < t < T/2$$

$$= 0 \qquad \text{elsewhere} \tag{1.13}$$

The average power is defined as

$$P = \lim_{T \to \infty} \frac{1}{T} \int_{-T/2}^{T/2} |m(t)|^2 \, dt$$

$$= \lim_{T \to \infty} \frac{1}{T} \int_{-\infty}^{\infty} |m_T(t)|^2 \, dt \tag{1.14a}$$

By the use of Parseval's theorem (Mohanty, 1987), Eq. (1.13) gives

$$P = \lim_{T \to \infty} \frac{1}{T} \int_{-\infty}^{\infty} |M_T(f)|^2 \, df = \int_{-\infty}^{\infty} \lim_{T \to \infty} \frac{|M_T(f)|^2 \, df}{T} \tag{1.14b}$$

where $M_T(f)$ is the spectrum or Fourier transform of $m_T(t) \cdot |M_T(f)|^2$, called the *energy spectrum*.

The *autocorrelation function* of signal $m(t)$ is defined as

$$R_m(\tau) = \lim_{T \to \infty} \frac{1}{T} \int_{-T/2}^{T/2} m(t) m^*(t + \tau) \, dt \tag{1.15}$$

where τ is the *delay* or *time lag*. The *power spectral density* (PSD) of a signal $m(t)$ is defined as

$$S_m(f) = \int_{-\infty}^{\infty} [R_m(\tau) - (\bar{m})^2] e^{-i2\pi f \tau} \, d\tau \tag{1.16}$$

where

$$\bar{m} = \lim_{T \to \infty} \frac{1}{T} \int_{-T/2}^{T/2} m(t)\, dt$$

(PSD) is expressed in watts per hertz. It can be shown that (Mohanty, 1987)

$$S_m(f) = \lim_{T \to \infty} \left(\frac{|M_T(f)|^2}{T} \right) \tag{1.17}$$

where $M_T(f)$ is the Fourier transform of $m_T(t)$, defined by Eq. (1.13). Note that the average power is

$$P = \int_{-\infty}^{\infty} S_m(f)\, df \tag{1.18}$$

Let

$$m(t) = A \sin 2\pi f_c t \qquad -\infty < t < \infty \tag{1.19}$$

then

$$R_m(\tau) = \frac{A^2}{2} \cos 2\pi f_c \tau \qquad -\infty < \tau < \infty \tag{1.20}$$

$$\bar{m} = 0 \tag{1.21}$$

$$S_m(f) = \frac{A^2}{4} [\delta(f - f_c) + \delta(f + f_c)] \tag{1.22}$$

where $\delta(t)$ is the direct *delta function* or *impulse function*, defined as

$$\int_{-\infty}^{\infty} g(t)\delta(t - t_0)\, dt = g(t_0) \tag{1.23}$$

for any function $g(t)$ continuous at $t = t_0$, t_0 is arbitrary. The impulse function or delta function is also defined as

$$\delta(t) = \lim_{\Delta \to 0} \frac{1}{\Delta} \qquad \frac{-\Delta}{2} \le t \le \frac{\Delta}{2}$$
$$= 0 \qquad \text{elsewhere} \tag{1.24}$$

The average power for the signal defined in Eq. (1.19) is

$$P = \int_{-\infty}^{\infty} \frac{A^2}{4} [\delta(f + f_c) + \delta(f - f_c)]\, df$$
$$= \frac{A^2}{2}$$

using Eqs. (1.18) and (1.22).

A signal $x(t)$ can be represented by

$$x(t) = \lim_{N \to \infty} \sum_{j=-N}^{N} a_j \phi_j(t) \tag{1.25}$$

where $\phi_j(t)$, $j = 0, \pm 1, \ldots, \pm N$, is an *orthonormal signal*; that is,

$$\int_{-\infty}^{\infty} \phi_j(t) \phi_m^*(t) \, dt = 1 \qquad j = m$$
$$= 0 \qquad j \neq m \tag{1.26}$$

where $\phi_m^*(t)$ is the complex conjugate of $\phi_m(t)$. For example,

$$\phi_n(t) = 1 \qquad (n-1)T \leq t \leq nT$$
$$= 0 \qquad \text{elsewhere}$$

Two signals $x(t)$ and $y(t)$ are called *orthogonal* if

$$\int_{-\infty}^{\infty} x(t) y^*(t) \, dt = 0 \tag{1.27}$$

where

$$\int_{-\infty}^{\infty} |x(t)|^2 \, dt = \xi_1$$

$$\int_{-\infty}^{\infty} |y(t)|^2 \, dt = \xi_2$$

If $\xi_1 = \xi_2 = 1$, then $x(t)$ and $y(t)$ have equal energy, and hence they are orthonormal.

It can be shown that

$$a_j = \int_{-\infty}^{\infty} x(t) \phi_j^*(t) \, dt \qquad j = 0, \pm 1, \pm 2, \pm 3, \ldots \tag{1.28}$$

using Eqs. (1.25) and (1.26). Examples of orthonormal signals are as follows:

1. $x_n(t) = e^{i 2\pi f_c n t}$ $\quad -\infty \leq t \leq \infty$, $n = 0, \pm 1, \pm 2, \ldots$

   Note that $x_n(t)$ is a complex signal.

2. $x(t) = A \cos 2\pi f_c t$ $\quad 0 \leq t \leq T$

   $y(t) = A \sin 2\pi f_c t$ $\quad 0 \leq t \leq T$

   where $f_c = n/T$, n is some integer.

If the signal $x(t)$ can be represented by

$$x(t) = \sum_{j=1}^{N} a_j e^{i 2\pi f j t} \qquad -\infty \leq t \leq \infty \tag{1.29}$$

then $a_1, a_2, \ldots, a_N$ are called the coordinates of the signal $x(t)$. For example, let

$$x(t) = A \cos[2\pi f_c t + \phi(t)] \tag{1.30}$$

where

$$\phi(t) = 0, \pi/2, 3\pi/2, \pi$$

In Eq. (1.30), $x(t)$ can be written as follows:

$$\begin{aligned}
x(t) &= A \cos 2\pi f_c t \cos \phi - A \sin 2\pi f_c \sin \phi \\
&= A(\cos \phi)\phi_1(t) - A(\sin \phi)\phi_2(t) \\
&= a_1 \phi_1(t) + a_2 \phi_2(t)
\end{aligned}$$

where

$$\phi_1(t) = \cos 2\pi f_c t$$

$$\phi_2(t) = \sin 2\pi f_c t$$

$$a_1 = A \cos \phi(t)$$

$$a_2 = -A \sin \phi(t)$$

Note that a_1 and a_2 are the coordinates of $x(t)$, and $\phi_1(t)$ and $\phi_2(t)$ are ortho-normal functions.

Let $x(t)$ be a periodic signal with period T_0 and of finite energy. The signal $x(t)$ can be represented by the Fourier series

$$x(t) = \sum_{n=-\infty}^{\infty} a_n \exp[i2\pi f_0 nt] \tag{1.31}$$

where $f_0 = 1/T_0$, and

$$a_n(f_0) = \frac{1}{T_0} \int_0^{T_0} x(t) e^{-i2\pi n f_0 t} \, dt \qquad n = 0, \pm 1, \pm 2, \ldots \tag{1.32}$$

Note that $\phi_n(t) = e^{ni2\pi f_0 t}$ and $\phi_n^*(t) = \exp(-in2\pi f_0 t)$, $n = 0, \pm 1, \pm 2, \ldots$. In general, the a_n coordinates, $n = 0, \pm 1, \pm 2$, are complex. If $x(t)$ is real, then

$$a_n(f_0) = [a_{-n}(f_0)]^* \tag{1.33}$$

where $[a_{-n}(f_0)]^*$ is the complex conjugate of $a_{-n}(f_0)$, $n = 0, \pm 1, \pm 2, \ldots$. If $x(t)$ is real, then a_n is real. The spectrum of the signal $x(t)$ defined in Eq. (1.31) is

$$X(f) = \text{Fourier transform of } x(t)$$

$$= \sum_{n=-\infty}^{\infty} a_n[\delta(f - nf_0)] \tag{1.34}$$

where a_n, $n = 0, \pm 1, \ldots$, are the spectral components at $f = nf_0$, $n = 0, \pm 1, \pm 2, \ldots$, where $\delta(f)$ is a Dirac delta function. It can be shown, using Parseval's theorem, that the power of the signal $x(t)$ is given by

$$P = \lim_{T \to \infty} \frac{1}{T} \int_{-T/2}^{T/2} |x(t)|^2 \, dt$$

$$= \frac{1}{T_0} \int_0^{T_0} |x(t)|^2 \, dt$$

$$= \sum_{n=-\infty}^{\infty} |a_n|^2 \tag{1.35}$$

Therefore, $|a_n|^2$ can be considered as a power component at frequency $f = nf_0$, $n = 0, \pm 1, \pm 2, \ldots$. Adding all the power components in the frequency domain, the power of the periodic finite energy signal is obtained.

A channel (propagation path) or filter is called *linear* if $x_1(t)$ and $x_2(t)$ are two inputs and $y_1(t)$ and $y_2(t)$ are corresponding outputs, then input $a_1 x_1(t) + a_2 x(t)$ will yield output $a_1 y_1(t) + a_2 y_2(t)$ for all a_1 and a_2.

The output $y(t)$ and input $x(t)$ for a time-variant linear channel are related as follows:

$$y(t) = \int_{-\infty}^{\infty} h(t, \tau) x(t - \tau) \, d\tau \qquad -\infty < t < \infty \tag{1.36}$$

where $h(t, \tau)$ is the channel (filter) impulse response function at time t with delay τ. If the linear channel (filter) is *time-invariant*,[a] the output

$$y(t) = \int_{-\infty}^{\infty} h(\tau) x(t - \tau) \, d\tau \qquad -\infty < t < \infty$$

$$= \int_{-\infty}^{\infty} h(t - \tau) x(\tau) \, d\tau \triangleq x(t) \otimes h(t) \tag{1.37}$$

where $x(t)$ is the input to the channel (filter) or the transmitted signal, and $\otimes$ stands for convolution integral. The output transfer function for a time-invariant channel (filter) is given by

$$Y(f) = H(f) X(f) \tag{1.38}$$

where $H(f)$ and $X(f)$ are the Fourier transforms of $h(t)$ and $x(t)$, $h(t)$ is the impulse response function of the time-invariant channel (filter) and $H(f)$ is called the *channel transfer function*. The channel (filter) transfer function can be written as

$$H(f) = |H(f)| \exp[i\psi(f)] \tag{1.39}$$

where $|H(f)|$ is the amplitude of the spectrum time-invariant channel (filter) $h(t)$ and $\psi(f)$ is the phase spectrum. The *group delay* is defined as

---

[a] A channel is called a time-invariant channel if when the input is delayed by T_0 seconds, then the output is delayed by T_0 seconds.

$$g(f) = -\frac{1}{2\pi}\frac{d\psi}{df}$$ (1.40)

If

$$|H(f)| = K_1 \text{ (constant)}$$
$$g(f) = K_2 \text{ (constant)} \qquad \text{for all } f$$ (1.41)

then the channel is called a distortionless channel.

A channel is called *physically realizable* and stable if

$$\int_0^\infty |h(t)|\, dt < \infty$$ (1.42)

Let $h(t)$ be real and physically realizable, and

$$x(t) = A \cos 2\pi f_c t \qquad -\infty < t < \infty$$ (1.43)

then the output is

$$y(t) = \int_{-\infty}^\infty h(t - \tau)A\cos(2\pi f_c \tau)\, d\tau$$
$$= |H(f_c)|A\cos[2\pi f_c t + \psi(f_c)] \qquad -\infty < t < \infty$$ (1.44)

where

$$H(f) = \int_{-\infty}^\infty h(t)\exp(-i2\pi ft)\, dt$$
$$= |H(f)|\exp[i\psi(f)]$$

For a distortionless channel (filter) the output $y(t)$ for the input given by Eq. (1.42) is

$$y(t) = AK_1\cos(2\pi f_c t - 2\pi K_2 f) \qquad -\infty < t < \infty$$

where $|H(f)| = K$, $\psi(f) = 2\pi K_2 f$.

The bandwidth (absolute) of the time-invariant linear channel $h(t)$ is B if $|H(f)| = 0$ for $|f| > B/2$, where $H(f)$ is the channel transfer function.

The *3-dB bandwidth* of the channel for which $\max_f |H(f)| = A$, or $|H(f_m)| = A$, is the region in which $|H(f)| = (A/\sqrt{2})$ The equivalent bandwidth of the channel is defined as

$$B_{\text{eq}} = \int_{-\infty}^\infty \frac{|H(f)|^2\, df}{\max|H(f)|^2}$$ (1.45)

The unit for bandwidth is Hertz. All real channels are *band-limited*, i.e.,

$$|H(f)| = 0 \qquad \text{for } |f| > B_{\text{CH}}$$ (1.46)

where B_{CH} is the bandwidth (absolute) of the channel $h(t)$. Let the signal's bandwidth be B_S. For a distortionless output,

$$B_{CH} > B_S$$

and

$$B_{CH} = K \text{ (constant)}$$

$$(1.47)$$

for the region $|f| \le B_S$. If $B_{CH} < B_S$, then the signal is distorted. This distortion is called the *intersymbol interference*.

A filter is called a *low-pass filter* if

$$|H(f)| = \text{constant} \qquad \text{for } |f| \le B/2$$

$$= 0 \qquad \text{elsewhere}$$

A filter is called a *high-pass filter* if

$$|H(f)| = 0 \qquad \text{for } |f| \le B/2$$

$$= \text{constant} \qquad \text{elsewhere}$$

A filter is called a *band pass* if its $|H(f)|$ is symmetric about $\pm f_c$ and

$$|H(f)| \neq 0 \qquad f \in (f_c - B/2, f_c + B/2) \text{ and}$$

$$f \in (-f_c - B/2, -f_c + B/2)$$

where f_c is the carrier frequency.

Let the channel (filter) impulse function be

$$h(t) = \frac{1}{\pi t} \tag{1.48}$$

The output $y(t)$ is given by

$$y(t) = \frac{1}{\pi} \int_{-\infty}^{\infty} \frac{X(\tau)}{t - \tau} d\tau \triangleq \hat{x}(t) \tag{1.49}$$

The Fourier transform of $y(t)$ given by Eq. (1.49) is

$$Y(f) = \begin{cases} -iX(f) & f > 0 \\ 0 & f = 0 \\ iX(f) & f < 0 \end{cases} \tag{1.50}$$

The output $y(t)$, defined in Eq. (1.49), is called the *Hilbert transform*.

If

$$x(t) = m(t)(\cos 2\pi f_c t + i \sin 2\pi f_c t) \tag{1.51}$$

$$y(t) = m(t)(\cos 2\pi f_c - i \sin 2\pi f_c t) \tag{1.52}$$

where $m(t)$ is a message signal,

$$M(f) = 0 \qquad \text{for } f < f_c - f_m \text{ or } f > f_c + f_m \text{ and } f_c > f_m \tag{1.53}$$

it can be shown that

$$X(f) = M(f - f_c) \qquad f_c - f_m \leq f \leq f_c + f_m$$
$$= 0 \qquad \text{elsewhere} \tag{1.54}$$

$$Y(f) = iM(f - f_c) \qquad f_c - f_m \leq f \leq f_c + f_m$$
$$= 0 \qquad \text{elsewhere} \tag{1.55}$$

Let

$$z(t) = x(t) + iy(t) = x(t) + i\hat{x}(t)$$

where $\tag{1.56}$

$$y(t) = \frac{1}{\pi} \int_{-\infty}^{\infty} \frac{x(\tau)}{t - \tau} d\tau = \hat{x}(\tau)$$

then

$$Z(f) = X(f) + iY(f) = X(f) + i\hat{X}(f)$$
$$= 2X(f) \qquad f > 0$$
$$X(0) \qquad f = 0 \tag{1.57}$$
$$0 \qquad f < 0$$

The analytical signal $z(t)$ has a spectrum only for positive frequencies; it is also called *preenvelope* of $x(t)$.

Note further that if $\hat{x}(t)$ is Hilbert transform of $x(t)$, then

$$\int_{-\infty}^{\infty} x(t)\hat{x}(t) \, dt = 0$$

that is, $\hat{x}(t)$ is orthogonal to $x(t)$. Let

$$x(t) = a(t) \cos[2\pi f_c t + \phi(t)]$$
$$= a(t) \cos 2\pi f_c t - b(t) \sin 2\pi f_c t$$
$$= \text{Re}[u(t)e^{i2\pi f_c t}] \tag{1.58a}$$

where $a(t) = A(t) \cos \phi(t)$

$$b(t) = A(t) \sin \phi(t) \tag{1.58b}$$

$$u(t) = A(t) \exp(i\phi(t)) = a(t) + ib(t) \tag{1.58c}$$

$u(t)$ is called the *complex envelope* of $x(t)$. The $a(t)$ and $b(t)$ values given in Eq. (1.58b) are called *in-phase* and *quadrature* components. If the spectra of $a(t)$ and $b(t)$ are such that $|A(f)| = 0$ for $|f| > f_m$ and $|B(f)| = 0$ for $|f| > f_m$ and $f_c \gg f_m$, then $x(t)$ defined by Eq. (1.58) is called a *narrowband signal*. The spectrum $x(t)$ is

$$X(f) = \int_{-\infty}^{\infty} x(t)e^{-i2\pi ft}\,dt = \int_{-\infty}^{\infty} \mathrm{Re}[u(t)e^{i2\pi f_c t}]e^{-i2\pi f_c t}\,dt$$

$$= \frac{1}{2}\int_{-\infty}^{\infty} [u(t)e^{i2\pi f_c t} + u^*(t)e^{-i2\pi f_c t}]e^{-i2\pi f_c t}\,dt$$

$$= \frac{1}{2}[U(f - f_c) + U^*(-f - f_c)] \tag{1.59a}$$

where $U(f)$ is the Fourier transform of $U(t)$ [$u(t)$ is also called the *equivalent low-pass signal* because $U(f)$ is concentrated near $f = 0$].

A signal $x(t)$ is called *baseband* if its the amplitude spectrum $|X(f)|$ is symmetric about zero or some small frequency and $|X(f)| = 0$ for $|f| > B$. A signal $X(t)$ is called *bandpass* if its amplitude spectrum $|X(f)|$ is symmetric about some carrier frequency and $|X(f)| = 0$ for $|f - f_c| > B$. The PSD of $x(t)$ is

$$S_x(f) = \tfrac{1}{2}[S_x(f + f_c) + S_u(-f - f_c)] \tag{1.59b}$$

1.3 RANDOM SIGNAL AND NOISE

The signal $x(t, \alpha)$ is a function of α, a random variable, where $\alpha \in \Omega$, and $t \in (-\infty, \infty)$ or $t = \pm n\,\Delta t$, $n = 0, 1, 2, \ldots$, Δt is some interval, and Ω is the sample space. If α is fixed, then $x(t)$ is called a *sample function*. If t is fixed, $X(t, \alpha)$ is a random variable. If t is fixed and α is fixed, $x(t, \alpha)$ is a number. Let us denote $x[t, \alpha(\omega)] = x(t, \omega) \triangleq x(t)$. The signal $x(t)$ is also defined as an indexed family of random variables. Let $t = t_i$, then

$$X(t_i, \omega) \triangleq X(i, \omega) \triangleq X_i \qquad i = 0, \pm 1, \pm 2, \ldots$$

The random variable $\{X_i\}$ can be discrete or continuous if x takes a finite set of values or continuous values. Examples of discrete random variables include binomial, Poisson, and geometric; that is,

$$P(X = k) = \binom{n}{k} P^k(1 - p)^{n-k} \qquad k = 0, 1, 2, \ldots, n \qquad \text{(binomial)}$$

$$P(X = k) = \frac{\lambda^k e^{-\lambda}}{k!} \qquad k = 0, 1, 2, \ldots, n, \ldots, \infty \qquad \text{(Poisson)}$$

$$P(X = k) = (1 - p)^{k-1}p \qquad k = 1, 2, 3, \ldots \qquad \text{(geometric)}$$

Examples of continuous random variables uniform and Gaussian distributions [see Eqs. (1.65a) and (1.65b).]

The nth moments of X are given by

$$E(X^n) = \sum_i (x_i)^n P(X = x_i) \qquad \text{if } X \text{ is discontinuous}$$

$$= \int_{-\infty}^{\infty} x^n f_x(x)\,dx \qquad \text{if } X \text{ is continuous} \tag{1.60}$$

where $f_n(x)$ is the *probability density function* (PDF). Note that the probability function $P(X = k)$ has the following properties:

1. $P(X = k) \geq 0$

2. $\sum_k P(X = k) = 1$

(1.61)

The probability density function has the following properties:

1. $f_x(x) \geq 0$

2. $\int_{-\infty}^{\infty} f_x(y)\,dy = 1$

3. $\int_{-\infty}^{x} f_x(y)\,dy = F_x(x)$

(1.62)

4. $\int_{a}^{b} f_x(y)\,dy = F_x(b) - F_x(a)$

$F_x(x)$ is called the probability distribution function. It satisfies the following conditions:

1. $F_x(x)$ is a monotomically increasing function; that is,

$$x_1 < x_2 = F_x(x_1) \leq F_x(x_2)$$

(1.63a)

2. $F_x(x)$ is right continuous; that is,

$$F_x(x) = \lim_{h \to 0} F_x(x + h) \qquad h > 0$$

(1.63b)

3. $F_x(-\infty) = 0, F_x(\infty) = 1$

(1.63c)

where

$$F_x(x) = P[\omega: X(\omega) \leq x] \qquad x \text{ is a real number, } -\infty < x < \infty$$

(1.6.3d)

If $n = 1$ in Eq. (1.60), $E(X)$ is called the *statistical average, ensemble average, mean*, or *first moment*. If $n = 2$ in Eq. (1.60), $E(X^2)$ is called the *average power*. The *variance* of X is defined as

$$\sigma_x^2 = E(X^2) - [E(X)]^2$$
$$= E[(X - m)^2]$$

(1.64)

where $m = E(X)$. The square root of the variance, σ_x, is called the *standard deviation*. Note that $\sigma_x \geq 0$.

As discussed previously, examples of continuous random variables are the uniform and the Gaussian (normal).

If X is a *uniform* random variable, the probability density function is given by

$$f_x(x) = \frac{1}{2a} \qquad a \le x < a$$

$$= 0 \qquad \text{elsewhere} \tag{1.65a}$$

It can be shown that

$$E(X) = 0$$

$$E(X^2) = a^2/3$$

Variance of $X = \text{var}(X) = \sigma_x^2 = a^2/3$. Hence the standard deviation $\sigma_x = a/\sqrt{3}$.

X is a *Gaussian* random variable if the probability density function is given by

$$f_x(x) = \frac{1}{\sqrt{2\pi\sigma^2}} \exp\left[-\frac{(x-m)^2}{2\sigma^2} \right] \tag{1.65b}$$

where

$$E(X) = m$$

$$\text{var}(X) = \sigma^2$$

Note: If $m = 5$, $\sigma^2 = 16$, then

$$E(X^2) = \sigma^2 + m^2 = 16 + 25 = 41$$

The characteristic function of a random variable is defined as

$$\phi_x(v) = E(e^{ixv})$$

$$= \sum_i e^{ix_i v} P(x = x_i) \qquad \text{if } x \text{ is discrete} \tag{1.66}$$

$$= \int_{-\infty}^{\infty} e^{ixv} f_x(x)\, dx \qquad \text{if } x \text{ is continuous}$$

If X is a Gaussian random variable with

$$f_x(x) = \frac{1}{\sqrt{2\pi\sigma^2}} \exp\left[\frac{-(x-m)^2}{2\sigma^2} \right] \tag{1.67}$$

then the characteristic function of a Gaussian random variable is

$$\phi_x(v) = \exp\left[imv - \frac{v^2\sigma^2}{2} \right] \tag{1.68}$$

The distribution function for a Gaussian (normal random variable defined in Eq. (1.67) is

$$F_x(x) = \int_{-\infty}^{x} \frac{1}{\sqrt{2\pi\sigma^2}} \exp\left[\frac{-1}{2\sigma^2}(y-m)^2 \right] dy = 1 - \int_{(x-m)/\sigma}^{\infty} \frac{1}{\sqrt{2\pi}} e^{-t^2/2}\, dy$$

$$= 1 - Q\left(\frac{x - m}{\sigma}\right)$$

$$= \frac{1}{2}\left[1 + \mathrm{erf}\left(\frac{x - m}{\sigma\sqrt{2}}\right)\right] \tag{1.69}$$

where

$$\mathrm{erf}(t) = \frac{2}{\sqrt{\pi}}\int_0^t e^{-y^2}\,dy$$

$$Q(t) = \frac{1}{2}\mathrm{erfc}\left(\frac{t}{\sqrt{2}}\right)$$

$$\mathrm{erfc}(t) = \frac{2}{\sqrt{\pi}}\int_t^\infty e^{-y^2}\,dy \tag{1.70}$$

$$\mathrm{erfc}(t) = 2Q(\sqrt{2}t)$$

The *error function* is denoted by $\mathrm{erf}(t)$. The complementary error function $\mathrm{erfc}(t)$ is defined as follows:

$$\mathrm{erfc}(t) = 1 - \mathrm{erf}(t) = \frac{2}{\sqrt{\pi}}\int_t^\infty e^{-y^2}\,dy \tag{1.71}$$

Note that

$$\mathrm{erf}(-t) = -\mathrm{erf}(t) \tag{1.72}$$

If x is a Gaussian random variable with mean m and variance σ^2, then

$$P[x > T] = \frac{1}{2}\mathrm{erfc}\left(\frac{T - m}{\sqrt{2}\sigma}\right)$$

Let $X_1, X_2, \ldots, X_n$ be n independent Gaussian random variables and

$$\chi^2 = X_1^2 + X_2^2 + \cdots + X_n^2 \tag{1.73}$$

where $E(X_i) = 0$ and $\sigma_{x_i}^2 = 1$ for all i, then the probability density function known as chi-squared (χ^2) is given by

$$f_{\chi^2}(x) = \frac{1}{2^{n/2}\Gamma(n/2)}e^{-(x^2/4)}(x)^{(n/2)-1}$$

$$\left(\Gamma(n + 1) = \int_0^\infty e^{-x}x^n\,dx\right) \tag{1.74}$$

where n is the number of degrees of freedom of the distribution and $\Gamma(\cdot)$ is the gamma function (see Mohanty, 1987).

If X is a log normal random variable, then the probability density function is given by

$$f_x(x) = \frac{1}{\sqrt{2\pi\sigma^2}} \frac{1}{x} \exp\left\{-\frac{1}{2}\left[\frac{(\ln x - m)^2}{\sigma}\right]\right\} \tag{1.75}$$

and

$$F_X(x) = \frac{1}{2}\left[1 + \text{erf}\left(\frac{\ln x - m}{\sqrt{2\sigma^2}}\right)\right] \tag{1.76}$$

where

$$E(X) = \exp\left(m + \frac{\sigma^2}{2}\right)$$

$$\sigma_x = \exp\left(m + \frac{\sigma^2}{2}\right)[\exp(\sigma^2) - 1]^{1/2} \tag{1.77}$$

Let $y = g(x)$, x is a random variable with PDF $f_x(x)$, then

$$E(y) = \int_{-\infty}^{\infty} g(x)f_x(x)\,dx$$

The *log normal* density is often used to represent rainfall rate distribution for low- and medium-rainfall regions.

Two random variables X_1 and X_2 are called *orthogonal* if

$$E(X_1 X_2) = 0 \tag{1.78}$$

and are *uncorrelated* if

$$E(X_1 X_2) = E(X_1)E(X_2) \tag{1.79}$$

Two random variables X_1 and X_2 are called *independent* if

$$P[X_1 \leq x_1, X_2 \leq x_2] = P[X_1 \leq x_1]P(X_2 \leq x_2) \tag{1.80}$$

If X_1 and X_2 are two *continuous, independent* random variables, then the joint PDF of x_1, x_2 is

$$f_x(x_1, x_2) = f_{X_1}(x_1)f_{X_2}(x_2) \tag{1.81}$$

where $\mathbf{X} = (X_1, X_2)$. If X_1 and X_2 are two independent random variables, then they are uncorrelated. Let X_1 and X_2 be two Gaussian random variables with joint probability density function

$$f_x(x_1, x_2) = \frac{1}{2\pi|C|^{1/2}} \exp\left[-\frac{1}{2}(\mathbf{X} - m)C^{-1}(\mathbf{X} - m)\right] \tag{1.82}$$

where

$$\mathbf{X} = \begin{pmatrix} X_1 \\ X_2 \end{pmatrix} \quad \text{and} \quad \mathbf{m} = \begin{bmatrix} E(X_1) \\ E(X_2) \end{bmatrix}$$

Also, in Eq. (1.82),

$$C = \begin{bmatrix} E[(X_1 - m_1)^2] & E[(X_1 - m_1)(X_2 - m_2)] \\ E[(X_1 - m_1)(X_2 - m_2)] & E[(X_2 - m_2)^2] \end{bmatrix}$$

where $m_i = E(X_i)$, $i = 1, 2$. Moreover, in Eq. (1.82), $|C|$ = determinant of matrix C. C is called the *covariance* matrix of vector X.

If X_1 and X_2 are uncorrelated, then

$$E[(X_1 - m_1)(X_2 - m_2)] = 0$$

The joint density functions of X_1 and X_2 are given by

$$f_x(x_1, x_2) = \frac{1}{2\pi\sigma_1\sigma_2} \exp\left[\frac{-(X_1 - m_1)^2}{2\sigma_1^2}\right] \exp\left[\frac{-(X_2 - m_2)^2}{2\sigma_2^2}\right] \tag{1.83}$$

where

$$\sigma_i^2 = E(X_i - m_i)^2 \qquad i = 1, 2$$

If x_1 and x_2 are uncorrelated, gaussian random variables, then x_1 and x_2 are independent. Let

$$m_1 = 0 \quad m_2 = 0 \quad \sigma_1 = \sigma_2 = \sigma^2 \tag{1.84}$$

and

$$X_1 = R\cos\theta$$

$$X_2 = R\sin\theta$$

Hence

$$R^2 = X_1^2 + X_2^2$$

$$\theta = \tan^{-1}(X_2/X_1)$$

Substitution of values of the parameter given by Eq. (1.84) in Eq. (1.83) yields

$$f(R, \theta) = \frac{1}{2\pi\sigma^2} \exp\left[\frac{-R^2}{2\sigma^2}\right]\frac{1}{R} \tag{1.85}$$

using the fact that (Mohanty, 1987)

$$f(X_1, X_2) = \frac{f(R, \theta)}{\begin{bmatrix} \partial X_1/\partial R & \partial X_2/\partial R \\ \partial X_1/\partial\theta & \partial X_2/\partial\theta \end{bmatrix}} \tag{1.86}$$

The *marginal density*[b] of R is given by

$$f_R(r) = \int_0^{2\pi} f(R, \theta)\, d\theta$$

$$= \frac{r}{\sigma^2} \exp\left(\frac{-r^2}{2\sigma^2}\right) \qquad r \geq 0 \tag{1.87}$$

---

[b] $f(x) = \int f(x, y)\, dy$, $f(x, y)$, is called the joint PDF, and $f(x)$ is called the marginal PDF.

The marginal density of θ is

$$f(\theta) = 1/2\pi \qquad 0 \le \theta \le 2\pi$$
$$= 0 \qquad \text{elsewhere} \tag{1.88}$$

The probability density function of R given by Eq. (1.87) is called the *Rayleigh density*. The probability density function of phase θ is a *uniform density*.

The probability density of R^2 is given by

$$f_{R^2}(r) = (e^{-r/2\sigma^2})(1/2\sigma^2) \qquad r \ge 0$$
$$= 0 \qquad r < 0 \tag{1.89}$$

$$E(R^2) = 2\sigma^2 \qquad\qquad \text{var}(R^2) = 4\sigma^4 \tag{1.90}$$

The PDF given by Eq. (1.89) is called *exponential density*.

The *Nakagami-m density* function is given by

$$f_x(x) = \frac{2m^m x^{2m-1} e^{-(m/\sigma)}}{\Gamma(m)\sigma^{2m}} \tag{1.91}$$

where

$$\Gamma(m) = \int_0^\infty e^{-t} t^{(m-1)}$$

$$E(X^2) = \sigma^2$$

When $m = 1$, the Nakagami density becomes the Rayleigh density. When $m = 1/2$, we get one-sided Gaussian density.

The *Gamma density function* is given by

$$f_x(x) = \frac{\alpha^n x^{n-1} e^{-\alpha x}}{\Gamma(n)} \tag{1.92}$$

where $E(X) = n/\alpha$, $\sigma_x^2 = n/\alpha^2$, and

$$\Gamma(n) = \int_0^\infty x^{(n-1)} e^{-x} \, dx$$

The *conditional probability density* of x given y is

$$f(x|y) = \frac{f(x, y)}{f(y)} = \frac{f(x, y)}{\int f(x, y) \, dx} \tag{1.93a}$$

The *conditional probability density* (*a posteriori*) of y given x is

$$f(y|x) = \frac{f(x|y)f(y)}{f(x)} = \frac{f(x|y)f(y)}{\int f(x, y) \, dy} \tag{1.93b}$$

Let a random signal be

$$Y(t) = \sum_{i=1}^t X_i \qquad t = 1, 2, \ldots, N \tag{1.94a}$$

where each X_i is a zero-mean random variable with variance σ_x^2 for all i.

$$E[Y(t)] = 0$$
$$\text{var}[Y(t)] = t\,\text{var}(X_i) \tag{1.94b}$$
$$= t\sigma_x^2$$

Note that $\text{var}[Y(t)]$ varies with t.

Let a random signal be

$$Y(t) = A\cos(2\pi f_c t + \phi) \qquad 0 \le t \le T \tag{1.95a}$$

where A is a Gaussian random variable with mean m and variance σ^2. The mean of $y(t)$ is

$$E[y(t)] = E(A\cos(2\pi f_c t + \phi))$$
$$= \cos(2\pi f_c t + \phi)E(A)$$
$$= m\cos(2\pi f_c t + \phi) \tag{1.95b}$$

The average power is given by

$$E(y^2) = E(A^2\cos^2(2\pi f_c t + \phi))$$
$$= \cos^2(2\pi f_c t + \phi)E(A^2)$$
$$= (\sigma^2 + m^2)\cos^2(2\pi f_c t + \phi) \tag{1.95c}$$

The variance is as follows:

$$\text{var}[y(t)] = \sigma^2\cos^2(2\pi f_c t + \phi) \tag{1.95d}$$

Note that both mean and variance of $y(t)$ vary with t. A random signal whose mean, variance, or other statistic varies with time t is called a *nonstationary* signal.

The *correlation function* of a random signal is defined as

$$R_x(t_1, t_2) = E(X(t_1)X^*(t_2)) \tag{1.96}$$

$X^*(t)$ is the complex conjugate of $X(t)$.

A random signal is called *covariance stationary* or *wide-sense stationary* (WSS) if

$$E[X(t)] = \text{constant}$$
$$E(|X(t)|^2) < \infty \tag{1.97}$$
$$R_x(t_1, t_2) = R_x(|t_1 - t_2|)$$
$$= R_x(\tau)$$

The autocorrelation function $R_x(\tau)$ is an even function and $R_x(\tau) \le R_x(0)$, where $R_x(0) = E(x^2)$, the average power of the signal.

The *covariance function* for a stationary signal is defined as

$$C_x(\tau) = R_x(\tau) - |m|^2 \tag{1.98}$$

where $m(t) = E[X(t)]$. If $m = 0$, $C_x(0) = R_x(0) = E[|x(t)|^2] \geq 0$.

The *power spectral density (PSD)* for a stationary signal is defined as

$$S_x(f) = \int_{-\infty}^{\infty} C_x(\tau) e^{-i2\pi f \tau} \, d\tau \tag{1.99}$$

Note that $S_x(f) \geq 0$, $S_x(f) = S_x(-f)$, $C_x(0) = \int_{-\infty}^{\infty} S_x(f) \, df$.

The covariance is

$$C_x(\tau) = \int_{-\infty}^{\infty} S_x(f) e^{i2\pi f \tau} \, d\tau$$

If

$$S_x(f) = N_0/2 \qquad \text{for all } f$$

then

$$C_x(\tau) = \frac{\delta(\tau) N_0}{2} \tag{1.100}$$

If the PSD $S_x(f)$ is constant, then $X(t)$ is called *white*.

Let a random signal be

$$X(t) = A \cos(2\pi f_c t + \phi) \qquad 0 \leq t \leq T$$

where ϕ is uniform, that is,

$$f_\phi(\phi) = 1/2\pi \qquad -\pi \leq \phi \leq \pi$$
$$= 0 \qquad \text{elsewhere}$$

The mean of $X(t)$ is

$$E[X(t)] = E[A \cos(2\pi f_c t + \phi)]$$
$$= A \cos(2\pi f_c t) E \cos\phi - A \sin(2\pi f_c t) E \sin\phi$$
$$= A \cos(2\pi f_c t) \int_{-\pi}^{\pi} \cos\phi \left(\frac{1}{2\pi}\right) d\phi - A \sin(2\pi f_c t) \int_{-\pi}^{\pi} \sin\phi \left(\frac{1}{2\pi}\right) d\phi$$
$$= 0$$

The average power of $X(t)$ is

$$E[X^2(t)] = E[A^2 \cos^2(2\pi f_c t + \phi)]$$
$$= E\left\{\frac{A^2}{2}[1 + \cos(4\pi f_c t + 2\phi)]\right\}$$

$$= \frac{A^2}{2} + \frac{A^2}{2} E[\cos(4\pi f_c t + 2\phi)]$$

$$= \frac{A^2}{2} + \frac{A^2}{2} \cos(4\pi f_c t) E(\cos 2\phi) - \frac{A^2}{2} \sin(4\pi f_c t) E(\sin 2\phi)$$

$$= A^2/2$$

The variation is as follows:

$$\text{var}[X(t)] = \frac{A^2}{2}$$

The autocorrelation function of $X(t)$ is

$$E[X(t_1)X(t_2)] = E[A^2 \cos(2\pi f_c t_1 + \phi)\cos(2\pi f_c t_2 + \phi)]$$

$$= \frac{A^2}{2} \cos[2\pi f_c(t_1 - t_2)]$$

$$+ \frac{A^2}{2} \cos[2\pi f_c(t_1 + t_2)] E(\cos 2\phi)$$

$$- \frac{A^2}{2} \sin[2\pi f_c(t_1 + t_2)] E(\sin 2\phi)$$

$$= \frac{A^2}{2} \cos(2\pi f_c \tau)$$

The autocorrelation of $X(t)$ is a function of $|t_1 - t_2|$. The power spectral density of $X(t)$ is

$$S_x(f) = \frac{A^2}{4} [\delta(f + f_c) + \delta(f - f_c)]$$

Therefore $X(t)$ is a stationary signal, but it is not a white noise.

The *cross correlation* of two stationary signals is given by

$$R_{xy}(\tau) = E[X(t_1)Y^*(t_2)] \qquad \tau = t_2 - t_1 \tag{1.101}$$

The *cross covariance* of $X(t)$ and $Y(t)$ is given by

$$C_{xy}(\tau) = R_{xy}(\tau) - m_x m_y^* \tag{1.102}$$

where

$$m_x = E[x(t)] \quad \text{and} \quad m_y = E[y(t)]$$

The *cross-power spectral density* of $x(t)$ and $y(t)$ stationary signals is

$$S_{xy}(f) = \int_{-\infty}^{\infty} C_{xy}(\tau) e^{-i2\pi f \tau} \, d\tau \tag{1.103}$$

Let $X(t)$ be a real zero-mean stationary signal and $h(t)$ be a stable time-invariant filter (channel) such that

$$\int_{-\infty}^{\infty} |h(t)| \, dt < \infty$$

and let $Y(t)$ be the output signal of the filter. Then

$$Y(t) = \int_{-\infty}^{\infty} h(\tau) X(t - \tau) \, d\tau \tag{1.104}$$

$$E[Y(t)] = \int_{-\infty}^{\infty} h(r) E[X(t - r)] \, dr$$

$$= \int_{-\infty}^{\infty} h(r) \cdot 0 \cdot dr$$

$$= 0 \tag{1.105a}$$

Note that

$$E(x(t - r)) = E[x(t)] = 0$$

because $E[x(t)] = 0$ for all t. Multiplying $X(t - \tau)$ on both sides of Eq. (1.104) and taking the average value, we get

$$E[Y(t)X(t - \tau)] = \int_{-\infty}^{\infty} h(r) E[X(t - r)X(t - \tau)] \, dr$$

$$= \int_{-\infty}^{\infty} h(r) R_{XX}(\tau - r) \, dr = R_{YX}(\tau)$$

or

$$R_{YX}(\tau) = \int_{-\infty}^{\infty} h(r) R_{XX}(r + \tau) \, d\tau = R_{XY}(-\tau) \ldots \tag{1.105b}$$

Taking the Fourier transform of Eq. (1.105b), we get

$$S_{YX} = H(f) S_{XX}(f)$$

or

$$S_{XY}(f) = H^*(f) S_{XX}(f) \tag{1.106}$$

where

$$H(f) = \int_{-\infty}^{\infty} h(t) e^{-i2\pi ft} \, dt$$

Using Eq. (1.104), we get

$$E[Y(t)Y(t + \tau)] = \int_{-\infty}^{\infty} h(r) R_{XY}(\tau - r) \, dr = R_{YY}(\tau) \tag{1.107a}$$

Taking the Fourier transform of Eq. (1.107a), we get

$$S_{YY}(f) = H(f)S_{XY}(f)$$
$$= H^*(f)H(f)S_{XX}(f)$$
$$= |H(f)|^2 S_{XX}(f) \tag{1.107b}$$

using Eq. (1.106).

We conclude that $Y(t)$ is a stationary with the PSD given by

$$S_{YY}(f) = |H(f)|^2 S_{XX}(f) \tag{1.108}$$

If $S_{XX}(f) = N_0/2$ (white), then the output $Y(t)$ is not white, because

$$S_{YY}(f) = \frac{N_0}{2}|H(f)|^2$$

$S_{YY}(f) = KN_0/2$ if and only if

$$|H(f)| = K \text{ (constant)}$$

Equation (1.108) is true even when $E[X(t)] = 0$. The average power of $Y(t)$ with mean zero is given by

$$E[Y^2(t)] = \int_{-\infty}^{\infty} S_{YY}(f)\,df$$
$$= \int_{-\infty}^{\infty} |H(f)|^2 S_{XX}(f)\,df$$

The mean of $Y(t)$ is

$$E[Y(t)] = m \int_{-\infty}^{\infty} h(t)\,dt$$
$$= mH(0) \tag{1.109}$$

where

$$H(f) = \int_{-\infty}^{\infty} h(t)e^{-i2\pi ft}\,dt$$

Let us denote the noise voltage as

$$n(t) = A(t)\cos[2\pi f_c t + \phi(t)] \tag{1.110a}$$
$$= n_c(t)\cos(2\pi f_c t) - n_s(t)\sin(2\pi f_c t) \tag{1.110b}$$

The preenvelope or analytical signal is given by

$$Z(t) = n(t) + iy(t) = u(t)e^{i2\pi f_c t}$$
$$= n(t) + i\hat{n}(t) \tag{1.111}$$

where

$$\hat{n}(t) = y(t) = \frac{1}{\pi} \int_{-\infty}^{\infty} \frac{n(\tau)}{(t - \tau)} d\tau \qquad (1.112)$$

the Hilbert transform of $n(t)$.

The *complex envelope* of $Z(t)$ is

$$u(t) = n_c(t) + in_s(t) \qquad (1.113)$$

It can be shown that

$$\hat{n}(t) = y(t) = n_c(t)\cos(2\pi f_c t) + n_s(t)\sin(2\pi f_c t) \qquad (1.114)$$

Assume that $n_c(t)$ and $n_s(t)$ are WSS processes; that is,

$$R_c(\tau) = E[n_c(t)n_c(t + \tau)]$$
$$R_s(\tau) = E[n_s(t)n_s(t + \tau)]$$
$$R_{cs}(\tau) = E[n_c(t)n_s(t + \tau)]$$
$$R_{sc}(\tau) = E[n_s(t)n_c(t + \tau)] \qquad (1.115)$$
$$R_c(\tau) = R_s(\tau)$$
$$R_{cs}(\tau) = -R_{sc}(\tau)$$

Note then that

$$R_{cs}(0) = R_{sc}(0) = E[n_c(t)n_s(t)] = 0$$

where $n_c(t)$ and $n_s(t)$ are orthogonal processes. It can be shown that

$$R_n(\tau) = R_c(\tau)\cos(2\pi f_c \tau) + R_{cs}\sin(2\pi f_c \tau)$$

The autocorrelation of the complex envelope is

$$R_u(\tau) = \frac{1}{2} E[u(t)u^*(t + \tau)]$$

$$= R_c(\tau) + iR_{cs}(\tau)$$

Note that

$$n(t) = \frac{1}{2}[z(t) + z^*(t)]$$

$$= \left(\frac{1}{2}\right)[u(t)e^{i2\pi f_c t} + u^*(t)e^{-i2\pi f_c t}]$$

$$R_n(\tau) = \left(\frac{1}{2}\right)[R_u(\tau)\exp(i2\pi f_c t) + R_u^*(\tau)\exp(-i2\pi f_c \tau)]$$

The power spectral density of the noise is

$$S_n(f) = \left(\frac{1}{2}\right)[S_u(f - f_c) + S_u^*(-f - f_c)]$$

But $S_u(f)$ is a real function. Hence,

$$S_n(f) = \left(\frac{1}{2}\right)[S_u(f - f_c) + S_u(-f - f_c)] \tag{1.116a}$$

The average noise power is

$$E[N(t)]^2 = R_c(0) = R_x(0)$$

The representation of noise by Eq. (1.110b) is called a *narrowband noise* if $S_{n_c}(f)$ and $S_{n_s}(f)$ and nonzero in $|f \pm f_c| < B, f_c \gg B$, and zero elsewhere.

It can be shown that the PSD of $n_c(t)$ and $n_s(t)$ are given by

$$S_c(f) = S_s(f) = S_n(f + f_c) + S_n(f - f_c) \tag{1.116b}$$

Note that $S_c(f) = N_0$ if $S_n(f) = N_0/2$ and $E(n_c^2) = N_0 B$, where B is the bandwidth of the filter.

Let the received signal be (see Fig. 1.3)

$$x(t) = s(t) + n(t) \qquad -\infty < t < \infty$$

where $s(t)$ is a deterministic signal and $n(t)$ is a zero-mean stationary noise. We would like to have a filter that will yield the maximum signal to noise power. Let $h(t)$ be the impulse response function of a time-invariant filter. Let $y(t)$ be the output of the filter with input signal $x(t)$:

$$y(t) = \int_{-\infty}^{\infty} x(\tau)h(t - \tau)\,d\tau$$

$$= \int_{-\infty}^{\infty} [s(\tau) + n(\tau)]h(t - \tau)\,d\tau$$

$$= S_0(t) + n_0(t) \qquad -\infty < t < \infty \tag{1.117}$$

where

$$S_0(t) = \int_{-\infty}^{\infty} S(\tau)h(t - \tau)\,d\tau \tag{1.118}$$

$$N_0(t) = \int_{-\infty}^{\infty} n(\tau)h(t - \tau)\,d\tau \tag{1.119}$$

The PSD of the output noise $n_0(t)$ is

$$S_{n_0}(f) = |H(f)|^2 S_n(f) \tag{1.120}$$

using Eq. (1.108).

Therefore the output noise power is

$$E[N_0^2(t)] = \int_{-\infty}^{\infty} S_n(f)|H(f)|^2\,df \tag{1.121}$$

The output signal power is

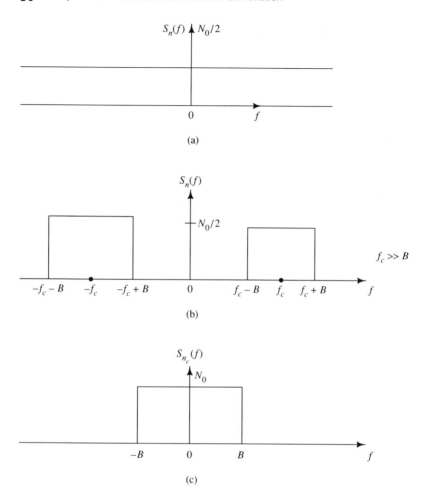

FIGURE 1.3   White noise power spectral density. (a) Additive white noise. (b) Narrowband white noise for the ideal band-pass filter. (c) In-phase or quadrature phase white noise for ideal band-pass filter.

$$S_0^2(t) = \left[\int_{-\infty}^{\infty} S(\tau)h(t - \tau)\, d\tau \right]^2$$

$$= \left[\int_{-\infty}^{\infty} S(f)H(f)e^{i2\pi ft}\, df \right]^2 \qquad -\infty < t < \infty \qquad (1.122)$$

using Parseval's theorem (see Mohanty, 1987). Therefore, the output signal power at $t = T$ is

$$S_0^2(T) = \left(\int_{-\infty}^{\infty} S(f)H(f)e^{i2\pi fT}\, df \right)^2 \qquad (1.123)$$

The output power signal to noise ratio (SNR) is

$$\text{SNR} = \frac{\left(\int_{-\infty}^{\infty} S(t)H(f)e^{i2\pi fT}\,df\right)^2}{\int_{-\infty}^{\infty} S_n(f)|H(f)|^2\,df} \tag{1.124}$$

using the Schwartz inequality; that is (see Mohanty, 1987),

$$\left|\int_{-\infty}^{\infty} X(f)Y(f)\,df\right|^2 \leq \int_{-\infty}^{\infty} |X(f)|^2\,df \int_{-\infty}^{\infty} |Y(f)|^2\,df \tag{1.125}$$

Therefore,

$$\text{SNR} = \frac{\left|\int_{-\infty}^{\infty} \frac{S(f)e^{i2\pi fT}}{\sqrt{S_n(f)}} \cdot \sqrt{S_n(f)}H(f)\,df\right|^2}{\int_{-\infty}^{\infty} S_n(f)|H(f)|^2\,df}$$

$$\leq \frac{\left|\int_{-\infty}^{\infty} \frac{|S(f)|^2}{\sqrt{S_n(f)}}\,df \int_{-\infty}^{\infty} S_n(f)|H(f)|^2\,df\right|}{\int_{-\infty}^{\infty} S_n(f)|H(f)|^2\,df} \tag{1.126}$$

$$\leq \int_{-\infty}^{\infty} \left[\frac{|S(f)|^2}{S_n(f)}\right]df \tag{1.127}$$

The equality holds when each of the two functions in the integral in Eq. (1.126) is equal to a constant times the complex conjugate of the other. This implies

$$\sqrt{S_n(f)} \cdot H(f) = K\left[\frac{S(f)e^{i2\pi Tf}}{\sqrt{S_n(f)}}\right]^* \tag{1.128}$$

simplifying Eq. (1.128), we get the optimum filter as

$$H^0(f) = K\left[\frac{S^*(f)e^{-i2\pi fT}}{S_n(f)}\right] \tag{1.129}$$

If the input noise $n(t)$ is white, with the power spectral density $S_n(f) = N_0/2$, the optimum filter is given by

$$H^0(f) = \frac{KS^*(f)e^{-i2\pi f T}}{N_0/2}$$

using Eq. (1.129).

Neglecting the constants, the optimum filter in the presence of white noise is given by

$$H^0(f) = S^*(f)e^{-i2\pi fT} \tag{1.130a}$$

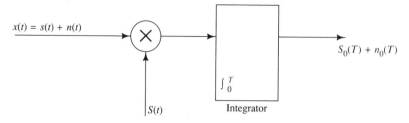

FIGURE 1.4   Matched-filter implementation.

Therefore the optimum filter response function is

$$h(t) = S(T - t) \qquad 0 \le t \le T$$
$$= 0 \qquad\qquad \text{elsewhere}$$

(1.130b)

The output $y(t)$ for the optimum filter is given by

$$y(t) = \int_{-\infty}^{\infty} y(\tau)S(T - t + \tau)\,d\tau \qquad -\infty < t < \infty$$

The output at $t = T$ is

$$y(T) = \int_{-\infty}^{\infty} y(\tau)S(\tau)\,d\tau$$

The optimum filter in the presence of white noise is a correlator.

The filter is called the *matched filter*. The matched-filter operation consists of multiplication of the output waveform with the exact replica of the transmitted waveform $S(t)$ and integrating for T seconds. This implementation is shown in Fig. 1.4.

The signal-to-noise (S/N) power is given in terms of $10\log_{10}(S/N)$—that is, in decibels referred to 1 watt (dBW).

Let the received signal be

$$x(t) = s(t) + n(t)$$

(1.131)

where

$$s(t) = A\cos(2\pi f_c t)$$

and the noise is a narrowband noise given by

$$n(t) = n_c(t)\cos(2\pi f_c t) - n_s(t)\sin(2\pi f_c t)$$

(1.132)

where $n_c(t)$ and $n_s(t)$ are uncorrelated Gaussian processes with zero mean and variance σ^2.

Equation (1.131) combined with Eq. (1.132) can be written

$$x(t) = [A + n_c(t)]\cos(2\pi f_c t) - n_s(t)\sin(2\pi f_c t)$$
$$\triangleq x_c(t)\cos(2\pi f_c t) + x_s(t)\sin(2\pi f_c t)$$

(1.133)

where

$$X_c(t) = A + n_c(t)$$
$$X_s(t) = -n_s(t)$$

Hence,

$$E(X_c) = A$$
$$\text{var}(X_c) = \sigma^2$$
$$E(X_s) = 0 \tag{1.134}$$
$$\text{var}(X_s) = \sigma^2$$

Let $X_c = R\cos\theta$ and $X_s = R\sin\theta$. Then

$$R^2 = X_c^2 + X_s^2$$
$$\theta = \tan^{-1}(X_s/X_c)$$
$$f(X_c, X_s) = \frac{1}{2\pi\sigma^2} \exp\left\{\frac{-[(X_c - A)^2 + X_s^2]}{2\sigma^2}\right\}$$

Proceeding as in Eq. (1.84) through Eq. (1.87), we get

$$f(R, \theta) = \frac{r}{2\pi\sigma^2} \exp\left[-\sigma^2 - 2Ar\cos\theta + \frac{A^2}{2\sigma^2}\right]$$

The marginal density R is given by

$$f_R(r) = \frac{r}{\sigma^2} \exp\left[-r^2 + \frac{A^2}{2\sigma^2}\right] I_0\left(\frac{Ar}{\sigma^2}\right) \qquad r \geq 0$$

$$f_\phi(\theta) = \frac{1}{2\pi} \qquad\qquad\qquad 0 \leq \theta \leq 2\pi \tag{1.135a}$$

$$= 0 \qquad\qquad\qquad\qquad \text{elsewhere}$$

where

$$I_0(x) = (1/2\pi) \int_0^{2\pi} e^{x\cos\theta}\, d\theta \tag{1.135b}$$

$I_0(\cdot)$ is the *modified Bissel function* of zero order. Equation (1.135a) is known as the *Rice probability density*. If $A = 0$, Eq. (1.135a) becomes

$$f_R(r) = \frac{r}{\sigma^2} \exp\left[\frac{-\sigma^2}{2\sigma^2}\right] \qquad r \geq 0$$

which is the *Rayleigh density*.

The input (received) signal-to-noise power ratio is given by

$$\text{SNR} = \frac{A^2/2}{\sigma^2} = \frac{A^2}{2\sigma^2}$$

(X_k) are called wide-sense stationary sequences if

$$E(X_k) = m \text{ (constant)}$$
$$E(|X_k|^2) < \infty \qquad \text{for all } k \tag{1.136a}$$

The autocorrelation of (X_k) is

$$E(X_k X_{k+j}^*) = R_x(j) \qquad \text{for all } k \text{ and } j$$

The *covariance* is defined as

$$C_x(j) = R_x(j) - |m|^2 \tag{1.136b}$$

The PSD of $\{x_k\}$ is defined as

$$S_x(f) = \sum_{j=-\infty}^{\infty} C_x(j) e^{-i2\pi f j} \tag{1.136c}$$

Let the message waveform be

$$m(t) = \sum_{n=-\infty}^{\infty} a_n p(t - nT) \qquad -\infty < t < \infty \tag{1.137}$$

T is the bit interval and $p(t)$ is a pulse of duration T, $\{a_n\}$ is a stationary random process with $E(a_n) = 0$, and

$$E[a_n a_m^*] = R_a(m - n) \tag{1.138}$$

The transmitted signal is

$$X(t) = m(t) \cos(2\pi f_c t) \qquad -\infty < t < \infty \tag{1.139}$$

where $m(t)$ is given in Eq. (1.137). The average value

$$\begin{aligned}
E(X(t)) &= E[m(t)] \cos(2\pi f_c t) \\
&= \sum_n E(a_n) p(t - nT) \cos(2\pi f_c t) \\
&= 0 \tag{1.140}
\end{aligned}$$

$$\begin{aligned}
E[X(t)X^*(t + \tau)] &= \sum_n \sum_m E(a_n a_m^*) p(t - nT_p) p(t + \tau - mT) \\
&\quad \cdot \cos(2\pi f_c t) \cos[(2\pi f_c(t + \tau)] \\
&= \frac{1}{2} [\sum \sum R_a(m - n) p(t - nT_p) p(t + \tau - mT)] \\
&\quad \cdot [\cos(2\pi f_c \tau) + \cos(4\pi f_c t + 2\pi f_c \tau)] \tag{1.141}
\end{aligned}$$

Note that $m(t)$ is a cyclostationary process; that is, the mean and second moments are periodic. We apply Wiener-Kinchin theorem (Mohanty, 1987) to estimate the PSD.

$$S_x(f) = F\left[\lim_{T \to \infty} \frac{1}{T} \int_{-T/2}^{T/2} R_x(t, t + \tau) \, dt \right] \tag{1.142}$$

where

$$F(\cdot) = \int_{-\infty}^{\infty} (\cdot)e^{i2\pi f\tau}\,df$$

It can be shown that

$$S_x(f) = \left(\frac{1}{4}\right)[S_p(f + f_c) + S_p(f - f_c)]S_a(f) \tag{1.143}$$

where

$$S_p(f) = \left(\frac{1}{T}\right)|P(f)|^2$$

$$P(f) = \int_{-\infty}^{\infty} p(t)e^{-i2\pi ft}\,dt \tag{1.144}$$

$$S_a(f) = \sum_{l=-\infty}^{\infty} R_a(l)e^{-2\pi flT}$$

is the PSD of sequence $\{a_k\}$.

Let

$$p(t) = A \qquad |t| < T/2$$
$$= 0 \qquad \text{elsewhere} \tag{1.145}$$

$$P(f) = AT\left(\frac{\sin \pi fT}{\pi fT}\right)$$

Hence

$$S_p(f) = \frac{A^2}{T}\left(\frac{\sin \pi fT}{\pi fT}\right)^2 \tag{1.146}$$

$$E(a_n a_m^*) = R_a(m - n) = 1 \qquad m = n$$
$$= 0 \qquad m = n \tag{1.147}$$

The PSD of $\{a_k\}$ is

$$S_a(f) = \sum_{j=-\infty}^{\infty} R_a(j)e^{-i\pi fjT}$$
$$= 1 \tag{1.148}$$

Therefore, the PSD of $X(t)$, using Eqs. (1.147), (1.146), and (1.142), is

$$S_x(f) = \frac{A^2}{4}T\left[\left\{\frac{\sin[\pi T(f - f_c)]}{\pi T(f - f_c)}\right\}^2 + \left\{\frac{(\sin[\pi T(f + f_c)]}{\pi T(f + f_c)}\right\}^2\right] \tag{1.149}$$

where T is the bit duration. This signaling scheme is known as *binary phase-shift keying* (BPSK). The PSD of QPSK or OQPSK and MSK signals are given by Eq. (1.359) in Section 1.4. See Fig. 11.13.

1.4 MODULATION

Modulation is a process in which transmitted signal parameters vary with message signal (baseband signal) in such a manner that the message can be extracted from the transmitted signals in the reversible process. The reversible process is known as *demodulation*. If the transmitted signal is a pulse whose amplitude, duration, or position varies with the message, then the process is called *pulse modulation*, and the transmitted signal is called *baseband signaling*. If the transmitted signal is a sinusoidal signal, also called a *continuous-wave* (CW) signal, such that the amplitude, phase, or frequency varies with message, then the modulation process is known as *bandpass modulation* and the transmitted signal is *radio-frequency signaling*. Examples of pulse modulations are telegraph and teletype. Examples of bandpass modulations are radio and satellite communications. Most input message signals, whether continuous or discrete, as they come from the transducer cannot be transmitted to a distant place. The message signals are converted to electric signals and transmitted as an electromagnetic wave from the antenna. Efficient electromagnetic radiation requires antennas whose physical dimensions are at least $1/10$ of wavelength λ. The wavelength is defined as c/f_c, where c in the velocity of light, and f_c is the transmitted signal (carrier) frequency. The unit of frequency is the hertz (Hz) and the unit for wavelength is the meter (m). Many baseband signals such as audio signals have frequency components down to 100 Hz or lower, which would require an antenna of height 300 km to radiate the electromagnetic wave directly. By transmitting the signal using a CW signal of radio frequency (RF) at a higher frequency than the maximum frequency content of the baseband signal, the antenna height can be reduced. Modulation can also be called a frequency translation process. The frequency assignment is made by an international organization known as International Telecommunication Union (ITU). One of the units of ITU, the International Radio Consultative Committee (CCIR) is concerned with standards, architecture, or coordination of communication and transmission systems of various national organization. The electromagnetic frequency ranges from 0 Hz to 10^{15} Hz. The frequency range from 0 to 5×10^{11} is called *ratio frequency* (RF). The subsection range from 10^{10} Hz to 5×10^{11} is called *microwave* (millimeter) *frequency*. The range from 5×10^{11} Hz to 5×10^{14} Hz is called the *infrared* (IR) region and the remaining range is called *visible optics*. The frequency designations and bands are given in Table 1.1.

Modulation provides several advantages to transmission of signals. Among them are: (1) frequency translation for ease of radiation (smaller antenna); (2) suppression of noise, interference, and fading; (3) frequency assignment to several users for simultaneous transmission; (4) multiplexing, combining several signals in one carrier; (5) operations at appropriate frequency range of devices such as amplifiers, path-related frequency, low frequency (UHF) for less rain attenuation, and high frequency (EHF) for less nuclear detonated channel degradation; and (6) larger bandwidth in the GHz, region—i.e, a very large number of users can be accommodated.

TABLE 1.1 RF Terminology

3–30 MHz	High frequency (HF)
30–300 MHz	Very high frequency (VHF)
300–3000 MHz	Ultrahigh frequency (UHF)
3–30 GHz	Superhigh frequency (SHF)
30–300 GHz	Extremely high frequency (EHF)

Note: kHz = 10^3 Hz, MHz = 10^6 Hz, GHz = 10^9 Hz.

In this chapter we will consider bandpass modulation or continuous-wave transmission where messages are transmitted varying the appropriate parameter of a sinusoidal signal known as the *carrier*. If the message is continuous, the continuous-wave modulation is called *analog modulation*. If the message is discrete or digitized at the source and the discrete values of message modulates appropriate parameter(s) of the carrier, the modulation is called *digital modulation*.

1.4.1 Analog Modulation

Let the carrier be

$$S(t) = A(t)\cos\theta(t) \tag{1.150}$$

where

$$\theta(t) = 2\pi f_c t + \phi(t)$$
$$= \omega_c t + \phi(t) \tag{1.151}$$

$A(t)$ is called the *instant amplitude*, $\theta(t)$ is called the *instant phase*, f_c is called the *carrier frequency*, and $\phi(t)$ is called the *phase*.

$$\frac{d\theta}{dt} = 2\pi f_c + \frac{d\phi}{dt}$$
$$= \omega_c + \frac{d\phi}{dt} \tag{1.152}$$

is called the *instant frequency*.

$$\frac{1}{2\pi}\frac{d\phi}{dt}$$

is called the *frequency deviation*.

If the amplitude varies with message

$$A(t) = A_c[1 + k_a m(t)]$$

A_c is called the amplitude of the carrier and k_a, $0 \le k_a \le 1$, is called the *amplitude modulation index* ($k_a = 1$ is called 100 percent modulation). Let ϕ_c

be the carrier phase. The transmitted signal, Eq. (1.150), is

$$S(t) = A_c[1 + k_a m(t)] \cos(2\pi f_c t + \phi_c)$$
$$= A_c \cos(2\pi f_c t + \phi_c) + A_c k_a m(t) \cos(2\pi f_c t + \phi_c) \quad 0 \le t \le T$$

(1.153)

is called the *amplitude-modulated* (AM) signal. If $\phi(t)$ in Eq. (1.151) varies with the message, that is,

$$\phi(t) = k_p m(t)$$

(1.154a)

then the transmitted signal with $A(t) = A_c$ in Eq. (1.50) is

$$S(t) = A_c \cos[\omega_c t + k_p m(t)] \quad 0 \le t \le T$$

(1.154b)

is called the *phase-modulated* (PM) signal, where A_c is the amplitude, f_c is the carrier frequency, k_p is the phase modulation index, and $\omega_c = 2\pi f_c$. If $d\phi(t)/dt$ varies with the message, that is,

$$\frac{d\phi}{dt} = k_f m(t)$$

(1.155)

then $\phi(t) = k_f \int_0^t m(\tau)\,d\tau$, where k_f is the *frequency modulation index*. The transmitted signal with $A(t) = A_c$ in Eq. (1.150) is

$$S(t) = A_c \cos\left[2\pi f_c t + k_f \int_0^t m(\tau)\,d\tau \right] \quad 0 < t < T$$

(1.156)

is called the *frequency-modulated* (FM) signal, where A_c is the amplitude and f_c is the carrier of the signal.

AM Signal

Let us assume that the dc value (average) of the message $m(t)$ is zero and $f_c = n/T$, where n is an integer and T is the duration of the signal. It can be shown from Eq. (1.53) that the AM signal power is

$$P_{AM} = P_c(1 + P_m)$$

$$P_c = \frac{A_c^2}{2}$$

(1.157)

$$P_m = \lim_{T \to \infty} k_a^2 \frac{1}{T} \int_{-T/2}^{T/2} m^2(t)\,dt$$

where P_c and P_m are the power of the carrier and message. Let the message spectrum $M(f)$, the Fourier transform of the message $m(t)$ is band-limited to f_m and $f_c \gg f_m$. It can be shown that the amplitude spectrum of the signal is given by

$$|S(f)| = \frac{A_c}{2}[\delta(f + f_c) + \delta(f - f_c)] + \frac{A_c}{2}k_a[M(f + f_c) + M(f - f_c)]$$

(1.158)

The amplitude spectrum of the AM signal $s(t)$ is given in Fig. 1.5.

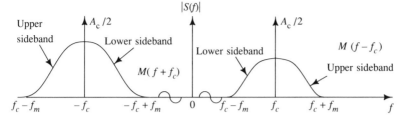

FIGURE 1.5   Spectrum of an AM signal.

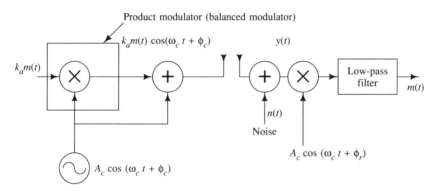

FIGURE 1.6   Amplitude modulation and demodulation.

The shape of the message spectrum is arbitrary. It is shown in Fig. 1.5, that at, $f = \pm f_c$, there is a delta function $(A_c/2)\delta(f)$ and message spectrum $M(f)$. The spectrum in the range of f_c to $f_c + f_m$ is called the *upper sideband*, the range of spectrum in $f_c - f_m$ to f_c is called the *lower sideband*. The transmitted signal bandwidth is $2f_m$, where f_m is the message bandwidth. We are counting $2f_m$ bandwidth only for the positive frequency. The spectrum of the transmitted signal in the negative frequency centered at $f = -f_c$ is the mirror image of the spectrum at the frequency at $f = f_c$.

The message can be extracted by using a coherent demodulator. The modulation and demodulation of an AM signal is shown in Fig. 1.6.

Let the received signal be

$$y(t) = [A_c + k_a A_c m(t)]\cos(2\pi f_c t + \phi_c) + n(t) \tag{1.159}$$

where $n(t)$ is a zero-mean white Gaussian noise with power $N_0/2$. The received signal is multiplied by $\cos(\omega_c t + \phi_r)$, where ϕ_r is the receiver oscillator phase. The output is passed through a low-pass filter (LPF) and an appropriate filter that removes the bias term $A_c/2$. The output signal power of LPF is

$$P_{S(AM)} = P_c P_m \cos^2(\phi_c - \phi_r) \tag{1.160}$$

where P_m is power of the message and output narrowband noise power is

$$P_n = \frac{N_0}{2} \times 2 \times 2f_m = 2N_0 f_m$$

Therefore the output signal-to-noise power for the AM signal is

$$(SNR)_{AM} = \frac{P_c P_m \cos^2(\phi_c - \phi_r)}{2N_0 f_m} \tag{1.161a}$$

$$= \frac{P_c P_m}{2N_0 f_m} \quad \text{if } \phi_c = \phi_r \tag{1.161b}$$

The efficiency of the AM signal is given by

$$\eta = \frac{\text{transmitted power}}{\text{message power}}$$

$$= \frac{P_c(1 + P_m)}{P_m} \tag{1.162}$$

The transmitted bandwidth is

$$B_{AM} = 2f_m = 2(\text{message bandwidth}) \tag{1.163}$$

Note that the message is proportional to the envelope of the transmitted signal. Therefore the message can be extracted by passing the received signal through a band-pass filter followed by an envelope detector.

The noise in the band-pass filter can be modeled as

$$n(t) = n_c(t)\cos(\omega_c t + \phi_c) - n_s(t)\sin(\omega_c t + \phi_c) \tag{1.164}$$

Therefore, the received signal can be written as

$$y(t) = u(t)\cos[\omega_c t + \phi_c + \psi(t)]$$

where

$$u(t) = \{(A_c[1 + k_a m(t)] + n_c(t))^2 + n_s^2(t)\}^{1/2}$$

$$\psi(t) = \tan^{-1}\left[\frac{n_s(t)}{A_c(1 + k_a m(t))} + n_c(t)\right] \tag{1.165}$$

The detector output is given

$$u(t) = [A_c + k_a m(t)]\left\{\frac{1 + 2n_c(t)}{A_c + k_a m(t)} + \frac{n_c^2(t) + n_s^2(t)}{[A_c(1 + k_a m(t))]^2}\right\}^{1/2} \tag{1.166}$$

when

$$A_c[1 + k_a m(t)] \gg \sqrt{E[n_c^2 + n_s^2(t)]} \tag{1.167}$$

where E is the expected operator, Eq. (1.166) gives

$$u(t) = A_c[1 + k_a m(t)] + n_c(t) \tag{1.168}$$

where $E[n_c^2(t)] = N_0$. The output signal-to-noise ratio is

$$(SNR)_{AM} = \frac{P_c P_m}{2f_m N_0} \tag{1.169}$$

which is the same as the output SNR for the coherent detector. Equation (1.167) does not apply when $A_c[1 + k_a m(t)] \ll \sqrt{E(n_c^2 + n_s^2)}$; the output signal-to-noise ratio decreases very rapidly. This transition behavior is usually called the *threshold effect*, usually 7 dB.

It should be noted that the efficiency η of the AM signal is not 100 percent because of the presence of the term $A_c \cos(2\pi f_c t + \phi)$. If this term is not present, then the signal is

$$s(t) = A_c m(t) \cos(\omega_c t + \phi_c) \tag{1.170}$$

where $k_a = 1$. This signal given by Eq. (1.170) is called *double-sideband suppressed carrier* (DSBSC). The amplitude spectrum of the DSBSC is given by

$$|S(f)| = \frac{A_c}{2}[M(f + f_c) + M(f - f_c)]. \tag{1.171}$$

The transmitted signal bandwidth

$$B_{\text{DSBSC}} = 2f_m = 2(\text{message bandwidth}) \tag{1.172}$$

and transmitted power is

$$P_{\text{SDBSSC}} = P_c P_m \tag{1.173}$$

The efficiency η for the DSBSC is

$$\eta = 100 \text{ percent}$$

The DSBSC signal can be demodulated by using a coherent demodulator with extra circuitry to extract the carrier from the received signal. The demodulation process is more complex and costly than AM demodulation. The output signal-to-noise ratio is

$$(\text{SNR})_{\text{DSBSC}} = \frac{P_c P_m}{2N_0 f_m} \tag{1.174}$$

where P_m is the power of the transmitted DSBSC signal and $N_0/2$ is the PSD of the noise present in the receiver.

It should be noted that the upper and lower sidebands are symmetrical. In order to preserve the bandwidth in a band-limited channel, it is necessary to send only the upper sideband or the lower sideband of the message and the other side can be reconstructed from the symmetry.

These signals are called *single-sideband suppressed carrier* (SSBSC); see Fig. 1.7. These signals are given by

$$s(t) = m(t)A_c \cos(2\pi f_c t + \phi_c)$$

$$\pm \hat{m}(t)A_c \sin(2\pi f_c t + \phi_c)$$

$$= \text{Re}[z(t)\exp(\pm i(2\pi f_c t + \phi_c)]A_c \tag{1.175}$$

where $z(t) = m(t) + i\hat{m}(t)$ is an analytical signal, $\hat{m}(t)$ is the Hilbert transform

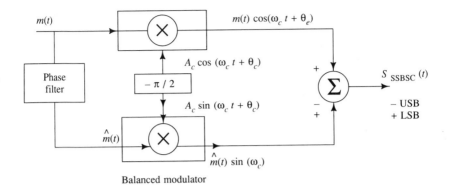

Balanced modulator

FIGURE 1.7   Generation of single-sideband suppressed-carrier signal.

of $m(t)$, and $m(t)$ and $\hat{m}(t)$ are orthogonal and have the same energy. The amplitude spectrum [see also Eq. (1.59)] of $s(t)$ is

$$|S(f)| = \frac{A_c}{2}[Z^*(-f - f_c) \pm Z(f - f_c)]^\epsilon$$

The implementation of SSBSC is shown in Fig. 1.7. The phase shifter filter has an impulse response function $h(t) = 1/\pi t$. The output $\hat{m}(t)$ of the phase shifter filter is the Hilbert transform of the message $m(t)$.

The spectrum of the SSBSC for the upper side band and lower side band is shown in Fig. 1.8. The transmitted signal bandwidth is

$$B_{SSBSC} = f_m \tag{1.176}$$

which is the same as the message bandwidth. The output signal-to-noise ratio of LPF for SSBSC is

$$(SNR)_{SSBSC} = \frac{P_m P_c}{N_0 f_m} \tag{1.177}$$

If we denote $(SNR)_i$ as the signal to the noise ratio of the input (received) signal and $(SNR)_0$ as the output SNR of the demodulated signal, then we have the following comparison:

$$\frac{(SNR)_0}{(SNR)_i} = \begin{cases} P_m/(1 + P_m) & \text{for AM} \\ 2 & \text{for DSBSC} \\ 1 & \text{for SSBSC} \end{cases}$$

$^\epsilon$For the negative sign

$$S(f) = A_c \begin{cases} M(f - f_c) & f > f_c + A_c \\ 0 & f < f_c \end{cases} \begin{cases} 0 & f > -f_c \\ M(f + f_c) & f < -f_c \end{cases}$$

for $\hat{M}(f) = -iM(f) \quad f > 0 \qquad \hat{M}(f) = iM(f)$ if $f < 0$.

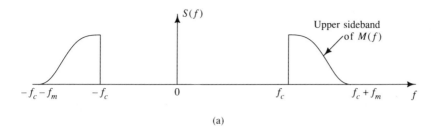

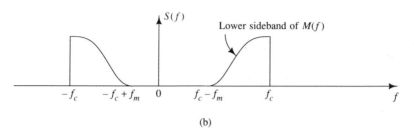

FIGURE 1.8   Spectrum of the single-sideband suppressed carrier.
(a) Upper-sideband SSBSC. (b) Lower-sideband SSBSC.

where

$$P_m = \lim_{T \to \infty} \frac{1}{T} \int_{-T/2}^{T/2} m^2(t)\, dt \tag{1.178}$$

The bandwidth comparison is as follows:

$$B = \begin{cases} 2f_m & \text{for AM} \\ 2f_m & \text{for DSBSC} \\ f_m & \text{for SSBSC (upper or lower)} \end{cases} \tag{1.179}$$

In certain applications, such as television, a trace of spectrum of the lower sideband is sent with the single-sideband suppressed-carrier (upper-sideband) transmitted signal. This modulation is called *vestigial sideband* (VSB) modulation. The transmitted power for the VSB is the same as the SSBSC signal and bandwidth of VSB is slightly larger than the SSBSC but much less than the DSB SC signal.

Angle Modulation

When the phase $\phi(t)$ or $d\phi/dt$ varies with the message, the modulation process is called *angle modulation* or *nonlinear modulation.*

If the message is a tone signal,

$$m(t) = \sin 2\pi f_m t$$

then the FM signal given by Eq. (1.156) is

$$s(t) = A_c \cos[2\pi f_c t + (k_f/2\pi f_m) \cos 2\pi f_m t] \qquad (1.180)$$

where $k_f/2\pi f_m$ is called the *frequency modulation index*. Using the identity

$$\cos(2\pi ft + a \sin \theta) = \sum_{n=-\infty}^{\infty} J_n(a) \cos(2\pi ft + n\theta)$$

where $J_n(\cdot)$ is the *n*th-order Bessel function, we can show that the FM signal given by Eq. (1.180) is

$$S(t) = A_c \sum_{n=-\infty}^{\infty} J_n(\beta) \cos(2\pi f_c t + n2\pi f_m t) \qquad (1.181)$$

where $\beta = k_f/2\pi f_m$. The amplitude spectrum of the FM given by Eq. (1.181) is

$$|S(f)| = \frac{A_c}{2} \sum_{n=-\infty}^{\infty} J_n(\beta)\{\delta[f + (f_c + nf_m)] + \delta[f - (f_c + nf_m)]\} \qquad (1.182)$$

The amplitude spectrum has an infinite number of discrete weights $A_c J_n(\beta)/2$ at $f = f_c + nf_m, n = \pm 1, \pm 2, \ldots$. The Bessel functions have the property that

$$J_{-n}(\beta) = (-1)^n J_n(\beta) \qquad \sum_{n=-\infty}^{\infty} J_n^2(\beta) = 1 \qquad (1.183)$$

Values of $J_n(\beta)$ decreases significantly for $n > \beta$. The significant amount (~ 98 percent) of power is contained in the bandwidth in B_{FM} where

$$B_{FM} = 2f_m(\beta + 1)$$
$$= 2(f_m\beta + f_m)$$
$$= 2(\Delta f + f_m) \qquad (1.184)$$

where Δf is called the *peak frequency deviation* for a sinusoidually modulated FM signal, $\Delta f = f_m\beta$. When $\beta < 1$, the FM signal is called *narrowband FM*. When $\beta > 1$, the signal is called *wideband FM*. In general, the bandwidth for FM signals is given by

$$B_{FM} = 2f_m(D + 1) = 2(\Delta f + f_m) \qquad (1.185)$$

when f_m is the message bandwidth and $\Delta f = Df_m$, $D = \max|m(t)|$. Equation (1.185) is called *Carson's rule*. Note that the bandwidth of the FM signal is greater than the bandwidth of the AM signal by an amount $2\Delta f$.

The generation and demodulation of FM signal is shown in Fig. 1.9. The FM signal is superheterodyned from RF to an intermediate frequency (IF), say 70 MHz, by using a mixer. The mixer gain is taken as unity. The IF filter bandwidth is larger than the FM bandwidth so that the IF filter will not distort the FM signal. The limiter restores the amplitude fluctuations, and the band-pass filter eliminates any harmonics. The discriminator is a slope network with an envelope detector. The output of the discriminator is proportional to the message. The output of the message is passed through a deemphasis filter to deattenuate the high-frequency components which were attenuated at the

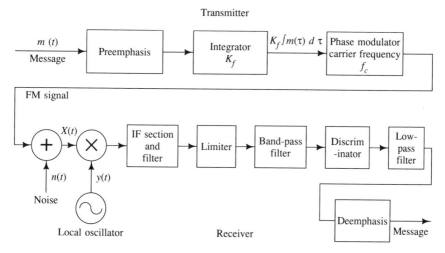

FIGURE 1.9   FM transmitter and receiver.

transmitter before the modulation. Finally, the message is recovered from the deemphasis filter. Let the received signal at the receiver be

$$x(t) = s(t) + n(t)$$

$$= A_c \cos\left[2\pi f_c t + k_f \int_0^t m(\tau) \, d\tau + \phi_c \right]$$

$$+ n_c \cos(\omega_c t + \phi_c) - n_s(t) \cos(\omega t_c + \phi_c) \qquad (1.186)$$

where $n_c(t)$ and $n_s(t)$ are uncorrelated narrowband white Gaussian noise in-phase and quadrature phase components.

The received signal can also be written as

$$x(t) = A_c \cos\left[2\pi f_c t + k_f \int_0^t m(\tau) \, d\tau + \phi_c \right] + r_n(t) \cos(2\pi f_c t + \phi_c + \phi_n) \qquad (1.187)$$

where $r_n(t)$ is a Rayleigh process and ϕ_n is uniform. The noise term can be written as

$$r_n \cos\left[2\pi f_c t + k_f \int_0^t m(\tau) \, d\tau + \phi_c + \phi_n - k_f \int_0^t m(\tau) \, d\tau \right]$$

$$= r_n \cos[\phi_n(t) - \phi(t)] \cos[2\pi f_c t + \phi_n + \phi(t)]$$

$$- r_n(t) \sin[(\phi_n(t) - \phi(t)] \sin[\omega_c t + \theta + \phi(t)] \qquad (1.188)$$

where

$$\phi(t) = k_f \int_0^t m(\tau) \, d\tau$$

Combining Eqs. (1.187) and (1.188), we get

$$x(t) = \{A_c + r_n(t)\cos[\phi_n(t) - \phi(t)]\}\cos[\omega_c t + \phi_c + \phi(t)]$$
$$- r_n(t)\sin[\phi_n(t) - \phi(t)]\sin[\omega_c t + \phi_c + \phi(t)]$$
$$= R(t)\cos[\omega_c t_c + \phi_c + \phi(t) + \psi(t)] \tag{1.189}$$

where

$$R(t) = [\{A_c + r_n\cos[\phi_n(t) - \phi(t)]\}^2 + \{r_n\sin[\phi_n(t) - \phi(t)]\}^2]^{1/2} \tag{1.190}$$

$$\psi(t) = \tan^{-1}\left\{\frac{r_n(t)\sin[\phi_n(t) - \phi(t)]}{A_c + r_n(t)\cos[\phi_n(t) - \phi(t)]}\right\} \tag{1.191}$$

Case 1. Assume that

$$E[A_c - r_n(t)]^2 \gg 0 \tag{1.192}$$

Under this assumption, Eq. (1.191) becomes

$$\psi(t) = \frac{r_n(t)}{A_c}\sin[\phi_n(t) - \phi(t)] \tag{1.193}$$

$R(t)$ in Eq. (1.189) remains constant at the output of the limiter, and the output of the discriminator is

$$y_D(t) = \frac{d}{dt}K_D\left\{\phi(t) + \left(\frac{r_n(t)}{A_c}\right)\sin[\phi_n(t) - \phi(t)]\right\}$$
$$= k_D k_f m(t) + k_d \frac{d}{dt}\left\{\left(\frac{r_n(t)}{A_c}\right)\sin[\phi_n(t) - \phi(t)]\right\}$$
$$\triangleq k_D k_f m(t) + n_0(t)$$
$$\triangleq s_0(t) + n_0(t) \tag{1.194}$$

where $s_0(t)$ and $n_0(t)$ are the output signal and noise at the discriminator with gain k_D. We assume that signal $s_0(t)$ and noise $n_0(t)$ are wide-sense stationary. In order for $n_0(t)$ to be stationary, $n_0(t)$ must be independent of message $m(t)$. Therefore we set $m(t) = 0$; that is, $\phi(t) = 0$. In this case,

$$n_0(t) = \left(\frac{1}{A_c}\right)k_d \frac{d}{dt}[r_n\sin\phi_n(t)]$$
$$= \frac{k_d}{A_c}\frac{d}{dt}[n_s(t)] \tag{1.195}$$

The discriminator transfer function is given as

$$H(f) = i2\pi f \tag{1.196}$$

The PSD of $n_0(t)$ (with no deemphasis) is

$$S_{n_0}(f) = k_D^2 |2\pi i f|^2 S_{ns}(f) \left(\frac{1}{A_c^2}\right) \qquad |f| \le B_{FM}$$

$$= k_D^2 \left(\frac{N_0}{A_c^2}\right) \cdot f^2 4\pi^2 \tag{1.197a}$$

Hence output noise power in the absence of deemphasis filter is

$$E(n_0^2) = \int_{-fm}^{fm} S_{n_0}(f)\, df = \frac{8 k_D^2 \pi^2 N_0 f_m^3}{3 A_c^2} \tag{1.197b}$$

The output signal power is

$$E[S_0^2(t)] = k_D^2 k_f^2 E[m^2(t)] = k_D^2 k_f^2 P_m$$

$$(SNR)_{OFM} = \frac{3 k_D^2 k_f^2 P_m A_c^2}{k_D^2 2 f_m^3 N_0 4\pi^2} \tag{1.198a}$$

where P_m is the power of the baseband signal, message $m(t)$. The output SNR can be written as

$$(SNR)_{OFM} = 3 \left(\frac{k_f^2 P_m}{4\pi^2 f_m^2}\right) \frac{A_c^2/2}{N_0 f_m} \tag{1.198b}$$

$$= 3\beta_1^2 \left(\frac{A_c^2/2}{N_0 f_m}\right) = 3\beta_1^2 (SNR)_B \tag{1.98c}$$

where $(SNR)_B$ is the carrier-to-noise ratio in the baseband, and

$$\beta_1 = \sqrt{P_m} k_f/(2\pi f_m) \tag{1.199}$$

If we assume

$$P_m = 1, \text{ where } P_m \text{ is the power of } m(t) \tag{1.200}$$

then

$$\beta_1 = k_f/(2\pi f_m) \tag{1.201}$$

which is the FM modulation index. The output SNR at the demodulator is

$$(SNR)_{OFM} = 3\beta^2 (SNR)_{baseband} \tag{1.202}$$

where $3\beta^2$ is called the FM *improvement factor*. The signal-to-noise ratio at the baseband is as follows:

$$(SNR)_B = \frac{A_c^2/2}{N_0 f_m}$$

$$= \left[\frac{A_c^2/2}{N_0^2 (\beta + 1) f_m}\right] 2(\beta + 1)$$

$$= \left(\frac{A_c^2/2}{N_0 B_{FM}} \right) 2(\beta + 1)$$

$$= (CNR)_{IF} 2(\beta + 1) \tag{1.203}$$

where

$$B_{FM} = 2(\beta + 1)f_m$$

Therefore, the output SNR can be written as

$$(SNR)_{OFM} = 6\beta^2(\beta + 1)(CNR)_{IF} \tag{1.204}$$

where $(CNR)_{IF}$ is the carrier-to-noise power ratio at the IF filter. From Eq. (1.204) it is seen that the output SNR can be increased by increasing β. The increase of β also increases the FM signal bandwidth. This increase of β also leads to a decrease of $(CNR)_{IF}$. This decrease of $(CNR)_{IF}$ may not satisfy the condition $E[A_c - r_n(t)]^2 \gg 0$. This requires

$$(CNR)_{IF} = \frac{A_c^2/2}{N_0 B_{FM}} \geq \text{Th} \tag{1.205}$$

where Th is the threshold required to maintain the approximation

$$\psi(t) \approx \frac{r_n(t)}{A_c} \sin[\phi_n(t) - \phi(t)]$$

The threshold value is about 12 to 16 dB. Below the threshold value, the FM detector produces clicks.

   Case 2. If $A_c \ll r_n(t)$, then

$$\psi(t) > \frac{r_n(t)}{A_c} \sin[\phi_n(t) - \phi(t)]$$

The noise parameter $\phi_n(t)$ dominates; that is, $(CNR)_{IF}$ is very weak. The message $m(t)$ is lost. The message can be recovered if $(CNR)_{IF}$ is above the threshold. The threshold extension can be done by using feedback tracking, which will be discussed later.

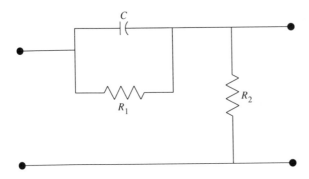

FIGURE 1.10   Preemphasis filter.

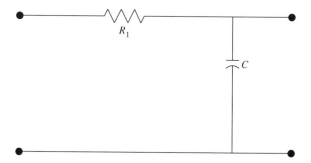

FIGURE 1.11   Deemphasis filter.

It is seen that the noise power increases with frequency. The frequency discriminator noise will attenuate the high-frequency terms. Therefore at the preemphasis the high-frequency terms of the message is attenuated by using a preemphasis network. A typical preemphasis filter is given by the transfer function

$$H_p(f) = \frac{if + 1/(2\pi R_1 C)}{if + (R_1 + R_2)/R_1 R_2 2\pi C} \tag{1.206}$$

where R_1 and R_2 and C are given in Fig. 1.10. The modified message from the preemphasis filter is restored by using an inverse filter known as deemphasis filter. A typical deemphasis filter transfer function is given by

$$H_d(f) = \frac{1/(\pi R_1 C)}{if + 1/(2\pi R_1 C)} = \frac{f_1}{if + f_1} \tag{1.207}$$

where

$$f_1 = \frac{1}{2\pi R_1 C}$$

and R_1 and C are given in Fig. 1.11.

The output noise $n_0(t)$ power with a deemphasis filter is

$$E[n_0^2(t)]_{\text{with deemphasis}} = N_0 f_1^2 \left(\frac{K_d^2}{A_c^2}\right)\left[f_m - f_1 \tan^{-1}\left(\frac{f_m}{f_1}\right)\right](2\pi)^2 \tag{1.208}$$

Therefore the ratio of output noise with and without deemphasis filter is as follows:

$$\frac{E[n_0^2(t)]_{\text{with deemphasis}}}{E[n_0^2(t)]_{\text{without deemphasis}}} = 3\left[\left(\frac{f_1}{f_m}\right)^2 - \left(\frac{f_1}{f_m}\right)^3 \tan^{-1}\left(\frac{f_m}{f_1}\right)\right]^d \tag{1.209}$$

using Eqs. (1.208), (1.197a), and (1.197b).

$$^d \int_{-f_m}^{f_m} S_{n_s}(f)|H_d(f)|^2 \, df = 2\int_0^{f_m} k_d^2 \frac{N_0 f^2}{A_c^2}\left[\frac{f_1^2}{f_1^2 + f^2}\right]df$$

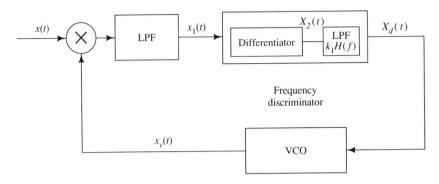

FIGURE 1.12   FM demodulation with feedback (FMFB) tracking.

If $f_1 = 2.1$ kHz and $f_m = 15$ kHz, then this ratio is equal to 0.047. This implies that the insertion of preemphasis and deemphasis filters yields a noise reduction of more than 12 dB for these frequencies.

We will discuss FM threshold extension by using feedback loops. The FM feedback demodulator is shown in Fig. 1.12. The feedback path has a voltage-controlled oscillator (VCO) whose output frequency deviation is proportional to the input voltage. The output of the VCO is given by

$$X_v(t) = \sin(2f_c \pi t + \theta_v(t)) \qquad \omega_c T = r, \text{ an integer}$$

where

$$\frac{d\theta_v}{dt} = k_D X_d(t) \tag{1.210}$$

where k_D is a constant and $X_d(t)$ is the output of the FM discriminator. The received signal $x(t)$ is multiplied by the output VCO signal $X_v(t)$ and is passed through a low-pass filter. The received signal is

$$X(t) = A_c \cos\left(2\pi f_c t + k_f \int_0^t m(\tau)\, d\tau\right) + n(t)$$

where $n(t)$ is a narrowband white Gaussian noise. The output of the low-pass filter is

$$X_1(t) = k_m A_c \sin[\theta_e(t)] + n_m(t) \tag{1.211}$$

where the double-frequency terms have been neglected,

$$\theta_e(t) = \theta(t) - \theta_v(t) \tag{1.212a}$$

$$\theta(t) = k_f \int_0^t m(\tau)\, d\tau \tag{1.212b}$$

k_m = multiplier gain

and $n_m(t)$ is the output noise of the mixer. Let us assume that $\theta_e(t)$, the phase difference between the received signal and the output VCO signal, is very small; that is,

$$|\theta_e(t)| < \frac{\pi}{6} \qquad (1.213)$$

Under this assumption, Eq. (1.211) becomes

$$X_1(t) = k_m A_c \theta_e(t) + n_m(t) \qquad (1.214)$$

Differentiating Eq. (1.214), we get

$$\frac{dX_1}{dt} = k_m A_c \left(\frac{d\theta}{dt} - \frac{d\theta_v(t)}{dt} \right) + \frac{dn_m(t)}{dt} \qquad (1.215a)$$

since

$$\frac{dX_1}{dt} = X_2(t) = A_c k_m \left[\frac{d\theta}{dt} - \frac{d\theta_v(t)}{dt} \right] + \frac{dn_m}{dt} \qquad (1.215b)$$

Taking the Fourier transform of both sides of Eq. (1.215b) and simplifying, we get

$$X_2(f) = A_c k_m \mathrm{i} f(\theta(f) - \theta_v(f)) + [N_m(f)]\mathrm{i} f \qquad (1.216)$$

The equivalent baseband model is shown in Fig. 1.13. Further,

$$X_d(f) = k_1 H(f) X_2(f) \qquad (1.217)$$

where $H(f)$ is the transfer function of the LPF and k_D is the discriminator gain. From Eqs. (1.216) and (1.217) we get

$$X_d(f) = k_1 H(f) \left\{ \mathrm{i} f A_c k_m \left[\theta(f) - \frac{k_D X_d(f)}{\mathrm{i} f} \right] \right\}$$
$$+ k_1 H(f) \mathrm{i} f N_m(f) \qquad (1.218)$$

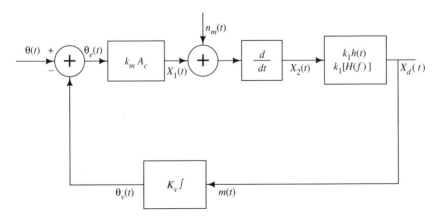

FIGURE 1.13   Equivalent baseband FMFB system.

Using Eqs. (1.4-62) and (1.4-68), we get

$$X_d(f) = \left[\frac{A_c k_m k_1 H(f)}{1 + kH(f)A_c}\right] k_f M(f) + \left[\frac{k_1 H(f)}{1 + kH(f)A_c}\right] if N_m(f) \qquad (1.219)$$

where we have used

$$\theta(f) = \frac{K_f M(f)}{if}$$

$$k = k_1 K_m k_D$$

$M(f)$ is the Fourier transform of $m(t)$. Let

$$H(f) = 1 \qquad \text{for } |f| \le B_{IF}$$
$$= 0 \qquad \text{elsewhere} \qquad (1.220)$$

Equations (1.219) and (1.220) yield

$$X_d(f) = \left(\frac{A_c k_m k_1}{1 + A_c k}\right)\left[k_f M(f) + if\frac{N_m(f)}{A_c k_m}\right]$$
$$= H_1(f)M(f) + H_2(f)N_m(f)$$
$$= S_0(f) + N_0(f) \qquad (1.221)$$

where

$$H_1(f) = \frac{A_c k_m k_1 k_f}{1 + A_c k} \qquad (1.222)$$

$$H_2(f) = \frac{A_c k_m k_1 if}{A_c k_m(1 + A_c k)} \qquad (1.223)$$

The output signal power is

$$E[S_0(t)]^2 = k_f^2 \left(\frac{A_c k_m k_1}{1 + A_c k}\right)^2 P_m$$

where P_m is the message power. The output noise power is

$$E[n_0(t)^2] = \int_{-f_m}^{f_m} \frac{A_c k_m k_1 f^2}{A_c^2 k_m^2(1 + A_c k_m)^2} \cdot \frac{N_0}{2} df$$

$$E[n_0(t)]^2 = \frac{N_0}{A_c^2}\left(\frac{A_c k_m k_1}{1 + A_c k}\right)^2 \frac{2f_m^3}{3} \qquad (1.224)$$

The output signal-to-noise power is

$$(\text{SNR})_0 = \frac{3A_c^2 k_f^2}{2f_m^3 N_0} P_m \qquad (1.225a)$$

$$= \frac{3}{2}\frac{k_f^2 A_c^2}{f_m^3 N_0} \qquad \text{if } P_m = 1 \qquad (1.225b)$$

which is same as Eq. (1.198b), the frequency discriminator demodulator SNR. However, in this derivation, we do not have to assume

$$E(A_c - N_0)^2 \gg 0$$

that is, we do not have a threshold of 10 to 14 dB for (CNR)IF. However, in this derivation we have assumed that the tracking error $\theta_e(t)$ is very small, and hence

$$\sin \theta_e(t) \approx \theta_e(t)$$

When the tracking error is large, the loop gain varies with the carrier amplitude. The carrier amplitude can be kept constant by using a limiter or automatic gain control (AGC) at the input. Therefore, the feedback tracking loop provides a signal-to-noise ratio of about 10 to 14 dB.

We will consider transmission of several baseband signals simultaneously in a channel. Simultaneous transmission of several baseband signals over a single channel by frequency or time translation with an appropriate separation is known as *frequency division* or *time division multiplexing* (FDM or TDM). We consider first frequency division multiplexing (FDM) in which the available bandwidth is divided into N number of nonoverlapping frequency slots and each message is assigned to a frequency slot. In order to avoid any crosstalk, any frequency overlapping, there is a guard band between every two messages. Let $m_1(t), m_2(t), \ldots, m_N(t)$ be messages with bandwidths $B_1, B_2, \ldots,$ B_N. Each message is passed through a low-pass filter to eliminate the residual frequencies and each assigned to a modulator (say, SSB) and a subcarrier. Then the modulated signals are summed; that is,

$$x(t) = \sum_{i=1}^{N} x_i(t) \qquad (1.226)$$

where $x_i(t) = m_i(t)\cos(2\pi f_i t + \phi_i) + f_i, i = 1, 2, \ldots, N$, is called the subcarrier.

The signal $x(t)$ is known as multiplexed baseband signal. The baseband signal $x(t)$ is modulated either by AM or FM. The transmitted signal for SSBSC modulation is

$$x_c(t) = A_c x(t)\cos(2\pi f_c t + \phi) + \hat{x}(t)\sin(2\pi f_c t + \phi) \qquad (1.227a)$$

where $\hat{x}(t)$ is the Hilbert transform of $x(t)$. For the FM signal,

$$x_c(t) = A_c \cos\left[2\pi f_c t + k_f \int_0^t x(\tau)\,d\tau \right] \qquad (1.227b)$$

The received signal is first carrier-demodulated. The demodulated signal $x(t)$ is then passed through a bank of N bandpass filters to extract appropriate $x_i(t)$ and then the message $m_i(t)$, $i = 1, 2, \ldots, N$, is recovered.

Figure 1.14 shows the amplitude spectrum of three baseband signals and the multiplexed SSBSC signal, whereas Fig. 1.15 shows frequency multiplexing and demultiplexing for amplitude modulation.

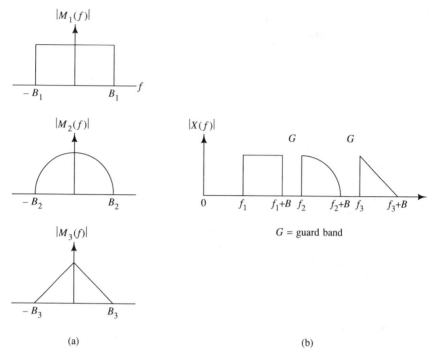

(a)   (b)

FIGURE 1.14   Multiplexing of three baseband signals. (a) Individual amplitude spectra of three baseband signals. (b) Amplitude spectrum of three baseband one-sided signals.

The minimum bandwidth of N message signals is

$$B = B_1 + G + B_2 + \cdots + B_N = (N-1)G + \sum_{i=1}^{N} B_i \qquad (1.228)$$

where G is the guard-band frequency, B_i is the message bandwidth of the ith message, $i = 1, 2, \ldots, N$, and N is the number of multiplexed messages. FDM is widely used in telephone systems for transmitting a large number of voice signals through a common carrier. Each message bandwidth is 3.4 kHz, and a bandwidth of 6 kHz is used. The Bell System uses a supermaster group to transmit 3600 voice channels using a 17,000-kHz bandwidth in the frequency range of 500 to 17,500 kHz.

In the CCIR *Handbook on Satellite Communication* (1985) the multiplexing process is recommended as follows:

1. Each telephone channel transmits audio frequencies from 0.3 to 3.4 kHz, the baseband signals being in the form of SSBSC with suppressed carriers at 4-kHz spacing.

2. Twelve telephone channels are frequency-converted to compose a basic group in the frequency range of 60 to 108 kHz.

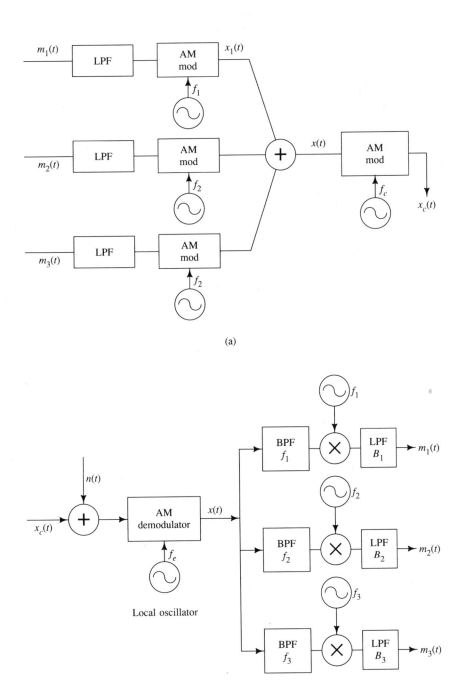

FIGURE 1.15 (a) Multiplexing and (b) demultiplexing of three baseband signals.

3. Five basic groups are again frequency-converted to compose a basic supergroup in the frequency range from 312 to 552 kHz.

4. Frequency conversion of this basic supergroup can be realized to multiplex many further telephone channels.

The basic relationship between the carrier-to-noise and signal-to-noise ratio in frequency division multiplexing FM (FDM-FM) system can be expressed as follows:

$$SNR = \frac{3(f_{tt})^2}{f_2^3 - f_1^3} B_{RF} p W(CNR) \tag{1.229}$$

where

SNR = the ratio of test tone (1 mW) at the point of zero relative level to the psophometrically weighted noise power in the highest telephone channel

CNR = the carrier-to-noise ratio in the radio-frequency bandwidth

f_2 = the upper frequency band of the passband of the highest baseband channel (Hz)

f_1 = the lower frequency band of the passband of the highest baseband channel (Hz)

W = the psophometric (picowatt, or 10^{-12} watt) weighting factor (pWp or, in dB above a 1 pWp reference, abbreviated dBp. If p is an absolute power level, then

$$p \text{ in dBp (unweighted)} = 10 \log_{10}(p \text{ in pWp}) + 2 \cdot 5$$

$$p \text{ in dBm (unweighted)} = p \text{ in dBp} - 90$$

Assuming a 0-dBm signal level,

$$SNR(\text{unweighted}) = p(\text{dBm}) = 87.5 - 10 \log_{10}(p \text{ in pWp})$$

$$SNR(\text{weighted}) = 90 - 10 \log_{10}(p \text{ in pWp})$$

$$f_{tt} = \text{test tone RMS deviation per channel (Hz)}$$

$$p = \text{preemphasis improvement factor}$$

$$B_{RF} = 2(\Delta f + f_m) \text{ (Carson's rule)}$$

where

$\Delta f = f_{tt} g l$

g = peak to RMS (numerical ratio)

= 10 to 13 dB (3.16 to 4.47)

l = loading factor

= $10^{(-15+10 \log N)/20}$ if $N \geq 240$

= $10^{(-1+4 \log N)/20}$ if $N < 240$

For one *single-channel-per-carrier (SCPC)* FM demodulator, the relationship between the CNR and SNR is given by

$$\text{SNR} = \frac{(f_{tt})^2}{(f_m)^2 + b^2/12}(B_{RF}/b)pW(\text{CNR}) \tag{1.230a}$$

where $b = f_2 - f_1$. If $b > 4f_m = 2(f_1 + f_2)$,

$$\text{SNR} = \left(\frac{f_{tt}}{f_m}\right)^2 (B_{RF}/b)pW(\text{CNR}) \tag{1.230b}$$

and other parameters are defined in Eq. (1.79), and $p = 2.5$ dB and $W = 4$ dB. Then

$$B_{RF} = 2(\Delta f + f_2)$$
$$\Delta f = f_{tt}gl \tag{1.230c}$$
$$l = 10^{(m+.115\beta^2)/20}$$

where m and β are the mean (dBm0) and standard deviation of the statistical distribution of the speech level in dBm0. If $m = -16$ dBm0 and $\beta = 5.8$ dB, then $l = 0.248$. In SCPC systems, an individual carrier is assigned to each voice channel. An individual voice band signal modulates a single sinusoidal carrier.

1.4.2 Baseband Processing

Some analog messages such as voice, music, and pictures are converted to binary digits 0 and 1. These 0's and 1's are assigned either a rectangular pulse or sinusoidal waveform. When these are assigned a pulse waveform, the process is called *pulse modulation*. For extracting the message the transmission of binary digits 0 and 1 or two waveforms $s_0(t)$ and $s_1(t)$ is preferable to the original analog message. The binary waveforms are detected at the receiver (demodulator) by comparing with a threshold. For example, $s_0(t)$ is a rectangular pulse with height A and duration T_b and $s_1(t) = -s_0(t)$. These pulses are sent through a channel. The received signal is sampled every T_b seconds and if the sampled value exceeds the threshold value 0, then the pulse representing 1 is present, otherwise the pulse representing 0 is present. This detection process is rugged to noise and channel impairment. These binary digits are added with redundant binary digits for error correction and encryption for message privacy and security. The advantage of binary communication is obtained at the cost of bandwidth expansion and complexity of circuits for sampling and synchronization of pulse timing. The generation of binary bits from analog signals prior to modulation and the generation of analog message from the binary bits after the demodulation is termed as *baseband processing*. At the transmit side, this process involves three stages: sampling, quantization, and encoding.

Sampling

Let $m(t)$ be an analog message band-limited to f_m; that is, the amplitude spectrum $M(f) = 0$ for $|f| > f_m$. It is assumed further that message amplitude spectrum is symmetric about the origin and the average value of $m(t)$ is zero. The message $m(t)$ is multiplied with another periodic signal $p(t)$ (see Figs. 1.16 and 1.17). The periodic signal is given by

$$p(t) = \sum_{n=\infty}^{\infty} \delta(t - nT_s) \tag{1.231}$$

where T_s is the period of the signal $p(t)$ and $\delta(t)$ is a Dirac delta function. The product signal is

$$x(t) = m(t)p(t)$$

$$= m(t) \sum_{n=-\infty}^{\infty} \delta(t - nT_s) \tag{1.232}$$

Since $p(t)$ is a periodic signal, it can be expressed by Fourier series as

$$p(t) = \sum_{n=-\infty}^{\infty} c_n e^{in2\pi f_s t} \qquad f_s = \frac{1}{T_s} \tag{1.233}$$

where the Fourier coefficient

$$c_n = \frac{1}{T_s} \int_0^{T_s} p(t)e^{-i2\pi n f_s t} \, dt$$

$$= \frac{1}{T_s} \sum_n \int_0^{T_s} \delta(t - nT_s)e^{-i2\pi n f_s t} \, dt$$

$$= \frac{1}{T_s} \int_0^{T_s} \delta(t)e^{-i2\pi n f_s t} \, dt$$

$$= \frac{1}{T_s} \qquad \text{for all } n \tag{1.234}$$

the other terms are ignored because these terms are outside the interval $(0, T_s)$. Therefore

$$p(t) = \frac{1}{T_s} \sum_{n=-\infty}^{\infty} \delta(t - nT_s) \tag{1.235}$$

The Fourier transform (FT) of $p(t)$ is

$$P(f) = \frac{1}{T_s} \int_{-\infty}^{\infty} \sum_{n=-\infty}^{\infty} \delta(t - nT_s)e^{-i2\pi f t} \, dt$$

$$= \frac{1}{T_s} \sum_{n=-\infty}^{\infty} \int_{-\infty}^{\infty} \delta(t - nT_s)e^{-i2\pi f t} \, dt$$

$$= \frac{1}{T_s} \sum_{n=-\infty}^{\infty} \delta(f - nf_s) \tag{1.236}$$

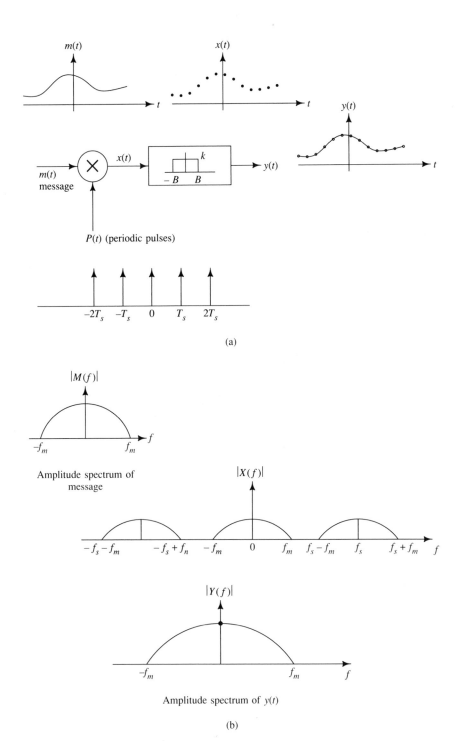

FIGURE 1.16 Sampling process. (a) Multiplying the message signal. (b) Output amplitude spectra of $x(t)$ and $y(t)$ if $f_s \geq 2f_m$.

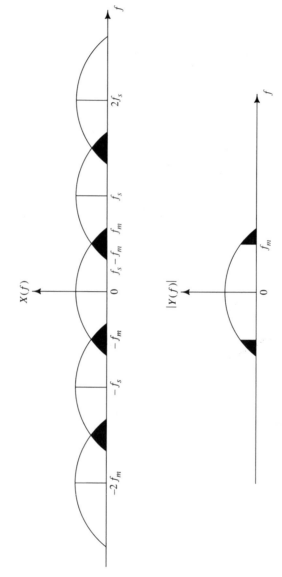

FIGURE 1.17 Output amplitude spectra of $x(t)$ and $y(t)$ if $f_s \leq 2f_m$.

The Fourier transform (FT) of $x(t)$, spectrum of sampled $m(t)$, is

$$X(f) = M(f) \otimes P(f) \text{ (convolution)}$$

$$= \int_{-\infty}^{\infty} M(f_1) P(f - f_1) \, df_1$$

$$= \int_{-\infty}^{\infty} M(f_1) \frac{1}{T_s} \sum_{n=-\infty}^{\infty} \delta(f - f_1 - nf_s) \, df_1$$

$$= \frac{1}{T_s} \sum_{n=-\infty}^{\infty} \int_{-\infty}^{\infty} M(f_1) \delta(f - f_1 - nf_s) \, df_1$$

$$= \frac{1}{T_s} \sum_{n=-\infty}^{\infty} M(f - nf_s) \qquad f_s = \frac{1}{T_s} \tag{1.237}$$

The output spectrum is a sum of infinite number of message spectra $M(f)$ located at $f = nf_s$, $n = 0, \pm 1, \pm 2, \ldots$. These spectra will not overlap if $f_s - f_m \geq f_m$. Therefore, it is necessary for the periodic signal period T_s to satisfy the condition

$$f_s \geq 2f_m \tag{1.238}$$

where f_s is called the sampling rate. If

$$f_s = 2f_m \tag{1.239}$$

f_s is called *Nyquist sampling rate*, where f_m is the maximum frequency content of the message. If the product signal is passed through a low-pass filter (LPF) with a bandwidth B such that

$$f_s - f_m \leq B \leq f_s \tag{1.240a}$$

and

$$H(f) = k \qquad |f| \leq B$$
$$= 0 \qquad \text{elsewhere} \tag{1.240b}$$

we can filter out the spectra located at $f = nf_s$, $n = \pm 1, \pm 2, \ldots$. The spectrum of the output of the LPF is

$$Y(f) = H(f) X(f) \tag{1.241}$$

$$= \frac{K}{T_s} M(f) \qquad |f| \leq B$$
$$= 0 \qquad \text{elsewhere} \tag{1.242}$$

where K is the gain of the LPF.

The output signal of the LPF has the same spectrum as that of the original signal $m(t)$ except for the term K/T_s. Taking the inverse Fourier transform of Eq. (1.241), we get

$$y(t) = \int_{-\infty}^{\infty} h(\tau)x(t - \tau)\, d\tau$$

$$= \int_{-\infty}^{\infty} h(\tau)[m(t - \tau)] \sum_{n=-\infty}^{\infty} \delta(t - \tau - nT_s)\, d\tau$$

$$= \sum_{n=-\infty}^{\infty} \int_{-\infty}^{\infty} h(\tau)m(t - \tau)\delta(t - \tau - nT_s)\, d\tau$$

$$= \sum_{n=-\infty}^{\infty} m(nT_s)h(t - nT_s) \tag{1.243}$$

Note that $h(t)$ is obtained by taking the inverse Fourier transform of Eq. (1.240b).

$$h(t) = \int_{-B}^{B} k e^{i2\pi f t}\, df$$

$$= \frac{k[e^{i2\pi Bt} - e^{-i2\pi Bt}]}{i2\pi t}$$

$$= k\left(\frac{\sin 2\pi Bt}{\pi t}\right) \tag{1.244}$$

Substitution of Eq. (1.244) in (1.243) yields

$$y(t) = \sum_{n=-\infty}^{\infty} m(nT_s)\frac{k\sin[2\pi B(t - nT_s)]}{\pi(t - nT_s)}$$

Equation (1.242) reveals that $y(t) = m(t)$, provided that $k = T_s$. Hence

$$m(t) = T_s \sum_{n=-\infty}^{\infty} m(nT_s)\frac{\sin[2\pi B(t - nT_s)]}{\pi(t - nT_s)} \tag{1.245}$$

For Nyquist sampling rate

$$T_s = \frac{1}{f_s} = \frac{1}{2f_m} \tag{1.246}$$

choosing $B = f_m$ and setting $T_s = 1/(2f_m)$. Equation (1.245) gives

$$m(t) = \sum_{n=\infty}^{} m\left(\frac{n}{2f_m}\right)\frac{\sin[\pi(2f_m t - n)]}{\pi(2f_m t - n)} \tag{1.247}$$

Equation (1.247) is known as the *sampling theorem*; that is, $m(t)$ can be recovered from the samples of $m(t)$, $t = nT_s$, $n = 0, \pm 1, \ldots$, provided that $T_s \le 1/(2f_m)$, where f_m is the bandwidth (one-sided) of message $m(t)$. The function

$$h(t) = \frac{\sin 2\pi f_m t}{2\pi f_m t}$$

is called the *sample function*. Note that

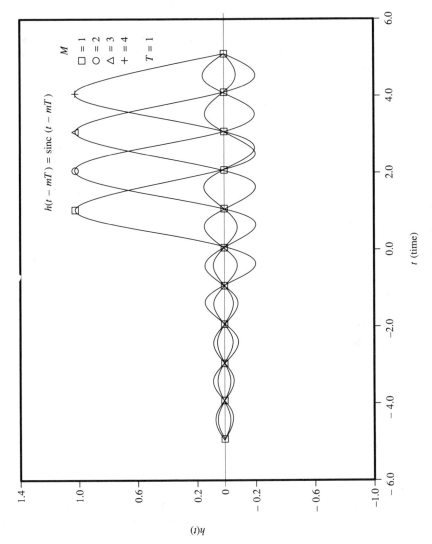

FIGURE 1.18 Sampling functions and delayed sampling functions.

$h(t - mT) = \text{sinc } (t - mT)$

M

□ = 1
○ = 2
△ = 3
+ = 4

$T = 1$

t (time)

$h(t)$

$$h(0) = 1 \qquad t = 0$$

$$\quad\;\; = 0 \qquad t = n/2f_m \qquad n \neq 0$$

as can be seen in Fig. 1.18. When $t = kT_s = k/2f_m$, from Eq. (1.247) we get

$$m\left(\frac{k}{2f_m}\right) = m\left(\frac{k}{2f_m}\right) + \sum_{n=k} m\left(\frac{n}{2f_m}\right) \operatorname{sinc}(k - n) \qquad (1.248a)$$

If $f_s < 2f_m$ or $H(f) \neq k, |f| < 2f_m$, then the second term in the right-hand side of Eq. (1.248a) is not equal to zero. In the frequency domain, the spectra of the sampled function $x(t)$ will overlap; see Fig. 1.17, where the overlapped spectra are shown in the shaded areas. This phenomena in sampling is known as *aliasing*. Aliasing distorts the output spectrum $y(t)$ and occurs if $f_s < 2f_m$. Note that even when $f_s \geq 2f_m$ but

$$h\left(\frac{n}{2f_m}\right) \neq 0 \qquad n = \pm 1, \pm 2 \pm \cdots \qquad (1.248b)$$

then Eq. (1.248a) gives

$$m\left(\frac{k}{2f_m}\right) = m\left(\frac{k}{2f_m}\right) + I$$

where I is called the *intersymbol interference*, given by

$$I = \sum_{n=k} m\left(\frac{n}{2f_m}\right) \operatorname{sinc}(k - n)$$

If the bandwidth of the transmitter channel or receive filter is such that

$$|H_c(f)| \neq k \qquad |f| \leq f_m$$

then the regenerated signal will suffer from intersymbol interference noise. The regenerated signal is passed through an equalizer to reduce the intersymbol interference. We will discuss the phenomenon of equalization later.

Quantization

Like the analog message, the nth sample $m(n/2f_m) = m(n)$ can take an infinite number of values. We have to transmit all such values. The advantages of digital communication will be lost if we transmit an infinite number of sampled values. We transform this infinite number to a finite number of Q values.

Let us denote

$$\max_n \; [m(n/2f_m] = V_b$$

$$\min_n \; [m(n/2f_m)] = V_a$$

Then

$$V_{pp} = V_b - V_a = Q\Delta$$

where Q is a positive integer and Δ is step size of quantized value. Usually Q is chosen as 2^q, where q is a positive integer. For example, for television, the illumination brightness is quantized to 256 levels; that is

$$Q = 256 = 2^8$$

$$q = 8$$

Speech signals have a maximum frequency of 4 kHz. It is sampled at a rate $f_s \geq -2 \times 4\,\text{kHz} = 8\,\text{kHz}$. The sampled values are quantized to eight values; thus

$$Q = 8 = 2^3$$

$$q = 3$$

In this quantization, the sampled value is equal to the nearest rounded value (see Fig. 1.19a).

Let $\hat{m}(n)$ denote the quantized value. We express the sample value as

$$m(n) = \hat{m}(n) + e(n) \qquad |e(n)| \leq \Delta/2 \qquad (1.249)$$

The value $e(n)$ is the *quantization error* of the nth sample $m(n)$; see Fig. 1.19b. This quantization process is called the *uniform quantization*. The probability density of error variable $e(n)$ is given by uniform probability density (Fig. 1.19c) as follows:

$$f_n(e) = 1/\Delta \qquad |e(n)| \leq \Delta/2$$
$$= 0 \qquad \text{elsewhere} \qquad (1.250)$$

Therefore, with this assumption, the mean of the quantization error is

$$E[e(n)] = 0 \qquad \text{for all } n$$

The power of the quantization error is

$$E[e(n)]^2 = \int_{-\Delta/2}^{\Delta/2} e^2 f_n(e)\,de$$

$$= \int_{-\Delta/2}^{\Delta/2} \frac{1}{\Delta} e^2\,de$$

$$= \frac{1}{\Delta} \cdot \frac{\Delta^3}{12} = \frac{\Delta^2}{12} \qquad \text{for all } n \qquad (1.251)$$

Let the mean and variance of the sample value be

$$E[m(n)] = 0 \qquad \text{for all } n$$

$$\text{var}[m(n)] = \sigma_m^2$$

The signal to quantization noise power ratio is

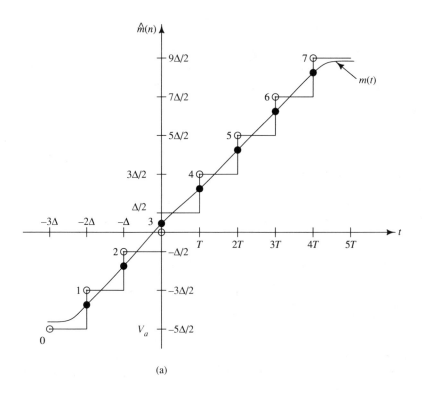

(a)

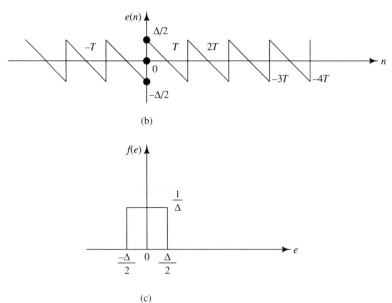

(b)

(c)

FIGURE 1.19 Quantization process with $Q = 8$. (a) Actual samples $m(n)$ are marked with a point, whereas quantized values $\hat{m}(n)$ are marked with circles. (b) Error sample at each step. (c) Probability density of quantization error variables.

$$(\text{SNR})_Q = \frac{\sigma_m^2}{\Delta^2/12}$$

$$= \frac{12\sigma_m^2}{\Delta^2} \tag{1.252}$$

The signal-to-noise ratio is improved by making Δ small. Small Δ makes the quantized value very close to the actual sample value. But this Q is very large in order to keep V_{pp} fixed. If we denote

$$\min[m(n)] = 0$$

$$\max[m(n)] = Q - 1$$

then

$$0 \le m(n) \le Q - 1$$

By means of quantization, the samples values takes only Q number of finite values. These Q values are called *symbols*.

In Fig. 1.19a, for $Q = 8$, there are eight quantization values:

$$V_a = -5\Delta/2$$

$$V_b = 9\Delta/2$$

$$V_{pp} = 9\Delta/2 + 5\Delta/2 = 7\Delta$$

$$m(-3) = 0 \qquad \text{(minimum value is } -5\Delta/2\text{)}$$

$$m(-2) = 1$$

$$m(-1) = 2$$

$$m(0) = 3$$

$$m(1) = 4$$

$$m(2) = 5$$

$$m(3) = 6$$

$$m(4) = 7 \qquad \text{(maximum value is } 9\Delta/2\text{)}$$

The quantized signal can be expressed as

$$x(t) = \sum_{n=-\infty}^{\infty} a_n P(t - nT_s) \tag{1.253a}$$

where

$$a_n \in [0, Q - 1]$$

and

$$P(t) = 1 \qquad 0 \le t \le T_s$$

$$= 0 \qquad \text{elsewhere} \tag{1.253b}$$

The signal $x(t)$ in Eq. (1.253a) is called the *pulse amplitude modulation* (PAM). Here the pulse amplitude takes 0 to $Q - 1$ values (symbols) and T_s is called the symbol duration.

Encoding

The encoding process converts Q values to binary digits. We will describe first natural binary coding. Let us denote

$$a_n = b_1 2^0 + b_2 2^1 + b_3 2^2 + \cdots + b_n 2^{q-1}$$

where b_1 is called the *least significant bit* (LSB) and b_n the *most significant bit* (MSB).

If

$$Q = 64 \qquad q = 6$$

$$0 = 00000$$

$$1 = 00001$$

$$\vdots$$

$$63 = 11111$$

Now each symbol has q bits. The encoded signal is given by

$$y(t) = \sum_{n=\infty}^{\infty} b_n p(t - nT_b)$$

where

$$b_n \in (0, 1) \tag{1.254}$$

where $p(t)$ is defined in Eq. (1.253b) and T_b is called the bit duration and

$$T_s = qT_b \tag{1.255a}$$

The signal $y(t)$ given by Eq. (1.254) is called the *pulse code modulation* (PCM). The bandwidth for PCM waveform transmission is

$$B = 2f_m \log_2 Q \tag{1.255b}$$

where f_m is the bandwidth of the baseband signal. The generation of a PCM signal is shown in Fig. 1.20.

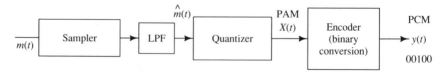

FIGURE 1.20   PMC signal generation (binary signals).

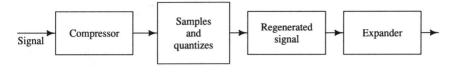

FIGURE 1.21   Nonuniform quantization.

In natural binary coding, two consecutive levels may differ in more than one place. For example, $14 = 1110$ and $13 = 1101$. They differ in two places. It is undesirable in some digital communications. The desired encoding with only one variation is given by *Gray encoding*. The Gray code is given by

$$g_k = b_q \qquad \text{if } k = q$$
$$b_{k+1} \oplus b_k \qquad \text{if } k < q \tag{1.256}$$

where $\oplus$ stands for mod 2 addition.

Uniform quantization yields poor signal-to-noise ratio when the dynamic range of the signal values is very large. For speech signals the dynamic range, (from loud talk to silence) $= 1000$. The quantization level should be very small for the weak-signal part and large (coarse) for the strong-signal part. Such quantization is called *nonuniform quantization* (Fig. 1.21). This can be accomplished by passing the signal through a *compander* before it is uniformly quantized; then the uniformly quantized signal is passed through an *expander* to get back the original signal.

Let $c(x)$ denote the compressor such that

$$c(x_1) \le c(x_2) \qquad \text{if } x_1 \le x_2$$

and

$$c(x) = -c(x)$$

where x is the input to the compressor and expander $e(x)$ is given by

$$e(x) = c^{-1}(x)$$

and therefore

$$e(x)c(x) = 1$$

Compressor and expander combination will leave the signal undistorted. There are two kinds of compressors.

1. μ-law companding:

$$\frac{c(|x|)}{x_{\max}} = \frac{\ln(1 + \mu|x|/x_{\max})}{\ln(1 + \mu)} \qquad 0 \le \frac{|x|}{x_{\max}} \le 1 \tag{1.257a}$$

where μ is the compression parameter

2. *A*-law companding:

$$\frac{c(|x|)}{x_{max}} = \begin{cases} \dfrac{A|x|/x_{max}}{1 + \ln A}, & 0 \le \dfrac{|x|}{x_{max}} \le \dfrac{1}{A} \\ \dfrac{1 + \ln(A|x|/x)_{max}}{1 + \ln A}, & \dfrac{1}{A} \le \dfrac{|x|}{x_{max}} \le 1 \end{cases} \qquad (1.257b)$$

where a sample of the input x is bounded by range $-x_{max}$ to x_{max}. CCIR recommends $\mu = 255$ and $A = 87.6$.

When signals are sampled at higher rate than the Nyquist rate, there will be oversampling. Samples are correlated and bandwidth of the signal is very large. To reduce the correlation or the bandwidth, we will filter the PCM signal. We can predict the estimated value of the samples from the past values and then take the difference of the incoming signal value and the predicted value. The difference value will have a small dynamic range such that Q will be small, which implies that q is small. We quantize the error signal and then encode it. The encoded signal is called a *differential PCM* (DPCM). The generation of DPCM and demodulation is shown in Fig. 1.22.

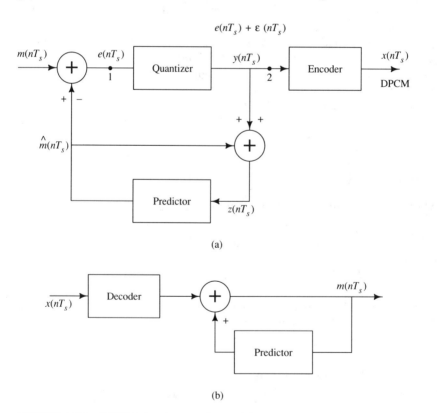

FIGURE 1.22   DPCM signal generation and recovery. (a) DPCM transmitter. (b) DPCM receiver.

Note that

$$y(nT_s) = f[e(nT_s)]$$
$$= e(nT_s) + \varepsilon(nT_s)$$
$$e(nT_s) = m(nT_s) - \hat{m}(nT_s) \tag{1.258}$$
$$z(nT_s) = y(n(T_s) + \hat{m}(nT_s))$$
$$\hat{m}(nT_s) = \sum_{m=0}^{n-1} a_m Z(mT_s)$$

where a_m are some constants, and $z(0)$, $z(T_s)$, $z[(n-1)T_s]$ are past values of $z(nT_s)$. It can be shown that

$$z(nT_s) = m(nT_s) + \varepsilon(nT_s) \tag{1.259}$$

where $\varepsilon(nT_s)$ is the quantization error. Let σ_m^2, σ_e^2, and σ_ε^2 denote variance of $m(nT_s)$, $e(nT_s)$, and quantization error $\varepsilon(nT_s)$. The ratio of signal to quantization error is

$$(\text{SNR})_T = \frac{\sigma_m^2}{\sigma_\varepsilon^2}$$

$$= \frac{\sigma_m^2}{\sigma_e^2} \cdot \frac{\sigma_e^2}{\sigma_\varepsilon^2} \tag{1.260a}$$

At node 1, the signal to prediction error is

$$(\text{SNR})_1 = \frac{\sigma_m^2}{\sigma_e^2} \tag{1.260b}$$

At node 2, the square of the prediction error to quantization error is

$$(\text{SNR})_2 = \frac{\sigma_e^2}{\sigma_\varepsilon^2} \tag{1.260c}$$

Equation (1.260a) reveals that by prediction methods,

$$(\text{SNR})_T = (\text{SNR})_1 \left(\frac{\sigma_e^2}{\sigma_\varepsilon^2}\right)$$

$(\text{SNR})_T$ is increased if $(\text{SNR})_1 > 1$. The coefficients $a_0, a_1, \ldots, a_{n-2}$ can be obtained by minimizing

$$E[m(nT_s) - \hat{m}(nT_s)]^2$$

We will consider this minimization when we discuss adaptive equalization. We now consider a special case when

$$\hat{m}(nT_s) = m[(n-1)T_s]$$
$$a_0 = a_1 = \cdots = a_{n-2} = 0, \, a_{n-1} = 1 \tag{1.261}$$

and

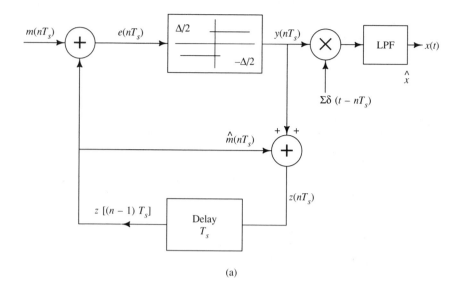

(a)

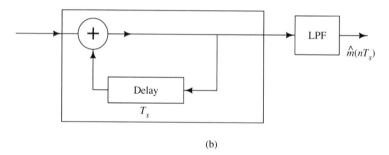

(b)

FIGURE 1.23   (a) Generation and (b) recovery of delta-modulated signal.

$$y(nT_s) = \frac{\Delta}{2} \triangleq 1 \qquad \text{if } e(nT_s) \geq 0$$

$$= \frac{-\Delta}{2} \triangleq 0 \qquad \text{if } e(nT_s) < 0$$

Therefore $Q = 2$, $q = 1$.

The expression $y(nT_s)$ is called a *delta-modulated* (DM) signal. It has only two values. The outputs are 0 and 1. There is no need for encoding. However, $y(nT_s)$ is passed through a low-pass filter to filter out any out-of-band noise. The generation and recovery of a DM signal are shown in Fig. 1.23.

The DM waveform can be written

$$x(t) = \sum_{n=-\infty}^{\infty} y(nT_s)\delta(t - nT_s) \qquad (1.262a)$$

$$= \sum_{n=-\infty}^{n=\infty} \frac{\Delta}{2} \text{sign} [m(nT_s) - \hat{m}(nT_s)] \delta(t - nT_s) \tag{1.262b}$$

$$\begin{aligned} \text{sign}(x) &= 1 & x \geq 0 \\ &= -1 & x < 0 \end{aligned} \tag{1.262c}$$

In delta modulation, quantization error arises from one of two cases. The first case is when the quantization slope

$$\frac{\Delta/2}{T_s} \gg \text{maximum of the slope of the message} = \max \left| \frac{d}{dt} [m(t)] \right| \tag{1.263a}$$

This condition is called *granular noise*, in which Δ is very large compared to slope of the signal. Step-size approximation is a very large approximation to a flat-type curve. In the second case,

$$\frac{\Delta/2}{T_s} \ll \max \left| \frac{dm(t)}{dt} \right| \tag{1.263b}$$

That is, the staircase-type approximation is far below the message curve. This noise is called *slope overhead*. Therefore, the choice step size Δ will enhance noise depending on the change of slope of the message signal. The step size Δ should be very small if the signal is constant or nearly constant and should be large when the slope of the signal changes rapidly. In *adaptive delta modulation* (ADM) the variable step size increases during a steep segment and decreases during a slowly varying segment of the input signal. It should be noted that the signal-to-noise ratio for PCM is better than that of DM when channel bandwidth is much larger than the baseband (message) bandwidth. Otherwise, DM is superior to PCM.

Time Division Multiplexing

Time division multiplexing (TDM) is a transmission scheme in which several digital signals are transmitted simultaneously in the same time period by interleaving several signal samples in time. This is a process in which a channel bandwidth is efficiently utilized by several users, such as FDM systems. Let $m_1(t), \ldots, m_M(t)$ be M number signals with bandwidth $B_1, B_2, \ldots, B_M$. In T seconds, there are $2B_1 T, 2B_2 T, \ldots, 2B_M T$ samples from M PCM signals. These samples are serially collected by a commutation with total samples

$$N = 2T \sum_{i=1}^{M} B_i$$

The bandwidths of the multiplexed M PCM signals are

$$B = \sum_{i=1}^{M} B_i$$

Let m_1, m_2, and m_3 have bandwidths $B, 2B, 4B$. The multiplexed PCM signals

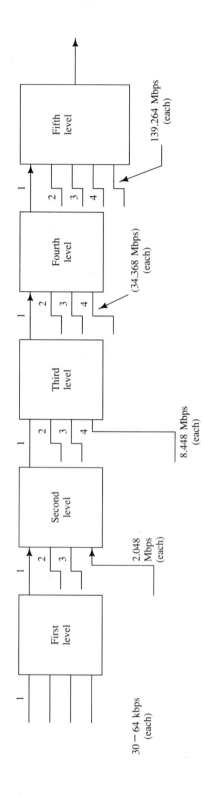

FIGURE 1.24   CCITT recommendations for TDM of 7680 voice frequency.

will be of the form $m_1m_4m_4m_2m_4m_4m_2$. One rotation of the commutators is called a *frame*. In one frame there is only one sample from m_1 but there are two samples from m_2 and four samples from m_4. This is due to the fact the sampling rate of m_4 is four times higher than that of m_1, and sampling rate of m_2 is two times that of m_1.

The frequency of the rotation of the commutators is equal to or greater than $2B_{max}$, where B_{max} is the bandwidth of the signal with the maximum bandwidth.

There are two types of TDM. In *synchronous TDM* each bit is clocked and controlled by a master clock and synchronizing data are transmitted along with message (baseband) data. In *asynchronous TDM*, the receiver clock starts and stops with a certain character's beginning bit and end bit. The CCITT digital TDM scheme is shown in Fig. 1.24.

1.4.3 Digital Modulation

In digital modulation discrete or digitized messages (data) are transmitted through a sinusoidal carrier whose amplitude, phase, frequency, or some combination thereof, varies with discrete values of the message. Digitized or encoded messages in the form of 0 and 1 occur at a rate of 1 bit per T_b seconds. These binary bits are converted to bipolar (nonreturn to zero) or Manchester type (biphase) pulses, or any of several other forms of pulses (see Fig. 1.25). These data are passed through driver amplifiers, a low-pass filter, and a mixer for appropriate amplitude, phase, or frequency modulations. The transmitted signals representing binary data zero and one will be denoted by $S_0(t)$ and $S_1(t)$ signals, respectively. For the purpose of power or bandwidth efficiency, these binary data are segmented to words of size k bits. Since each bit is a zero or one, there are 2^k distinct words. We denote $M = 2^k$. Each of these words, sometimes called *code words or symbols*, are assigned a sinusoidal signal. There are then M different signals, $S_0(t), S_1(t), \ldots, S_M(t)$. Each symbol duration is T_s seconds. Since each word has k bits, $T_s = kT_b$. The reciprocal of T_s is called the *symbol or baud rate*. If E_b is the energy of the binary bit, then the symbol energy $E_s = kE_b$. Note that if $M = 2$, then $E_s = E_b$, $T_s = T_b$ because $k = 1$. When $M = 2$, the transmission is called *binary transmission* and when $M > 2$, the transmission is known as *M-ary transmission*. The detection of these signals is based on whether or not the receiver has the exact knowledge of the phase information of the signal. When the receiver has the phase information, the receiver is called a *coherent* receiver. The detection scheme is called *coherent detection*. The receiver is a matched filter or *correlator detector*. When the receiver does not have the phase information, the receiver is known as a *noncoherent receiver*. The noncoherent receiver is a band-pass filter followed by an envelope detector. The envelope detector is a suboptimal receiver, whereas a matched filter is an optimal filter that yields the maximum signal-to-noise ratio, as discussed in Section 1.3. The choice of modulation and demodulation is based on system (mission) objectives.

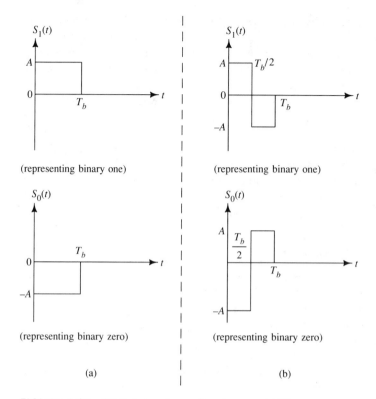

FIGURE 1.25 Digital signal-encoding formats. (a) Nonreturn-to-zero (bipolar) pulses. (b) Manchester (biphase) pulses.

There are various objectives:

1. Maximum throughput data rate, where cost of transmission is important.

2. Maximum resistance to jamming, where survivability of the signals is important; bandwidth expansion is not critical.

3. Minimum power or bandwidth; satellite channels are power-limited but not-band limited.

4. Minimum probability of bit error when the reliability of messages is most critical.

5. Least delay when messages are hopped satellite to satellite to avoid jammer detection or fading (nuclear scintillation).

6. Technological constraint and development cost.

The detection scheme described in this section is based on the minimum error probability. Let the received signal be

$$r(t) = S_i(t, \phi) + n(t) \qquad 0 \le t \le T_b \qquad i = 0, 1 \tag{1.264}$$

where $S_i(t, \phi)$, $i = 0$, 1, is the transmitted signal of phase ϕ and $n(t)$ is the zero-mean white Gaussian noise, referred to as *additive white Gaussian noise* (AWGN). $S_0(t)$ is transmitted with probability π_0 and $S_1(t)$ is transmitted $(1 - \pi_0)$, $0 < \pi_0 < 1$. The decision rule is based on the condition that

$$P[r|S_1(\phi)](1 - \pi_0) > P[r|S_1(\phi)]\pi_0$$

that is,

$$f[r|S_1(\phi)](1 - \pi_0) > f[r|S_0(\phi)]\pi_0 \qquad (1.265)$$

where $f[r|S_i(\phi)]$ is the conditional probability density function of $r(t)$ given that $S_i(t)$ is transmitted and phase ϕ is known, $i = 0$, 1. π_0 and π_1 are called *a priori probabilities* of signal $S_0(t)$ and $S_1(t)$. Equation (1.265) can be written as

$$\frac{f[r|S_1(\phi)]}{f[r|S_0(\phi)]} > \frac{\pi_0}{1 - \pi_0} \qquad (1.266)$$

when $S_1(t)$ is decided as present. The left side of Eq. (1.266) is known as *likelihood function*. Because of the assumption of AWGN, it is convenient to take logarithms of both sides to make a decision. That is,

$$\log\left[\frac{f[r|S_1(\phi)]}{f[r|S_0(\phi)]}\right] > \log\left[\frac{\pi_0}{1 - \pi_0}\right] \qquad \text{when } S_1 \text{ is present}$$

$$< \log\frac{\pi_0}{1 - \pi_0} \qquad \text{when } S_1 \text{ is absent } (S_0 \text{ is present}) \qquad (1.267)$$

The left-hand side of Eq. (1.267) is called the *log likelihood function*. Without loss of generality, assume that binary data 0 and 1—that is, $S_0(t)$ and $S_1(t)$— are transmitted with equal probability. In this case, $\pi_0 = \pi_1 = 1/2$. Equation (1.267) can be written as

$$\log\left[\frac{f[r|S_1(\phi)]}{f[r|S_0(\phi)]}\right] > 0 \qquad \text{when } S_1(t) \text{ is present}$$

$$< 0 \qquad \text{when } S_1(t) \text{ is absent } (S_0 \text{ is present}) \qquad (1.268)$$

For noncoherent detection, when phase information is unavailable, the decision rule is as follows:

$$\log\left[\frac{E_\phi f[r|S_1(\phi)]}{E_\phi f[r|S_0(\phi)]}\right] > 0 \qquad \text{if } S_1(t) \text{ is present}$$

$$< 0 \qquad \text{if } S_1(t) \text{ is absent (i.e., } S_0 \text{ is present)} \qquad (1.269)$$

where E_ϕ denotes the expected value of $f[r|S_i(\phi)]$ with respect to ϕ, $i = 0$, 1.

Coherent Binary Detection

Let $h(t)$ be the impulse function of the receiver. It is assumed that phase information is available. The output of the receiver (see Fig. 1.26) is given by

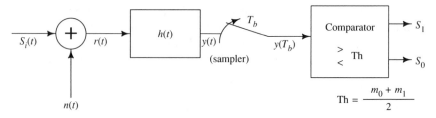

FIGURE 1.26   Detection of binary signals.

$$y(t) = \int_0^\infty h(\tau)r(t - \tau)\,d\tau \tag{1.270}$$

where we assume that $h(\tau)$ is stable and physically realizable; Since $n(t)$ is AWGN, $y(t)$ is also Gaussian. $y(T_b)$, which is sampled every T_b seconds, is a Gaussian random variable. Let us denote the mean and variance of $y(T_b)$ as follows:

$$m_i = E[y(T_b)|S_i(\phi)] \qquad i = 0, 1$$
$$\sigma_i^2 = \text{var}[Y(T_b)|S_i(\phi)] \qquad i = 0, 1 \tag{1.271}$$

Since $n(t)$ is AWGN,

$$E[n(t)] = 0$$
$$E[n(t)n(t + \tau)] = \left(\frac{N_0}{2}\right)\delta(\tau) \tag{1.272a}$$

where $N_0/2$ is the two-sided height of the PSD of $n(t)$. Therefore,

$$E[y(T_b)|S_i(\phi)] = \int_0^\infty h(\tau)S_i(T_b - \tau)\,d\tau = m_i$$
$$\text{var}[Y(T_b)|S_i(\phi)] = \frac{N_0}{2}\int_{-\infty}^\infty |H(f)|^2\,df = \sigma_i^2 \qquad i = 0, 1 \tag{1.272b}$$

using Eqs. (1.264), (1.270), and (1.272a) and

$$S_Y(f) = |H(f)|^2 S_x(f)$$

[see also Eq. (1.107b)].

The condition PDF of $y(T)$ is

$$f(Y(T)|S_i(\phi)) = \frac{1}{\sqrt{2\pi\delta_i^2}}\exp\left[\frac{-(Y(T) - m_i)^2}{2\delta^2}\right] \qquad i = 0, 1 \tag{1.273}$$

where m_i and δ_i are given by Eq. (1.273).

Note that

$$\sigma_1^2 = \sigma_0^2 \tag{1.274}$$

The likelihood ratio in Eq. (1.266) becomes

$$\frac{f[Y(T_b)|S_1(\phi)]}{f[Y(T_b)|S_0(\phi)]} = \frac{\dfrac{1}{\sqrt{2\pi\sigma_0{}^2}}\exp\left\{\dfrac{-1}{2\sigma_0^2}[y(T_b) - m_1]^2\right\}}{\dfrac{1}{\sqrt{2\pi\sigma_0^2}}\exp\left\{\dfrac{-1}{2\sigma_0^2}[y(T_b) - m_0]^2\right\}} \gtrless 1 \qquad (1.275)$$

using Eq. (1.273). The log likelihood function is therefore

$$[y(T_b) - m_0]^2 - [y(T_b) - m_1]^2 \gtrless 0$$

using Eqs. (1.268) and (1.275). The decision rule is therefore as follows: If

$$y(T_b) > \frac{m_0 + m_1}{2}, \text{ then } S_1 \text{ is present}$$

$$< \frac{m_0 + m_1}{2}, \text{ then } S_0 \text{ is present} \qquad (1.276)$$

$(m_0 + m_1)/2$ is called the *threshold*. It is important to note that $m_0 \neq m_1$; otherwise, it will not be possible to identify $S_0(t)$ and $S_1(t)$. There are two kinds of probability of errors:

P_{01} = Probability of deciding $S_0(t)$ is present when $S_1(t)$ is really present

$\quad = P[y(T_b) < \text{Th}|S_1(\phi)]$ \qquad (1.277a)

P_{10} = Probability of deciding S_1 is present when S_0 is the
transmitted signal \qquad (1.277b)

$\quad = \Pr[y(T_b) < \text{Th}|S_1(\phi)]$

The *probability of bit error* (see Fig. 1.27) is

$$P_e = \pi_0 P_{10} + (1 - \pi_0)P_{01} \qquad (1.277c)$$

where π_0 and $(1 - \pi_0)$ are the a priori probabilities of $S_0(t)$ and $S_1(t)$ and Th $= (m_0 + m_1)/2$. Hence

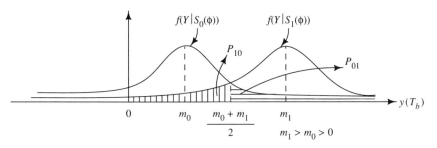

FIGURE 1.27   Error probabilities.

$$P_{10} = \int_{\text{Th}}^{\infty} f[Y(T_b)|S_0(\phi)]\,dy$$

$$= \int_{\text{Th}}^{\infty} \frac{1}{\sqrt{2\pi\sigma_0^2}} \exp\left\{\frac{-1}{2\sigma_0^2}[y(T_b) - m_0]^2\right\}$$

$$\triangleq Q\left(\frac{\text{Th} - m_0}{\sigma_0}\right) = Q\left(\frac{m_1 - m_0}{2\sigma_0}\right)$$

$$= \frac{1}{2}\text{erfc}\left(\frac{\text{Th} - m}{\sqrt{2}\delta_0}\right) = \frac{1}{2}\text{erfc}\left(\frac{m_1 - m_0}{2\sqrt{2}\delta_0}\right) \qquad (1.278a)$$

$$P_{10} = \int_{-\infty}^{\text{Th}} \frac{1}{\sqrt{2\pi s_1^2}} \exp\left\{\frac{-1}{2\sigma_1^2}[y(T_b) - m_1]^2\right\}dy$$

$$= 1 - Q\left(\frac{m_0 - m_1}{2\sigma_1}\right)$$

$$= 1 - Q\left(\frac{m_0 - m_1}{2\sigma_0}\right) \qquad (1.278b)$$

since $\sigma_1^2 = \sigma_0^2$. If $\pi_0 = 1/2$, then

$$P_e = P_{10}\pi_0 + P_{01}(1 - \pi_0)$$

$$= \frac{1}{2}Q\left(\frac{m_1 - m_0}{2\sigma_0}\right) + \frac{1}{2}\left[1 - Q\left(\frac{m_0 - m_1}{2\sigma_0}\right)\right] = Q(d)$$

$$= \frac{1}{2}\text{erfc}\left(\frac{d}{\sqrt{2}}\right) \qquad (1.279a)$$

where

$$d = \frac{m_1 - m_0}{2\sigma_0} \qquad (1.279b)$$

using Eqs. (1.277c), (1.278a), and (1.278b). The probability of error is minimum if d is maximum. Therefore,

$$\min(P_e) = \max_d Q(d)$$

$$d^2 = \frac{|m_1 - m_0|^2}{4\sigma_0^2}$$

$$= \frac{\left|\int_{-x}^{\infty} h(\tau)S_1(T_b - \tau)\,d\tau - \int_0^{\infty} h(\tau)S_0(T_b - \tau)\,d\tau\right|^2}{4\dfrac{N_0}{2}\int_{-\infty}^{\infty}|H(f)|^2\,df}$$

$$= \frac{\left| \int_{-\infty}^{\infty} h(\tau) [S_1(T_b - \tau) - S_0(T_b - \tau)] \, d\tau \right|^2}{4 \frac{N_0}{2} \int_{-\infty}^{\infty} |H(f)|^2 \, df}$$

using Eq. (1.273) and $\sigma_1^2 = \sigma_0^2$. Since $n(t)$ is white noise, d^2 is maximum if

$$h(t) = S_1(T_b - t) - S_0(T_b - t) \qquad 0 \le t \le T_b$$
$$= 0 \qquad\qquad\qquad\qquad\qquad \text{elsewhere}$$

See the matched-filter derivation in Eq. (1.130b). The receiver is shown in Figure 1.28. Therefore, the maximum is given by

$$d^2 = \frac{\int_0^T [S_1(T_b - t) - S_0(T_b - t)]^2 \, dt}{2N_0} \qquad (1.280a)$$

Let us denote

$$E_i = \int_0^{T_b} S_i^2(t) \, dt \qquad\qquad\qquad (1.280b)$$

$$P = \frac{1}{\sqrt{E_1 E_2}} \int_0^{T_b} S_0(t) S_1(t) \, dt$$

the crosscorrelation of the two signals. Therefore, the maximum d is given by

$$d = \left(\frac{E_1 + E_2 - 2\rho\sqrt{E_1 E_2}}{2N_0} \right)^{1/2} \qquad (1.281)$$

Using Eqs. (1.279a) and (1.281) we obtain the probability of bit error as

$$P_e = Q \left\{ \left[\frac{E_1 + E_2 - 2\rho\sqrt{E_1 E_2}}{2N_0} \right]^{1/2} \right\} \qquad (1.282a)$$

when $S_1(t)$ and $S_0(t)$ are transmitted with equal probability. When $E_1 = E_0 = E_b$, Eq. (1.282a) gives

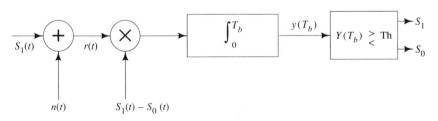

FIGURE 1.28   Detection of coherent binary signals.

$$P_e = Q\left\{\left[\frac{E_b}{N_0}(1-\rho)\right]^{1/2}\right\}$$

$$= \frac{1}{2}\text{erfc}\left\{\left[\frac{E_b(1-\rho)}{2N_0}\right]^{1/2}\right\}$$

(1.282b)

where

$$\text{erfc}(x) = \int_x^\infty \frac{2}{\sqrt{\pi}} e^{-t^2}\, dt$$

(1.282b)

Case 1. Orthogonal signals, $\rho = 0$. Let

$$\left.\begin{array}{l} S_0(t) = A\cos 2\pi f_c t \\ S_1(t) = A\sin 2\pi f_c t \end{array}\right\} \quad 0 \le t \le T_b$$

where $f_c = n/T_b$ and n is an integer.

These signals can also be written as

$$\left.\begin{array}{l} S_i(t) = A\cos(2\pi f_c t + \theta_i) \\ \theta_0 = 0 \\ \theta_1 = \pi/2 \end{array}\right\} \quad 0 \le t \le T_b$$

These signals are *binary coherent phase-shift keying* (BPSK). Note further that

$$E_1 = E_0 = \frac{A^2 T_b}{2} = E_b$$

$$\rho = 0$$

Therefore, the probability of bit error for *orthogonal coherent* BPSK, using Eq. (1.282b), is

$$P_e = \frac{1}{2}\text{erfc}\left(\sqrt{\frac{E_b}{2N_0}}\right)$$

(1.283a)

where E_b/N_0 is called the *signal energy-to-noise ratio*.

Case 2. Antipodal signals, $\rho = -1$. Let

$$\left.\begin{array}{l} S_0(t) = A\cos(2\pi f_c t) \\ S_1(t) = A\cos(2\pi f_c t) \end{array}\right\} \quad 0 \le t \le T_b$$

or

$$\left.\begin{array}{l} S_i(t) = A\cos(2\pi f_c t + \theta_i) \\ \theta_0 = 0 \\ \theta_1 = \pi \end{array}\right\} \quad 0 \le t \le T_b$$

In this case,

$$E_1 = E_0 = \frac{A^2 T}{2}$$

$$\rho = -1$$

The probability of bit error for *antipodal coherent binary signal* is

$$P_e = \frac{1}{2} \text{erfc}\left(\sqrt{\frac{E_b}{N_0}}\right) \qquad \text{(antipodal)} \tag{1.283b}$$

when signals $S_0(t)$ and $S_1(t)$ are equally likely and have equal energy. For $P_e = 10^6$, $d = 10.3$ dB in the case of antipodal signals and $d = 13.6$ dB in case of orthogonal signals. See Fig. 1.31b.

When the oscillators have some instabilities, the probabilities of error for coherent binary signals are given by

$$P_e = \frac{1}{2} \text{erfc}\left(\sqrt{\frac{d}{2}} \cos \phi\right) \qquad \text{(orthogonal signals)} \tag{1.284a}$$

$$P_e = \frac{1}{2} \text{erfc}(\sqrt{d} \cos \phi) \qquad \text{(antipodal signals)}$$

$$\tag{1.284b}$$

$$d = E_b/N_0$$

where ϕ is the difference of phase between the transmitter and receiver oscillators. When $\phi = 0$, Eqs. (1.283) are the same as Eqs. (1.283). In some systems, the phase difference ϕ is a random variable. The phase error ϕ is obtained by using a phase lock loop, as described earlier. The probability density function of the phase error in some systems is given by

$$f_\phi(\phi) = \frac{\exp[\alpha \cos \phi]}{2\pi I_0(\alpha)} \qquad |\phi| \leq \pi \tag{1.285}$$

where $I_0(\alpha)$ is the modified Bessel function of zero order and

$$\alpha = \frac{4A}{KN_0}$$

where K is the combined loop gain of the phase lock loop (see Mohanty, 1986, p. 83).

The probability of bit error is obtained by averaging the probability of bit error in Eqs. (1.284a and b) over ϕ. Hence the probability of bit error when there is an imperfect correlation is given by

$$P_e = E_\phi\left[\frac{1}{2} \text{erfc}\left(\left(\frac{\sqrt{d}}{\sqrt{2}}\right) \cos \phi\right)\right]$$

$$= \frac{1}{2\pi I_0(\alpha)} \int_{-\pi}^{\pi} \frac{1}{2} \text{erfc}\left(\frac{\sqrt{d}}{\sqrt{2}} \cos \phi\right) \exp(\alpha \cos \phi) \, d\phi \tag{1.286a}$$

(for orthogonal signal)

$$= \frac{1}{2\pi I_0(\alpha)} \int_{-\pi}^{\pi} \frac{1}{2} \mathrm{erfc}(\sqrt{d} \cos \phi) \exp(\alpha \cos \phi) \, d\phi$$

$$(1.286b)$$

(for antipodal signal)

$$d = \frac{A^2 T_b}{2N_0}$$

$$\mathrm{erfc}(x) = \frac{2}{\sqrt{\pi}} \int_x^{\infty} e^{-y^2} \, dy$$

We can choose two signals as follows:

$$S_i(t) = A \cos(2\pi f_i t) \qquad 0 \le t \le T_b \qquad i = 0, 1$$

$$f_1 = f_c + f_d$$

$$f_0 = f_c - f_d$$

$$(1.287a)$$

$$f_d = \frac{n}{2T_b}$$

$$f_c \ge \frac{2}{T_b}$$

$$(1.287b)$$

where n is an integer. When $n = 1$, the frequency separation is minimum. The signal scheme is called the *minimum shift keying (MSK)*. We note that

$$E_i = \int_0^{T_b} S_i^2(t) \, dt = \frac{A^2 T_b}{2} = E_b \qquad i = 0, 1$$

$$\rho = \frac{1}{E_b} \int_0^{T_b} S_0(t) S_1(t) \, dt$$

$$= \frac{\sin 2\pi(f_1 - f_0) T_b}{2\pi(f_1 - f_0) T_b} = 0 \qquad \text{if } (f_1 - f_0) T_b = \frac{1}{2}$$

The signals scheme described in Eq. (1.287a, b) is called the *orthogonal binary frequency shift keying* (BFSK). Therefore, the probability of bit error for BFSK with no phase error is

$$P_e = \frac{1}{2} \mathrm{erfc}\left(\sqrt{\frac{d}{2}}\right)$$

$$(1.288a)$$

where $d = E_b/N_0$.

Let us consider BFSK signals

$$S_i(t) = A \cos(2\pi f_i t) \qquad 0 \le t \le T_b$$

$$f_0 = f_c - f_d \qquad f_c \ge \frac{2}{T_b}$$

$$f_1 = f_c + f_d$$

$$(1.288b)$$

It can be shown that

$$E_1 = E_2 = E_b = \frac{A^2 T}{2}$$

$$\rho = \frac{1}{E_b} \int_0^{T_b} A^2 \cos(2\pi f_1 t) \cos(2\pi f_2 t)\, dt$$

$$= \frac{\sin(2\pi f_d T_b)}{2\pi f_d T_b}$$

The minimum value of $\rho = -0.22$ if $2\pi f_d T_b = 3\pi/2$. This signaling scheme given by Eq. (1.288b) is called the *coherent binary frequency-shift keying* (BFSK) with $2\pi f_d T_b = 3\pi/2$. The probability of error for coherent BPFK modulation is

$$P_e = \frac{1}{2} \text{erfc}\left(\sqrt{\frac{1.22 E_b}{N_0}} \right)$$

$$= \frac{1}{2} \text{erfc}(\sqrt{1.22d}) \tag{1.288c}$$

The *coherent PSK* modulation is about 0.86 dB better than orthogonal PSK given by Eq. (1.288b). See Fig. 1.31b.

Noncoherent Detection

Binary Frequency-Shift Keying (BFSK). Let a received signal be

$$\left. \begin{array}{l} y(t) = A \cos(2\pi f_c t + \psi_1 + \theta) + n(t) \quad \text{with probability } 1/2 \\ = A \cos(2\pi f_c t + \psi_0 + \phi) + n(t) \quad \text{with probability } 1/2 \end{array} \right\} 0 \le t \le T_b \tag{1.289a}$$

where $\psi_1 = \pi t/2T_b$, $\psi_0 = -\psi_1$, $n(t)$ is white Gaussian noise with mean zero and spectral density $N_0/2$ and A and ψ_i are the amplitude and phase, respectively, of $S_i(t)$, $t = 0, 1$. The probability densities of θ and ϕ are given by

$$f(\theta) = \frac{1}{2\pi} \qquad -\pi < \theta < \pi$$

$$= 0 \qquad \text{elsewhere}$$

$$f(\phi) = \frac{1}{2\pi} \qquad -\pi < \phi < \pi$$

$$= 0 \qquad \text{elsewhere}$$

The conditional probability densities of $S_0(t)$ and $S_1(t)$ are given by

$$f\left(\frac{y}{S_i}\right) = \frac{1}{\sqrt{N_0 \pi}} \exp\left[-\frac{1}{N_0} \int_0^T (y(t) - A\cos(2\pi f_c t - \psi_i + \theta))^2 \right] dt$$

$$i = 0, 1 \tag{1.289b}$$

The likelihood ratio is given by

$$
\Lambda = \frac{E_\theta f(Y|S_1, \theta)}{E_\phi f(Y|S_0, \phi)}
$$

$$
= \frac{\dfrac{1}{2\pi} \displaystyle\int_{-\pi}^{\pi} \exp\left[-\dfrac{1}{N_0} \int_0^T (y(t) - A\cos(2\pi f_c t + \psi_1 + \theta))^2 \, dt \right] d\theta}{\dfrac{1}{2\pi} \displaystyle\int_{-\pi}^{\pi} \exp\left[-\dfrac{1}{N_0} \int_0^T (y(t) - A\cos(2\pi f_c t + \psi_0 + \phi))^2 \, dt \right] d\phi}
$$

$$
= \frac{\displaystyle\int_{-\pi}^{\pi} \exp\left[\dfrac{2}{N_0} \int_0^T y(t) A \cos(2\pi f_c t + \psi_1 + \theta) \, dt \right] d\theta}{\displaystyle\int_{-\pi}^{\pi} \exp\left[\dfrac{2}{N_0} \int_0^T y(t) A \cos(2\pi f_c t + \psi_0 + \phi) \, dt \right] d\phi}
$$

canceling terms containing integrals of y^2 and also canceling integrals of S_1^2 and S_0^2, since the power S_1 is the same as S_0, from both the numerator and denominator. Let

$$
\left.
\begin{aligned}
X_i &= \int_0^T y(t) A \cos(2\pi f_c t + \psi_i) \, dt \\[2mm]
Z_i &= \int_0^T y(t) A \sin(2\pi f_c t + \psi_i) \, dt
\end{aligned}
\right\} \quad i = 0, 1 \tag{1.289c}
$$

The likelihood ratio function can be written as

$$
\Lambda(y) = \frac{\displaystyle\int_{-\pi}^{\pi} \exp\left[\dfrac{2A}{N_0} (\cos\theta X_1 - \sin\theta Z_1) \right] d\theta}{\displaystyle\int_{-\pi}^{\pi} \exp\left[\dfrac{2A}{N} (\cos\phi X_0 - \sin\phi Z_0) \right] d\phi}
$$

Let

$$
\left.
\begin{aligned}
q_i &= \sqrt{X_i^2 + Z_i^2} \\[1mm]
\eta_i &= \arctan(Z_i/X_i)
\end{aligned}
\right\} \quad i = 0, 1 \tag{1.289d}
$$

The likelihood function is

$$
\Lambda(y) = \frac{\displaystyle\int_{-\pi}^{\pi} \exp\left[\dfrac{2A_1}{N_0} q_1 \cos(\theta + \eta_1) \right] d\theta}{\displaystyle\int_{-\pi}^{\pi} \exp\left[\dfrac{2A_0}{N_0} q_0 \cos(\phi + \eta_0) \right] d\phi}
$$

$$
= \frac{2\pi I_0\left(\dfrac{2A_1}{N_0} q_1 \right)}{2\pi I_0\left(\dfrac{2A_0}{N_0} q_0 \right)} \tag{1.290a}
$$

where

$$I_0(\alpha) = \frac{1}{2\pi} \int_{-\pi}^{\pi} e^{\alpha \cos \theta} \, d\theta \qquad (1.290b)$$

is the modified Bessel function of the first kind of order 0. If

$$\frac{\log \left[I_0 \left(\frac{2A_1}{N_0} q_1 \right) \right]}{\log \left[I_0 \left(\frac{2A_0}{N_0} q_0 \right) \right]} \geq 0 \qquad S_1 \text{ is selected}$$

$$< 0 \qquad S_0 \text{ is selected}$$

Therefore the decision rule is that if

$$\log I_0 \left(\frac{2A_1}{N_0} q_1 \right) \geq \log I_0 \left(\frac{2A_0}{N_0} q_0 \right)$$

then S_1 is decided. The equivalent decision rule is that if

$$q_1 - q_0 \geq 0$$

then S_1 is decided. This is because the inequality is valid for an inverse modified Bessel function. Therefore the detection scheme is as shown in Fig. 1.29. This detection scheme is called *noncoherent detection*; X_i and Z_i are called *in-phase* and *quadrature* components, $i = 0, 1$.

$$h_i(t) = A \cos(\omega_c(T - t) + \theta + \psi_i) \qquad 0 \leq t \leq T$$

$$= 0 \qquad \text{elsewhere}; \ i = 0, 1$$

The output of the matched filters is

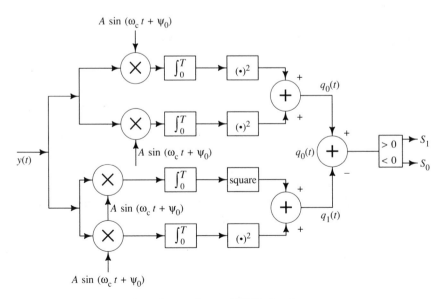

FIGURE 1.29   Detection of noncoherent BPSK signals.

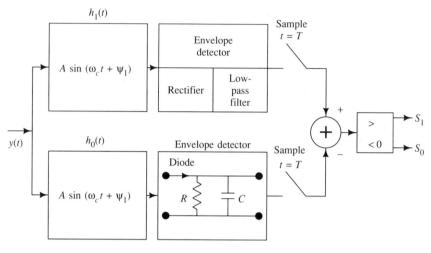

FIGURE 1.30   Envelope detection of BPSK signals.

$$r_i(t) = \int_{-\infty}^{t} y(\tau)h_i(t-\tau)\,d\tau \qquad i = 0, 1$$

$$r_1(t) = \Lambda \cos(\omega_c(T-t) + \theta) \int_0^t y(\tau)\cos(\omega_c\tau + \psi_1)\,d\tau$$

$$+ A\sin(\omega_c(T-t) + \theta) \int_0^t y(\tau)\sin(\omega_c\tau + \psi_1)\,d\tau$$

The envelope of $r_1(t)$ evaluated at $t = T$ is given by

$$\left\{ \left(\int_0^T y(t)A\cos(\omega_c t + \psi_1)\,dt \right)^2 + \left(\int_0^T y(t)A\sin(\omega_c t + \psi_1)\,dt \right)^2 \right\}^{1/2} = q_1$$

Similarly, the envelope of $r_0(T)$ is

$$\left\{ \left(\int_0^T y(t)A\cos(\omega_c t + \psi_0)\,dt \right)^2 + \left(\int_0^T y(t)A\sin(\omega_c t + \psi_0)\,dt \right)^2 \right\}^{1/2} = q_0$$

We can implement the detection scheme by using a parallel matched filter followed by envelope detectors. Therefore this detection process is called *envelope detection*, and it is shown in Fig. 1.30. The envelope detector is a rectifier and is followed by a low-pass device. It is usually used in practice to demodulate AM signals. X_1 and Z_1, defined in Eq. (1.289c) are Gaussian random variables. The conditional mean of X_1 is

$$E[X_1/S_1(\theta)] = E\left[\int_0^T y(A)\cos(\omega_c t + \psi_1)\,dt | S_1(\theta) \right]$$

$$= \int_0^T A^2 \cos(\omega_c\tau + \psi_1 + \theta)\cos(\omega_c\tau + \psi_1)\,d\tau$$

$$= \frac{A^2}{2} \left[\int_0^T [\cos\theta + \cos(2\omega_c t + 2\psi_1 + \theta)] \, dt \right]$$

$$= \frac{A^2 T}{2} \cos\theta = E_1 \cos\theta \tag{1.291}$$

where $E_1 = A^2 T/2$. The contribution to the second term is zero assuming $\omega_c T = K\pi$, where K is an integer.

The conditional variance of X_1 is

$$\text{var}[X_1|S_1(\theta)] = E\left\{ \left(X_1 - \frac{A^2 T}{2} \cos\theta \right) \middle| S_1 \right\}^2$$

$$= E\left(\int_0^T n(t) A \cos(\omega_c t + \psi_1) \, dt \right)^2$$

$$= A^2 E\left[\int_0^T \int_0^T n(t) n(\tau) \cos(\omega_c t + \psi_1) \cos(\omega_c \tau + \psi_1) \, dt \, d\tau \right]$$

$$= A^2 \int_0^T \int_0^T E[n(t)n(\tau)] \cos(\omega_c t + \psi_1) \cos(\omega_c \tau + \psi_1) \, dt \, d\tau$$

$$= A^2 \int_0^T \int_0^T \frac{N_0}{2} \delta(t - \tau) \cos(\omega_c t + \psi_1) \cos(\omega_c \tau + \psi_1) \, dt \, d\tau$$

$$= \frac{N_0}{2} A^2 \int_0^T \cos^2(\omega_c t + \psi_1) \, dt$$

$$= \frac{N_0}{4} A^2 \int_0^T [1 + \cos(2\omega_c t + 2\psi_1)] \, dt$$

$$= \frac{N_0 T}{4} A^2 \tag{1.292}$$

Similarly,

$$E(Z_1|S_1(\theta)) = \frac{A^2 T}{2} \sin\theta = \mathscr{E}_1 \sin\theta \tag{1.293}$$

$$\text{var}[Z_1|S_1(\theta)] = \frac{A^2 N_0 T}{4} = \sigma^2 \tag{1.294}$$

The joint condition density of X_1 and Z_1 is

$$f(X_1, Z_1|\theta) = \frac{1}{2\pi\sigma^2} \exp\left[\frac{-[(X_1 - \mathscr{E}_1 \cos\theta)^2 + (Z_1 - \mathscr{E}_1 \sin\theta)^2]}{2\sigma^2} \right] \tag{1.295a}$$

using Eqs. (1.293) and (1.294), $\mathscr{E}_1 = A^2 T/2$, and $\sigma^2 = A^2 N_0 T/4$, and substituting

$$\begin{aligned} X_1 &= q_1 \cos\delta_1 & q_1 &\geq 0 \\ Z_1 &= q_1 \sin\delta_1 & -\pi &\leq \delta_1 \leq \pi \end{aligned} \tag{1.295b}$$

in Eq. (1.295a). The joint conditional density of q_1 and δ_1 is

$$f(q_1, \delta_1 | \theta) = \frac{q_1}{2\pi\sigma^2} \exp\left[-\frac{1}{2\sigma^2}[q_1^2 + \mathscr{E}_1^2 - 2\mathscr{E}_1 q_1 \cos(\theta - \delta_1)] \right] \quad (1.296)$$

where $\delta_1 = \arctan(Z_1/X_1)$, and the Jacobian $|J| = q_1$.

The marginal density of q_1, when S_1 is present, is

$$\begin{aligned}
f(q_1 | \theta) &= \int_{-\pi}^{\pi} f(q_1, \delta_1 | \theta) \, d\delta_1 \\
&= \frac{q_1}{\sigma^2} \exp\left[-\frac{1}{2\sigma^2}(q_1 + \mathscr{E}^2) \right] \frac{1}{2\pi} \int_{-\pi}^{\pi} \exp\left[\frac{\mathscr{E}_1 q_1}{\sigma^2} \cos(\theta - \delta) \right] d\delta \\
&= \frac{q_1}{\sigma^2} \exp\left[-\frac{1}{2\sigma^2}(q_1^2 + \mathscr{E}_1^2) \right] \cdot I_0\left(\frac{\mathscr{E}_1 q_1}{\sigma^2} \right) \quad q_1 \geq 0 \quad (1.297)
\end{aligned}$$

where $I_0(\cdot)$ is the modified Bessel function of the first kind of order 0 and is given by Eq. (1.290b).

The probability density function of $f(q_1 | S_1)$ is called the *Rician density*. The nth moment of the Rician random variable is

$$\begin{aligned}
E(q_1^n | S_1) &= \int_0^\infty \frac{q_1^{n+1}}{\sigma^2} \exp\left[-\left(\frac{q_1^2}{2\sigma^2} + \frac{\mathscr{E}_1^2}{2\sigma^2} \right) \right] I_0\left(\frac{\mathscr{E}_1 q_1}{\sigma^2} \right) dq_1 \\
&= (2\sigma^2)^{n/2} \Gamma\left(\frac{n}{2} + 1 \right) {}_1F_1\left(\frac{-n}{2}, 1, \frac{\rho_1^2}{2} \right) \quad (1.298)
\end{aligned}$$

where ${}_1F_1(a, b, x)$, the *confluent hypergeometric function*, is defined by

$${}_1F_1(a, b, x) = 1 + \frac{a}{b}x + \frac{a(a+1)}{b(b+1)}\frac{x^2}{2!} + \frac{a(a+1)(a+2)}{b(b+1)(b+2)}\frac{x^3}{3!} + \cdots \quad (1.299)$$

and $\rho_1^2 = \mathscr{E}_1/\sigma^2$ is the signal-to-noise ratio. The *gamma function* is

$$\Gamma(n) = \int_0^\infty e^{-x} x^{(n-1)} \, dx = (n-1)!$$

Similarly, the PDF of q_0 when S_0 is present is

$$f(q_0 | S_0) = \frac{q_0}{\sigma^2} \exp\left[-\frac{1}{2\sigma^2}(q_0^2 + \mathscr{E}_0^2) \right] I_0\left(\frac{\mathscr{E}_0}{\sigma^2} q_0 \right) \quad q_0 \geq 0 \quad (1.300)$$

where $\mathscr{E}_0 = A_0^2 T/2$ and $\sigma^2 = A_0^2 N_0 T/4$.

If $A = 0$, then Eq. (1.300) takes the form of

$$f(q_0 | S_0 = 0) = \frac{q_0}{\sigma^2} \exp\left[-\frac{1}{2\sigma^2} q_0^2 \right] \quad q_0 \geq 0 \quad (1.301)$$

This density is known as a *Rayleigh* PDF.

Using Eqs. (1.289a) and (1.289c), the conditional mean of X_0 when $S_1(\theta)$ is transmitted is

$$E(X_0|S_1) = E\left[\int_0^T [S_1(\theta) + n(\tau)] A \cos(\omega_c\tau + \psi_0)\, d\tau\right]$$

$$= E\left[\int_0^T AS_1(\theta)\cos(\omega_c\tau + \psi_0)\, d\tau\right]$$

$$= E\left[\int_0^T A^2 \cos(\omega_c\tau + \theta + \psi_1)\cos(\omega_c\tau + \psi_0)\, d\tau\right]$$

$$= 0$$

Similarly, the conditional density of Z_0, when $S_1(\theta)$ is transmitted, is

$$E(Z_0|S_1(\theta)) = 0$$

The variance of X_0 and Z_0 when $S_1(\theta)$ is transmitted is

$$\mathrm{var}(X_0|S_1) = \mathrm{var}(Z_0|S_1) = \frac{N_0 TA^2}{4}$$

The joint density of X_0 and Z_0 when S_1 is transmitted is

$$f(X_0, Z_0|S_1) = \frac{1}{2\pi\sigma^2}\exp\left[-\frac{1}{2\sigma^2}(X_0^2 + Z_0^2)\right] \qquad (1.302a)$$

Substitute in Eq. (1.302a)

$$\begin{aligned} X_0 &= q_0\cos\delta_0 & q_0 &\geq 0 \\ Z_0 &= q_0\cos\delta_0 & -\pi &\leq \delta_0 \leq \pi \end{aligned} \qquad (1.302b)$$

and note that the Jacobian $|J| = q_0$. Then Eq. (1.302a) yields

$$f(q_0, \delta_0|S_1) = \frac{1}{2\pi\sigma^2} q_0 \exp\left[-\frac{q_0^2}{2\sigma^2}\right]$$

The marginal density of

$$f(q_0|S_1) = \int_{-\pi}^{\pi} f(q_0, \delta_0|S_1)\, d\delta_0$$

$$= \frac{1}{\sigma^2} q_0 \exp\left[-\frac{q_0^2}{2\sigma^2}\right] \qquad q_0 \geq 0 \qquad (1.302c)$$

which is the same as Eq. (1.301) and is known as the Rayleigh PDF. The distribution function of the Rayleigh random variable is

$$P[q_0 \leq q] \triangleq F_{q_0}(q) = 1 - e^{-q^2/2\sigma^2} \qquad q \geq 0 \qquad (1.303)$$

The distribution of a Rician random variable is given by

$$F_{q_1}(x) = \int_0^x f_{q_1}(q)\,dq$$

$$= \int_0^x \frac{q_1}{\sigma^2}\exp\left[\frac{q_1^2 + \mathscr{E}_1^2}{-2\sigma^2}\right]I_0\left(\frac{\mathscr{E}_1 q_1}{\sigma^2}\right)dq_1$$

$$= \int_0^{x/\sigma} q\exp\left[\frac{q^2 + d_1^2}{-2}\right]I_0(d_1 q)\,dq \qquad (1.304)$$

substituting $q_1/\sigma = q$. Hence

$$F_{q_1}(x) \triangleq 1 - Q(y, d_1)$$

$$v = x/\sigma \qquad (1.305)$$

$$d_1 = \mathscr{E}_1/\sigma$$

$$Q(v, \rho) = \int_0^\infty \exp\left[\frac{-q^2 + \rho^2}{2}\right]I_0(\rho q)\,dq$$

is called the *Marcum Q function*.

The probability of error is

$$P_e = P[q_1 - q_0 < 0 | S_1 \text{ is true}]$$

$$= P[q_1 < q_0 | S_1(\theta) \text{ is true}]$$

$$= P[q_0 > q_1 | S_1(\theta)]$$

$$= \int_0^\infty P[q_0 > q_1 = q' | S_1]f_{q_1}(q^1)\,dq^1$$

$$P[q_0 > q_1 = q^1] = \int_{q_1}^\infty f_{q_0}(q_0 | q_1, S_1)\,dq_0$$

$$P_e = \int_0^\infty \int_{q_1}^\infty f_{q_1}(q_1 | S_1)f_{q_0}(q_1, S_1)\,dq_1\,dq_0$$

$$= \int_0^\infty \int_{q_1}^\infty f(q_0, q_1 | S_1)\,dq_0\,dq_1$$

$$= \int_0^\infty \int_{q_1}^\infty f_{q_0}(q_0 | S_1)f_{q_1}(q_1 | S_1)\,dq_0\,dq_1$$

since q_1 and q_0 are independent random variables. Note that

$$f_{q_0}(q_0 | q_1, S_1) = f_{q_0}(q_0 | S_1) = \frac{1}{\sigma^2}q_0\exp[-q_0^2/2\sigma^2]$$

$$P_e = \int_0^\infty dq_1 f_{q_1}(q_1)\left[\int_{q_1}^\infty q_0\exp[q_0^2/2\sigma^2]\,dq_0\right]$$

$$= \int_0^\infty \int_{q_1}^\infty \frac{q_0 q_1}{\sigma^4}\exp\left[\frac{q_0^2 + q_1^2 + \mathscr{E}_1^2}{-2\sigma^2}\right]I_0\left(\frac{q_1\mathscr{E}_1}{\sigma^2}\right)dq_0\,dq_1$$

$$= \int_0^\infty \frac{q_1}{\sigma^2} \exp\left[\frac{2q_1^2 + \mathscr{E}_1^2}{-2\sigma^2}\right] I_0\left(\frac{q_1 \mathscr{E}_1}{\sigma^2}\right) dq_1$$

$$= \frac{1}{2} \exp\left[-\frac{A^2 T}{2 \cdot 2N_0}\right]$$

$$= \frac{1}{2} \exp\left[-\frac{\delta}{2N_0}\right] \tag{1.306}$$

where $\mathscr{E} = (A^2 T/2)$ is the energy of the signal. Therefore the probability of error for *noncoherent FSK* signals is

$$P_e = \frac{1}{2} \exp\left[-\frac{\mathscr{E}}{2N_0}\right] \qquad \mathscr{E} = \frac{A^2 T}{2}$$

These noncoherent BPSK signals are orthogonal ($\int_0^T S_0(t)S_1(t)\,dt = 0$) and of equal energy ($\mathscr{E}_1 = \mathscr{E}_0$) and are equal likely ($\pi_0 = \pi_1$). For the same signal set

$$P_e = 0.5 \operatorname{erfc}\left(\sqrt{\frac{\mathscr{E}}{2N_0}}\right) \qquad \mathscr{E} = \frac{A^2 T}{2}$$

for coherent BFSK.

To achieve $P_e = 10^{-4}$, the envelope detector for a noncoherent BFSK signal needs approximately 4 dB more than for the matched-filter detection of orthogonal, equally likely, equal energy coherent BFSK signals. See Fig. 1.31b.

Differential Phase-Shift Keying (DPSK). In this scheme, the transmitted signals are

$$A \cos(\omega_c t + \theta_k^e) \qquad (k-1)T \le t \le kT$$

where

$$\left.\begin{array}{l} \theta_k^e = 0 \quad \text{if } \theta_k - \theta_{k-1} \ne 0 \\ \quad = \pi \quad \text{if } \theta_k - \theta_{k-1} = 0 \end{array}\right\} \quad k \ge 1, \theta_0 = \pi \tag{1.307a}$$

where

$$\theta_k = \pi \qquad \text{when } S_1 \text{ is transmitted}$$

$$\quad = 0 \qquad \text{when } S_0 \text{ is transmitted}$$

$$\omega_c T = 2\pi$$

θ_k^e is called *differential encoding* of the phase. The received signal is

$$y(t) = A \cos(\omega_c t + \theta_k^e + \phi) + n(t) \qquad 0 < \phi < \pi, 0 \le t \le T$$

where $n(t)$ is a zero-mean white narrowband Gaussian noise with $\sigma^2 = N_0/2$. The received signal phase is differentially decoded as follows:

$$\theta_k^d = 0 \quad \text{if } (\theta_k^e + \phi) - (\theta_{k-1}^e + \phi) \ne 0 \quad \text{that is,} \quad \theta_k^d - \theta_{k-1}^d \ne 0, \theta_0 = \pi$$

$$\pi = \quad \text{if } (\theta_k^e + \phi) - (\theta_{k-1}^e + \phi) = 0 \quad \text{that is,} \quad \theta_k^d - \theta_{k-1}^d = 0$$

$$\tag{1.307b}$$

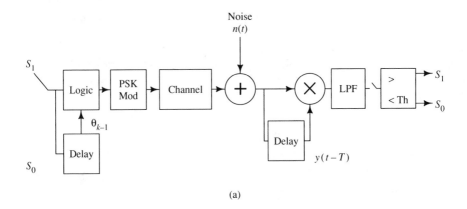

(a)

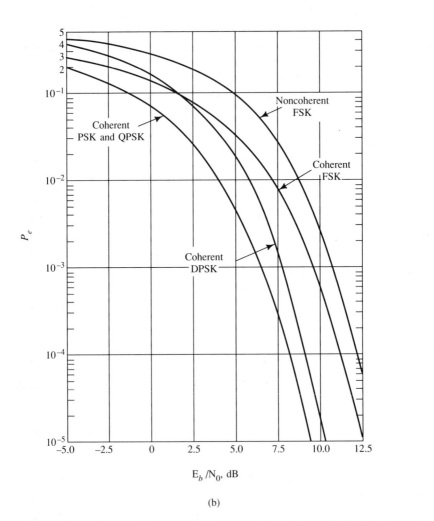

(b)

FIGURE 1.31   (a) DPSK system. (b) P_e versus E_b/N_0. (From H. L. Van Trees, Ed., *Satellite Communications*, New York: IEEE Press. © 1979 IEEE.)

TABLE 1.2 Differential Encoding and Decoding

k	1	2	3	4	5	6	7	8	9	10
$\theta_k,\ \theta_0 = \pi$	π	π	0	π	0	0	0	π	π	0
θ_k^e	π	π	0	0	π	0	π	π	π	0
$\theta_k^d,\ \theta_0 = \pi$	π	π	0	π	0	0	0	π	π	0

The differential encoding and decoding scheme is shown in Fig. 1.31a. In terms of circuit implementation, the received signal is delayed by T seconds, is multiplied by the original delayed signal, and the product signal is multiplied by passing through a low-pass filter.[c]

In the absence of any additive noise, the encoded and decoded phase sequence is given by Table 1.2 for a given sequence of θ_k.

Notice that $\theta_k^d = \theta_k^d$; the phase delay ϕ (constant) has not caused any error in the decoded phase sequence.[e]

The input to the low-pass filter (LPF) is

$$Z(t) = y(t)y(t - T) \qquad (k - 1)T \le t < kT$$

$$= \{A \cos(\omega_c t + \theta_k^e + \phi) + \bar{n}(t)\}$$

$$\times \{A \cos[\omega_c(t - T) + \theta_{k-1}^e + \phi] + \bar{n}(t - T)\}$$

where $\bar{n}$ is output of the LPF and is a narrowband Gaussian noise. The simplification of the above equation yields

$$Z(t) = A \cos(\omega_c t + \theta_k^e + \phi)\bar{n}(t - T) + A \cos(\omega_c(t - T) + \theta_{k-1}^e + \phi)\bar{n}(t)$$

$$+ \tfrac{1}{2}[A^2 \cos(\omega_c T + (\theta_k^e - \theta_{k-1}^e))$$

$$+ A^2 \cos(2\omega_c t - \omega_c T + (\theta_k^e + \theta_{k-1}^e + 2\phi)] + \bar{n}(t)\bar{n}(t - T)$$

The output of the LPF (integrator) is

$$q(T) = \int_0^T \left\{ \frac{A^2}{2} \cos(\omega_c T + \theta_k^d) + A \cos(\omega_c t + \theta_k^e + \phi)[\bar{n}(t - T) + \bar{n}(t)] \right.$$

$$\left. + \bar{n}(t)\bar{n}(t - T) \right\} dt \tag{1.308}$$

We write the narrowband Gaussian noise output $\bar{n}(t)$ as

$$\bar{n}(t) = n_c(t) \cos \omega_c t - n_s(t) \sin \omega_c t \tag{1.309}$$

---

[e] The encoding and decoding operation can also be written as

$$\theta_k^e = \theta_k \oplus \theta_{k-1}$$

$$\theta_k^d = \theta_k^e \oplus \theta_{k-1}^e$$

where $\oplus$ stands for exclusive-OR operation, $\theta_k = 0$ or 1.

and

$$\bar{n}(t - T) = n_c(t - T)\cos\omega_c(t - T) - n_s(t - T)\sin\omega_c(t - T)$$

$$\approx n_c(t - T)\cos\omega_c t - n_s(t - T)\sin\omega_c t$$

$$E[n_c(t)] = E[n_s(t)] = 0, \quad E(n_c^2) = E(n_s^2) = \sigma^2 = N_0/2$$

$$E[n_c(t)n_s(t)] = 0 \tag{1.310}$$

Also, $\omega_c T = 2\pi$.

Substitution of Eqs. (1.309) and (1.310) in Eq. (1.308) yields

$$q(T) = \int_0^T \left\{ \frac{A^2}{2}\cos(\theta_k^d + \omega_c T) \right.$$

$$+ \frac{A}{2}\cos(\theta_k^e + \phi)[n_c(t) + n_c(t - T)]$$

$$- \frac{A}{2}\sin(\theta_k^e + \phi)[n_s(t) + n_s(t - T)]$$

$$\left. + \frac{1}{2}n_c(t)n_c(t - T) + \frac{1}{2}n_s(t)n_s(t - T) \right\} dt \tag{1.311}$$

neglecting the double-frequency terms.

When $\theta_k = 0$, then $\theta_k^d = 0$ and

$$q(T) = \frac{A^2 T}{2} + \frac{AT}{2}\cos\phi[\bar{n}_c(T) + \bar{n}_c(T - 1)]$$

$$- \frac{AT}{2}\sin\phi[\bar{n}_s(t) + \bar{n}_s(T - 1)]$$

$$+ \frac{1}{2}T\bar{n}_c(T)\bar{n}_c(T - 1) + \frac{T}{2}\bar{n}_s(T)\bar{n}_s(T - 1)$$

where

$$\bar{n}_c(T) \triangleq \int_0^T n_c(t)\,dt \triangleq \bar{n}_c$$

$$\bar{n}_c^1 \triangleq \int_0^T n_c(t - T)\,dt \triangleq \bar{n}_c(T - 1) \quad \text{etc.}$$

If $q(T) > 0$, then S_1 is present. Therefore, the probability of error is

$$P_e = [q(T) \le 0 | S_1 \text{ is present}]$$

Let

$$\alpha_1 = A\sqrt{T} + \tfrac{1}{2}[(\bar{n}_c + \bar{n}_c^1)\cos\phi - (\bar{n}_s + \bar{n}_s^1)\sin\phi]$$

$$\beta_1 = \tfrac{1}{2}[(\bar{n}_c + \bar{n}_c^1)\sin\phi + (\bar{n}_s + \bar{n}_s^1)\cos\phi]$$

$$\gamma_1 = \tfrac{1}{2}[(\bar{n}_c - \bar{n}_c^1)\cos\phi - (\bar{n}_s - \bar{n}_s^1)\sin\phi]$$

$$\delta_1 = \tfrac{1}{2}[(\bar{n}_c - \bar{n}_c^1)\sin\phi + (\bar{n}_s - \bar{n}_s^1)\cos\phi]$$

$$q_1 = \alpha_1^2 + \beta_1^2$$

$$q_0^2 = \gamma_1^2 + \delta_1^2$$

and so

$$q(T) = \frac{T}{2}(\alpha_1^2 + \beta_1^2) - (\gamma_1^2 + \delta_1^2)$$

$$= \frac{T}{2}(q_1^2 - q_0^2)$$

Therefore

$$P_e = P[q_1 \le q_0 | S_1 \text{ is present}]$$

The conditional densities of q_1 and q_0 are

$$f(q_1|S_1) = \frac{q_1}{\sigma^2}\exp\left[-\frac{(q_1^2 + A^2)}{2\sigma^2}\right]I_0\left(\frac{Aq_1}{\sigma^2}\right) \qquad q_1 > 0$$

$$f(q_0|S_1) = \frac{q_0}{\sigma^2}\exp\left[\frac{-q_0^2}{2\sigma^2}\right] \qquad q_0 > 0$$

where $\sigma^2 = 2N_0/T$.

Proceeding as in noncoherent FSK detection, we get

$$P_e = \frac{1}{2}\exp\left[-\frac{\mathscr{E}}{N_0}\right]$$

where $\mathscr{E} = A^2 T/2$. This shows that DPSK requires 1 dB more ($\mathscr{E}/N_0$) than coherent PSK at $P_e = 10^{-4}$, but it needs 3 dB less ($\mathscr{E}/N_0$) than the noncoherent FSK for the same $P_e = 10^{-4}$. See Fig. 1.31b. We now consider the case when the amplitude of the signal is a random variable.

Slow Rayleigh Fading. The received FSK signal is given by

$$y(t) = A_1\cos(\omega_1 t + \theta_1) + n(t) \qquad \text{under } H_1$$

$$= A_0\cos(\omega_0 t + \theta_0) + n(t) \qquad \text{under } H_0 \qquad (1.312a)$$

$$0 \le t \le T$$

where $n(t)$ is a zero-mean white Gaussian noise with PSD $N_0/2$, the a priori probabilities π_0 and π_1 are equal, and

$$f(A_i) = \frac{A_i}{\sigma_0^2}\exp\left(-\left[\frac{A_i^2}{2\sigma_0^2}\right]\right) \qquad i = 0, 1$$

$$A_i \ge 0 \qquad (1.312b)$$

$$f(\theta_i) = \frac{1}{2\pi} \qquad -\pi \leq \theta_1 \leq \pi$$

$$= 0 \qquad \text{elsewhere} \qquad i = 0, 1$$

where

$$E(A_i^2) = 2\sigma_0^2, \qquad i = 0, 1$$

$$E(A_i) = \sigma_0\sqrt{\pi/2}$$

and θ_i and A_i are independent. It is assumed further that $A_i(t)$ and $\theta_i(t)$ are slowly changing with time. The envelopes remain essentially constant during $(0, T)$. This phenomenon is known as *slow fading*.[f]

The conditional densities are given by

$$f(y|S_i, A_i, \theta_i) = \frac{1}{\sqrt{2\pi\sigma^2}} \exp\left\{ \int_0^T \left[-\frac{(y(t) - A_i\cos(\omega_i t + \theta_i))^2}{2\sigma^2} \right] dt \right\}$$

$$i = 0, 1$$

where $\sigma^2 = N_0/2$. The likelihood ratio is

$$\Lambda = \frac{E[f(y|S_1, A, \theta_1)]}{E[f(y|S_0, A_0, \theta_0)]}$$

$$= \frac{\dfrac{1}{2\pi} \displaystyle\int_{-\pi}^{\pi} d\theta_1 f(\theta_1) \int_0^\infty f(A_1)f(y|S_1, A_1, \theta_1)\, dA_1}{\dfrac{1}{2\pi} \displaystyle\int_{-\pi}^{\pi} d\theta_0 f(\theta_0) \int_0^\infty f(A_0)f(y|S_0, A_0, \theta_0)\, dA_0} \qquad (1.312c)$$

From Eqs. (1.312a,b,c), we get

$$\Lambda = \frac{\displaystyle\int_0^\infty \frac{A_1}{\sigma_0^2} \exp\left[-\frac{A_1^2}{2}\left(\frac{1}{\sigma_0^2} + \frac{T}{N_0}\right) \right] I_0\left(\frac{2A_1 q_1}{N_0}\right) dA_1}{\displaystyle\int_0^\infty \frac{A_0}{\sigma_0^2} \exp\left[-\frac{A_0^2}{2}\left(\frac{1}{\sigma_0^2} + \frac{T}{N_0}\right) \right] I_0\left(\frac{2A_0 q_0}{N_0}\right) dA_0}$$

$$= \frac{\left(\dfrac{N_0}{N_0 + T\sigma_0^2}\right) \exp\left[\dfrac{2\sigma_0^2 q_1^2}{N_0(N_0 + T\sigma_0^2)}\right]}{\left(\dfrac{N_0}{N_0 + T\sigma_0^2}\right) \exp\left[\dfrac{2\sigma_0^2 q_0^2}{N_0(N_0 + T\sigma_0^2)}\right]}$$

$$= \exp\left\{ \left[\frac{2\sigma_0^2}{N_0(N_0 + T\sigma_0^2)}\right](q_1^2 - q_0^2) \right\}$$

where q_i, $1 = 0, 1$, are defined in Eq. (1.289b), with $\mathscr{E}_i = A_i^2 T/2$, $i = 0, 1$. Therefore, S_1 is selected if

$$\log_e \Lambda > \log\frac{\pi_0}{\pi_1}$$

---

[f] The rate of fading is much less than the reciprocal of the highest frequency of the carrier signal.

So S_1 is selected if

$$q_1^2 - q_0^2 > \frac{N_0(N_0 + T\sigma_0^2)}{2\sigma_0^2} \cdot \log \frac{\frac{1}{2}}{\frac{1}{2}} = 0$$

In other words, S_1 is selected if

$$q_1 > q_0$$

The probability of error of the BFSK signal [see Eq. (1.306)] is

$$P_e(\delta_1) = \frac{1}{2}\exp\left[\frac{-\mathscr{E}_1}{2N_0}\right] \qquad \mathscr{E}_1 = \frac{A_1^2 T}{2} \tag{1.313}$$

The probability of error is a random variable because $\mathscr{E}_1$ is a function of random amplitude A_1. Therefore, the average error probability is

$$P_e = \int_0^\infty P_e(A_1)f(A_1)\,dA_1 = \int_0^\infty \frac{A_1}{2\sigma_0^2}\exp\left[-\frac{A_1}{2}\left(\frac{1}{\sigma_0^2} + \frac{T}{2N_0}\right)\right]dA_1$$

Using Eqs. (1.313) and (1.302b). Hence

$$P_e = \frac{1}{2 + \sigma_0^2 T/N_0} = \frac{1}{2 + (E_{ave}/N_0)}$$

where the average energy is $E_{ave} = \sigma_0^2 T$. For the error probability $P_e = 10^{-4}$, coherent BFSK needs 10 dB, the noncoherent BFSK needs 12.5 dB, and noncoherent slow fading FSK needs a 40-dB signal-to-noise ratio. See Fig. 11.12.

Multiple Signals

In radar, sonar, and electrooptical sensor systems, it is often required to detect multiple objects in the presence of a cluttered background. In radioastronomy, many space objects have to be identified and classified. In binary digital communications, instead of transmitting binary digits bit by bit, a block of length k is transmitted. This technique is known as *source coding*. There are then $M = 2^k$ number of binary words of length k. For $k = 2$, $M = 4$, there are four coded binary words given as follows:

$$b_1 = 00, \quad b_2 = 01, \quad b_3 = 10, \quad b_4 = 11 \tag{1.314a}$$

Each symbol is transmitted by a waveform $S_i(t)$, $0 \le t \le T_s$, $i = 1, 2, 3, 4$. T_s is the symbol duration. For example,

$$S_i(t) = A\cos(2\pi f_c + \phi_i) \qquad 0 \le t \le T_s$$
$$\phi_i = \frac{\pi}{2}(i - 1) \qquad\qquad i = 1, 2, 3, 4 \tag{1.314b}$$

$T_s = 2T_b$, where T_b is the bit duration.

When $k = 3$, $M = 8$. In this case, three consecutive bits are assigned as symbols. We have here eight such symbols, as follows:

$$b_1 = 000 \quad b_4 = 011 \quad b_7 = 110$$

$$b_2 = 001 \quad b_5 = 100 \quad b_8 = 111$$

$$b_3 = 010 \quad b_6 = 101$$

These eight symbols are transmitted by using eight different waveforms, $S_i(t)$, $i = 1, 2, 3, 4, 5, 6, 7, 8, 0 \leq t \leq T_s$.

In Gray coding [see Eq. (1.256)], the assignment of k binary bits is made such that the adjacent symbols differ only by one bit. In such a case, eight symbols in Gray coding are 000, 001, 011, 010, 110, 111, 101, 100. As a result, only a single bit error occurs in the k-bit sequence in the decision space. The signaling scheme of Eq. (1.255b) is called *quadrature phase-shift keying* (QPSK). The block encoding scheme conserves bandwidth when used with *multiple PSK* signaling (MPSK) schemes and conserves power when used with *multiple frequency-shift keying* (MFSK) schemes. The received signal waveform is

$$y(t) = S(t) + n(t) \quad 0 \leq t \leq T_s, \quad T_s = kT_b$$

where $n(t)$ is an additive white Gaussian noise and $S(t) = S_i(t)$ with probability π_i, $i = 1, 2, \ldots, M$, $\sum_{i=1}^{M} \pi_i = 1$.

The decision rule is that S_k is selected if

$$\pi_k f(Y|S_k) \geq \pi_j f(Y|S_j) \quad \text{for all } j \neq k$$

and S_k is not selected otherwise.

The equivalent decision rule is that if

$$\pi_k f(Y/S_k) = \max_{1 \leq j \leq M} \{\pi_j f(Y/S_j)\} \tag{1.315}$$

then S_k is selected.

Dividing both sides of Eq. (1.315) by $\sum_{i=1}^{M} P(Y|S_i)\pi_i$ and using the fact that $P[y_1 \leq y \leq y_1 + dy_1] = f(y_1)dy_1$, we get

$$P[S_k|Y] = \max_{1 \leq j \leq M} P[S_j|Y] \tag{1.316}$$

when S_k is selected. This is known as *MAP detection*. MAP detection is equivalent to the maximum likelihood (ML) detection when $f(S_j)$ is independent of j.

We note that

$$\int_0^{T_s} S_i^2(t) \, dt = \mathscr{E}_i \quad i = 1, \ldots, M$$

where $\mathscr{E}_i$ is the energy of the ith symbol or signal.

As before, we assume that the additive noise $n(t)$ is a zero-mean white Gaussian noise. The conditional density is

$$f(Y/S_i) = \frac{1}{(2\pi)^{N/2}(\sigma^2)^{N/2}} \exp\left[\frac{-(Y - S_i)'(y - S_i)}{2\sigma^2}\right] \quad i = 1, 2, \ldots, M \tag{1.317}$$

assuming $E[n(t)] = 0$, $E[n^2(t)] = \sigma^2$ of additive noise $n(t)$, and

$$Y = [y_1, y_2, \ldots, y_N]'$$

$$S_i = [S_{i1}, S_{i2}, \ldots, S_{iN}]' \qquad i = 1, 2, \ldots, M$$

$$n = [n_1, n_2, \ldots, n_N]', \qquad E(n) = 0, \; E(nn') = \sigma^2 I$$

The decision rule is that S_k is present if

$$\log \pi_k + \log f(Y|S_k) \geq \log \pi_j + \log f(Y|S_j) \qquad \text{for all } j$$

or

$$\log \pi_k - \frac{\mathscr{E}_k}{2\sigma^2} + \frac{Y'S_k}{\sigma^2} \geq \log \pi_j - \frac{\mathscr{E}_j}{2\sigma^2} + \frac{Y'S_j}{\sigma^2} \qquad \text{for all } j \neq k \qquad (1.318)$$

Therefore S_k is decided, in the case when $\pi_j = 1/M$, $\mathscr{E}_j = \mathscr{E}$ for all j, if

$$Y'S_k = \max_j \{Y'S_j\} \qquad 1 \leq j \leq M$$

or if

$$\int_0^{T_s} y(t)S_k(t)\,dt = \max_j \left\{ \int_0^{T_s} y(t)S_j(t)\,dt \right\} \qquad 1 \leq j \leq M \qquad (1.319)$$

The optimal receiver for the multiple-signal case in the presence of white Gaussian noise is a bank of M matched filters coupled with a comparator. The receiver is shown in Fig. 1.32.

Let us denote

$$Z_j = \int_0^{T_s} y(t)S_j(t)\,dt \qquad 1 \leq j \leq M \qquad (1.320)$$

When S_k is present, the conditional average of Z_j is

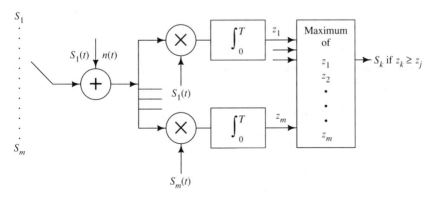

FIGURE 1.32   Optimum receiver for multiple-signal detection.

$$E(Z_j|S_k) = E\left(\int_0^{T_s} [S_k(t) + n(t)]S_j(t)\,dt\right)$$

$$= \int_0^{T_s} S_k(t)S_j(t)\,dt$$

$$\triangleq r_{kj} \tag{1.321}$$

The conditional covariance Z_j is

$$E[(Z_i - r_{ik})(Z_j - r_{jk})|S_k] = E\left(\int_0^{T_s}\int_0^{T_s} n(t)S_i(t)n(\tau)S_j(\tau)\,dt\,d\tau\right)$$

$$= \sigma^2\left(\int_0^{T_s}\int_0^{T_s} \delta(t - \tau)S_i(t)S_j(\tau)\,dt\,d\tau\right)$$

$$= \sigma^2\left(\int_0^{T_s} S_i(t)S_j(t)\,dt\right)$$

$$= \sigma^2 r_{ij} \tag{1.322}$$

Equation (1.322) establishes the fact that even though noise is white, Z_j is not uncorrelated.

For the sake of simplicity we will assume that the signals are orthogonal to each other, of equal energy, and are given by

$$r_{ij} = \mathscr{E} \qquad i = j$$

$$= 0 \qquad i \neq j$$

In this case Z_j is uncorrelated. Each Z_j is a Gaussian random variable with mean $\mathscr{E}$ and variance $\sigma^2\mathscr{E}$ when S_j is present. Note that

$$f(Z_k|S_k) = \frac{1}{\sqrt{2\pi\sigma^2\mathscr{E}}}\exp\left[\frac{-(Z_k - \mathscr{E})^2}{2\sigma^2\mathscr{E}}\right]$$

and

$$f(Z_k|S_i) = \frac{1}{\sqrt{2\pi\sigma^2\mathscr{E}}}\exp\left[\frac{-(Z_k)^2}{2\sigma^2\mathscr{E}}\right]$$

The probability of symbol (signal) detection when S_k is present is

$$P(D_k|S_k) = P[Z_k \geq Z_j, j \neq k|S_k]$$

$$= P[Z_k \geq Z_1, Z_k \geq Z_2, \ldots, Z_k \geq Z_M|S_k]$$

$$= P[Z_1 \leq Z_k, Z_2 \leq Z_k, \ldots, Z_M \leq Z_k|S_k]$$

$$= \prod_{i \neq j} P(Z_j \leq Z_k|S_k)$$

The conditional probability of Z_j when S_k is present is

$$P[Z_j \leq Z_k | S_k] = \int_{-\infty}^{Z_k} \frac{1}{\sqrt{2\pi\sigma^2 \mathscr{E}}} \exp\left[\frac{-(Z_j)^2}{2\sigma^2 \mathscr{E}}\right] dZ_j$$

$$= 1 - 0.5\,\mathrm{erfc}\left[\frac{Z_k}{\sqrt{2\sigma^2 \mathscr{E}}}\right] \qquad j \neq k \qquad (1.323a)$$

Hence

$$P(D_k | S_k) = (P(Z \leq Z_k | S_k])^{M-1}$$

$$= \left[1 - 0.5\,\mathrm{erfc}\,\frac{Z_k}{\sqrt{2\sigma^2 \mathscr{E}}}\right]^{M-1} \qquad \text{for all } k \qquad (1.323b)$$

and therefore

$$PD_k = \int_{-\infty}^{\infty} (PD_k | S_k) f(Z_k | S_k)\, dZ_k = \int_{-\infty}^{\infty} (P[Z \leq Z_k])^{M-1} f(Z_k | S_k)\, dZ_k$$

$$= \int_{-\infty}^{\infty} \left(1 - 0.5\,\mathrm{erfc}\left(\frac{Z}{g}\right)\right)^{M-1} \frac{1}{\sqrt{2\pi\sigma^2 \mathscr{E}}} e^{\frac{-(Z - \mathscr{E})^2}{2\sigma^2 \mathscr{E}}}\, dZ \qquad \text{for all } k$$

$$(1.324a)$$

Put $(Z - \varepsilon)/\sqrt{g} = t$ and $g = 2\sigma^2\delta$. Then

$$PD_k = \int_{-\infty}^{\infty} (1 - 0.5\,\mathrm{erfc}(\mathscr{E} + t\sqrt{g}))^{M-1} \frac{1}{\sqrt{\pi}} e^{-t^2}\, dt \qquad \text{for all } k$$

Hence the probability of symbol (signal) error for orthogonal coherent signals is

$$PE_k = 1 - \int_{-\infty}^{\infty} (1 - 0.5\,\mathrm{erfc}(\mathscr{E} + t\sqrt{g}))^{M-1} \frac{1}{\sqrt{\pi}} e^{-t^2}\, dt \qquad \text{for all } k$$

and the total probability of symbol error for orthogonal coherent signals is

$$PE_M = \frac{1}{M} \sum_{k=1}^{M} PE_k$$

$$(1.324b)$$

$$= 1 - \int_{-\infty}^{\infty} (1 - 0.5\,\mathrm{erfc}(\mathscr{E} + t\sqrt{g}))^{M-1} \frac{1}{\sqrt{\pi}} e^{-t^2}\, dt$$

$$\leq \left(\frac{M-1}{2}\right) \mathrm{erfc}[\sqrt{\mathscr{E}/2N_0}) \qquad (1.324c)^g$$

---

[g] Note that $PE_M \leq (M - 1)P_e$, where P_e is the bit error probability for a binary PSK signal. This bound is known as the *union bound*.

Multiple Frequency-Shift Signals. Signals employing MFSK are given by

$$S_i(t) = \sqrt{\frac{2\mathscr{E}_1}{T_s}} \sin\left(\omega_c t + \frac{2\pi i}{T_s} t + \theta\right) \qquad 1 \le t \le T_s, i = 1, 2, 3, \ldots, M$$

where θ is a uniform random variable in $[-\pi, \pi]$ and T_s is the symbol or signal duration. Find the probability of symbol error when the signals are equally likely and the additive noise $n(t)$ is white Gaussian with mean zero and variance σ^2.

The signals are of equal energy because

$$\int_0^{T_s} S_i^2(t)\,dt = \frac{2\mathscr{E}_1}{T_s} \int_0^{T_s} \sin^2\left(\omega_c t + \frac{2\pi i}{T} t + \theta\right) dt \triangleq \mathscr{E}(\theta)$$

The average energy is

$$E_\theta\left(\int_0^{T_s} S_i^2(t)\,dt\right) = E_\theta[\mathscr{E}(\theta)] = \bar{\mathscr{E}} \qquad \text{for all } i$$

$$f(y|S_j) = \int f(y|S_j, \theta) f(\theta)\,d\theta$$

$$= \frac{1}{2\pi} \int_{-\pi}^{\pi} f(y|S_j, \theta)\,d\theta$$

when θ is uniform in $(-\pi, \pi)$.

The decision rule is to decide S_k if

$$\frac{1}{2\pi} \pi_k \int_{-\pi}^{\pi} d\theta \exp\left\{-\frac{1}{2\sigma^2} \int_0^{T_s}\left[y(t) - \sqrt{\frac{2\mathscr{E}_1}{T}} \sin\left(\omega_c t + \frac{2\pi k t}{T} + \theta\right)\right]^2 dt\right\}$$

$$\ge \pi_j \frac{1}{2\pi} \int_{-\pi}^{\pi} d\theta \exp\left\{-\frac{1}{2\sigma^2} \int_0^{T_s}\left[y(t) - \sqrt{\frac{2\mathscr{E}}{T}} \sin\left(\omega_c t + \frac{2\pi j t}{T} + \theta\right)\right]^2 dt\right\}$$

since $\pi_k = 1/M$, $\bar{\mathscr{E}}_k = \bar{\mathscr{E}}$ for all k, decide S_k if

$$\log I_0\left(\frac{R_k}{\sigma^2}\right) \ge \log I_0\left(\frac{R_j}{\sigma^2}\right) \qquad j \ne k$$

where

$$R_i^2 \sim \left[\int_0^{T_s} y(t) \cos\left(\omega_c t + \frac{2\pi i t}{T}\right) dt\right]^2 + \left[\int_0^{T_s} y(t) \sin\left(\omega_c t + \frac{2\pi i t}{T}\right) dt\right]^2$$

$$1 \le i \le M$$

where $I_0(\cdot)$ is the modified Bessel function.

The detection scheme for noncoherent MFSK signals is shown in Fig. 1.33. The conditional density of the envelope is given by

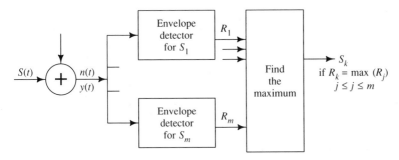

FIGURE 1.33   Noncoherent MFSK receiver.

$$f(R_k|S_k) = \frac{R_k}{\sigma^2} \exp\left[\frac{-(R_k^2 + \sigma^4\rho^4)}{2\sigma^2}\right] I_0\left(R_k \frac{\bar{\mathscr{E}}}{\sigma^2}\right)$$

$$f(R_k|S_j) = \frac{R_k}{\sigma^2} \exp\left(-\frac{R_k^2}{2\sigma^2}\right) \qquad k \neq j$$

(1.325a)

where $\rho^2 = \bar{\mathscr{E}}/\sigma^2 = \bar{\mathscr{E}}/\eta r_b$, r_b is the data rate, $\eta/2$ is the noise power spectral

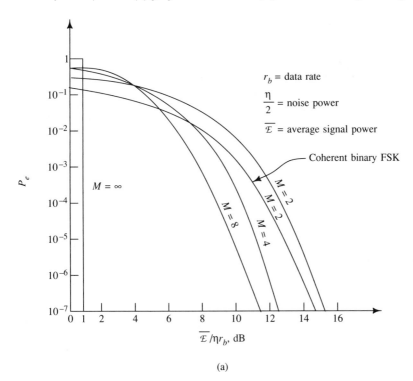

(a)

FIGURE 1.34   (a) Multiple FSK signal error probability. (From H. L. Van Trees, Ed., *Satellite Communications*, New York: IEEE Press. © 1979 IEEE.)

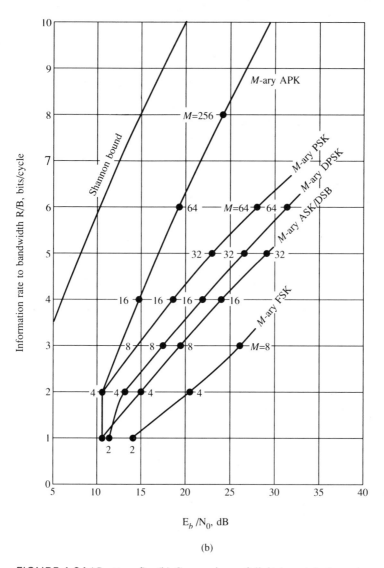

FIGURE 1.34 (*Continued*) (b) Comparison of digital modulation schemes. (From H. L. Van Trees, Ed., *Satellite Communications*, New York: IEEE Press. © 1979 IEEE.)

density, and $\bar{\mathscr{E}}$ is the average signal power. Therefore, the probability of symbol error is given by

$$P_e = 1 - P_d = 1 - \int_0^\infty f(R_k|S_k)(P[R < R_k])^{M-1} dR_k$$

$$= 1 - \int_0^\infty \left\{ U e^{-(U^2+P^2)/2} I_0(UP) \left[\int_0^U t e^{-t^2/2} dt \right]^{M-1} \right\} dU$$

$$= \sum_{n=1}^{M-1} (-1)^{n+1} \binom{M-1}{n} \frac{1}{n+1} \exp\left[\frac{-nr_s}{n+1} \right] \qquad (1.325b)$$

where r_s is the symbol energy-to-noise ratio. Note that if θ is nonrandom, then symbol error for Mary FSK is given by Eq. (1.324c).

For various values of M, the probability of error is shown in Fig. 1.34a. For fixed data rate r_b and $\eta/2$ when M increases, the required signal-to-noise ratio for a fixed P_e decreases. When $M \to \infty$,

$$P_e = 1 \qquad \text{if } \bar{\mathscr{E}}/\eta r_b < 0.7$$

$$= 0 \qquad \text{if } \bar{\mathscr{E}}/\eta r_b > 0.7$$

The bandwidth of the signal set increases as M increases. If the data rate $r_b \leq (\bar{\mathscr{E}}/0.7\eta)$ and is less than the capacity of the channel, the probability of error can be made arbitrarily small.

It should be mentioned that the bit energy E_b, bit rate T_b, and bit error probability P_b are related to the symbol energy E_s, symbol time T_s, and symbol error P_s in the following manner

$$\left. \begin{array}{l} E_b = \dfrac{E_s}{\log_2 M} = \dfrac{E_s}{k} \\[2mm] T_s = k T_b \end{array} \right\} \qquad (1.326)$$

$$P_b = \frac{M}{2(M-1)} P_s \qquad \text{(for orthogonal signals)}$$

Detection of Biorthogonal Signals

A set of M signals are called *biorthogonal* if half of signals $S_j(t)$, $j = 1,\dots,$ $M/2$ are orthogonal and the other half signals are $-S_j(t)$, $j = 1,2,\dots,M/2$, $0 \leq t \leq T_s$, $M = 2^k$. For $M = 4$, the biorthogonal signals are $S_1(t)$, $S_2(t)$, $-S_1(t)$, $-S_2(t)$, $0 \leq t \leq T_s$. These signals are received in the presence of a zero mean white Gaussian noise $n(t)$ with PSD $N_0/2$. Let us denote the received signal by

$$y(t) = \pm S_m(t) + n(t) \qquad 0 \leq t \leq T_s, m = 1,\dots,M/2 \qquad (1.327)$$

Let us assume that the signals are of equal energy and are equally likely.

Therefore the MAP decision is equivalent to finding

$$\max_{1 \leq j \leq M/2} \left\{ \left| \int_0^{T_s} y(t) S_j(t)\, dt \right| \right\} = \max_{t \leq j \leq M/2} \{|z_j|\}$$

The detection scheme is to compute the correlation detector output and then finding the maximum of the outputs.

Note that when S_1 is transmitted,

$$\begin{aligned} z_j &= \varepsilon + \int_0^{T_s} n(t) S_1(t)\, dt & j &= 1 \\ &= \int_0^{T_s} n(t) S_j(t)\, dt & j &= 2, \ldots, M/2 \end{aligned}$$

(1.328)

Therefore,

$$E(z_1) = \varepsilon, \quad E(z_j) = 0 \qquad j = 2, \ldots, M/2$$

where

$$\varepsilon = \int_0^{T_s} S_j^2(t)\, dt \qquad j = 1, \ldots, M/2$$

Note that z_1 is Gaussian with mean ε and z_m, $m = 2, \ldots, M/2$, are Gaussian random variables with mean zero.

When S_1 is transmitted,

$$\begin{aligned} E[z_j^2] &= \int_0^{T_s} \int_0^{T_s} S_j(t) S_j(\tau) E[(n(t)n(\tau)]\, dt\, d\tau \\ &= \frac{N_0}{2} \int_0^{T_s} \int_0^{T_s} - S_j(t) S_j(\tau) \frac{N_0}{2} \delta(t - \tau)\, dt\, d\tau \\ &= \frac{N_0}{2} \varepsilon \quad 2 \leq j \leq M/2 \end{aligned}$$

(1.329)

$$E(z_1^2) = \varepsilon^2 + \frac{\varepsilon N_0}{2}$$

Hence

$$\operatorname{var}(z_j) = (N_0/2)\varepsilon \qquad j \geq 1$$

(1.330)

when $S_1(t)$ is transmitted, $[z_m, m = 1, \ldots, M/2]$ are independent Gaussian random variables with mean and variance given by Eqs. (1.329) and (1.330). S_1 is selected if

$$z_1 > \max_{2 \leq j \leq M/2} |z_j| \qquad z_1 \geq 0$$

The probability of detection of S_1 is

$$P[z_1 > |z_2|, \ldots, |z_{M/2}| z_1 > 0] = \prod_{j=2}^{M/2} P[z_1 > |z_j| | z_1 > 0]$$

$$= \prod_{j=2}^{M/2} P[|z_j| \le z_1 | z_1 > 0]$$

$$= \prod_{j=2}^{M/2} P[-z_1 < z_j < z_1 | z_1 > 0]$$

$$= \prod_{j=2}^{M/2} \int_{-z_1/2}^{z_1/2} \frac{1}{\sqrt{2\pi\sigma^2}} \exp\left[\frac{-z_j^2}{2\sigma^2}\right] dz_j$$

$$= \left[\int_{-z_1/2}^{z_1/2} \frac{1}{\sqrt{2\pi\sigma^2}} \exp\left(\frac{-z_j^2}{2\sigma^2}\right) dz_j\right]^{(M/2)-1}$$

where $\sigma^2 = (N_0/2)\varepsilon$.

Therefore, the probability of detecting S_1 is

$$P_d = \int_0^\infty \left(\int_{-z_1/2}^{z_1/2} \frac{1}{\sqrt{2\pi\sigma^2}} \exp\left[\frac{-z_j^2}{2\sigma^2}\right] dz_j\right)^{(M/2)-1} f(z_1)\, dz_1$$

$$= \int_0^\infty \left[\left(\int_{-z_1/2}^{z_1/2} \frac{1}{\sqrt{2\pi\sigma^2}} \exp\left[\frac{-z_j^2}{2\sigma^2}\right] dz_j\right)^{(M/2)-1}\right.$$

$$\left. \times \frac{1}{\sqrt{2\pi\sigma}} \exp\left[\frac{(z_1 - \varepsilon)^2}{2\sigma^2}\right]\right] dz_1 \qquad (1.331)$$

Therefore, the probability of symbol error for the M-ary biorthogonal signal is

$$P_s = 1 - \frac{1}{\sqrt{2\pi}} \int_{-d}^\infty \left(e^{-z_1^2/2}\left(\frac{1}{\sqrt{2\pi}} \int_{-(z+d)}^{z+d} e^{-v^2/2}\, dv\right)^{(M/2)-1}\right) dz_1 \qquad (1.332)$$

where $d^2 = 2\varepsilon/N_0$, the signal-to-noise ratio.

Biorthogonal MPSK Signals. The received signal is

$$y(t) = S_j(t) + n(t) \qquad (1.333)$$

where

$$S_j(t) = A\cos(\omega_c t + \theta_j) \qquad \omega_c T_s = 2\pi, \; j = 1, 2, \ldots, M$$

$$= \frac{\pi}{M}(2j - 1) \qquad 0 \le t \le T_s \qquad (1.334)$$

where T_s is the signal (symbol) duration and $n(t)$ is a zero-mean white Gaussian noise. $S_j(t)$ are called *M-ary polyphase or phase-shift keying* (MPSK) signals.

The MPSK signals can be expressed as

$$S_j(t) = A\cos\omega_c t \cos\theta_j - A\sin\omega_c t \sin\theta_j$$

$$\triangleq a_j A \cos\omega_c t + b_j A \sin\omega_c t \qquad 1 \le j \le M, 0 \le t \le T_s, \omega_c T_s = 2\pi \qquad (1.335)$$

These signals are on a circle of radius A and are separated by $2\pi/M$. (a_j, b_j), $j = 1, \ldots, M$ are the coordinates of $S_j(t), j = 1, \ldots, M$, in the signal space with basis functions $\phi_1(t) = A \cos \omega_c t$ and $\phi_2(t) = A \sin \omega_c t$. $\{a_j\}$ and $\{b_j\}$ are called in-phase and quadrature components. When $M = 4$, $\{S_j(k)\}$ are biorthogonal and are also called *QPSK signals*.

The coherent optimum detector is a bank of matched filters. The received signal is multiplied with $S_j(t)$ and integrated and sampled at $t = T_s$. The output at the matched filters is given by

$$z_j = \int_0^{T_s} y(t) S_j(t) \, dt$$

$$= A \cos \theta_j \int_0^{T_s} y(t) \cos \omega_c t \, dt - A \sin \theta_j \int_0^{T_s} y(t) \sin \omega_c t \, dt$$

$$= A \cos \theta_j R_1 - A \sin \theta_j R_2 \tag{1.336}$$

$$= R \cos(\phi - \theta_j), \qquad 1 \le j \le M \tag{1.337}$$

where

$$R_1 = \int_0^{T_s} y(t) A \cos \omega_c t \, dt$$

$$R_2 = \int_0^{T_s} A y(t) \sin \omega_c t \, dt \tag{1.338}$$

$$R = \sqrt{R_1^2 + R_2^2}$$

$$\phi = \arctan(R_2/R_1)$$

When $S_1(t)$ is present,

$$E(R_1) = \mathscr{E} = A^2 T_s/2$$

$$\text{var}(R_1) = N_0 \varepsilon/2$$

$$E(R_2) = 0$$

$$\text{var}(R_2) = N_0 \varepsilon/2$$

Note that R_1 and R_2 are Gaussian random variables. If S_1 is transmitted, the joint probability density function of R_1 and R_2 is

$$f(R_1, R_2 | S_1) = (2\pi\sigma^2)^{-1} \exp\left[-\frac{(R_1 - \varepsilon)^2 + R_2^2}{2\sigma^2} \right] \tag{1.339}$$

where $\varepsilon = A^2 T_s/2$ and $\sigma^2 = N_0 \varepsilon/2$.

If z_1 exceeds $z_j, j = 2, \ldots, M$, then S_1 is selected; $z_1 = \max\{z_j\}$. z_j is maximum when the angular distance $|\phi - \theta_j|$ is minimum. Select S_1 if

$$|\phi - \theta_1| \le \{|\phi - \theta_j|\} \qquad 2 \le j \le M$$

The joint PDF of R, ϕ given S_1 is present, is

$$f(R, \phi | S_1) = \frac{R}{2\pi\sigma^2} \exp\left(-\frac{R^2 + \varepsilon^2 - 2\varepsilon R \cos\phi}{2\sigma^2}\right) \qquad R \geq 0$$

$$= 0 \qquad\qquad\qquad \text{elsewhere} \tag{1.340}$$

using Eqs. (1.338) and (1.339); see Mohanty (1986, p. 140).

Therefore, the PDF of ϕ when S_1 is present is given by

$$f(\phi | S_1) = \int_0^\infty f(R, \phi | S_1) \, dR$$

$$= \frac{1}{2\pi} e^{-d^2/2} \left(1 + \sqrt{2\pi d^2} \cos\phi \, e^{-d^2/2 \cos^2\phi} \frac{1}{\sqrt{2\pi}} \int_{-\infty}^{d\cos\phi} e^{-v^2/2} \, dv\right) \tag{1.341}$$

where $d^2 = 2\varepsilon/N_0$, the signal-to-noise ratio. Since the signals are equally at a distance $2\pi/M$ on the circle, the error is made if ϕ is outside $[-\pi/M, \pi/M]$. Therefore, the probability of symbol error for MPSK signals is

$$P_s = 1 - \int_{-\pi/M}^{\pi/M} f(\phi | S_1) \, d\phi \tag{1.342}$$

where $f(\phi | S_1)$ is given by Eq. (1.341).

Assuming that $d^2 \gg 1$,

$$f(\phi | S_1) \approx \sqrt{\frac{d^2}{2\pi}} \cos\phi \exp\left[-\frac{d^2}{2} + \frac{d^2}{2} \cos^2\phi\right] \tag{1.343}$$

The probability of symbol error for M-ary PSK signals is given by

$$P_s \approx \mathrm{erfc}[(d/\sqrt{2})\sin(\pi/M)] \qquad M \geq 4 \tag{1.344}$$

using Eqs. (1.342) and (1.343), and

$$\mathrm{erfc}(x) = \frac{2}{\sqrt{\pi}} \int_x^\infty e^{-v^2} \, dv$$

The *bandwidth efficiency* η is defined as R/B, where R and B are the rate and bandwidth of a signal respectively.

It should be noted that the *bandwidth efficiency* for M-ary PSK is

$$\eta_{\mathrm{MPSK}} = R/B = \log_2 M \tag{1.345}$$

and the bit error probability is related to symbol error as

$$P_b = (1/\log_2 M) P_s \tag{1.346a}$$

where P_s is the M-ary PSK symbol error. The bandwidth efficiency for M-ary FSK signals is

$$\eta_{\mathrm{MFSK}} = (2\log_2 M)/M \tag{1.346b}$$

When $M \rightarrow \infty$, $\eta_{MFSK} \rightarrow 0$, but the MFSK symbol error P_s decreases. On the other hand, when $M \rightarrow \infty$, η_{MPSK} increases by a factor k but the M-ary PSK symbol error P_s increases.

The probability of symbol error for M-ary differentially coded phase shift keying (M-ary DPSK) is given by

$$P_s = \tfrac{1}{2}\mathrm{erfc}[d\sqrt{2}\sin(\pi/2M)] \qquad M > 4^h$$

When $M = 4$, Eq. (1.335) can be written as

$$
\begin{aligned}
S_j(t) &= a_j A \cos \omega_c t + b_j A \sin \omega_c t \\
&= A[m_1(t)\cos \omega_c t + m_2(t)\sin \omega_c t] \qquad \omega_c T = 2\pi \\
&= m_1(t)\phi_1(t) + m_Q(t)\phi_2(t)
\end{aligned}
\qquad (1.347)
$$

where

$$
\begin{aligned}
m_I(t) &= a_j p(t) & a_j &= \cos \theta_j \\
m_Q(t) &= b_j q(t) & b_j &= \sin \theta_j \\
q(t) &= p(t) = 1 & 0 &\leq t \leq T_S \\
&= 0, & &\text{elsewhere}
\end{aligned}
\qquad (1.348)
$$

$$\phi_1(t) = A_1 \cos \omega_c t$$

$$\phi_2(t) = A_2 \sin \omega_c t \qquad A_1 = \sqrt{\frac{2}{T_s}} = A_2$$

$m_I(t)$ and $m_Q(t)$ can be considered as two independent data streams. $s(t)$ is called *quadrature multiplexing* (QM). $m_1(t)$ and $m_\theta(t)$ are called in-phase and quadrature phase data.

The received signal is

$$y(t) = s_j(t) + n(t) \qquad (1.349)$$

where $n(t)$ is a white Gaussian random variable with variance equal $N_0/2$. $m_I(t)$ can be recovered by multiplying the received signal $y(t)$ with $\cos \omega_c t$ and passing through a low-pass filter (see Fig. 1.35). The output of the LPF is compared with a decision device to determine whether $a_j = 1$ or -1. Similarly, $m_Q(t)$ is recovered by multiplying the received signal by $\sin \omega_c t$, passing through an LPF. The output of the LPF is compared with a decision device to determine whether $b_j = 1$ or -1. Let us denote the output of the LPFs as follows:

---

[h] See Table 12.5 for a summary of probability symbol errors for various modulations in Chapter 12.

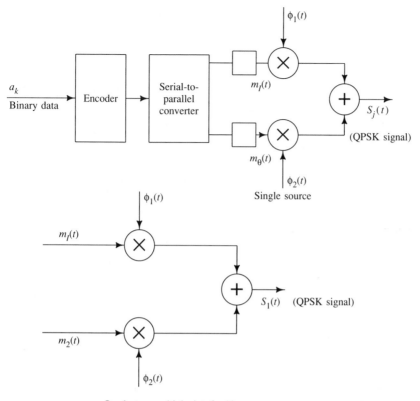

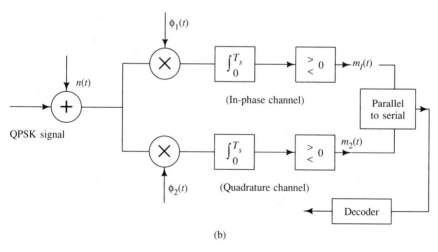

FIGURE 1.35   QPSK signal modulation and demodulation. (a) Generation of SPSK signal. (b) Demodulation of QPSK signal. (c) QPSK signal structure.

113

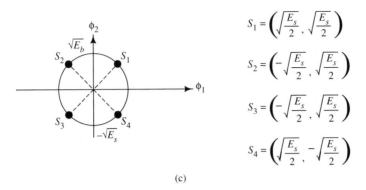

$$S_1 = \left(\sqrt{\frac{E_s}{2}}, \sqrt{\frac{E_s}{2}} \right)$$

$$S_2 = \left(-\sqrt{\frac{E_s}{2}}, \sqrt{\frac{E_s}{2}} \right)$$

$$S_3 = \left(-\sqrt{\frac{E_s}{2}}, \sqrt{\frac{E_s}{2}} \right)$$

$$S_4 = \left(\sqrt{\frac{E_s}{2}}, -\sqrt{\frac{E_s}{2}} \right)$$

(c)

FIGURE 1.35   (*Continued*)

$$
\begin{aligned}
x_1 &= \int_0^{T_s} y(t) \cos \omega_c t \, dt \\
&= a_j + \int_0^{T_s} n(t) \cos \omega_c t \, dt \\
&\triangleq a_j + w_1 \qquad j = 1, 2, 3, 4
\end{aligned}
\tag{1.350a}
$$

$$
\begin{aligned}
x_2 &= \int_0^{T_s} y(t) \sin \omega_c t \, dt \\
&= b_j + \int_0^{T_s} u(t) \sin \omega_c \, dt \\
&\triangleq d_j + w_2 \qquad j = 1, 2, 3, 4
\end{aligned}
\tag{1.350b}
$$

x_1 and x_2 are uncorrelated. It can be shown that the probability densities of x_1 and x_2 are given by

$$f_{x_i}(x_i) = \frac{1}{\sqrt{\pi N_0}} \exp\left[\frac{-(x_i - \sqrt{A^2 T_s/2})^2}{N_0} \right] \qquad i = 1, 2 \tag{1.351}$$

The joint probability density of x_1 and x_2 is

$$f_x(x_1, x_2) = f_{x_1}(x_1) f_{x_2}(x_2) \tag{1.352}$$

Let us assume that $a_j > 0$, $b_j > 0$. The probability of correct decision of s_1 is

$$
\begin{aligned}
P_d &= P[x_1 > 0, x_2 > 0 \,|\, a_j > 0, b_j > 0] \\
&= \int_0^\infty \int_0^\infty f_x(x_1, x_2) \, dx_1 \, dx_2 \\
&= \left\{ \int_0^\infty \frac{1}{\sqrt{\pi N_0}} \exp\left[\frac{-(x_1 - \sqrt{A^2 T_s/2})^2}{N_0} \right] dx_1 \right\}^2
\end{aligned}
$$

$$= \left[1 - \frac{1}{2} \text{erfc}\left(\sqrt{\frac{E_s}{2N_0}} \right) \right]^2$$

$$= 1 - \text{erfc}\left(\sqrt{\frac{E_s}{2N_0}} \right) + \frac{1}{4} \text{erfc}^2\left(\sqrt{\frac{E_s}{N}} \right)$$

where

$$E_s = \frac{A^2 T_s}{2} \tag{1.353}$$

The probability of symbol error for the QPSK signal is

$$P_{e(s)} = 1 - P_d = \text{erfc}\left(\sqrt{\frac{E_s}{2N_0}} \right) - \frac{1}{4} \text{erfc}^2\left(\sqrt{\frac{E_s}{2N_0}} \right)$$

$$= \text{erfc}\left(\sqrt{\frac{E_b}{N_0}} \right) - \frac{1}{4} \text{erfc}^2\left(\sqrt{\frac{E_b}{N_0}} \right) \tag{1.354}$$

where $E_b = A^2 T_b/2$, $T_b = T_s/2$, and $E_b/N_0 \gg 1$.
The probability of symbol error for s_1 is given by

$$P_{e(s)} = \text{erfc}\left(\sqrt{\frac{E_b}{N_0}} \right) \tag{1.355}$$

If we substitute $M = 4$ and $E_s = 2E_b$ in Eq. (1.344), we get Eq. (1.355). Probabilities of detection of s_1, s_2, s_3, and s_4 are symmetrical. Hence, symbol errors are equal. The transmitter and receiver details for a QPSK system are given in Fig. 1.35.

We note that for high signal-to-noise ratio, the symbol error for QPSK is same as bit error for BPSK, but the symbol time T_s is twice the bit time T_b. Hence, the *symbol rate* (also called *baud rate* is half the bit rate. For a given bit rate, a QPSK system requires half the transmission bandwidth of the BPSK system. Therefore, a QPSK system carries twice as many data as a BPSK system for a given bit error probability.

In QPSK system, the carrier phase changes every $2T_b$ seconds. The phase change is equal to 0, $\pm \pi/4$, or π, depending on the values of $m_I(t)$ and $m_Q(t)$. When the phase change is π, the envelope of the QPSK signal is 0, which yields AM/PM distortion (out-of-band interference) when a QPSK signal passes through a traveling-wave-tube amplifier (TWTA) amplifier. It is desired to have a constant envelope input to the TWTA. In order to reduce this AM/PM distortion, the maximum phase change should be $\pi/4$. This happens if $m_I(t)$ and $m_Q(t)$ do not change at the same time. The modified QPSK signal is

$$S_j(t) = m_I(t)\phi_1(t) + m_Q(t)\phi_2(t) \qquad 0 \le t \le T_s$$

$$m_I(t) = \sqrt{E_s/2} U_j p(t) \qquad\qquad j = 1, 2, 3, 4 \tag{1.356a}$$

$$m_Q(t) = \sqrt{E_s/2} V_j q(t) \qquad\qquad U_j, V_j = \pm 1$$

where

$$p(t) = 1 \qquad -T_s/2 \leq t \leq T_s/2$$
$$\quad = 0 \qquad \text{elsewhere} \tag{1.356b}$$

$$q(t) = p(t + T_b)$$
$$\quad = 1 \qquad 0 \leq t \leq T_s$$
$$\quad = 0 \qquad \text{elsewhere}$$

This signaling system given by Eq. (1.356a) is called *offset QPSK* (OQPSK); it is also called *staggered QPSK* (SQPSK). The quadrature channel is offset T_b from the in-phase channel. The envelope of OQPSK is not zero at any time. The recovery OQPSK is the same as QPSK except that the integration for the in-phase channel is from $-T_s/2$ to $T_s/2$. Since $m_I(t)$ and $m_Q(t)$ are independent as in QPSK, the symbol error for OQPSK is same as QPSK, given by Eq. (1.354).

Even though the envelope of the OQPSK is not zero, the change of phase in OQPSK system is not smooth. The smooth change can be obtained by using nonrectangular continuous waveforms. Let us define

$$p(t) = \cos(\pi t/T_s) \qquad -T_s/2 \leq t \leq T_s/2$$
$$q(t) = \sin(\pi t/T_s) \qquad 0 \leq t \leq T_s \tag{1.357a}$$

The multiplexed signal is given by

$$s_j(t) = \sqrt{2/T_s}[m_I(t)\cos \omega_c t + m_Q(t)\sin \omega_c t] \tag{1.357b}$$

where

$$m_I(t) = \sqrt{E_s/2}U_j p(t) = U_j \cos(\pi t/2T_b)\sqrt{E_s/2} \qquad U_j = \pm 1$$
$$m_Q(t) = \sqrt{E_s/2}V_j q(t) = V_j \sin(\pi t/2T_b)\sqrt{E_s/2} \qquad V_j = \pm 1$$

Equation (1.357b) can be written as

$$S_j(t) = U_j \phi_1(t) + V_j \phi_2(t) \qquad j = 1, 2, 3, 4 \tag{1.358a}$$

where

$$\phi_1(t) = \left(\sqrt{\frac{2}{T_s}}\right)\cos(\pi t/T_s)\cos \omega_c t \qquad -T_s/2 \leq t \leq T_s/2$$

$$\phi_2(t) = \left(\sqrt{\frac{2}{T_s}}\right)\sin(\pi t/T_s)\sin \omega_c t \qquad 0 \leq t \leq T_s, \omega_c = 2\pi f_c$$

Equation (1.358) can be written as

$$S_j(t) = \frac{A}{\sqrt{2}}\cos\left[2\pi\left(f_c - \frac{U_j V_j}{4T_b}\right)t + \phi_j\right] \qquad 0 < t < T_s$$

where

$$U_j, V_j = \pm 1$$

$$\phi_j = 0 \qquad \text{if } U_j = 1 \tag{1.358b}$$

$$= \pi \qquad \text{if } U_j = -1$$

The signal given by Eq. (1.358) is called 4-ary *minimum shift keying* (MSK). The demodulation of MSK is similar to demodulation of OQPSK. Hence, the symbol error for MSK is the same as QPSK signal. The MSK signal has a constant and its phase transition is not abrupt. MSK can be generated by two FSK carriers with $f_1 = f_0 + 1/(4T)$ and $f_2 = f_0 - 1/(4T)$. The frequency separation between these two tones is $1/(2T_b)$.

It can be shown that PSD of the QPSK or OQPSK and MSK signals are given by (see Fig. 12.13)

$$S_x(f) = 2A^2 T_b \{\sin^2 c[2T_b(f + f_c)]$$

$$+ \sin^2 c[2T_b(f - f_c)]\} \qquad \text{(QPSK)} \tag{1.359}$$

$$S_x(f) = \frac{16A^2 T_b}{\pi^2} \left\{ \frac{\cos^2 2\pi T_b(f + f_c)}{[1 - (4T_b(f + f_c))^2]^2} \right.$$

$$+ \left. \frac{\cos^2 2\pi T_b(f - f_c)}{[1 - (4T_b(f - f_c))^2]^2} \right\} \qquad \text{(MSK)}$$

The efficiency of modulation system is given by

$$\eta = \frac{\text{bit rate}}{\text{bandwidth}} = \frac{r_B}{B} \text{ (bits/Hz)} \tag{1.360}$$

The capacity of band-limited white Gaussian channel, in bits per second (bps), is given by Mohanty (1986) as follows:

$$C = B \log_2(1 + S/N)$$

$$= B \log_2(1 + S/BN_0)$$

$$= B \log_2\left(1 + \frac{E_b}{T_b B N_0}\right)$$

$$= B \log_2(1 + r_b/B \cdot E_b/N_0) \qquad r_b = 1/T_b \tag{1.361}$$

where B is bandwidth of the channel and S is the signal power and N_0 is the noise power.

$$C/B = [\log_2(1 + r_b/B \cdot E_b/N_0)]$$

Hence

$$E_b/N_0 \geq \frac{(2^\eta - 1)}{\eta} \tag{1.362}$$

where the spectral bit rate $\eta = r_b/B$, $C \geq r_b$, $B \to \infty$,

$$E_b/N_0 \geq -1.6 \text{ dB} \tag{1.363}$$

which is known as *Shannon bound* regardless of modulation and coding schemes. A link operating with $r_b < B$ is said to be *power-limited*. This is the case when $\eta < 1$. The satellite channels are wide, whereas satellite powers are limited. For a given bit error probability, E_b/N_0 can be reduced by expanding the bandwidth including coding schemes. On the other hand, if the system is bandwidth-limited, the capacity (data rate) can be increased by increasing power for a given bandwidth to maintain the certain probability bit error. The case $\eta > 1$ is called the *bandwidth-limited case*. E_b/N_0 increases with η in this case. The bandwidth efficiency, information rate to bandwidth, is shown in Fig. 1.34b.

1.5 SYNCHRONIZATION

Synchronization is a process of generating carrier, phase, symbol, word, and network time information at the receivers for such purposes as correlation and sampling in demodulation and decoding. In some systems, these types of information are sent along with information signals. For example, in AM systems the unmodulated carrier is sent with modulated signals. These synchronization signals use extra bandwidth and power. In some systems, these synchronization signals (parameters) are extracted from the received signals. Even though these parameters are deterministic, the received signal is a random process because of the presence of noise and propagation path. The extraction method of these parameters involves the study of estimation theory. We will describe here the maximum-likelihood estimation method. Let us denote the received signal as

$$y(t) = s(t, \alpha) + n(t) \qquad 0 \leq t \leq T$$

where $n(t)$ is a zero-mean white Gaussian noise, $s(t, \alpha)$ is the transmitted signal, α represents the unknown parameters, $\alpha = (\alpha_1, \alpha_2, \ldots, \alpha_n)$, and T is the duration of the symbol. It can be shown that the conditional density of $y(t)$ given $s(t, \alpha)$ is

$$f[y(t)|s(t,\alpha)] = \frac{1}{\sqrt{2\pi\sigma^2}} \exp\left\{\frac{-1}{2\sigma^2} \int_0^T [y(t) - s(t,\alpha)]^2 \, dt\right\} \tag{1.364}$$

where $\sigma^2 = E[n^2(t)]$; see Mohanty (1986).

The conditional density given by Eq. (1.364) is also called the *likelihood function*, the estimate of α has to be chosen such that $f[y(t)|S(t,\alpha)]$ is maximum. It is convenient sometimes to seek the maximum of $\log f[y(t)|s(t,\alpha)]$, inasmuch as the log function will yield the same maximum value as the original function. Hence, the *maximum-likelihood estimates* of α are given by

$$\frac{\partial}{\partial \alpha_i}\left\{\log[f(y|s(t,\alpha)]\right\} \qquad i = 1, 2, \ldots, n \tag{1.365}$$

An estimator $\hat{\alpha}$ is called an unbiased estimator if

$$E(\hat{\alpha}) = \alpha \qquad (1.366)$$

The lower bound for the covariance of an *unbiased estimator* α is given by

$$\text{cov}(\alpha - \hat{\alpha}) \geq [E_\alpha\{[\nabla_\alpha f_y(Y|s(\alpha))\nabla_\alpha f_y(Y|s(\alpha))]\}]^{-1} \qquad (1.367)$$

where

$$\nabla_\alpha = \left(\frac{\partial}{\partial \alpha_1}, \ldots, \frac{\partial}{\partial \alpha_n}\right)'$$

$$Y = [y(t_1), \ldots, y(t_N)]'$$

$$t_1 < t_2 < \cdots \leq t_N$$

$$S(\alpha) = [S(t_1, \alpha), \ldots, S(t_N, \alpha)]'$$

Here, the "prime" symbol is used to indicate a transpose. Equation (1.367) is known as the Cramer-Rao bound (Mohanty, 1986). An estimator α is called an *efficient estimator* if Eq. (1.367) is satisfied with equality.

Let the received signal be

$$y(t) = A\cos[\omega_c t + \theta + \phi(t)] + n(t) \qquad 0 \leq t \leq T \qquad (1.368)$$

where θ is the unknown phase and $n(t)$ is a zero-mean white Gaussian noise with various σ^2. The conditional density or the likelihood function is given by

$$f[y(t)|S(t,\theta)] = \frac{1}{\sqrt{2\pi\sigma^2}}\exp\left[\frac{-1}{2\sigma^2}\int_0^T \{y(t) - A\cos[\omega_c t + \theta + \phi(t)]\}^2\, dt\right]$$

$$(1.369)$$

The log likelihood function is given by $\log f[y(t)|S(t,\theta)]$:

$$\log f[y(t)|S(t,\theta)] = -\frac{1}{2\sigma^2}\int_0^T \{y(t) - A\cos[\omega_c t + \phi + \theta(t)]\}^2\, dt$$

$$-\frac{1}{2}\log(2\pi\sigma^2) \qquad (1.370)$$

Differentiating Eq. (1.370) with respect to ϕ and equating the zero, we get

$$\int_0^T \{y(t) - A\cos[\omega_c t + \hat{\phi} + \theta(t)]A\sin[\omega_c t + \hat{\phi} + \theta(t)]\}\, dt = 0 \qquad (1.371)$$

which is a nonlinear equation. We cannot obtain the maximum-likelihood, optimal estimator for ϕ. In most parameters of interest, the estimation problem will be nonlinear. We will describe here some suboptimal procedures to derive carrier, phase, and time estimation. Estimation of these parameters are known as carrier, phase, and symbol synchronization.

We will now consider carrier and symbol time synchronization.[i] The process is shown in Fig. 1.36. The carrier, phase, and time signal are obtained

---

[i] Frame (TDMA) synchronization is dealt with in Section 1.9.

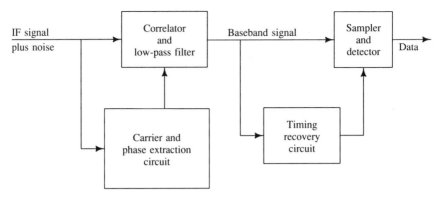

FIGURE 1.36   Synchronization scheme.

by converting the modulated signal to unmodulated carrier and baseband signals.

1.5.1 Carrier Synchronization

The extraction of carrier frequency for an M-ary PSK system is shown in Fig. 1.37. The received signal is given by

$$y(t) = S(t) + n(t) \qquad 0 \le t \le T \tag{1.372a}$$

$$= A\cos[2\pi f_c t + \theta + (2\pi/M)(m-1)] + n(t) \qquad m = 1, 2, \ldots, M \tag{1.372b}$$

where $n(t)$ is a zero-mean white Gaussian noise. The received signal is passed through an Mth power device and bandpass filter tuned to Mf_c such that the output from the Mth harmonic generating circuit is $\cos(2\pi Mf_c + M\theta)$ plus terms containing noise.

For simplicity, we consider a binary system, $M = 2$. We can express the signal as

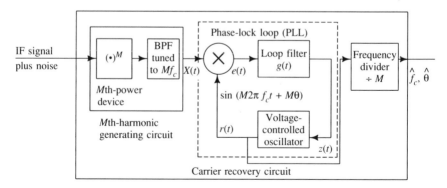

FIGURE 1.37   Carrier recovery circuit for M-ary PSK system.

$$S(t) = Aa(t)\cos(2\pi f_c t + \theta)$$
$$= A\cos(2\pi f_c t + \pi(m - 1) + \theta) \qquad m = 1, 2 \tag{1.373}$$

where $a(t) = \pm 1$. The harmonic frequency generating circuit is a squaring loop. The output of the bandpass filter is $A^2 \cos(4\pi f_c t + 2\theta)$ plus noise terms. The output of the voltage-controlled oscillator is

$$r(t) = A_V \sin(\omega_c t + \hat{\phi}) \qquad \omega_c = 4\pi f_c$$

$$\hat{\phi} = 2\hat{\theta} = -k_v \int_{-\infty}^{t} z(\tau) \, d\tau \tag{1.374}$$

where k_v is the VCO sensitivity $(\text{rad} \cdot \text{s}^{-1} \cdot \text{V}^{-1})$ and $z(t)$ is the input to the VCO. Let us write the input to the phase-locked loop (PLL) as

$$x(t) = A_c \cos(\omega_c t + \phi) + n_e(t) \tag{1.375}$$

where $\omega_c = 4\pi f_c$, $\phi = 2\theta$, $A_c = A^2/2$, and

$$n_c(t) = n_c(t)\cos\omega_c t - n_s(t)\sin\omega_c t$$

$n_c(t)$ and $n_s(t)$ are independent Gaussian random processes. The output of the phase detector is

$$e(t) = x(t)r(t)$$
$$= \tfrac{1}{2}k_m A_v A_c \sin(\phi - \hat{\phi})$$
$$+ \tfrac{1}{2}k_m A_v A_c \sin(2\omega_c t + \phi + \hat{\phi})$$
$$+ \tfrac{1}{2}k_m A_v [n_c(t)\sin\hat{\phi} - n_s\cos\hat{\phi}]$$
$$+ \tfrac{1}{2}k_m A_v [n_c(t)\sin(2\omega_c t + \hat{\phi}) + n_s(t)\cos(2\omega_c t + \hat{\phi})] \tag{1.376}$$

The low-pass filter eliminates the double-frequency term. The output

$$e(t) = k_1 \sin\phi_e + k_1 n^1 \tag{1.377}$$

where

$$k_1 = \tfrac{1}{2}k_m A_v A_c$$
$$\phi_e = \phi - \hat{\phi} = 2(\theta - \hat{\theta})$$
$$n'(t) = \frac{n_c(t)}{A_c}\sin\hat{\phi} - \frac{n_s(t)}{A_c}\cos\hat{\phi}$$

The output of the loop filter is

$$z(t) = \int_{-\infty}^{\infty} e(\tau)g(t - \tau) \, d\tau \tag{1.378}$$

where $g(t)$ is the impulse response function of the loop filter. From Eqs. (1.377), (1.378), and (1.374), we get

$$\hat{\phi} = -k_v \int_{-\infty}^{t} \left[\int_{-\infty}^{\infty} e(r)g(t-r)\,dr \right] dt \tag{1.379}$$

Differentiating Eq. (1.379), we obtain

$$\frac{d\phi}{dt} = +\frac{d\phi_e}{dt} - k \int_{-\infty}^{\infty} [\sin(\phi_e(\tau)) + n'(\tau)]g(t-\tau)\,d\tau \tag{1.380}$$

where

$$k = \tfrac{1}{2}k_v k_m A_v A_c$$

is the loop gain.

Assuming that the error ϕ_e is small (i.e., $|\phi_e| \leq \pi/6$), we can use the approximation

$$\sin \phi_e \approx \phi_e \qquad |\phi_e| \leq \pi/6 \tag{1.381}$$

Equations (1.380) and (1.381)

$$\frac{d\phi}{dt} = +\frac{d\phi_e}{dt} - k \int_{-\infty}^{\infty} [\phi_e(\tau) + n'(\tau)]g(t-\tau)\,d\tau \tag{1.382}$$

The linearized PLL model is shown in Fig. 1.38.

Taking the Laplace transform of Eq. (1.382), we get

$$s\phi(s) = +s\phi_e(s) - kG(s)\phi_e(s)$$

$$= +[s - kG(s)]\phi_e(s)$$

$$= -[s - kG(s)][\hat{\phi}(s) - \phi(s)] \tag{1.383}$$

where

$$n(t) = 0$$

$$\phi(s) = \int_{0}^{\infty} \phi(t)e^{-ts}\,dt$$

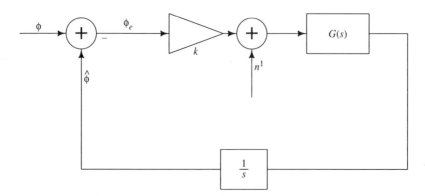

FIGURE 1.38   Linearized PLL model.

Hence

$$\hat{\phi}(s) = \frac{kG(s)}{kG(s) + s}\phi(s) \qquad (1.384)$$

The closed-loop transfer function[j] is

$$H(s) = \frac{kG(s)}{s + kG(s)} \qquad (1.385)$$

We consider a second-order low-pass filter such that

$$G(s) = \frac{1 + \tau_2 s}{1 + \tau_1 s} \qquad \tau_1 \gg \tau_2 \qquad (1.386)$$

The closed-loop transfer function is

$$H(s) = \frac{1 + \tau_2 s}{1 + \left(\tau_2 + \dfrac{1}{k}\right)s + \dfrac{\tau_1}{k}s^2} \qquad (1.387a)$$

$$= \frac{(2\xi\omega_n - \omega_n^2/k)s + \omega_n^2}{s^2 + 2\xi\omega_n S + \omega_n^2} \qquad (1.387b)$$

where the *natural frequency*

$$\omega_n = \sqrt{k/\tau_1} \qquad (1.387c)$$

and the loop damping factor

$$\xi = \frac{\tau_2 + 1/k}{2} \cdot \omega_n \qquad (1.387d)$$

The noise equivalent bandwidth (one-sided) is

$$B_L = B_{eq} = \frac{1}{\max|H(f)|^2}\int_0^\infty |H(f)|^2\,df \qquad (1.388)$$

$$= \frac{1 + (\tau_2\omega_n)^2}{8\xi\omega_n} \qquad (1.389a)$$

substituting Eq. (1.387b) in Eq. (1.388).

The variance of the phase estimate is

$$E(\hat{\phi}^2) = \int_{-\infty}^\infty |H(f)|^2 S_{n^1}(f)\,df$$

$$\qquad (1.389b)$$

$$= 4\frac{1 + \frac{1}{2}\rho}{(B_{bp}/B_{eq})\rho} = \sigma_{\hat{\phi}}^2$$

---

[j]For a first-order loop, $G(s) = 1$.

where $\rho = (A_c^2/2)/N_0 B_{eq}$, B_{bp} is the bandwidth of the bandpass filter and ρ is the input signal-to-noise ratio. Therefore, the variance of estimate of $\hat{\theta}$ is

$$E(\hat{\theta}^2) = E(\hat{\phi}^2)/4 \qquad (1.390)$$

The signal-to-noise ratio is

$$\rho = \frac{A_c^2/2}{N_0 B_{eq}} \qquad (1.391)$$

and the *squaring loss* is defined as

$$S_L = \frac{1 + (B_{bp}/2B_{eq})}{\rho} \qquad (1.392)$$

B_{bp} is the bandwidth of the bandpass filter following the squaring loop and $B_{bp} \gg B_{eq}$. The probability density function of the phase error is given by

$$f(\phi_e) = \frac{1}{2\pi I_0(\rho)} \exp(\rho \cos \phi_e) \qquad |\phi_e| < \pi/6 \qquad (1.393)$$

where (see Mohanty, 1986, p. 83)

$$I_0(\rho) = \frac{1}{2\pi} \int_{-\pi}^{\pi} e^{\rho \cos \phi} \, d\phi$$

where $I_0(\cdot)$ is the modified Bessel function of zero order. The probability of error for a BPSK signal with a phase error ϕ_e is

$$P_{BPSK}(\phi_e) = Q(\sqrt{\rho} \cos \phi_e)$$

where

$$Q(x) = \frac{1}{\sqrt{2\pi}} \int_{x}^{\infty} e^{-u^2/2} \, du$$

The probability of error for a BPSK signal is

$$P_{BPSK} = \frac{1}{2\pi} \int_{-\pi}^{\pi} \frac{\exp(\rho \cos \phi_e)}{I_0(\rho)} Q(\sqrt{\rho} \cos \phi_e) \, d\phi_e \qquad (1.394)$$

It should be noted that the output of the frequency divider has a phase ambiguity of $360°/2 = 180°$, relative to the phase of the received signal phase.

For a QPSK signal, the carrier extraction is done by using a fourth-order ($M = 4$) harmonic frequency generator. The phase error variance[k] is given by

$$\sigma_\theta^2 = \frac{1 + (4.5)\rho^{-1} + 6(\rho)^{-2} + (1.5)\rho^{-3}}{2(B_{bp}/B_{eq})\rho} \qquad (1.395a)$$

---

[k] S. A. Butman and J. R. Lesh. The Effects of Bandpass Limiters on *n*-Phase Tracking Systems, *IEEE Trans. Commun.*, vol. COM-25, no. 6, pp. 569–576, June 1977.

where ρ is the input signal-to-noise ratio given in Eq. (1.391). In this case, the output of the frequency divider has a phase ambiguity of $360°/4 = 90°$ relative to the phase of the received signal phase. The symbol error for a QPSK signal is given by

$$P_{\text{QPSK}} = \int_{-\pi}^{\pi} \frac{1}{2\pi} I_0(\rho) \exp[\rho \cos \phi_e] P(\phi_e)\, d\phi_e \qquad (1.395b)$$

where

$$P(\phi_e) = \frac{1}{2}\{Q[\sqrt{\rho}(\cos \phi_e + \sin \phi_e)] + Q[\sqrt{\rho}(\cos \phi_e - \sin \phi_e)]\}$$

Another method of extraction of carrier and phase information of M-ary PSK is by using a decision feedback phase-lock loop. The decision feedback carrier loop is shown in Fig. 1.39. In the decision feedback case, the received

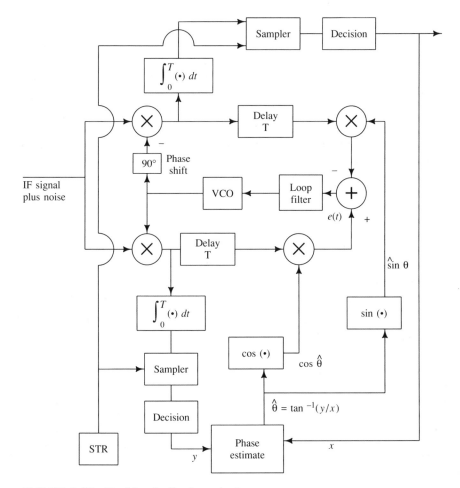

FIGURE 1.39   Decision feedback carrier loop.

signal is multiplied by $\cos(2\pi f_c t + \hat{\phi})$ and $\sin(2\pi f_c t + \hat{\phi})$ and delayed. The delayed signals are multiplied by $\sin\hat{\theta}$ and $\cos\hat{\theta}$ and are added. The filtered added signal drives the VCO; the VCO output is $\sin(2\pi f_c t + \hat{\phi})$. x and y are the decision parameters that are used to estimate θ and to compute $\cos\hat{\theta}$ and $\sin\hat{\theta}$. Note that the decision feedback loop requires the decisions x and y and symbol time recovery information. However, the VCO is driven by addition of in-phase and quadrature components rather than the product as in the Mth power loop. The carrier recovery for M-ary system has a phase ambiguity of $360°/M$ relative to the received signal. The decision feedback loop yields less phase error variance if correct decisions are made. It yields about four to ten times better than the squaring loop if the signal-to-noise ratio is large.

1.5.2 Symbol Synchronization

Symbol timing is required at the sampler and correlator to recover the binary data. The symbol timing recovery (STR) circuit is used to extract the transition time of bits. The received signal is multiplied by the extracted carrier loop signal discussed earlier. The product signal is passed through a tuned bandpass filter. The filtered signal is square law rectified and is passed through another bandpass filter tuned to $1/T_S$ where T_S is the symbol duration. The output signal is sinusoidal with frequency $1/T_S$. The signal will give the transition time of a bit. The symbol time recovery circuit is shown in Fig. 1.40. This method is known as *spectral-line method*. Let us write the input and output of the tuned bandpass filter as

$$x(t) = \sum_{k=-\infty}^{\infty} a_k q(t - kT_S) \tag{1.396a}$$

$$y(t) = \sum_{k=-\infty}^{\infty} a_k p(t - kT_S) \tag{1.396b}$$

where

$$p(t) = q(t) \otimes h_L(t) \quad \text{(convolution)} \tag{1.397}$$

in which $h_L(t)$ is the impulse response of baseband filter, and $q(t)$ is the transmitted baseband waveform, and $\{a_k\}$ represent the information symbols. The pulse waveform $p(t)$ is such that the amplitude spectrum is

$$|P(f)| = 0 \quad \text{for } |f| > 1/2T_S \tag{1.398}$$

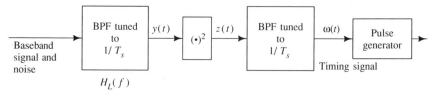

FIGURE 1.40   Symbol timing recovery circuit using spectral-line method.

where T_S is the symbol duration and $P(f) = Q(f)H_L(f)$.

We assume that

$$E(a_k a_m) = \sigma^2 \qquad k = m$$
$$= 0 \qquad k \neq m \tag{1.399}$$

We denote $z(t)$ as the output of the square-law rectifier. We can show that

$$E[z(t)] = E[y^2(t)] = \sigma^2 \sum_{k=-\infty}^{\infty} p^2(t - kT_S) \tag{1.400}$$

$z(t)$ is a cyclostationary process; the mean is periodic with period T_S. Hence we write

$$E[z(t)] = \sum_{n=-\infty}^{\infty} c_n e^{i2\pi f_0 tn} \qquad f_0 = \frac{1}{T_S} \tag{1.401a}$$

$$c_n = \frac{1}{T_S} \int_{-T_S/2}^{T_S/2} \left[\sigma^2 \sum_{k=-\infty}^{\infty} p^2(t - kT_S) e^{-i2\pi f_0 kt} \right] dt$$

$$c_n = \frac{\sigma^2}{T_S} \int_{-\infty}^{\infty} p^2(\tau) e^{-2\pi f_0 n\tau} d\tau$$

$$= \frac{\sigma^2}{T_S} G\left(\frac{n}{T_S}\right) \tag{1.401b}$$

where

$$G(f) = \int_{-\infty}^{\infty} P(f^1)P(f - f^1) df^1 \tag{1.401c}$$

$$E[z(t)] = \frac{\sigma^2}{T_S} \sum_{m=-\infty}^{\infty} G(m/T_S) \exp[i2\pi f_0 mt] \tag{1.401d}$$

If we set $g(t) = p^2(t)$ in Eq. (1.400), then from Eqs. (1.400) and (1.401d) we obtain

$$\sum_{n=-\infty}^{\infty} g(t - nT_S) = \frac{1}{T_S} \sum_{m=-\infty}^{\infty} G\left(\frac{m}{T_S}\right) e^{+i2\pi m f_0 t} \tag{1.402}$$

which is known as *Poisson sum formula*. Since $P(f) = 0$, for $|f| > 1/2T_S$,

$$G(m/T_S) = 0 \qquad m = \pm 2, \pm 3, \pm \ldots \tag{1.403}$$

Hence, Eqs. (1.401d) and (1.403) yield

$$E[z(t)] = \frac{\sigma^2}{T_S} G(0) + \frac{\sigma^2}{T_S} [G(-1)e^{-2\pi f_0 t} + G(1)e^{i2\pi f_0 t}]$$

$$= \frac{\sigma^2}{T_S} G(0) + \frac{\sigma^2}{T_S} G(1) \cos(2\pi t/T_S) 2 \tag{1.404}$$

since $G(-1) = G(1)$.

The signal $z(t)$ is passed through a bandpass filter and the output is denoted by $\omega(t)$.

The dc term $(\sigma^2/T_S)G(0)$ is not passed through the bandpass filter centered at $1/T_S$. Hence, $\omega(t)$ is a sinusoidal signal that will be used to generate a square wave at the frequency $1/T_S$. This frequency will be used as the sampling instant. The alternate zero crossing of the waveform $\omega(t)$ will yield the timing phase at the sampling instant.

It was shown, by Moeneclay (1982) that the variance of the symbol time recovery error τ is given by

$$\sigma_\tau^2 = B_L T_b^{\,3}(0.017 + 0.545d^{-1} + 0.318d^{-2}) \tag{1.405}$$

where

$$T_b = \text{bit duration} = \frac{3}{16}\cdot\frac{1}{f_3}$$

$f_3 = $ 3-dB bandwidth of the RC prefilter $H_L(f)$

$d = PT/N_0 = $ signal-to-noise ratio

$$B_L = \frac{1}{2}\int_{-\infty}^{\infty}\left|\frac{k\sqrt{P}\,\text{erf}(\sqrt{d})F(f)}{2\pi i f + k\sqrt{P}\,\text{erf}(\sqrt{d})F(f)}\right|^2 df$$

in which k is the loop gain, $F(f)$ is the loop transfer function, P is the power of the signal, and $N_0/2$ is the variance of the noise. It is assumed that $B_L T \ll 1$.

The second method we will describe is the absolute-value early-late gate synchronizer, which utilizes two matched filters in two separate areas, as shown in Fig. 1.41. If the baseband signal is a rectangular pulse of height A and duration T seconds, the matched filter output is maximum $A^2 T$ at $t = T$. The autocorrelation function is an even function. Let us denote the auto-correlation function as $R(\tau)$; if $R(T) = A^2 T, R(\tau) = R(-\tau)$. Hence, $R(T - \delta) = R(T + \delta)$, where δ is the delay. Let us denote $R(T + \delta) - R(T - \delta) = e, \delta < T$, $e = 0$, if the signals generated by local symbol generators are matched to the incoming baseband signals. The delay δ can be adjusted to make the error small. The error signal is passed through a loop filter. The output of the loop filter is a control voltage for a VCO whose output is used to generate symbol waveforms. It is shown by Lindsey and Simon (1973) that the variance of the symbol time recovery error τ is given by

$$\sigma_\tau^2 = \frac{B_{\text{eq}} T_b}{8(E_b/N_0)\,\text{erf}\left[\dfrac{\sqrt{E_b/N_0}}{2}\right]} \tag{1.406}$$

where

$$B_{\text{eq}} = \text{loop bandwidth}$$

$$T_b = \text{bit duration}$$

$$E_b/N_0 = \text{bit energy-to-noise ratio}$$

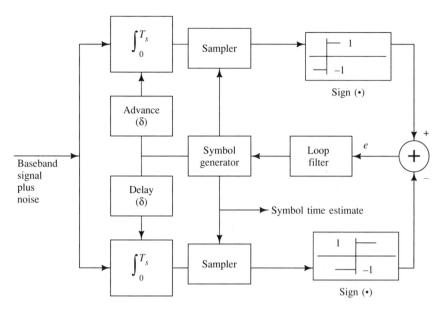

FIGURE 1.41   Absolute-value early-late gate time synchronizer.

1.6 IMPAIRMENTS AND INTERFERENCE EFFECTS

In addition to additive noise and jamming noise, there are various transmission-related impairments that cause signal degradation and loss of symbol energy. These transmission impairments arise when signals are passed through band-limited channels, band-limiting filters, and nonlinear amplifiers. Besides, earth stations may receive signals from the satellites adjacent to the desired satellites. The signals from the undesired satellites will cause interference at the receiver. We will refer to these signals as *intersatellite interference*. The impairments caused by nonlinear amplifiers will be called *intermodulation interference*. The impairments caused by band-limited filters and channels will be called *intersymbol interference*.

1.6.1 Intermodulation Interference

Consider a nonlinear amplifier whose input and output are given by

$$y(t) = a_1 x(t) + a_3 x^3(t) \tag{1.407}$$

where $x(t)$ is the input. Let

$$x(t) = A \cos(\omega_1 t) + B \cos(\omega_2 t) \tag{1.408}$$

It can be shown that the output of the amplifier is

$$y(t) = Aa_1\left[1 + \frac{3}{4}\left(\frac{a_3}{a_1}\right)A^2 + \frac{3}{2}\left(\frac{a_3}{a_1}\right)B^2\right]\cos\omega_1 t$$

$$+ Ba_1\left[1 + \frac{3}{4}\left(\frac{a_3}{a_1}\right)B^2 + \frac{3}{2}\left(\frac{a_3}{a_1}\right)A^2\right]\cos\omega_2 t$$

$$+ \frac{3}{4}ABa_3[A\cos(2\omega_1 - \omega_2)t + B\cos(2\omega_2 - \omega_1)t]$$

$$+ \frac{a_3}{4}[A^3\cos 3\omega_1 t + B^3\cos 3\omega_2 t + 3AB^2\cos(2\omega_1 + \omega_2)t$$

$$+ 3AB^2\cos(2\omega_2 + \omega_1)t] + \text{higher terms} \qquad (1.409)$$

assuming $2\omega_1 - \omega_2 \approx \omega_1$ and $2\omega_2 - \omega_1 \approx \omega_2$. The first and second terms in Eq. (1.409) give the fundamentals with amplitudes Ac_1 and Bc_2, where

$$c_1 = a_1\left[1 + \frac{3}{4}\left(\frac{a_3}{a_1}\right)A^2 + \frac{3}{2}\left(\frac{a_3}{a_1}\right)B^2\right]$$

$$c_2 = a_1\left[1 + \frac{3}{4}\left(\frac{a_3}{a_1}\right)B^2 + \frac{3}{2}\left(\frac{a_3}{a_1}\right)A^2\right] \qquad (1.410)$$

c_1 and c_2 are called *suppression factors*. The third term in Eq. (1.409) is called the *intermodulation product*. The fourth term in Eq. (1.409) is the higher harmonics which is usually filtered out in the bandpass filter. The third term in Eq. (1.409) is also called the *intermodulation interference noise* (IM).

In practice, the input and output relations are given by data. The output can be modeled as

$$y(t_i) = a_1 x(t_i) + a_3 x^3(t_i) + a_5 x^5(t_i) + \cdots + a_{2L+1} x(t_i)$$

$$L \geq 1, 1 \leq i \leq N \qquad (1.411)$$

Let us denote

$$Y = [y(t_1), \ldots, y(t_N)]'$$

$$X = \begin{bmatrix} (x(t_1) & x^3(t_1) & \cdots & x^{2L+1}(t_1) \\ \vdots & \vdots & & \vdots \\ x(t_N) & x^3(t_N) & \cdots & x^{2L+1}(t_N) \end{bmatrix}$$

$$a = (a_1, a_3, \ldots, a_{2L+1})'$$

Equation (1.411) can be written as

$$Y = Xa \qquad (1.412)$$

It can be shown (Mohanty, 1986) that the least-square fit of the data is given by

$$a = (X'X)^{-1}X'Y$$

where it is assumed that

$$\det[X'X] \neq 0 \qquad (1.413)$$

The input $x(t)$ is given by sum of n number of signals:

$$x(t) = \sum_{i=1}^{n} A \cos[\omega_i t + \theta_i(t)] \tag{1.414}$$

where it is assumed that n carriers have equal power. The total input power is

$$\rho_i = \frac{nA^2}{2} \tag{1.415}$$

where n is the number of input signals to an amplifier.

The output voltage $y(t)$ in Eq. (1.407) can be expressed as before as

$$y(t) \triangleq \sum_{i=1}^{n} B_i \cos[\omega_i t + \phi_i(t)] + IM + H \tag{1.416}$$

where IM = intermodulation

   H = higher harmonics of the fundamentals

   B_i = AM-to-AM conversion coefficients for the ith carrier

   $\phi_i(t)$ = AM-to-PM conversion coefficients for the ith carrier

The amplified signal contains intermodulation products that fall within the transponder bandwidth at frequencies

$$\omega_x = m_1 \omega_1 + m_2 \omega_2 + \cdots + m_n \omega_n$$

where $m_1, m_2, \ldots$, and m_n are integers. The order m of intermodulation product x is defined as

$$m = |m_1| + |m_2| + \cdots + |m_n|$$

When the center frequency of the transponder is large compared to the bandwidth of the transponder, only odd-order IMs fall within the useful frequency band and interfere with other signals. $2\omega_1 - \omega_2, \omega_1 + \omega_2 - \omega_3$, etc. are third order IM. There are $n(n-1)$ third-order IMs such as $2\omega_i - \omega_{i+1}, i = 1, 2, \ldots$, and $\frac{1}{2}n(n-1)(n-2)$ IM products at frequencies $\omega_i + \omega_{i-1} - \omega_{i+2}, i = 1, 2, \ldots$ Fifth-order IM products are produced at frequencies $3\omega_i - 2\omega_{i+1}$, $2\omega_i + \omega_{i+1} - 2\omega_{i+2}, 3\omega_i - \omega_{i+1} - \omega_{i+2}, 2\omega_i + \omega_{i+1} - \omega_{i+2} - \omega_{i+3}$, and $\omega_i + \omega_{i+1} + \omega_{i+2} - \omega_{i+3} - \omega_{i+4}$. There are $n(n-1)$, $n(n-1)(n-2)$, $\frac{1}{2}n(n-1)(n-2), \frac{1}{2}n(n-1)(n-2)(n-3)$, and $\frac{1}{2}n(n-1)(n-2)(n-3)(n-4)$ such fifth-order IM products. For $n = 5$, there are 230 IM products containing third-order and fifth-order products. Therefore these IM products can cause severe degradation to the output signals when the amplifier is operated at the saturation point. It has been shown by Pritchard and Sciulli (1986) that the carrier-to-IM ratio (C/IM) is

$$\left(\frac{C}{IM}\right) = \frac{1}{6}\left[\sqrt{\left(\frac{C}{IM}\right)_2} + 1\right]^2 \tag{1.417}$$

where

$$\left(\frac{C}{IM}\right)_2 = \frac{16}{9}\left(\frac{a_1^2}{a_3}\right)\frac{F_2^2}{\rho_i^2}$$

$$F_2 = 1 + \frac{9}{4}\left(\frac{a_3}{a_1}\right)\rho_i$$

$$\rho_i = \frac{\eta A^2}{2}$$

when n is large and a_5, a_7, ... are equal to zero. The output power per carrier versus total input power and output power per intermodulation product versus total input power are given in Figs. 1.42 and 1.43.

If we set peak single-carrier output power to 1 (that is, 0 dB), then the output backoff is

$$BO_0 = \frac{1}{nB_n^2} \tag{1.418}$$

where

$$B_n = a_1\sqrt{\frac{2\rho_i}{n}}\left[1 + 3\frac{a_3}{a_1}\left(\frac{\rho_i}{n}\right)\left(n - \frac{1}{2}\right)\right] \qquad \rho_i = \frac{nA^2}{2}$$

assuming that a_5, a_7, ... are equal to zero.

The output backoff to input backoff for one, three, and ten carriers is shown in Fig. 1.44. The third-order IM product power is also shown in the same figure.

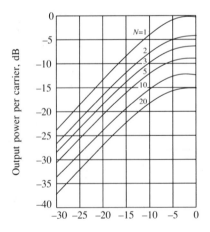

Total input power, dB

FIGURE 1.42   Output power per carrier versus total input power. (Courtesy of CCIR, ITV, Geneva, 1985.)

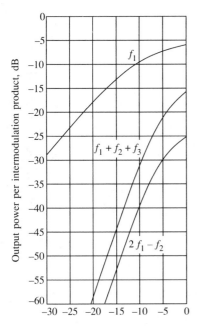

FIGURE 1.43   Intermodulation, with three carriers. (From CCIR, *Handbook on Satellite Communications*, 1985. Courtesy of ITU.)

1.6.2 Intersatellite Interference

Because of the positioning of several satellites close to each other in the geosynchronous and other orbits, the earth stations receive signals from the undesired satellite systems. Figure 1.45 shows the interference geometry between two satellite networks. Let us denote

C = wanted carrier power, dBW

E_W = EIRP (wanted), dBW

L_{dW} = space loss (downlink in the direction of the wanted satellite, dB

$G_4(0)$ = earth station gain in direction of wanted satellite, dB

$G_4(\theta)$ = earth station gains in direction of interfering satellite, dB

e_i = EIRP of interfering satellite, dBW

l_{di} = space loss (downlink) in direction of interfering satellite, dB

Y_d = polarization discrimination, dB

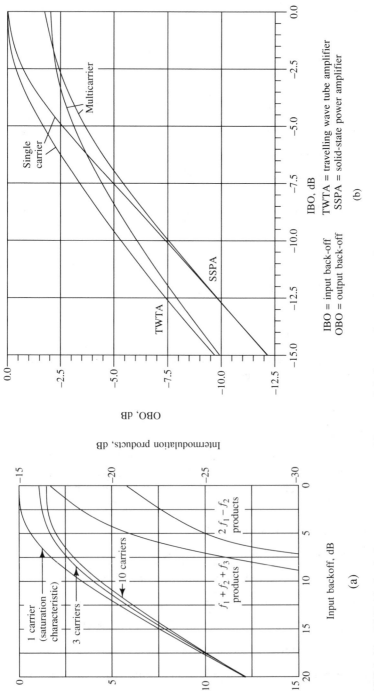

FIGURE 1.44 (a) Typical relative output power and intermodulation products of a traveling-wave amplifier. (From CCIR, *Handbook on Satellite Communications*, 1985. Courtesy of ITU, Geneva.) (b) Single-carrier and multicarrier power transfer characteristics of a TWTA and an SSPA. (From K. Jonalgadda and L. Schiff, "Improvements in Capacity of Analog Voice Multiplex Systems Carried by Satellite," *Proc. IEEE*, vol. 72, November 1984. © 1984 IEEE.)

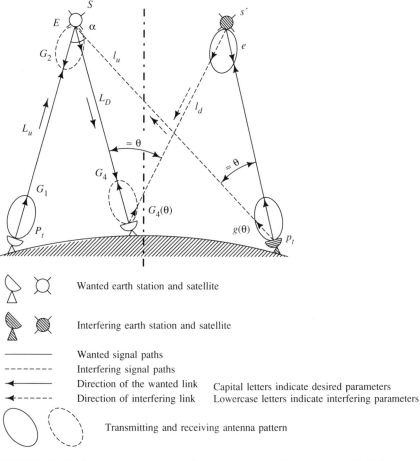

Wanted earth station and satellite

Interfering earth station and satellite

——————— Wanted signal paths
- - - - - - - Interfering signal paths
◄—————— Direction of the wanted link   Capital letters indicate desired parameters
◄- - - - - Direction of interfering link   Lowercase letters indicate interfering parameters

Transmitting and receiving antenna pattern

FIGURE 1.45   Interference geometry between two satellite networks. Uplink of wanted network shares frequencies with uplink of interfering network (and similarly in the downlink). (Courtesy of CCIR, ITU, Geneva, 1985.)

Therefore, the carrier power is

$$C = E_W G_4(0)/L_{dW}$$
$$= E_W - L_{dW} + G_4(0) \quad (dB) \tag{1.419}$$

The intersatellite interference power is

$$I_{sat} = e_i G_4(\theta)/L_{di} Y_d$$

$$= e_i - L_{di} + G_4(\theta) - Y_d \quad (dB)$$

Therefore, carrier-to-interference power for intersatellite power for the downlink

$$\left(\frac{C}{I}\right)_{sat_d} = E_W - e_i - (L_{dW} - L_{di}) + [G_4(0) - G_4(\theta)] + Y_d \tag{1.420}$$

Similarly, carrier-to-interference for intersatellite power for the uplink is

$$\left(\frac{C}{I_{sat}}\right)_u = P_t + G_1(0) - [p_t + g_1(\theta)]$$
$$- (L_{uW} - l_{ui}) + (G_{2W} - G_{2i}) + Y_u - M_u \qquad (1.421)$$

where P_t = earth station transmit power (wanted), dB

p_t = earth station transmit power (interfering), dB

$G_1(0)$ = wanted earth station transmit antenna gain in direction of wanted satellite, dB

$g_1(\theta)$ = interfering earth transmit antenna gains in direction of wanted satellite, dB

l_{ui} = space loss (uplink) in direction of interfering satellite, dB

L_{uW} = space loss (uplink) in direction of wanted satellite, dB

G_{2W} = wanted satellite receive antenna gain in direction of wanted earth station, dB

G_{2i} = wanted satellite receive antenna gain in direction of interfering earth station, dB

M_u = uplink margin

The total carrier-to-interference noise ratio from one adjacent satellite is found by reciprocal addition, as follows:

$$\left(\frac{C}{I_{sat}}\right)_T^{-1} = \left(\left(\frac{C}{I_{sat}}\right)_u^{-1} + \left(\frac{C}{I_{sat}}\right)_d^{-1}\right)^{-1} \qquad (1.422)$$

where

$$\left(\frac{C}{I_{sat}}\right)_u \quad \text{and} \quad \left(\frac{C}{I_{sat}}\right)_d$$

are given by Eqs. (1.420) and (1.421), respectively.

If there is more than one satellite interference and these interference signals are independent, the total interference satellite signal power can be given by

$$I_{sat} = \sum_j \{[e_j + G_4(\theta_j)] - L_j - Y_j\} \, dB \qquad (1.423)$$

The CCIR antenna sidelobe formulas are given by

$$G_4(\theta) = 32 - 25 \log \theta \qquad D/\lambda > 100$$
$$= 52 - 10 \log(D/\lambda) - 25 \log \theta \quad D/\lambda < 100 \qquad (1.424)$$

where D = antenna diameter, λ = wavelength, and θ = angle between main-lobe and sidelobe. θ, in this case, is called the *topocentric angle*[1]. See Fig. 1.54.

---

[1]The *orbital separation* is defined as the angle α subtended at the center of the earth by two satellites. This angle is also known as the *geocentric* angle. See Fig. 1.54.

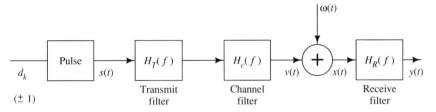

FIGURE 1.46   Baseband transmission.

1.6.3 Intersymbol Interference

We consider a transmission of a (pulse amplitude modulation) PAM system, as shown in Fig. 1.46. Let a_k be binary data $(1, -1)$, representing 1 volt for symbol 1 and -1 volt for symbol 0. Let the PAM signal be

$$S(t) = \sum_{k=-\infty}^{\infty} a_k g(t - kT_b) \tag{1.425}$$

where T_b is the bit duration, and $g(t)$ is the baseband pulse with $g(0) = 1$. Let $H_T(f)$, $H_c(f)$, and $H_R(f)$ be the transfer function of the transmit, channel, and receive filters. In Fig. 1.46, $\omega(t)$ is an additive white Gaussian noise. The output of the receive filter

$$y(t) = \sum_{k=-\infty}^{\infty} a_k h(t - kT_b) + n(t) \tag{1.426}$$

where

$$h(t) = g(t) \otimes h_T(t) \otimes h_c(t) \otimes h_R(t)$$

$$n(t) = \int h_R(\tau)\omega(t - \tau)\,d\tau$$

$$h(0) = 1$$

In addition, $h_T(t)$, $h_c(t)$, and $h_R(t)$ are, respectively, impulse response functions of the transmit, channel, and receive filters and $h(t)$ is the convolution of $g(t)$, $h_T(t)\,h_c(t)$ and $h_R(t)$. The sampling output at $t = mT_b$, $m = 0, \pm 1, \pm 2, \pm 3, \ldots$, is

$$y(mT_b) = \sum_{k=-\infty}^{\infty} a_k h(mT_b - kT_b) + n(mT) \tag{1.427}$$

For the sake of brevity, we denote

$$y(n) = a_n h(0) + \sum_{k \neq n} a_k h(n - k) + n(m)$$

$$= a_n + \sum_{k \neq n} a_k h(n - k) + n(m) \tag{1.428}$$

The second term is the *intersymbol interference*. Because of the band-limiting effects, the baseband pulse is dispersed in time and spread beyond T_b. The frequency component of the pulse is attenuated and phase-delayed. In the absence of the $n(m)$ and the intersymbol interference, the transmitted sequence is received correctly. Disregarding the receiver noise, the intersymbol interference can be made zero if

$$h[(n-k)T_b] = 1, \quad n = k$$
$$= 0, \quad n \neq k \tag{1.429}$$

Let us denote the sampled function

$$h_S(t) = h(t) \sum_{m=-\infty}^{\infty} \delta(t - mT_b)$$

$$= \sum_{m=-\infty}^{\infty} h(mT_b)\delta(t - mT_b) \tag{1.430}$$

The spectrum of $h_S(t)$, see Eq. (1.237), is

$$H_S(f) = \frac{1}{T_b} \sum_{m=-\infty}^{\infty} H\left(f - \frac{m}{T_b}\right) \tag{1.431}$$

Taking the Fourier transform of Eq. (1.430),

$$H_S(f) = \int_{-\infty}^{\infty} \sum_{m=-\infty}^{\infty} h(mT_b)\delta(t - mT_b)e^{-i2\pi ft}\, dt \tag{1.432}$$

$$= \int_{-\infty}^{\infty} h(0)\delta(t)e^{-i2\pi ft}\, dt$$

$$= h(0) = 1 \tag{1.433}$$

Therefore, Eqs. (1.431) and (1.433) yield

$$\frac{1}{T_b} \sum_{m=-\infty}^{\infty} H\left(f - \frac{m}{T_b}\right) = 1$$

or

$$\sum_{m=-\infty}^{\infty} H\left(f - \frac{m}{T_b}\right) = T_b \tag{1.434}$$

In order to have nonoverlapping spectrum, it is necessary to have

$$H_N(f) = T_b \quad |f| \leq 1/2T_b$$
$$= 0 \quad \text{elsewhere} \tag{1.435}$$

Therefore, a signal at a data rate $R_b (= 1/T_b)$ needs a bandwidth of $R_b/2$ Hz (one-sided). Equation (1.435) is called the *Nyquist condition*. The Nyquist filter is given by

$$h_N(t) = \frac{\sin(2\pi t/2T_b)}{(2\pi t/2T_b)} = \text{sinc}(t/T_b) \tag{1.436}$$

The filter described in Eq. (1.435) is an ideal, having sharp cutoff at $f = \pm 1/2T_b$. This is unrealizable. Further the function $h(t)$ decreases at the rate of $1/|t|$. If there is a sampling jitter, then the intersymbol term does not converge.

To overcome these disadvantages, the filter can be expanded from $1/2T_b$ to $1/T_b$ such that there will be a smooth roll-off. This can be achieved if

$$H\left(f - \frac{1}{T_b}\right) + H(f) + H\left(f + \frac{1}{T_b}\right) = T_b \qquad |f| \le \frac{1}{2T_b} \qquad (1.437)$$

The transfer function that satisfies Eq. (1.437), known as a *raised cosine filter*, is

$$H_{RC}(f) = \begin{cases} T_b, & |f| < f_1 \\[2mm] \dfrac{T_b}{2}\left\{1 + \cos\left[\dfrac{\pi(|f| - f_1)}{\dfrac{1}{T_b} - 2f_1}\right]\right\} & f_1 < |f| < \dfrac{1}{T_b} - f_1 \\[4mm] 0, & |f| > \dfrac{1}{T_b} - f_1 \end{cases}$$

$$(1.438)$$

This transfer function is shown in Fig. 1.47a. The impulse response function of the raised cosine filter (Fig. 1.47b) is

$$h_{RC}(t) = \text{sinc}\left(\frac{t}{T_b}\right)\left\{\frac{\cos\left[\pi\alpha\left(\dfrac{1}{T_b}\right)t\right]}{1 - 4\alpha^2\left(\dfrac{1}{T_b^2}\right)t^2}\right\} \qquad (1.439)$$

where $\alpha = 1 - 2f_1 T_b$, $0 \le \alpha < 1$, and is called the *roll-off factor*. If $\alpha = 0$, $f_1 = 1/2T_b$, $H_{RC}(f) = H_N(f)$, the Nyquist filter given by Eq. (1.436). The filter $h_{RC}(t) = 0$ when $t = nT_b$, n is an integer. Further, $h_{RC}(t)$ decreases at the rate of $1/|t|^2$. The intersymbol interference term goes to zero even when there is a sampling-time error. Since the pulse-shaping filter $G(f)$, the Fourier transform of $g(t)$, and the channel filter are given, the transmitting and receiving filter should be designed such that

$$G(f)H_c(f)H_T(f)H_R(f) = H_{RC}(f) \qquad (1.440a)$$

We could choose

$$H_T(f) = \left[\frac{H_{RC}(f)}{H_c(f)G(f)}\right]^{1/2}$$

$$= H_R(f) \qquad (1.440b)$$

to satisfy the condition of zero intersymbol interference in the absence of noise.

We proceed here to design the receive filter $H_R(f)$ such that the mean-square error in decoding (estimating) symbol is minimum. Let us represent the transmission system in Fig. 1.46. Let the observed signal be

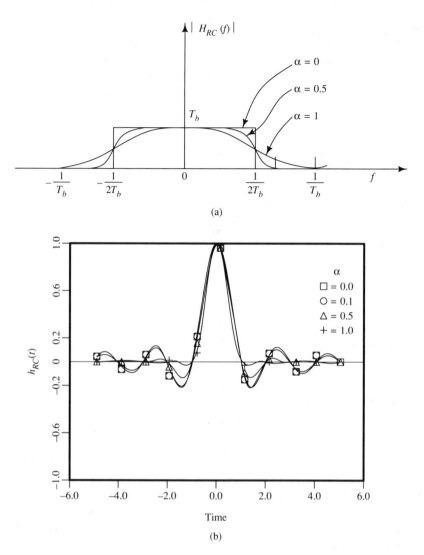

FIGURE 1.47   Raised cosine filter. (a) Transfer function. (b) Impulse response function.

$$x(t) = v(t) + \omega(t)$$

$$= \sum_{k=-\infty}^{\infty} a_k q(t - kT_b) + \omega(t) \qquad (1.441)$$

where

$$q(t) = g(t) \otimes h_T(t) \otimes h_c(t)$$

and $\omega(t)$ is a zero-mean white Gaussian noise with variance $N_0/2$. Let us denote the samples of the observed signal as

$$x_k = \sum_{j=-\infty}^{\infty} a_j q[(k-j)T_b] + \omega(kT_b) \qquad -N \le k < N \tag{1.442}$$

Let the estimate (decoded) of a_k be

$$a_k = \sum_j c_j x_{k-j} \qquad k = 0, \pm 1, \ldots, \pm L \tag{1.443}$$

where $c_j, j = 0, 1, 2, \ldots, 2L + 1$ represent the weights to be determined. Let the estimation error be

$$\varepsilon_k = a_k - \hat{a}_k \tag{1.444}$$

The c_j values are to be adjusted such that

$$\begin{aligned}
\zeta &= E[(a_k - \hat{a}_k)^2] \\
&= E(a_k^2 - c'\mathbf{X}_k \mathbf{X}_k' c - 2a_k \mathbf{X}_k' \mathbf{c})
\end{aligned} \tag{1.445}$$

is minimum where

$$\mathbf{c} = (c_0, c_1, \ldots, c_L)'$$

$$\mathbf{X}_k = [x_k, x_{k-1}, \ldots, x_{k-L}]'$$

Let us denote

$$R = E(\mathbf{X}_k \mathbf{X}_k') \tag{1.446}^m$$

$$\mathbf{P} = E(a_k \mathbf{X}_k)$$

It is known from the orthogonality principle (Mohanty, 1986) that

$$E(\varepsilon_k \mathbf{X}_m) = 0 \qquad m = k - 1, \ldots, k - N + 1 \tag{1.447}$$

It can be shown that ζ is minimum if the optimum weight is

$$C_0 = R^{-1}\mathbf{P} \tag{1.448}$$

and

$$\zeta_{\min} = E(a_k^2) - P'C_0 \tag{1.449}$$

The optimum filter (receive) is given by Eq. (1.448). It can be shown that

$$\zeta = \zeta_{\min} + (C - C_0)'R(C - C_0) \tag{1.450}$$

Let us denote

$$\mathbf{C} - \mathbf{C}_0 = \mathbf{V}$$

$$\begin{aligned}
R &= Q\Lambda Q^{-1} \\
&= Q\Lambda Q'
\end{aligned} \tag{1.451}$$

$$\Lambda = \text{diag}[\lambda_1, \ldots, \lambda_L]$$

---

^{m} We have assumed $E(X_k) = 0$.

$$Q = [z_1, \ldots, z_L]$$

$$Rz_i = \lambda_i z_i \qquad i = 1, \ldots, L$$

where z_i and λ_i are, respectively, the eigenvector and eigenvalue of the covariance matrix R. We have assumed that R is symmetric and positive definite. Q is an orthogonal matrix such that $Q^{-1} = Q'$. Equations (1.450) and (1.451) give

$$\zeta = \zeta_{\min} + V'RV$$

$$= \zeta_{\min} + V'Q\Lambda Q'V$$

$$= \zeta_{\min} + U'\Lambda U \tag{1.452a}$$

where

$$U = Q'V = [U_1, \ldots, U_L]' \tag{1.452b}$$

Hence,

$$\zeta = \zeta_{\min} + \sum_{i-1}^{L} \lambda_i U_i^2 \tag{1.453}$$

The gradient of ξ is

$$\nabla\xi = 2R(C - C_0)$$

$$= 2RV \tag{1.454}$$

using Eqs. (1.450) and (1.451).

The weights C_0 can be obtained recursively by using *linear mean square* (LMS) *algorithm*, given by

$$C_{k+1} = C_k + \mu(-\nabla_k) \tag{1.455}$$

where μ is a convergence factor, ∇_k is the gradient of ∇ at time $t = k$. The recursive weight is

$$C_{k+1} = C_k - 2\mu R(C_k - C_0)$$

$$= C_k + 2\mu RV \tag{1.456}$$

Subtracting C_0 from the both sides of Eq. (1.456), we get

$$V_{k+1} = V_k + 2\mu RV_k$$

$$= (I + 2\mu R)V_k \tag{1.457}$$

Substitution of Eq. (1.452) in Eq. (1.457) yields

$$U_{k+1} = (I - 2\mu\Lambda)U_k \tag{1.458}$$

The solution of Eq. (1.458) is

$$U_k = (I - 2\mu\Lambda)^k U_0 \tag{1.459}$$

where

$$(I - 2\mu\Lambda)^k = \begin{bmatrix} (1 - 2\mu\lambda_1)^k & & 0 \\ & \ddots & \\ 0 & & (1 - 2\mu\lambda_L)^k \end{bmatrix}$$

When $k \to \infty$,

$$(I - 2\mu\Lambda)^k \to 0 \tag{1.460}$$

if $|(1 - 2\mu\lambda_i)| \le 1$ for all i. Hence, when $k \to \infty$,

$$\mathbf{U}_k \to 0 \Rightarrow \mathbf{C}_k \to \mathbf{C}_0$$

provided that

$$\max|1 - 2\mu\lambda_i| < 1 \tag{1.461}$$

Therefore, $\mathbf{C}_k \to \mathbf{C}_0$ if

$$\frac{2}{\lambda_{max}} > \mu > 0 \qquad \lambda_{max} \text{ is the maximum of } \lambda_1, \dots, \lambda_L \tag{1.462}$$

Equation (1.450) can be written as

$$\xi_k = \xi_{min} + (C_k - C_0)'R(C_k - C_0) \tag{1.463}$$

Since $\mathbf{C}_k \to \mathbf{C}_0$ when $k \to \infty$,

$$\xi_k \to \xi_{min}$$

provided that μ satisfies Eq. (1.462). Note that

$$\xi_k = E(\varepsilon_k^2) = E(a_k - \mathbf{X}_k'\mathbf{C}_k)^2 \tag{1.464}$$

In the event that R is not known, the gradient of Λ_k has to be estimated. The estimate of the gradient is

$$\nabla\xi_k = \nabla(a_k - \mathbf{X}_k\mathbf{C}_k)^2 = \nabla\varepsilon_k^2$$
$$= 2\varepsilon_k\nabla\varepsilon_k$$
$$= -2\varepsilon_k\mathbf{X}_k \tag{1.465}$$

The *Widrow algorithm* for the LMS is given by

$$\mathbf{C}_{k+1} = \mathbf{C}_k + \mu(2\varepsilon_k\mathbf{X}_k)$$

using Eq. (1.455) and (1.465). The algorithm converges if $2/\lambda_{max} > \mu > 0$, where λ_{max} is the maximum eigenvalue of the covariance matrix R. An estimate correlation function is

$$r_m = E(X_k X_{k+m}) = \frac{1}{N} \sum_{k=-N+|m|+1}^{N-|m|-1} X_k X_{k+m}$$

The covariance matrix is a *Topelitz matrix* given by

$$R = \begin{bmatrix} r_0 & r_1 & & & r_L \\ r_1 & r_0 & & & r_{L-1} \\ & & \ddots & & \\ r_{L-1} & & & r_0 & r_1 \\ r_L, r_{L-1}, r_{L-2} & \cdots & r_1, & r_0 \end{bmatrix} \tag{1.466}$$

The performance of this equalizer (filter) can be improved by using previous decisions through a feedback loop. This equalizer is called a *linear-feedback equalizer*. Therefore, the estimate is given by

$$\hat{a}_k = \sum_{j=-L}^{0} c_j X_{k-j} + \sum_{j=1}^{L} c_j \hat{a}_{k-j} \tag{1.467}$$

where $\hat{a}_{k-j}, j = 1, \ldots, L$ are the previous L decision symbols. Let us denote

$$\mathbf{C} = (C_{-L}, \ldots, C_{-1}, C_0, \ldots, C_L)'$$
$$\bar{\mathbf{X}}_k = (X_{k+L}, \ldots, X_k, \hat{a}_{k-1}, \ldots, \hat{a}_{k-L})' \tag{1.468}$$

The recursive weights for a linear-feedback equalizer is given by

$$\mathbf{C}_{k+1} = \mathbf{C}_k + 2\mu\varepsilon_k \bar{\mathbf{X}}_k$$

where $\varepsilon_k = a_k - \hat{a}_k$.

Decision Feedback Loop Equalizer

Now we replace the feedback loop with input $\bar{a}_k$, where $\bar{a}_k$ is the decision of which symbol is present. We form the estimate of the transmitted symbol as

$$\hat{a}_k = \sum_{j=-L}^{0} c_j X_{k-j} + \sum_{j=1}^{L} c_j \bar{a}_{k-j} \tag{1.469a}$$

where $\bar{a}_{k-j}, j = 1, \ldots, L$, are the past decision symbols. Let us denote

$$\mathbf{C} = [c_{-L}, c_{-(L-1)}, \ldots, c_0, c_1, \ldots, c_L]'$$
$$\bar{\mathbf{X}}_k = (X_{k+L}, X_{k+L-1}, \ldots, X_k, \bar{a}_{k-1}, \ldots, \bar{a}_{k-L})' \tag{1.469b}$$

The weights are given by

$$C_0 = R^{-1}P$$

where

$$R = E(\bar{X}_k \bar{X}_k')$$
$$P = E[a_k \bar{X}_k] \tag{1.470}$$

The recursive weights are given by

$$c_{k+1} = c_k - 2\mu\varepsilon_k \bar{X}_k \tag{1.471}$$

where $\varepsilon_k = a_k - \hat{a}_k$.

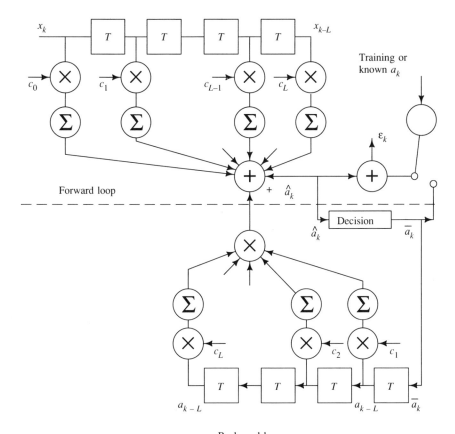

FIGURE 1.48   Decision feedback equalizer.

The decision feedback equalizer is not a linear equalizer unless $c_k = 0$, $k = 0, \ldots, L$–that is, unless the feedback coefficients are zero. The decision feedback loop is shown in Fig. 1.48.

1.7 EARTH SEGMENT

The earth segment includes several earth stations. Each earth station receives various incoming signals, such as telephone, television, and data transmission signals, from terrestrial links. It modifies these signals through a transmission system and transmits RF signals through its antenna system. It also receives signals from satellites through its antenna systems and extracts baseband signals through its receive system, and distributes these signals through its terrestrial links. A typical earth station has the following systems:

1. Antenna system
2. Transmit system

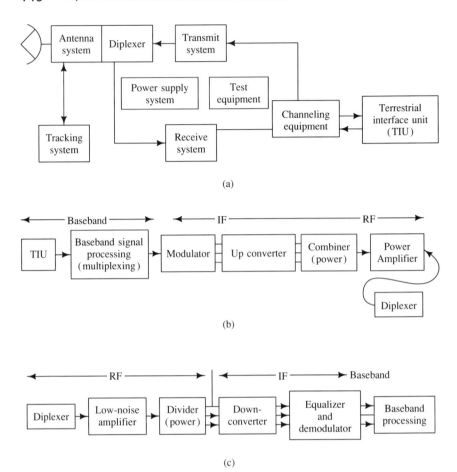

FIGURE 1.49   Earth station. (a) Functional elements. (b) Transmit system.
(c) Receive system.

3. Receive system
4. Terrestrial interface
5. Power system
6. Tracking system
7. Test equipment system (monitoring and control)

The interconnection of these systems is shown in Fig. 1.49. An earth station can be only of the transmit or receive type. The types of service include fixed, maritime, aeronautical, and broadcasting, among others.

1.7.1 The Antenna System

The antenna system is the most important element of the earth station. There are various kinds and sizes of antennas, the most common of which is the parabolic antenna. The diameter of such an antenna ranges from 3 to 33 meters. It is used both for transmission and reception of signals to and from satellites. Signals are transmitted and received at two different frequencies and are polarized to avoid interference. The antenna performance is given by G/T_S where G is the gain of the antenna and T_S is the system temperature. The *gain* is given by

$$G = \pi^2(D/\lambda)^2\eta$$

where D = diameter

$\lambda = c/f_c$ = wavelength

η = efficiency (0.5 to 0.8)

$T_S = T_e + T_R$

T_e = equivalent noise temperature of antenna (see Eq. 1.529)

T_R = noise temperature of the receiver

The value of the power radiated by the antenna to the satellite is equal to gain of the antenna and power of the *high-power amplifier* (HPA). The radiated power is called *equivalent isotropically radiated power* (EIRP). Transmitter power varies from few tens of watts to 400 kW. One of the most widely used HPAs is the *traveling-wave tube amplifier* (TWTA). It has a very wide bandwidth of 500 MHz or more. It is a nonlinear device. The simultaneous transmission of several signals in TWTA produces intermodulation noise, which increases as the operating point reaches saturation. The drive power at which output saturation occurs is called input saturation power. The decreasing input saturation is called the *input backoff*. The input backoff reduces the output power but does not create intermodulation noise. The output backoff (loss from the saturation point) is plotted versus input backoff in Fig. 1.50.

The output signal amplitude also varies nonlinearly with input signal amplitude. This conversion is called *amplitude-to-amplitude modulation conversion* (AM to AM) and the phase of the output signal varies nonlinearly with the input signal amplitude (power). This output signal phase suffers a conversion called amplitude-to-phase modulation conversion (AM to PM) as shown in Fig. 1.51. See also Eq. (1.416).

Another type of high-power amplifier is the klystron, a narrowband device in the range of 40 to 80 MHz. Klystrons are used normally with frequency-modulated multiple access systems. There are also solid-state amplifiers such as input diode amplifier and gallium arsenide (GaAs) field-effect transistor (FET) amplifiers, which are used for low-power applications.

The received signal from the satellite is very weak compared to the transmitted signal. The received signal is passed through a *low-noise amplifier* (LNA), which is kept very close to the diplexer of the antenna feed. LNAs are wideband, and most are parametric.

The operating temperatures for parametric LNAs are 300 K, 223 K, and 23 K for uncooled, thermoelectrically cooled, and cryogenically cooled ampli-

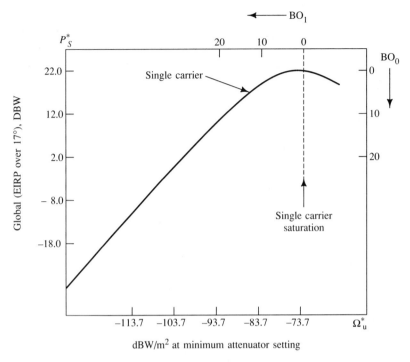

FIGURE 1.50   Input and output backoff for TWTA. (From H. L. Van Trees, Ed., *Satellite Communications*, New York: IEEE Press. © 1979 IEEE.)

fiers, respectively. GaAs FET LNAs are also available for 3- to 5-GHz links at the ambient operating temperature (290 K).

1.7.2 Transmit System

The transmit system may have multiple transmit chains depending on the number of separate chains, a number of up-converters from baseband to IF to RF and a number of parallel TWTA amplifiers in case of malfunctioning of any HPAs. Similarly the receive system may have a number of receiver chains with multiple down-converters and parallel LNAs for redundancy and reliability. These two systems differ in power, frequency, and polarization. Because of the satellite's limited power, antenna gain, the uplink carrier frequency and power is much higher than the received signal frequency and power.

1.7.3 The Terrestrial Interface

The terrestrial interface unit (TIV) provides, connects, and distributes telephones, video, and data to inland users. For fixed-service earth stations, transmission and reception to inland users are accomplished by the use of microwave and cable links. For *single-channel-per-carrier links* (SCPC), the interface

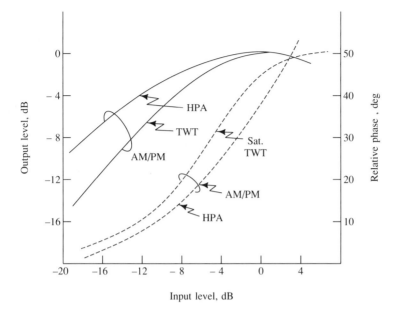

FIGURE 1.51   Phase output versus input. (From K. Jonnalgadda and L. Schiff, "Improvements in Capacity of Analog Voice Multiplex Systems Carried by Satellite," *Proc. IEEE*, vol. 72, November 1984. © 1984 IEEE.)

unit processes terrestrial time-division multiplexed signals to frequency-division multiplexed signal for transmission, and does the opposite for distribution to telephone lines.

1.7.4 The Power System

The power system unit depends on the nature of earth stations such as transmit-only, receive-only, and transmit-and-receive stations for users. The system size ranges from simple battery installations to huge power stations. During power outages, the system is equipped with diesel motor-generators or with batteries that can replace the electric power system without any interruption of service. A system equipped with such a switchover capability is called no-break power system.

1.7.5 The Tracking System

A tracking system (see Chapter 4) is necessary to align an earth station antenna's boresight axis (the direction of maximum power) with that of the satellite's (see Fig. 1.52) and also to switch from one satellite to another. The satellite's position, even though stationary to the observer, deflects from the satellite's orbit because of many disturbing forces. It is required to know azimuth, elevation, and ranging distance of the satellite (see Fig. 1.53). These equations are given by Eqs. (1.476) to (1.478).

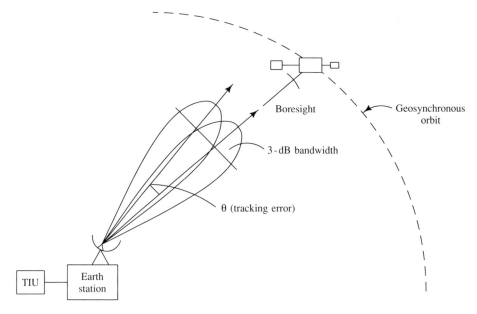

FIGURE 1.52   Earth station antenna pointing to orbiting satellite.

There are two main classes of *antenna steering*. The class that uses satellite beacon signals is called *autotracking*; the other class, called *program control tracking*, uses predicted orbit data of the satellite.

1.7.6 The Test Equipment System

The test equipment unit controls and monitors various signal, noise, and earth station parameters. It measures voltage, power, temperature, and many other variables. There are two important parameters to measure. They are *noise-to-power ratio* (NPR) and *gain over temperature* (G/T). The test equipment system also monitors on-line equipment failures and automatic switchover from the failed unit to standby units. The *reliability* of any unit is given by the reliability function

$$R(t) = 1 - F(t)$$

where $F(t)$ is the distribution of the failure time of the equipment (Mohanty, 1986, 1987). It is usually modeled as an exponential distribution. Thus

$$R(t) = \exp(-\lambda t) \qquad t \geq 0$$

where λ is the average failure rate. The *mean time to failure* (*MTTF*) is given by

$$\text{MTTF} = E(t) = \int_{0}^{\infty} R(t)\, dt = 1/\lambda \qquad (1.472)$$

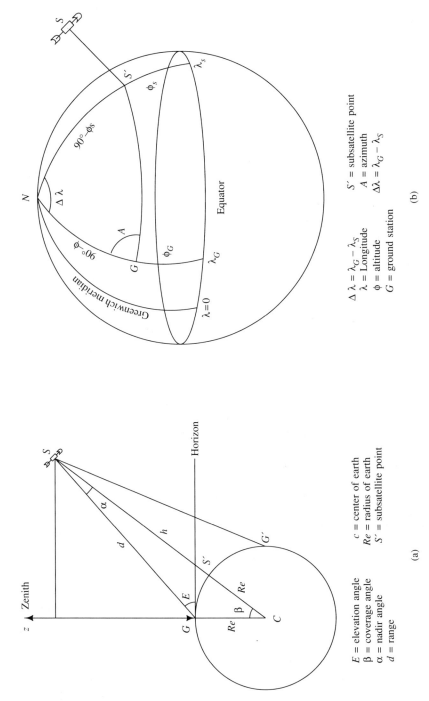

$\Delta \lambda = \lambda_G - \lambda_S$ $S' =$ subsatellite point
$\lambda =$ Longitude $A =$ azimuth
$\phi =$ altitude $\Delta \lambda = \lambda_G - \lambda_S$
$G =$ ground station

(b)

$E =$ elevation angle $c =$ center of earth
$\beta =$ coverage angle $Re =$ radius of earth
$\alpha =$ nadir angle $S' =$ subsatellite point
$d =$ range

(a)

FIGURE 1.53 Elevation, range, and azimuth of satellite. (a) Elevation angle E and range of satellite. (b) Azimuth of satellite.

If n components are cascaded, then the reliability function is given by

$$R_c(t) = \prod_{i=1}^{n} R_i(t) = \exp\left(-\sum_{i=1}^{n} \lambda_i t\right) \tag{1.473}$$

for exponential distributed failure rate components. If n components are connected in parallel, the reliability function of a redundant system is

$$R_p(t) = 1 - \prod_{i=1}^{n} (1 - R_i(t))$$

$$= 1 - \prod_{i=1}^{n} (1 - e^{-\lambda_i t}) \tag{1.474}$$

Let $MTTF_T$ and $MTTR_T$ denote mean time to failure and mean time to repair for the transmit side. Let $MTTF_R$ and $MTTR_R$ denote mean time to failure and mean time to repair at the receive side. The earth station *availability* is defined as

$$\text{Availability} = \frac{MTTF_R \cdot MTTF_T}{(MTTF_T + MTTR_T)(MTTF_R + MTTR_R)} \tag{1.475}$$

The earth stations' sites are selected at places with the least man-made or natural interference, including humidity, snow, rainfall, and earthquakes.

1.8 SPACE SEGMENT

Earth-orbiting satellites provide a significant amount of information about the earth and space by using appropriate sensors and signals. The satellites include communication satellites like Intelsat, navigational satellites such as global positioning satellites (GPS), weather satellites, defense meteorological satellites programs (DMSP), and Landsats. For communication purposes, satellites are used as a repeater and switchboard to connect various communication centers on the ground, in the air, and at sea. Communication satellites overcome the connecting problems encountered in terrestrial networks, and provide direct access to communication centers, and connect N number stations with line-of-sight (LOS) connections directly rather than having $N(N-1)/2$ LOS connections required in terrestrial connections. This capability of the satellites is due to a wide area coverage by satellite antenna. A satellite orbiting earth in a circular orbit in the equatorial plane at a distance of 35,786 km from the earth with a velocity 11,070 km/h can cover one-third of the earth stations if the satellite antenna beamwidth is 17.30°. In this case, only three equally spaced satellites of this type can cover the entire earth except for the polar region (see Fig. 1.54). Such satellites are called *geosynchronous* satellites, and their orbits are called *geosynchronous* or *geostationary* orbits. The *subsatellite point* on the earth, the projection of the satellite on the earth, will observe the satellite as stationary almost all the time. In addition to the wide coverage, there will be very few problems in tracking the satellite antennas

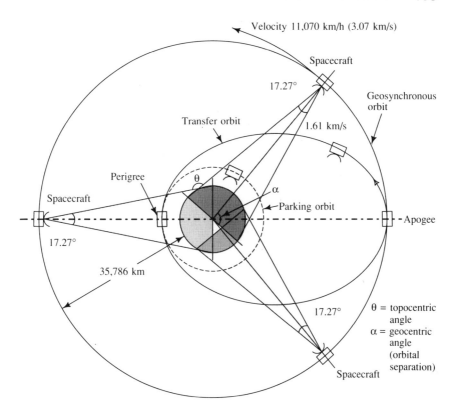

FIGURE 1.54   Coverage area of the earth, shown by shades, by three geosynchronous satellites at a height of 35,786 km above the earth.

for the LOS propagation. There are other types of satellites that move in elliptical orbits inclined to the equatorial planes, one of which is the Soviet satellite in Molyna. There are satellites with low altitudes, medium altitudes, and high altitudes covering a particular region for data collection, broadcasting, or other missions. Communication satellites have two main subsystems. One subsystem is called the *payload* and the other one is called *spacecraft bus*.

1.8.1 The Spacecraft Bus

The spacecraft bus has six subsystems: (1) attitude control, (2) propulsion, (3) power subsystem, (4) thermal control, (5) structure, and (6) *telemetry, tracking, and control* (TT&C). The attitude control system provides proper orientation for the spacecraft; the antenna points to the earth stations and solar cells point toward the sun and position in the space orbit. The system has sensors for attitude determination and activators for providing corrective torques against

(a)

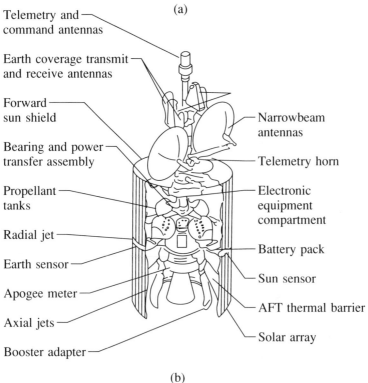

Telemetry and command antennas

Earth coverage transmit and receive antennas

Forward sun shield

Bearing and power transfer assembly

Propellant tanks

Radial jet

Earth sensor

Apogee meter

Axial jets

Booster adapter

Narrowbeam antennas

Telemetry horn

Electronic equipment compartment

Battery pack

Sun sensor

AFT thermal barrier

Solar array

(b)

FIGURE 1.55   Intelsat IV satellite. (a) Spacecraft. (b) Details. (c) Communication subsystem.

154

various disturbances due to pressures from the sun, moon, stars, and other sources. The propulsion subsystem injects the satellite into a desired orbit. The power system provides the necessary electric energy to all systems and payload. The power system has photovoltaic solar cells to collect solar energy from the sun, and has rechargeable nickel-hydrogen batteries to run the system in the event of an eclipse.

The thermal system controls the temperature of high-power amplifiers and other subsystem units. The structure system supports and sustains various mechanical units, antennas, and thruster jets. The Intelsat IV satellite is shown in Fig. 1.55.

The TT&C system deals with satellite management in the orbit, particularly orbit control, attitude control, on/off switching of redundant systems, monitoring satellite sensors, and ranging and tracking. The telemetry unit measures various sensor outputs, such as pressure, voltage, current, power, and temperatures. These outputs are converted to digital formats such as PCM, multiplexed and encrypted with PSK modulation at a frequency much lower than communication frequency. The tracking unit uses a beacon carrier, generated in the telemetry unit, to maintain contact with the earth station. The earth station antenna uses this signal to determine the angular measurements of the satellite. The TT&C also sends tone and pseudorandom noise (PRN) signals to the earth station to measure the satellite range by computing

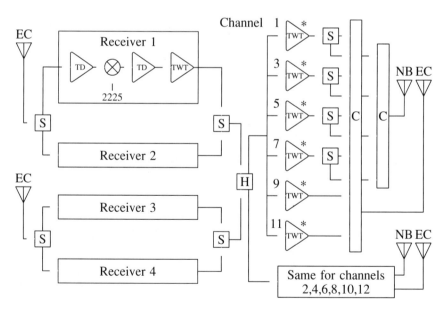

NB: Narrowbeam antenna
EC: Earth coverage antenna

*Redundant TWT for each channel not shown

(c)

FIGURE 1.55 (*Continued*)

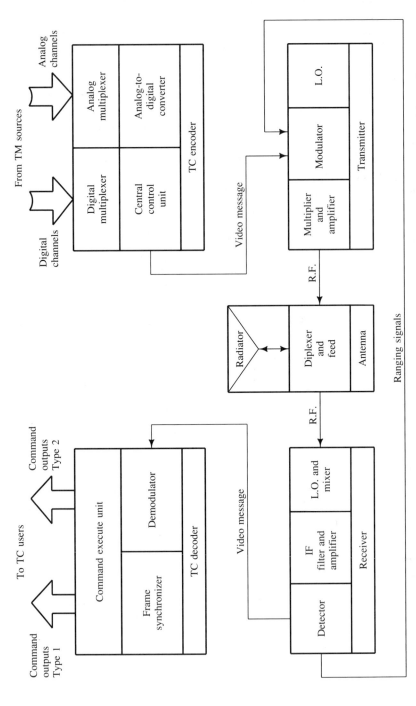

FIGURE 1.56   Example of TTC subsystem block diagram. (From R. Collette and B. L. Herdan, "Design Problems of Spacecraft for Communication Missions," *Proc. IEEE*, vol. 65, March 1977. © 1977 IEEE.)

round-trip phase shift or time delay. Commands are sent to the satellite from the earth stations to control apogee—the farthest point (see Fig. 1.54) on the satellite orbit from the ground—and to boost motor, attitude, orbit adjustment, and switching in communications systems. A typical TT&C unit is given in Fig. 1.56.

The azimuth is given by

$$Az = \tan^{-1}\left[\frac{\sin(\lambda_G - \lambda_S)}{\cos(\phi_G)\tan(\phi_s) - \sin(\phi_G)\cos(\lambda_G - \lambda_s)}\right] \tag{1.476}$$

where (see Fig. 1.53)

ϕ_G, ϕ_s = attitude of the earth stations and subsatellite point

λ_G, λ_s = latitude of the earth stations and subsatellite point

The slant range is given by

$$d = \frac{(R_e + h) - R_e\sin(\pi/2 - \beta)}{\cos(\pi/2 - \beta - E)}$$
$$= [(R_e + h)^2 + R_e^2 - 2R_e(R_e + h)\cos \beta]^{1/2} \tag{1.477}$$

where the elevation angle E is

$$E = \frac{\sin(\pi/2 - \beta) - R_e/(R_e + h)}{\cos(\pi/2 - \beta)} \tag{1.478a}$$

where 2β is the coverage angle (see Fig. 1.53a). The earth radius $R_e = 6370$ km. The coverage area is given by

$$A_{cov} = 2\pi R_e^2(1 - \cos \beta) \tag{1.478b}$$

If an earth station is in the coverage area, it can communicate with the satellite by pointing the earth station's antenna toward the satellite position. The satellite position, azimuth and elevation, is given by Eqs. (1.476) and (1.478a).

For geosynchronous (geostationary) satellites

$$h = 35,860 \text{ km}$$

$$\alpha \approx 17.3°$$

where α is sometimes referred to as the *look angle* or *global beam angle*. The angle illuminated on earth is

$$2\beta = 180° - 17.3° = 162.7°$$

The arc length illuminated on the earth is $GS'G' = 18,080$ km. The distance from the satellite to earth center is $R_e + h = 42,230$ km. The velocity of the satellite (see Fig. 1.57a) is given by

$$v = \left[2\mu\left(\frac{1}{r} - \frac{1}{2a}\right)\right]^{1/2} \qquad r = \frac{a(1 - e^2)}{1 + e\cos \phi} \tag{1.479}$$

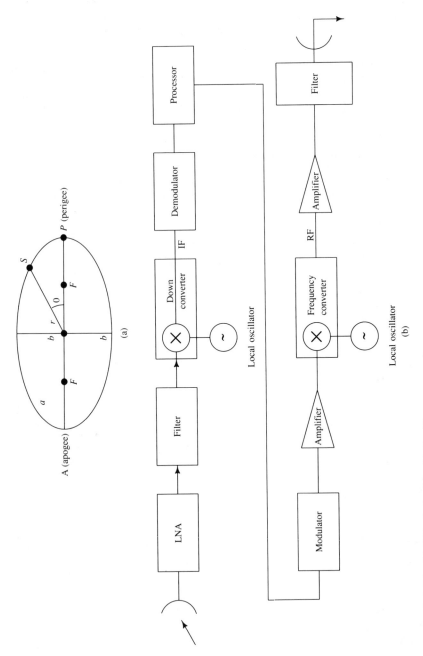

FIGURE 1.57 (a) Elliptical orbit. (b) Satellite transponder.

and the period T of the orbit is

$$T = \frac{2\pi a^{3/2}}{\sqrt{\mu}} \qquad \mu = 3.99 \times 10^{14} \text{ m}^3/\text{s}^2 \tag{1.480}$$

where a is the semimajor axis and

$\quad e = 0 \qquad$ for circular orbit

$\qquad = 1 \qquad$ for parabolic orbit

$\qquad > 1 \qquad$ elliptical orbit

For geosynchronous satellites $(e = 0)$,

$$r = (h + R_e) = 42{,}230 \text{ km}$$

$$v_{\text{syn}} = 11{,}070 \text{ km/h}$$

$$T_{\text{syn}} \approx 23 \text{ h } 56 \text{ m } 45 \text{ s}$$

$$\approx 24 \text{ h}$$

The minimum and maximum propagation times are as follows:

$$\tau_{\text{min}} = 2SS'/c = 0.239 \text{ s}$$

$$\tau_{\text{max}} = 2SG/c = 0.279 \text{ S}$$

where $SG = SG' = 42{,}230 \text{ km}$

$\qquad SS = 35{,}860 \text{ km}$

$\qquad c = 3 \times 10^8 \text{ m/s}$

S, S', and G are given in Fig. 1.53a. The transmission delay in general is given by

$$\tau = \frac{2(R_a + h) \sin \beta / \cos E}{c}$$

(see Fig. 1.54). The maximum latitude that can be served by the satellite is

$$\beta = \cos^{-1}\left(\frac{R_a \cos E}{R_a + h}\right) - E \tag{1.481}$$

When $E = 0$, the maximum north or south latitude is $81.25°$. Therefore, geosynchronous satellites cannot illuminate north or south poles. Equation (1.481) reveals that reasonable communication is possible if $E \geq 5°$.

The satellite is placed into geosynchronous orbit in two or three steps. Most expendable launch vehicles like Delta and Ariane put the satellite in an inclined elliptical orbit called a *transfer orbit* (Fig. 1.54), with an apogee at geosynchronous altitude and a 185.370-km perigee. At the transfer orbit apogee, the apogee kick motor puts the satellite into circular orbit with zero inclination to the equatorial plane. Some satellites are *spin stabilized*, main-

	I	II	III	IV	IV A	V	V A	VI
Intelsat Designation								
Year of First Launch	1965	1966	1968	1971	1975	1980	1983	1986
Prime Contractor	Hughes	Hughes	TRW	Hughes	Hughes	Ford Aerospace	Ford Aerospace	Hughes
Dimensions (Undeployed)								
Width, m	0.7	1.4	1.4	2.4	2.4	2.0	2.0	3.6
Height, m	0.6	0.7	1.0	5.3	6.8	6.4	6.4	5.3
Launch Vehicles	Thor Delta			Atlas Centaur		Atlas Centaur or Ariane 1.2		Sts or Ariane 4
Spacecraft Transfer Orbit Mass, kg	68	162	293	1,385	1,469	1,946	2,140	13,806 3750
Communications								
Payload Mass, kg	13	36	56	185	190	235	260	600
Power Eol Equinox	40	75	134	460	600	1,270	1,270	2,200
Design Lifetime, yr	1.5	3	5	7	7	7	7	10
Rated Voice Channel Capacity in AOR								
Half Circuits	480	480	2,400	8,000	12,000	25,000	30,000	80,000
Bandwidth, MHz	50	130	300	500	800	2,300	2,180	3,680
Antenna Beam Coverages								
C-Band	Toroidal Northern Only	Toroidal Almost Full Earth	Despun Earth Cov	Despun Earth Cov/and 2 Spots Steerable	Despun Earth Cov/and 2 Hemi	3 Axis Earth Cov 2 Hemi 2 Zone	3 Axis Dual Pol Earth Cov 2 Hemi 2 Zone 2 Spots	Despun Dual Pol Earth Cov 2 Hemi 4 Zone
Ku Band	NA	NA	NA	NA	NA	2 Spots Steerable	2 Spots Steerable	2 Spots Steerable
L Band	NA	NA	NA	NA	NA	Earth Cov	NA	NA

FIGURE 1.58 Evolution of Intelsat satellite. (From S. O. Bennett and D. J. Braverman, "Intelsat VI—Continuing Evolution," *Proc. IEEE*, vol. 72, November 1984. © 1984 IEEE.)

taining their orientation in space by means of a rotation of the body of the satellite along the axis parallel to the earth axis. Some satellites are *body stabilized*, maintaining their orientation by small jets and momentum wheels. Satellite weights range from 500 to 2000 kg.

1.8.2 The Satellite Payload

The satellite *payload* comprises communication *transponders* and antennas. The signal path from the receive satellite antenna to the transmit satellite path is called the transponder path. There are several transponders in a satellite. These transponders contain switching systems for transferring signals among multiple spot beams. A typical transponder consists of a low-noise amplifier, down-converter filter, high-power amplifier, and a filter. Some satellite transponders demodulate the signal and remodulate it and then amplify it before transmission to the ground. This process, called *on-board signal processing*, reduces amplitude, phase, and intermodulation distortion caused by nonlinearity of the amplifier and by ripples in the filter passband. A typical transponder is shown in Fig. 1.57b.

The received signal from the ground is weak and is passed through an LNA. These amplifiers have a flat band in a very wide band (500 MHz) and have a very low noise figure. They are usually GaAs FET amplifiers. The output is filtered through bandpass filters with a flat band over the signal bandwidth. Most practical filters such as Tchebychev or elliptical filters have ripples in the passband whereas the Butterworth filter has slow roll-off (6-dB/octave per pole) skirts; sharp-cutoff filters produce group delay distortion at the band edges. The signal is heterodyned by mixing with local oscillator, sometimes to baseband, for complete demodulation in case of regenerative repeaters. The signals are passed through HPAs such TWTAs and is transmitted to the earth using the downlink frequency. The transponder provides an amplification of 100–110 dB in several stages. An Intelsat transponder has a 500-MHz bandwidth in 12 channels. Each channel has a 36-MHz bandwidth, and there is a 4-MHz separation between the channels. With dual polarization, this transponder can provide 24 channels over the same 500-MHz bandwidth. The evolution of Intelsat system is shown in Fig. 1.58.

1.9 MULTIPLE-ACCESS METHODS

Since a satellite antenna beam can cover a large number of earth stations, these earth stations can simultaneously access the satellite for relaying their messages to other stations. These access methods can be fixed, demand assigned, or random. For fixed methods also known as preassigned or controlled, each earth station is assigned a frequency, time or code such that transmitted signals are orthogonal in terms of frequency, time or code. When the signals are transmitted at different RF bands at the same time, the access method is called the *frequency division multiple access* (FDMA). When the signals are transmitted at different time intervals in the same frequency, the access method is known as *time division multiple access* (TDMA). When the signals are assigned with different codes but with same frequency and time, the access method is known as the *code division multiple access* (CDMA). The CDMA system is also a random access method. In the demand assigned method, stations are assigned a frequency or time slot when a frequency or time slot is available. In random access or uncontrolled modes, earth stations access satellites at random time in the same frequency. The design goal of the multiaccess methods is the efficient use of satellite power and bandwidth. Other considerations are to accommodate as many users as possible, to provide connections in the least time, to adapt dynamically to new requirements, and to be reliable, survivable, and cost-effective.

1.9.1 Frequency Division Multiple Access

In FDMA, the transponder bandwidth is divided into several nonoverlapping bandwidths. Each earth station (user) is assigned a unique frequency and bandwidth. In order that signals from one earth station do not interfere with the signals from the other earth stations, there are frequency guard bands in between users' bands. These guard bands are kept large enough to avoid adjacent-channel interference but small enough not to waste the transponder bandwidth. Users send their messages to earth stations through a terrestrial network. At the earth station, these signals are multiplexed, either FDM or TDM. These multiplexed signals are modulated, say FM or PSK. These modulated signals are transmitted from various earth stations simultaneously at various frequencies. The transmitted signals can take various forms depending on the nature of multiplexing, modulation and multiple-access method. Various schemes are shown in Table 1.3. The most common configurations are DPCM-FDM-FM-FDMA, DM-TDM-QPSK–convolutional coded–Viterbi decoded–TDMA and PCM-FDM-FSK–half rate–coded SSMA systems. Two popular FDMA systems are *multiple channel per carrier* (MCPC) and *single channel per carrier* SCPC. The MCPC and SCPC systems can be analog or digital. The analog MCPC is SSB/FDM/FM/FDMA. The digital MCPC is usually ADPCM/TDM/PSK/FDMA. The frequency assignment for the MCPC system is shown in Fig. 1.59. The guard bands between adjacent frequencies are provided to prevent frequency drifts of the satellite and earth station oscillators.

In analog MCPC, individual voice channels are first single-sideband (SSB) modulated and then frequency-division multiplexed (FDM). The composite baseband signal is frequency-modulated (FM) and then propagated through an assigned FM carrier using FDMA transmission equipment. Several such FDMA signals are amplified through the satellite transponders at the same time. Since the satellite amplifiers are nonlinear, the output signals are degraded by intersymbol and adjacent channel noise. In order to reduce the intersymbol noise, the satellites are operated in the linear region. Satellite power is backed

TABLE 1.3 Transmission Systems Configurations

Baseband Processing (if applicable)	Multiplexing	Modulation	Coding (Optional)	Multiple-Access Method
PCM	FDM	SSB	Block	FDMA
DPCM			BCH	Fixed
ADPCM		FM	RS	Demand
DM	TDM		Golay	Assigned
ADM		PSK	Convolutional	
		DPSK	Concatenated	
		FSK	Rate	
		Coherent	(1/2, 1/3, 2/3, etc.)	TDMA
		FSK	Sequential decoding	Fixed Demand
		Noncoherent		Assigned
			Viterbi decoding	CDMA (SSMA)
				DS
				FH
				TH
				Random access

off from the output saturation point. Typically, this reduction of average output power is 50 percent or more. The amplified signals are transmitted to the ground stations (earth station) in separate downlink frequencies. The earth stations demodulate and demultiplex the desired signal.

In digital MCPC systems, the baseband signals are converted to adaptive pulse code modulation (ADPCM) and then the bit streams from the various

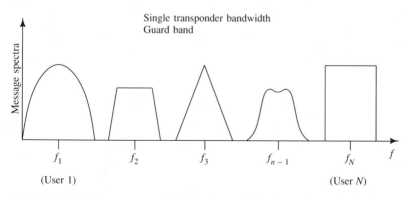

FIGURE 1.59 Frequency assignment for multiple-channel-per-carrier FDMA system: N earth stations with N distinct frequencies.

users are time-division multiplexed. The composite signal is further coded for error correction and privacy. The coded signal is modulated on a (four) phase-shift keying carrier. Several such signals are transmitted using FDMA equipment to the satellite. In an SCPC system, each signal (voice/data) is transmitted through an individual carrier. In this scheme, no multiplexing is done at the earth station. Typical single-channel-per-carrier (SCPC) systems use companded frequency modulation (CFM), delta modulation, or pulse code modulation with PSK on an RF carrier. In this type of system, the carrier is usually voice activated, saving 60 percent of the satellite power. A 36-MHz transponder can handle 800 to 1600 simultaneous SCPC channels. SCPC systems are preferable when there is only a small number of users per earth station. MCPC systems are preferable for a large number of users at each earth station.

The carrier-to-noise calculation for FDMA system is given below.

$$\left(\frac{C}{N}\right)_{\text{total}}^{-1} = \frac{1}{(C/N)_{\text{up}}} + \frac{1}{(C/N)_{\text{down}}} + \frac{1}{(C/N)_{\text{intermod}}} \tag{1.482}$$

where

$$\left(\frac{C}{N}\right)_{\text{up}} = F_s + (G/T)_s - 10\log_{10}\left(\frac{4\pi}{\lambda^2}\right) - 10\log_{10}k$$
$$- 10\log_{10}(B_{\text{IF}}) - BO_i \text{ (dB)}$$
$$= (\text{EIRP})_E - L_{\text{up}} + (G/T)_s - 10\log_{10}k$$
$$- 10\log_{10}(B_{\text{IF}}) - BO_i \text{ (dB)}$$

$$\left(\frac{C}{N}\right)_{\text{down}} = \text{EIRP}_S - L_{\text{down}} + (G/T)_E - 10\log_{10}k$$
$$- 10\log_{10}(B_{\text{IF}}) - BO_0 \text{ (dB)}$$

$$\left(\frac{C}{N}\right)_{\text{intermod}} = \text{carrier-to-intermodulation noise ratio}$$

F_s = single-carrier saturation flux density at the beam center (dBW/m^2)

$\quad = (\text{EIRP})_E/4\pi R_u^2$

$(G/T)_s$ = satellite gain over temperature (*figure of merit* of the satellite antenna)

k = Boltzmann's constant (dB)

B_{IF} = IF bandwidth (dB)

BO_i = input backoff power (dBW)

BO_0 = output backoff power (dBW)

$(G/T)_E$ = earth station antenna gain over temperature

R_u = uplink distance from the earth station to the satellite

L_{up} = uplink path loss = $20\log_{10}(4\pi R_u/\lambda_u)$

L_{down} = total path loss in the downlink

$EIRP_s$ = satellite EIRP (effective isotropic radiated power) for saturated single-carrier operation

λ = wavelength (m)

The *flux density F* is defined as

$$F = \frac{GP}{4\pi R^2} \qquad (1.483)$$

where

G = gain of the antenna [see Eq. (1.537)]

P = power of the transmitting station

R = distance between earth station and satellite

The carrier-to-noise ratio can be expressed as

$$C/N = \left(\frac{E_s}{N_0}\right)\left(\frac{R_s}{B}\right) \qquad (1.484)$$

where

E_s = symbol energy

R_s = symbol rate

N_0 = noise density

B = bandwidth

E_s/N_0 is known as symbol energy-to-noise ratio. For digital communication, this ratio is determined from the type of modulation for a given symbol error rate, see Section 1.4.

1.9.2 Time Division Multiple Access

In TDMA systems, each station is assigned a time slot in a transmission duration called a *frame*. Each station's transmission is known as *burst*, a finite set of binary bits. In TDMA systems, all the earth stations use a single uplink frequency to transmit their messages (data, voice, facsimile, etc.) and receive signals from the satellite on a single downlink frequency. At any time, only one signal is present in the satellite, and each station uses the transponder's whole bandwidth. Since the transponder amplifies only one signal during a time slot, there is no intermodulation noise. There is no need to bring the amplifier into the linear region. Hence in TDMA the satellites operate at the saturation point. The satellite capacity can be used efficiently in terms of power and bandwidth. In FDMA the system may have to back off 3 to 6 dB of the available power, whereas TDMA systems use 90 percent of the available power. However, in order to avoid any overlapping of time slots from various users, there are guard time bands between adjacent time slots. Since all the earth stations are dispersed over a large area and the satellite has certain anomalies, the coordination of these bursts in the right time slots requires stringent synchronization systems. A typical TDMA frame is shown in Fig. 1.60a.

A frame begins with reference burst(s) and ends with burst from the Mth station, and this pattern is repeated continuously. The *reference burst* (RB) consists of carrier and bit timing recovery bits (CBR); unique word (UW); teletype (TTY); service channel (SC); voice order wire (VOW); and control and delay channel bit (CDC). The CBR sequence provides the carrier and timing reference for the demodulation at the receiver. The UW is used to identify the starting position of the burst in the frame and the position of bits in the burst. The VOW bits are used for interstation communication. The TTY bits carry *parameter information* for network control and SC bits are used for error analysis when the system is in service. The CDC bits are used for transmitting information regarding network management such as acquisition, synchro- nization, and system monitoring. Each traffic burst has this information in addition to information bits. Non–information bits are called *preamble* or *overhead* bits. One of the earth stations is chosen to transmit *reference bursts* and other stations follow the reference burst in a controlled manner such that there will be no collision of various bursts emanating from the various stations.

The UW sequence determines the start of each frame and serves to remove phase ambiguity for quadrature phase-shift keying (QPSK) modulation. The detection of UW sequences is done by correlating the received UW sequence[n] with the stored UW sequence at the receiving station. If the UW sequence is received without any error, then the correlation output will be equal to the length of the sequence. Let N be the length of the sequence and n be the number of places the received sequence has errors. Let p be the error in a single place. The probability of n number of errors is given by

$$P(n \text{ errors}) = \binom{N}{n} p^n (1-p)^{N-n} \qquad 1 \le n \le N, 0 \le p \le 1$$

---

[n] See Chapter 2 for sequence with high main lobe and small sidelobe. Such sequences are Barker code and maximum length sequences. *Barker code* of length 13 is $(1, 1, 1, 1, 1, -1, -1, 1, 1, 1, -1, 1, -1, 1)$. Note that

$$\sum_{i=1}^{13-j} c_i c_{i+j} = 13 \qquad j = 0$$

$$= 0 \qquad j \neq 0$$

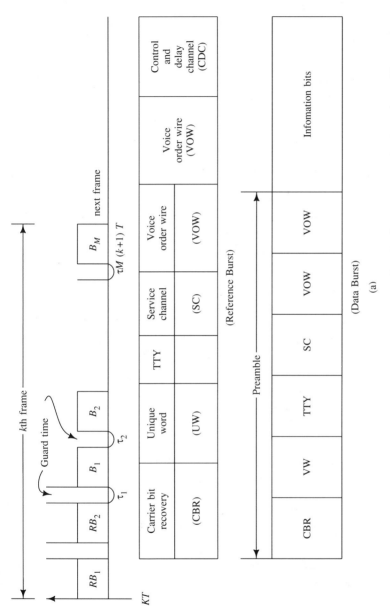

FIGURE 1.60 (a) TDMA frame structure. (b) Determination of D_N for frame synchronization. (c) Satellite-switched TDMA. (d) Transponder capacity for FDMA and TDMA versus number of accesses. (b, c, d from S. J. Campanella and J. V. Carrington, "Satellite Communications Networks, *Proc. IEEE*, vol. 72, November 1984. © 1984 IEEE.)

167

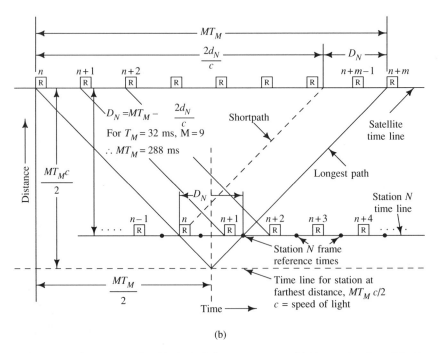

(b)

FIGURE 1.60 (*Continued*)   (a) TDMA frame structure. (b) Determination of D_N for frame synchronization. (c) Satellite-switched TDMA. (d) Transponder capacity for FDMA and TDMA versus number of accesses. (b, c, d from S. J. Campanella and J. V. Carrington, "Satellite Communications Networks, *Proc. IEEE*, vol. 72, November 1984. © 1984 IEEE.)

Let e be the total number of errors; then

$$P(n \le e) = P(\text{errors}) = \sum_{n=0}^{e} \binom{N}{n} p^n (1 - p)^{N-n} \tag{1.485}$$

The system is designed to have a maximum of E errors. Therefore, the probability of more than e' errors is given by

$$P(\text{no detection}) = 1 - \sum_{n=0}^{e'} \binom{N}{n} P^n (1 - P)^{N-n}$$

$$= \sum_{n=e'+1}^{N} \binom{N}{n} p^n (1 - p)^{N-n} \tag{1.486}$$

If the bits are equally likely to be erased, then

$$p = 1/2 \tag{1.487a}$$

Hence, the error in detecting the presence of frame is given by

$$P(\text{no detection}) = \frac{1}{2^N} \sum_{n=e'+1}^{N} \binom{N}{n} \tag{1.487b}$$

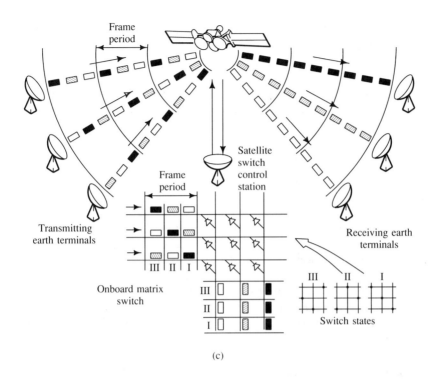

Frame period

Satellite switch control station

Frame period

Transmitting earth terminals

Receiving earth terminals

Onboard matrix switch

III II I

III

II

I

Switch states

(c)

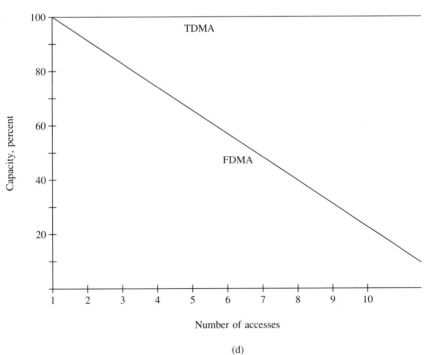

TDMA

FDMA

Capacity, percent

Number of accesses

(d)

FIGURE 1.60 (*Continued*)

It should be emphasized that the UW sequences for the reference burst and traffic burst B_m, $1 \leq m < M$, are all different.

It is assumed that the reference burst can be received by all earth stations. Each earth station sends an acquisition burst and correlates with the received acquisition burst to determine its local timing. The detection of the UW sequence and the delay time of the individual station's burst will determine the station's access time.

1.9.3 Burst Synchronization and TDMA Performance

The problem of controlling burst transmission timing at each access station, in order to prevent overlapping of bursts at the satellite transponder, is called *burst or frame synchronization*. The burst transmission at each station can be performed at Mth station by delaying a certain amount of time D_M after receiving the start of receiving frame (SORF) markers from the satellite such that twice the propagation time of SORF to the earth station plus the station delay D_M remain constant for all M stations. This constant is 288 ms for the Intelsat network. The D_M numbers are transmitted to the earth stations through the CDC channel in the reference bursts. This synchronization information is transmitted back to an earth station from a reference station if the earth station cannot listen to its own transmission. This scheme is known as *feedback closed-loop synchronization*. An earth station can listen to its own transmission when a satellite uses global beam to cover all the earth stations. In this case of global beam coverage, the burst synchronization is known as *direct closed-loop synchronization*. In the case where there is spot beam coverage and no reference station available to transmit D_M information, then the station uses a fixed, constant delay D_M. This scheme is known as *open-loop synchronization*. In the *open-loop synchronization* the constant delay D_M at such earth stations may be more than the guard time. The guard time has to be increased to accommodate the burst timing of these stations. The scheme is simple but it reduces the TDMA frame efficiency.

In Intelsat's TDMA architecture (see Fig. 1.60b) the delay time at the mth station is given by

$$D_m = k_1 T_f - \frac{2d_m}{c} \tag{1.488}$$

where

T_f = frame duration

k_1 = the least integer which will yield positive D_m.

$k_1 T_f$ = 228 ms for Intelsat Network

d_m = distance between the satellite and the transmitting earth station,

c = velocity of light = 2.99×10^8 m/s

k is selected such that delay D_m is the smallest, positive quantity.

In the above discussion for FDMA and TDMA, it is assumed that earth stations are covered by satellite's global antenna beam. However, some satellites use spot beams (multiple beams). The spot beams have two advantages: (1) higher gain and (2) frequency reuse. When spot beams are used, the various earth stations are connected by switching from the one beam to the other beam. This switching is done at the satellite. These systems are known *satellite switched (SS) FDMA or TDMA* (see Fig. 1.60c) depending upon the multiple-access systems. It is necessary to have enough beam isolation such that there will be no beam-to-beam interference.

To evaluate the performance of TDMA systems, we define the following:

$$
\left.
\begin{aligned}
b_g &= \text{number of bits in the guard band} \\
b_p &= \text{number of bits in the preamble in traffic burst} \\
b_r &= \text{number of bits in a reference burst} \\
b_T &= \text{total number bits in a frame} \\
T_F &= \text{frame duration} \\
M &= \text{number of earth stations} \\
R_T &= \text{TDMA bit rate} = b_T/T_F \\
R_b &= \text{bit rate} = k/T_F = 1/T_B \\
R_p &= \text{preamble bit rate} = b_p/T_F \\
R_g &= \text{guard bit rate} = b_g/T_F \\
R_r &= \text{reference bit rate} = b_r/T_F \\
K &= \text{number of bits in burst time } T_b \\
n_r &= \text{number of reference earth stations} \\
T_B &= \text{burst time}
\end{aligned}
\right\}
\tag{1.489}
$$

The efficiency η of a TDMA system is given by

$$
\eta = 1 - \frac{Mb_p + n_r b_r + (n_r + M)b_g}{R_T T_F}
\tag{1.490}
$$

The bit rate for information-oriented data is given by

$$
R_i = R_T - n_r(R_r + R_g) - M(R_p + R_g)
\tag{1.491}
$$

The voice channel capacity is given by

$$
C = \left[\frac{R_i}{R_v} \right]_{\text{int}}
\tag{1.492}
$$

where R_i is given by Eq. (1.491), R_v is the voice channel bit rate, and $[\,\cdot\,]_{\text{int}}$ denotes the largest integer value less than $[\ \ ]$.

1.9.4 Code Division Multiple Access

In the CDMA scheme, all earth stations transmit their signals in a single uplink frequency simultaneously and continuously, without sharing frequency or time with other users. Let us denote transmitted signals as $S_m(t)$, $0 \le t \le T$, $m = 1, \ldots, M$. In order for the desired signal to be extracted from the rest of the signal it is necessary that

$$\int_0^T S_j(t)S_m(t)\, dt = 0 \qquad j \ne m$$

$$= E \qquad j = m \tag{1.493}$$

where it is assumed that all transmitted signals have equal energy E. Let us denote the jth signal as

$$S_j(t) = m_j(t)c_j(t)\cos(\omega_c t + \phi) \tag{1.494a}$$

$$= \cos\{\omega_c[c_j(t)]t + \phi[m_j(t)]\} \qquad 1 \le j \le M \tag{1.494b}$$

where $m_j(t)$ is the baseband message, $c_j(t)$ is a codeword for the jth earth station, $1 \le j \le M$, $\omega_c = 2\pi f_c$, f_c is the carrier frequency, and ϕ is the phase. Let us denote R_b as the bit rate of the message $m_j(t)$ signal and R_c as the code (chip) rate for the code signal $c_j(t)$, for all j. Equation (1.494a) is called direct-sequence (DS) CDMA and Eq. (1.494b) is called the frequency-hopping (FH) CDMA. The message bandwidth is chosen to be much smaller than the code bandwidth; $R_c > R_b$. The RF signal in either Eq. (1.494a) or Eq. (1.494b) has a larger bandwidth than the bandwidth uncoded signal $m(t)\cos(\omega_c + \phi)$. For this reason, CDMA systems are also called spread spectrum multiple access (SSMA) systems (see Chapter 5). For FHCDMA, the frequency of the transmitted signal hops every T_c ($= 1/R_c$) seconds. Note that $T_c < T_b$, where T_b is the duration of the message signal. $N = T_b/T_c$ is called the number of hops for the FHCDMA. For the DSCDMA, the bandwidth is the convolution of message bandwidth with code bandwidth. The result of the multiplication of message and code sequences yields a larger bandwidth than the message bandwidth. At the receiver the signal is multiplied with the exact code word and is passed through a low-pass filter to strip off the code sequence. Let us denote the received signal as

$$y(t) = \sum_{j=1}^M S_j(t) + n(t) \tag{1.495}$$

The correlator output is given by

$$z = \int_0^T y(t)S_j(t)$$

$$= E + \sum_{m \ne j} \int_0^T S_j(t)S_m(t)\, dt + \int_0^T n(t)S_j(t)\, dt$$

$$\triangleq E + I + \int_0^T n(t)S_j(t)\, dt = E + I + n_1 \tag{1.496}$$

The power of n_1 is very small compared to the interference power of I and can be neglected. Let us denote B as the bandwidth of the CDMA signal. The noise density of the undesired signal and noise is given by

$$N_0 = \frac{I}{B} \tag{1.497}$$

Let P be the individual carrier power. For a binary system, the energy-per-bit noise power density is given by

$$\frac{E_b}{N_0} = \frac{PT_b}{N_0} = \frac{P}{R_b} \cdot \frac{B}{I} \tag{1.498}$$

where T_b is the bit duration.

The ratio of the undesired signal (sometimes called jammer) power to the desired signal power is

$$\frac{I}{P} = \frac{B/R_b}{E_b/N_0} \tag{1.499}$$

I/P is called the *jamming margin* and B/R_b is called the *processing gain*, G; that is,

$$G = B/R_b \tag{1.500}$$

When there are M users and each has same carrier power P,

$$I = (M - 1)P \tag{1.501a}$$

Therefore, the number of users is given by

$$M = 1 + \frac{B/R_b}{E_b/N_0} \tag{1.501b}$$

where E_b/N_0 is the desired bit energy to noise ratio, B is the CDMA bandwidth, which is equal to the transponder bandwidth, and R_b is the bit rate $(1/T_b)$. For $B = 36$ MHz and $R_b = 36$ kbps, the processing gain $G = 30$ dB. For a bit error rate of 10^{-5}, $E_b/N_0 = 9.6$ dB for BPSK modulation and $M = 110$. It can be shown that the probability of bit error for a direct sequence BPSK system, with code rate r_c and minimum distance $2t + 1$, is approximately given by

$$P_e = \frac{1}{2} \text{erfc} \left[\left(\frac{B/R_b}{I/P} r_c (2t + 1) \right)^{1/2} \right]$$

where

$$E_b = PR_b$$

$$\text{erfc } x = \frac{2}{\sqrt{\pi}} \int_x^{\infty} e^{-t^2} \, dt$$

1.9.5 Demand-Assigned Multiple Access

Fixed-assigned FDMA or TDMA systems are not efficient if volume of traffic is low for some earth stations. In demand-assigned multiple-access (DAMA) systems, frequency or time slots are assigned when there is a demand from the users and are taken back after the duration of the use. The *S*ingle channel per carrier *P*CM multiple-*A*ccess *D*emand *E*quipment (SPADE) is an example of FDMA where the 36-MHz bandwidth of the Intelsat system is assigned on demand. A *common signaling channel* (CSC) operating in a TDMA mode serves as an order wire for access request and assignment. Similarly the 14/12 GHz Telecom, the French public network, is an example of demand-assigned TDMA system. Also, the satellite business system (SBS) is demand-assigned TDMA in the US. A variation of DAMA is called a *priority-oriented demand-assigned access* (PODA) system. In PODA systems, the frequency or time slots are assigned on priority basis depending on users' influence or urgency.

1.9.6 Comparison

Let us assume, for comparison, the power of a satellite transponder is the same for TDMA and FDMA systems. Each system uses a digital modulation. The input bit rate is R_b for a FDMA system. Let the transmitted bit rate for the TDMA system be R_T; see Eq. (1.489). For the same E_b/N_0 and G/T, and other losses at the satellite for uplink transmission, the power of the earth station for TDMA and FDMA systems are given by

$$(\text{Power})_{\text{TDMA}} = (\text{Power})_{\text{FDMA}} + R_T - R_b \qquad (1.502)$$

The capacity of a 36 MHz transponder for a FDM-FM-FDMA is 400 telephones and for a PCM-QPSK-TDMA 1000 telephones. The transponder capacity for FDMA and TDMA is shown in Fig. 1.60d.

1.10 PROPAGATION EFFECTS AND PERFORMANCE EVALUATION

The transmitting antenna converts electric energy to electromagnetic energy. The information-bearing electromagnetic energy is radiated in the form of electromagnetic waveforms. These waves are traveling waves of the electric and magnetic fields that vary with time and distance in the direction of propagation. The electric field and magnetic field, which are perpendicular to each other, are orthogonal to the direction of propagation. If the z axis is taken as the direction of propagation, then electric field E is along the x axis and magnetic field is along the y axis. In free space E and H are related by a simple constant, $E = 120\pi H$. So we can represent the electromagnetic field by the E field alone. Let us denote it by

$$E = E_x i_x + E_y i_y$$

where i_x and i_y are unit vectors along the x and y axes. If $E_y = 0$, the wave is called *horizontally polarized*. If $E_x = 0$, then the wave is *vertically polarized*. These polarizations are known as *linearly polarized*. The same frequency can be used to transmit signals by the simultaneous use of horizontal and vertical polarizations. This process is known as *frequency reuse*. Another set of orthogonal polarizations consists of *right-hand circular polarization* (RHCP) and *left-hand circular polarization* (LHCP) and is represented by

$$E_{\text{RHCP}} = E_0 \sin(2\pi f_c t + \pi/2)i_x + E_0 \sin(2\pi f_c t)i_y$$

$$E_{\text{LHCP}} = E_0 \sin(2\pi f_c t)i_x + E_0 \sin(2\pi f_c t + \pi/2)i_y$$

where f_c is the carrier frequency and E_0 is the magnitude. A *wave front* is defined as a surface joining all points of equal phase. If an electromagnetic wave is radiated equally in all directions from a point source in free space, known as *isotropic* radiation, the energy received at any point in a spherical wave front is the same. If P_t is the transmitted power, then the power density P_d on a surface of a sphere of a radius R is given by $P_d = P_t/4\pi R^2$. If R increases, then P_d diminishes. In addition to this radiation loss, there are other losses.

1.10.1 Propagation Path Modeling

Ultraviolet (UV) radiation from the sun causes ionization of the air into free electrons and positive and negative ions. The ionization extends from 50 km to 2000 km above the earth. It varies with the sun's position and time. The atmosphere has three ionized regions: D, E, and F. The lower region D extends from ground level to a height of 50 to 95 km. The region is also called the *troposphere*. In this region the air is in permanent motion and meteorological phenomena take place. The E zone is above the D zone and extends from 95 to 150 km above the D zone. This region is also known as *stratosphere*. The temperature in this zone is constant, and humidity is absent. The F layer extends from 150 to 2000 km above the E layer. The E zone contains ionized particles and is called the *ionosphere*. Beyond the ionosphere is *free space*. The process of ionization is mainly the result of electromagnetic radiation from the sun, changed particles from the solar region, and galactic cosmic rays. The degree of ionization, the number of electrons per unit volume, is about 10^9 electron/m^3 in the region D, 10^{11} electrons/m^3 in region E and 10^{12} electrons/m^3 in the region F during the daytime. It fades away at night. Ionization depends mainly on the zenith angle x of the sun and is a function of $(\cos x)^4$. It is given by $N_S = \int_S n(s)\,ds$, where $n(s)$ is the electron density and N is the *total electron content* (TEC) at the position S in the space. The ionization causes scattering, absorption, multipath disturbances, polarization, and change of angle. The presence of free electrons along the path causes the signal to travel with a group velocity less than the in vacuo speed of light. This retardation suffered by the signal is commonly called the *group delay*.

The group delay is given by

$$\Delta\tau = (1.33 \times 10^{-7}/f_c^2)N_S \tag{1.503}$$

where f_c is the carrier frequency and N_S is the TEC. When the signal has a significant bandwidth, the propagation delay causes *dispersion*. The dispersion is inversely proportional to frequency cubed. The signal attenuation is given by

$$L_a = \frac{1.16 \times 10^{-3}N_S v}{f_c^2} \quad \text{(dB/km)} \tag{1.504}$$

where v is the collision frequency per second.

Scintillation of signals refers to the fluctuation of signal amplitude, phase, frequency, and angle of arrival. Scintillation of signals passing through the ionosphere is due mainly to field-aligned time-dependent irregularities along the transmission path. We model the scintillation channel as $h(t, \tau)$, which is the response of the channel at time t due to an impulse applied at $t - \tau$, and τ is the delay. The channel $h(t, \tau)$ is a complex-valued Gaussian random process. The transfer function of the channel is defined as

$$H(f, t) = \int_{-\infty}^{\infty} h(\tau, t)\exp(-2\pi f\tau)\,d\tau \tag{1.505}$$

If $u(t)$ is the complex envelope of the transmitted waveform, then the output envelope $z(t)$ is given by

$$\begin{aligned} z(t) &= \int_{-\infty}^{\infty} h(t - \tau, t)u(\tau)\,d\tau \\ &= \int_{-\infty}^{\infty} H(f, t)U(f)\exp(i2\pi ft)\,df \end{aligned} \tag{1.506}$$

where

$$U(f) = \int_{-\infty}^{\infty} u(t)e^{-i2\pi ft}\,dt$$

the spectrum of the transmitted signal envelope. The output (received) envelope can be written by

$$z(t) = x(t) + iy(t) \tag{1.507}$$

The joint probability distribution of x and y is given by

$$f(x, y) = \frac{1}{2\pi\bar{s}}\exp\left[\frac{-(x^2 + y^2)}{2\bar{s}}\right]$$

where

$$\bar{s} = E[\tfrac{1}{2}|z|^2] \tag{1.508a}$$

Let

$$R = \sqrt{(x^2 + y^2)}$$
$$G = R^2/2 \tag{1.508b}$$
$$\theta = \tan^{-1}(y/x)$$

The PDFs of R and θ are given by

$$f_R(r) = \frac{r}{s}\exp\left(\frac{-r^2}{2s}\right) \quad \text{(Rayleigh)} \qquad r \geq 0 \tag{1.508c}$$

$$f_\theta(\phi) = \frac{1}{2\pi} \qquad\qquad 0 \leq \phi \leq 2\pi$$
$$\qquad\qquad\qquad\qquad\qquad\qquad\qquad \tag{1.508d}$$
$$= 0 \qquad\qquad \text{elsewhere}$$

$$f_G(g) = \frac{1}{s}\exp\left(\frac{-g}{s}\right) \quad \text{(exponential)} \qquad 0 \leq g < \infty \tag{1.508e}$$

If the mean of x or y is nonzero, there is a dominant component known as *specular* or nonfading component S_0. The PDF of R is given by

$$f_R(r) = \frac{r}{S_f}\exp\left(\frac{-S_0}{S_f}\right)\exp\left(\frac{-r^2}{2S_f}\right)I_0\left(\frac{r\sqrt{2S_0}}{S_f}\right) \tag{1.508f}^o$$

where

$$I_0(\cdot) = \text{modified Bessel function}$$

$$S_f = \text{mean power in fading (\textit{diffuse}) component}$$

Let us assume that the channel impulse response function $h(t, \tau)$ is wide-sense stationary. We denote the covariance of the *channel impulse function* as

$$R_h(\tau_1, \tau_2, t_1, t_2) = \tfrac{1}{2}E[h^*(t_1, \tau_1)h(t_2, \tau_2)]$$
$$= R_h(\tau_1 - \tau_2, t_1 - t_2) \tag{1.509}$$

If the delays from various scatters are uncorrelated, then

$$R_h(\tau_1 - \tau_2, t_1 - t_2) = R_h(\tau_1, t_1 - t_2)\delta(\tau_1 - \tau_2) \tag{1.510}$$

If $t_1 - t_2 = 0$, then $R_h(\tau_1, 0)$ is average power of the channel at the delay τ_1. The range of values for which $R_h(\tau_1, 0)$ is nonzero is called the *multipath time spread* of the channel. This measures the pulse spreading when the pulses are transmitted through the channel. We denote this range by T_m. We define the *covariance function* of the channel transfer function as

$$R_H(f_1, f_2, t_1, t_2) = E[\tfrac{1}{2}H^*(f_1, t_1)H(f_2, t_2)]$$
$$= R_H(f_1 - f_2, t_1 - t_2) \tag{1.511}$$

---

^{o} Rice density, see Eq. (1.135a), $E(x) = S_0$, $\text{var}(x) = S_p = \bar{s}$

assuming that the channel is wide-sense stationary. If we set $t_1 = t_2$, then $R_H(f_1 - f_2, 0)$ is the frequency correlation function of the channel. The range of the frequencies for which $R_H(f_1 - f_2, 0)$ is nonzero is called the *coherence bandwidth* of the channel. It can be seen that

$$R_H(\Delta f, 0) = \int_{-\infty}^{\infty} R_h(\tau_1, 0) e^{-i2\pi f t_1} \, dt_1 \tag{1.512}$$

where $\Delta f = f_1 - f_2$. We denote the bandwidth of $R_H(\Delta f, 0)$ by $(\Delta f)_c$. It can be noted that this bandwidth is proportional to the reciprical of multipath time spread T_m. If the coherence bandwidth $(\Delta f)_c$ is greater than the bandwidth B of the transmitted signal, then the frequency components of the transmitted signal fade uniformly. Such fading is called *flat fading or nonselective fading*. In this case,

$$H(f_1, t) = H(f_2, t) = \alpha(t) \tag{1.513}$$

Using Eq. (1.513), Eq. (1.506) gives the output envelope as

$$z(t) = \alpha(t) u(t) \tag{1.514}$$

In this case, fading is multiplicative. If the coherence bandwidth $(\Delta f)_c$ is less than B, the bandwidth of the signal, then the signal is distorted severely. This fading is called the *frequency-selective fading* or *nonflat fading*. In this case, the channel injects intersymbol interference. Let us denote the power spectral density of the channel transfer function as

$$S_H(\Delta f, \lambda) = \int_{-\infty}^{\infty} R_H(\Delta f, \Delta t) \exp(-i2\pi\lambda \, \Delta t) \, d(\Delta t) \tag{1.515}$$

where $\Delta t = t_1 - t_2$, the Fourier transform of the covariance of the channel transfer function defined in Eq. (1.511). When $\Delta f = 0$, $S_H(0, \lambda)$ is called the *Doppler spread power spectrum*. The range of values for which $S_H(0, \lambda)$ is nonzero is called the *Doppler spread bandwidth*. It is denoted by B_d. Equation (1.515) shows that $S_H(0, \lambda)$ is the Fourier transform of $R_H(0, \Delta t)$, the channel transform time correlation function. The range of values for which $R_H(0, \Delta t)$ is nonzero is called the *coherence time or decorrelation time*. The coherence time $(\Delta t)_c$ is proportional to the reciprocal of the Doppler spread B_d. If the coherence time $(\Delta t)_c > T$, the symbol duration, the fading is called *slow*. When $(\Delta t)_c < T$, the fading is called *fast fading*. Define the *spread factor*

$$L = T_m B_d \tag{1.516}$$

where T_m is the multipath spread and B_d is the Doppler spread. When $L < 1$, the channel is called *unspread*. When $L > 1$, the channel is called *overspread*. When the channel is slow and fading is flat, $L < 1$. Taking the Fourier transform $R_h(\Delta \tau, \Delta t)$, the channel impulse response covariance function defined in Eq. (1.509), we get

$$S_h(\Delta\tau, \lambda) = \int_{-\infty}^{\infty} R_h(\Delta\tau, \Delta t) \exp(-i2\pi\lambda\,\Delta t)\,d(\Delta t)$$

where

$$\Delta\tau = \tau_1 - \tau_2 \quad \text{and} \quad \Delta t = t_1 - t_2 \tag{1.517}$$

It can be proved that

$$S_h(\Delta\tau, \lambda) = \int_{-\infty}^{\infty} S_H(\Delta f, \lambda) \exp(i2\pi\,\Delta\tau\,\Delta f)\,d(\Delta f)$$

$$= \int_{-\infty}^{\infty}\int_{-\infty}^{\infty} R_H(\Delta f, \Delta t) \exp[-i2\pi\,\Delta\tau]\exp[i2\pi\,\Delta\tau\,\Delta f]\,d(\Delta f)\,d(\Delta t) \tag{1.518}$$

using Eq. (1.515). $S_h(\Delta\tau, \lambda)$ is called the *scattering function* of the channel.[p]
Integrating the scattering function with respect to $\Delta\tau$, we get the *power spectrum*

$$V_h(\lambda) = \int_{-\infty}^{\infty} S_h(\Delta\tau, \lambda)\,d(\Delta\tau) \tag{1.519}$$

The *rate of Rayleigh fading* is proportional to the *RMS bandwidth* f_N of $V_h(\lambda)$.
The RMS bandwidth is given by

$$f_N = \left[\frac{\int_{-\infty}^{\infty} \lambda^2 V_h(\lambda)\,d\lambda}{\int_{-\infty}^{\infty} V_h(\lambda)\,d\lambda}\right]^{1/2} \tag{1.520}$$

For telephone fading channels, the Rummler (1979) model is given by

$$H(f) = \alpha\{1 - \beta\exp[-2\pi(f - f_m)\tau]\} \tag{1.521}$$

where α is the overall attention parameter, β is called the *diffuse parameter*, τ
is the time difference between the direct path and the multipath, and f_m is the
frequency when the fading is minimum. The measure of scintillation is given
by the *scintillation index*

$$S_X = \sigma_x/m_x \tag{1.522a}$$

where m_x and σ_x are, respectively, the mean and standard deviation of the
received signal. For tropospheric scintillation,

$$\sigma_x = [2.5 \times 10^{-2} \times f_c^{7/12}(\operatorname{cosec}\theta)^{0.85}(G(D)]^{1/2}\ dB \tag{1.522b}$$

where f_c is the carrier frequency, θ is the elevation angle, and $G(D)$ is an
antenna aperture averaging factor determined from the antenna diameter D.

---

[p] See Proakis (1983).

For a digital modulation, the probability of bit error rate is given by $P_b(r_b)$, where $r_b = \alpha^2 E_b/N_0$ and the PDF of α^2 is given by Eq. (1.508e). Therefore, the bit error rate is given by

$$P_e = \int_0^\infty P_b(r_b) f(r_b) \, dr_b \qquad (1.523a)$$

For DPSK modulation, the probability of the bit error is

$$P_e = \frac{1}{2(1 + \bar{S})} \qquad (1.523b)$$

for Rayleigh slow and flat fading.

1.10.2 Loss Due to Precipitation and Cloud (CCIR)

Let

R_p = point rainfall rate (mm/h)

h_0 = height above mean sea level (km)

θ = elevation angle (degrees)

l = latitude of earth station (degrees)

The isothermal height h_F is given by

$$h_F = 5.1 \cos(1.06l)$$

The rain height h_R is

$$
\begin{aligned}
h_R &= 0.6 h_F & l \leq 20° \\
&= 0.6 + 0.02(l - 20) & 20 \leq l \leq 40° \qquad (1.524a) \\
&= h_F & l \geq 40°
\end{aligned}
$$

The path length L_S is

$$
\begin{aligned}
L_S &= \frac{2(h_R - h_0)}{\left[\sin^2\theta + \dfrac{2(h_R - h_0)}{8500}\right]^{1/2} + \sin\theta} & \theta < 10° \\[2mm]
&= \frac{h_R - h_0}{\sin\theta} & \theta > 10° \qquad (1.524b)
\end{aligned}
$$

Let the slant length L_g be

$$L_g = L_S \cos\theta \qquad \text{km} \qquad (1.524c)$$

The point rainfall rate varies with location of earth station. The rainfall rate around the earth has been tabulated in the CCIR *Handbook on Satellite*

Communications (1985). The rainfall rate zone is shown in Fig. (1.61).[q] The corresponding rain rate exceeded (mm/h) is given in Table 1.4. Let

$$r_p = \frac{90}{90 + 4L_g} \tag{1.524d}$$

The attenuation for 0.01 percent of the time is given by

$$A_{01} = k(R_p)^\alpha L_S r_p \quad \text{dB} \tag{1.524e}$$

The attenuation for any particular time is

$$A = b(A_{01})p^{-c} \quad \text{dB} \tag{1.524f}$$

where

$$b = 0.22, c = 0.33 \quad \text{if } 0.001 \le p \le 0.01$$

$$b = 0.15, c = 0.41 \quad 0.01 < p \le 0.1$$

$$b = 0.12, c = 0.5 \quad 0.1 < p \le 1.0$$

where p stands for percentage of a mean year and [see Eq. (1.524e)]

$$\begin{aligned}
k &= 4.21 \times 10^{-5} \times f_c^{2.42} & 2.9 < f_c < 54 \text{ GHz} \\
&= 4.0 \times 10^{-2} \times f_c^{.699} & 54 \le f_c \le 160 \text{ GHz} \\
\alpha &= 1.4 f_c^{-.0779} & 8.5 \le f_c \le 25 \text{ GHz} \\
&= 2.63 f_c^{-.272} & 25 \le f_c \le 160 \text{ GHz}
\end{aligned} \tag{1.524g}$$

In addition to attenuation, atmospheric precipitation causes depolization of transmitted waves. In systems where dual polarization is used for frequency reuse, one polarization component may contain an unwanted component of another polarization. The ratio of the wanted polarization component (say RHCP) to the unwanted orthogonal component (say LHCP) is called the *cross-polarization discrimination* ratio (XPD) and is given by

$$\text{XPD} = U - V \log \text{CPA} \tag{1.525}$$

where

$$U = -10 \log \tfrac{1}{2}(1 - \cos 4\tau) - 40 \log \cos \theta + 3 \log f_c$$

$$V = 20, \text{ if } 8 < f_c < 15 \text{ GHz}$$

$$V = 23, \text{ if } 15 < f_c < 35 \text{ GHz}$$

$$\tau = \text{polarization tilt angle with respect to the horizontal plane}$$

$$\text{CPA} = \text{coplanar attenuation}$$

$$= 20 \log(E_0/E_c)$$

---

[q] See Chapter 3 and Figs. 3.14, 3.16, and 3.17.

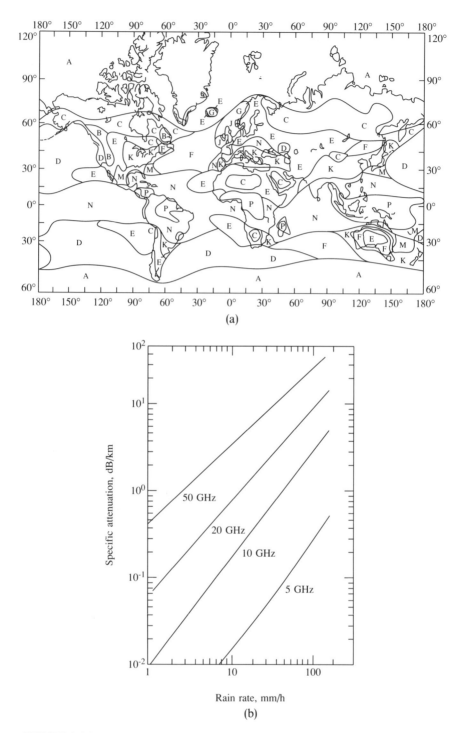

(a)

(b)

FIGURE 1.61   (a) Rain climatic zones. (From *NASA Propagation Effects Handbook*, 1983.) (b) Dependence of specific attenuation on rain rate for several frequencies. (From David Rogers, Propagation Considerations for Satellite Broadcasting above 10 GHz, *IEEE J. Selected Areas Commun.*, 1985. © 1985 IEEE.)

TABLE 1.4 Rainfall Intensity Exceeded (mm/h) for CCIR Rain Climatic Zones

% Time	A	B	C	D	E	F	G	H	J	K	L	M	N	P
1.0	—	1	—	3	1	2	—	—	—	2	—	4	5	12
0.3	1	2	3	5	3	4	7	4	13	6	7	11	15	34
0.1	2	3	5	8	6	8	12	10	20	12	15	22	35	65
0.03	5	6	9	13	12	15	20	18	28	23	33	40	65	105
0.01	8	12	15	19	22	28	30	32	35	42	60	63	95	145
0.003	14	21	26	29	41	54	45	55	45	70	105	95	140	200
0.001	22	32	42	42	70	78	65	83	55	100	150	120	180	250

Source: NASA Propagation Effects Handbook, 1983.

where E_0 is the field density of incident wave and E_c and E_x are, respectively, the field density copolar and cross polar waves leaving the rain layer. The cross-polarization discrimination ratio is then

$$XPD = 20\log(E_c/E_x)$$

1.10.3 Clear-Sky Attenuation

The attenuation due to oxygen and water vapor is given by

$$A_{O_2} = \left[\left(\frac{6.6}{f_c^2 + 0.33}\right) + \frac{9.8}{[(f_c - 57.5)^2 + 2.2]}\right] f_c^2 \cdot 10^{-3} \qquad f_c < 57.5 \text{ GHz}$$

$$= 14.7 \text{ dB} \qquad 57.5 < f_c < 62.5$$

$$= \left[\left(\frac{4.13}{(f_c - 62.5)^2 + 1.1}\right) + \frac{0.19}{(f_c - 118.7)^2 + 2}\right] f_c^2 \cdot 10^{-3}$$

$$62.5 \le f_c < 350 \text{ GHz}$$

(1.526a)

$$A_{H_2O} = \left[0.067 + \frac{2.4}{(f_c - 22.3)^2 + 6.6} + \frac{7.33}{(f_c - 183.5)^2 + 5}\right.$$

$$\left. + \frac{4.4}{(f_c - 323.8)^2 + 10}\right] f_c^2 \rho \cdot 10^{-4} \qquad (1.526b)$$

where ρ is the grams of water vapor per m^3 of air. For frequencies lower than 100 GHz, only the first two terms in Eq. (1.527) are taken into account. Equation (1.527) is valid for f_c in GHz. The atmospheric attenuation is shown in Fig. 1.62 for $\rho = 7.5$ g/m^3. The atmospheric loss is

$$L_a = A_{O_2} \times \text{path length} + A_{H_2O} \times \text{path length} \qquad (1.527)$$

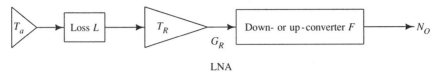

FIGURE 1.62   Receiving unit

1.10.4 Noise Power

The noise power in watts is given by

$$N = kT_S B_N \qquad (1.528)$$

where

$$B_N = \frac{\int_{\infty}^{\infty} |H(f)|^2 \, df}{\max |H(f)|^2} = \text{equivalent noise bandwidth}$$

$H(f) = $ transfer function of a two-port device

$k = $ total Boltzmann constant $ = -228$ dB

$T_S = $ temperature of the system

The noise caused by a receiver is also expressed by equivalent temperature T_e. The equivalent temperature is given by

$$T_e = (F - 1)T_0 \qquad (1.529)$$

where T_0 is environment temperature and F is usually called the *noise factor*. The noise factor (*noise figure*) is the ratio of the noise power at the input to the noise power at the output of a blackbody at ambient temperature of 290 K. It is expressed in dB. A noise factor of 0 dB implies that the two-port network has no noise.

A receiving unit is shown in Fig. 1.62. Let T_a be the temperature of the antenna, T_R be the effective noise temperature of the low-noise amplifier with gain G_R, and F be the noise figure of the frequency converter. The total noise is

$$T_s = T_a + LT_R + \frac{(L - 1)T_0}{G_R} + \frac{L(F - 1)T_0}{G_R} \qquad (1.530a)$$

where T_0 is the ambient noise temperature of the loss L. The antenna temperature T_a is

$$T_a = T_0 + T_{\text{sky}} + T_{\text{sun}} \qquad (1.530b)$$

where $T_0 = 290$ K, usually. The *sky temperature*[r] T_{sky} and *sun temperature* T_{sun};

---

[r] $T_{\text{sky}} = T_m(1 - 10^{-L_a/10})$ (no rain)

$\quad\quad\quad = T_m(1 - 10^{-A/10})$ (in presence of rain)

$T_m = (1.12 \times \text{surface temperature} - 50)$

L_a and A are defined in Eqs. (1.527) and (1.524f). See also Fig. 3.14 and Chapter 3.

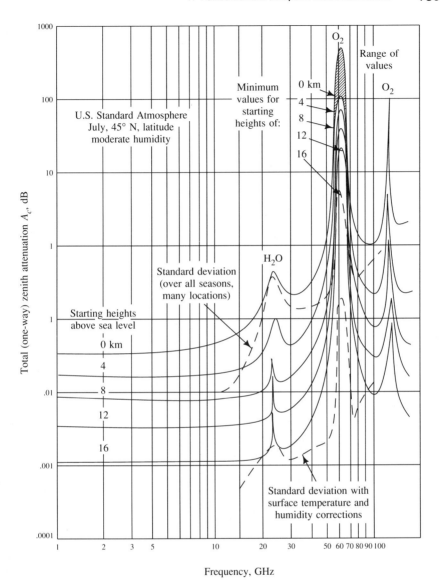

FIGURE 1.63 Total youth attenuation versus frequency. (From *NASA Propagation Effects Handbook*, 1983.)

are given in Figs. 1.63 and 1.64. Therefore, the noise power at the input is

$$N = kT_sB_N \qquad (1.531)$$

where T_s is given in Eq. (1.530a) and B_N is the equivalent bandwidth. The noise power spectral density is

$$N_0 = N/B_N$$

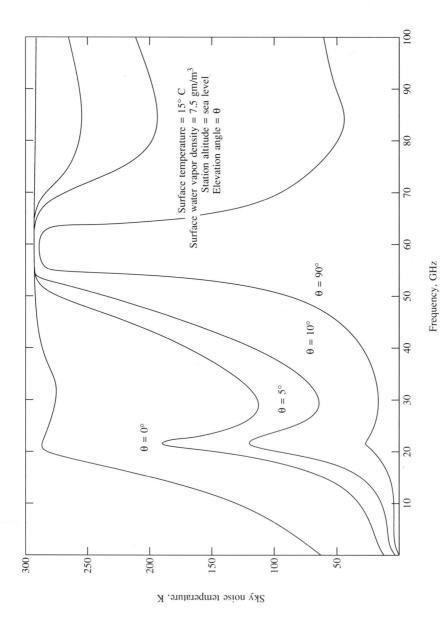

FIGURE 1.64   Sky noise temperature due to clear air. (From *NASA Propagation Effects Handbook*, 1983.)

186

1.10.5 Link Calculation

The link model is shown in Fig. 1.65. Let us first investigate the uplink calculation. The flux density at the satellite is given by

$$\Omega_u = \frac{P_{ET} G_{ET}}{\pi R_u^2} \tag{1.532}$$

where

P_{ET} = transmitted power of earth station (watts)

G_{ET} = earth station transmitter antenna gain

R_u = uplink distance between the earth station and the satellite (meters)

The flux density required at the receiving antenna to yield saturation of the travelling wave tube amplifier is called the *saturation flux density*. The received signal power at the satellite due to environment and device noise and degradation is

$$P_{su} = \Omega_u A_{su}$$

$$= \Omega_u [(G_{SU}/4\pi)\lambda_u^2] = \Omega_u \frac{G_{SU}}{4\pi} \cdot \frac{c^2}{f_c^2}$$

A_{su} = effective area of the satellite antenna

G_{SU} = satellite receive antenna gain (dB) $\qquad$ (1.533)

$c = 3 \times 10^8$ m/s, the velocity of light

f_c = transmitted signal frequency (GHz)

λ_u = wavelength of the signal

where Ω_u is given in Eq. (1.532).

The noise consists of thermal noise, rain-induced noise and sky noise. The signal-to-noise power is given by

$$(C/N)_u = P_{su}/N = P_{ET} G_{ET}(\lambda_u/4\pi R_u)^2 \left(\frac{G_{su}}{T_u}\right)\left(\frac{1}{L_{au}}\right)\left(\frac{1}{kB_u}\right) \tag{1.534}$$

Using Eqs. (1.534) and (1.533) and $N = K T_u B_u$, where B_u is the uplink channel bandwidth and T_u is the total uplink noise temperature, we can obtain the carrier-to-noise ratio as

$$(C/N)_u = 10\log(P_{ET} G_{ET}) - 20\log(4\pi R_u/\lambda_u)$$

$$+ 10\log(G_{su}/T_u)_{SR} - 10\log L_{au} - 10\log kB_u \tag{1.535}$$

The earth effective isotropic radiated power (EIRP) is defined as

$$\text{EIRP} = 10\log(P_{ET} G_{ET})\,\text{dB}$$

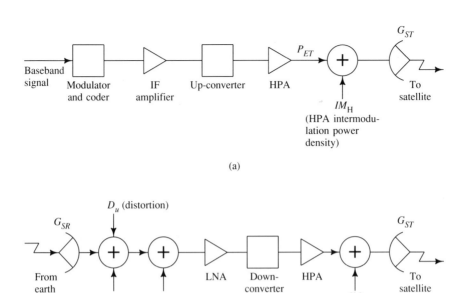

(a)

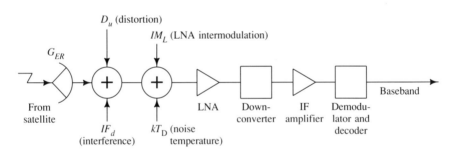

(b)

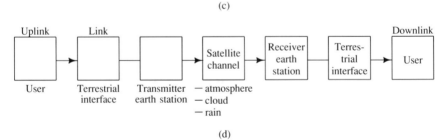

(c)

(d)

FIGURE 1.65   The satellite link. (a) Earth station transmitter. (b) Satellite transponder. (c) Earth station receiver. (d) User-to-user link.

The satellite antenna figure of merit is defined by

$$(G/T)_{SR} = 10\log_{10}(G_{su}/T_u)\,\text{dB}$$

T_u represent uplink total antenna noise temperature. Equation (1.535) can be written as

$$(C/N)_u = (\text{EIRP})_E + (G/T_u)_{SR} - (L_{pu} + L_{au}) - 10\log(kB_u) \qquad (1.536)$$

where

L_{pu} = path loss (uplink)

$\quad = 20\log(4\pi R_u/\lambda_u)\,\text{dB}$

$\quad = 183.5 + 20\log f_c$ for geostationary satellites

L_{au} = loss due to band-limiting, nonlinearity of the transponder
amplifiers, phase noise, cross polarization loss; antenna tracking
error, intermodulation, atmospheric, rain, galactic, star, and
other noise.

The gain of the antenna is given by

$$G = \eta(\pi D/\lambda_u)^2 \qquad (1.537a)$$

where

η = efficiency of the antenna = 0.55

D = antenna diameter for a parabolic antenna

λ_u = wavelength

A parabolic antenna with a diameter D is shown in Fig. 1.66. CO is the boresight pointing towards the antenna. The antenna gain is maximum at 0. A and B are two points in the main beam where the gain is 3 dB less than the gain at 0, and ϕ is the angle (*antenna beamwidth*) that AB subtends at the

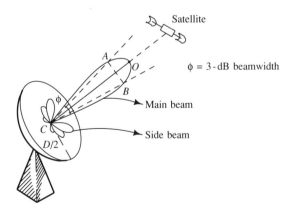

FIGURE 1.66   Antenna pattern.

center. The beamwidth is given by

$$\phi = 65(\lambda/D) \tag{1.537b}$$

For a multiple-access system the satellite power is reduced by input backoff power. In addition to the satellite antenna noise, there are intermodulation noise, adjacent satellite noise, and other noise. A modified equation for uplink carrier to noise is

$$(C/N)_u = (EIRP)_E + (G)_S - 10\log_{10} N - (L_{pu} + L_{au} + BO_i) \tag{1.538a}$$

where

EIRP = earth station EIRP

G_S = satellite antenna gain

BO_i = input backoff power (dB) (See Figs. 1.50 and 1.51.)

$N = kT_u B_u + IM_u + IF_u$

IM_u = uplink intermodulation noise power

IF_u = uplink interference noise power

Equation (1.538a) can also be written as

$$(C/N)_u = [(C/kT_u B_u)_u^{-1} + (C/IM_u)^{-1} + (C/IF_u)^{-1}]^{-1} \tag{1.538b}$$

where

$$(C/kT_u B_u) = (EIRP)_E + (G/T)_s - (L_{pa} + L_{au} + BO_i) - 10\log kB_u \tag{1.538c}$$

If $BO_i = 0$ dB, $IM_u = 0$ dB, and $IF_u = 0$ dB, then Eq. (1.538c) is the same as Eq. (1.536).

Let us now investigate the downlink calculation. It can be shown that the downlink's carrier to noise is as follows:

$$\left(\frac{C}{N}\right)_D = (EIRP)_S - L_{PD} - L_{aD} + (G_R/T)_E - 10\log_{10} k - 10\log_{10} B_d \tag{1.539}$$

where

$(EIRP)_S$ = satellite EIRP = $10\log_{10} P_{ST} G_{ST}$

P_{ST} = satellite transmitter power

G_{ST} = satellite transmit antenna gain

$(G_R/T)_E$ = earth station receive antenna figure of merit

L_{PD} = downlink free space path loss = $20\log_{10}(4\pi R_D/\lambda_D)$; here R_D is the downlink distance and λ_D is the downlink wavelength

B_D = downlink receiver noise bandwidth

k = Boltzmann's constant

The downlink antenuation L_{PD} is attributable to atmospheric and other effects, including tracking error, adjacent channel, cochannel, intermodulation, and antenna losses, phase noise, band-limiting distortion, implementation loss, synchronization loss, intersymbol interference, cross-polarization loss, as well as atmospheric, galactic, cloud, and other losses.

For the multiple-access system, the carrier-to-noise ratio is

$$(C/N)_D = (EIRP)_S - BO_0 + (G)_E - (L_{PD} + L_{aD}) - 10\log_{10} N \qquad (1.540a)$$

where

L_{aD} = downlink atmospheric and other loss

BO_0 = output backoff power (dB) (See Figs. 1.50 and 1.51.)

$N = kT_d B_d + IM_D + IF_D'$ (dBW)

T_d = downlink total noise

B_d = receiver (earth) bandwidth

IM_D = downlink intermodulation noise

IF_D = downlink interference total noise power

Equation (1.540a) can be written as

$$(C/N)_D = [(C/kT_d B_d)^{-1} + (C/IM_d)^{-1} + (C/IF_d)^{-1}]^{-1} \qquad (1.540b)$$

where

$$(C/kT_d B_d) = (EIRP)_S - (BO)_0 + (G/T)_E - 10\log kB_d - (L_{PD} + L_{aD})$$

The composite (total) *carrier-to-noise power* is

$$(C/N)_T^{-1} = [(C/N)_u^{-1} + (C/N)_D^{-1})]\,dB \qquad (1.541a)$$

The total *carrier-to-noise power density* is given by

$$(C/N_0)_T = [(C/N_0)_u^{-1} + (C/N_0)_D^{-1})]^{-1}\,dB \qquad (1.541b)$$

$$(C/N_0)_T = 10\log[10^{-(C/N_0)u/10} + 10^{-(C/N_0)D/10}] \qquad (1.541c)$$

where $N_0 = N/B$ dBW/Hz.

In digital communication, the ratio of bit energy to noise density is given by

$$(E_S/N_0) = (C/N_0)_T(1/R_S) \qquad (1.542)$$

where R_S is the symbol rate in bps, and E_S is the symbol energy. For analog communication using an FM system, the signal-to-noise ratio is given by

$$(S/N) = (C/N_0)_T(1/b)(f_r/f_m)^2 pwB \qquad (1.543)$$

where

B = Rf channel bandwidth [see Eq. (1.230)]

$p = 1.78$ (psophometric weighting factor) dB

TABLE 1.5 Uplink and Downlink Parameters

Parameter	Uplink	Downlink
1. Transmitter power, dBW	35⎫ earth	17⎫ satellite
2. Transmitting antenna gain, dBi	57⎭	23⎭
3. Receiving antenna gain, dBi	23	57
4. Noise power density (kt), dBW/Hz	199	208
5. Interference power density, dBW/Hz	0	2
6. HPA intermodulation power density, dBW/Hz	2	1
7. LNA intermodulation power density, dBW/Hz	1	1
8. Distortion power density, dBW/Hz	2	1
9. Path loss, dB	206 (14)	205 (12)
10. Atmospheric loss, dB	2	2
11. Rain loss, dB	6	4
12. Tracking loss, dB	1	1
13. Backoff	2	4
14. Bandwidth	78 (72 MHz)	78

w = 2.5 (preemphasis weighting factor) dB

b = channel bandwidth = 3.1 kHz

f_r = RMS test tone

f_m = maximum baseband frequency ($4.2 \times n_v$)

n_v = number of noise channel

The link parameters are given in Table 1.5.

In the parameters in Table 1.5 the uplink calculations are as follows:

Earth station EIRP = 35 + 57 − 2 = 90 dB   (1.544)

obtained by adding parameters 1 and 2 and subtracting parameter 13.

Loss = −(206 + 2 + 6 + 1) = −215 dB   (1.545)

obtained by adding parameters 9, 10, 11, and 12.

Note that the path loss is

$$20 \log(4\pi R_u/\lambda_u) = 183.5 + 20 \log(14 \times 10^9)$$

The noise power density is 198 dB.

$$N_0 = kT$$

$$T = 228 - 198 = 30 \text{ dB}$$

$$= 1000 \text{ K}$$

The satellite (G/T) is given by

$$G/T = G\,(\text{dB}) - T\,\text{dB}$$

$$= 23 - 30 = -7\,\text{dB} \tag{1.546}$$

using parameter 3 and $T = 30$ dB.

The total interference noise density is

$$IM_u + IF_u = -(2 + 1 + 2) = -5\,\text{dB} \tag{1.547}$$

adding parameters 6, 7, and 8.

The contribution of Boltzmann's constant is

$$-10\log_{10} k = 228\,\text{dB} \tag{1.548}$$

Adding Eqs. (1.544) through (1.548), we get the uplink carrier-to-noise power density

$$(C/N_0)_u = 91\,\text{dBW/Hz} \tag{1.549a}$$

Therefore, the carrier-to-noise power is

$$(C/N)_u = 91 - 78 = 13\,\text{dB}$$

using Eq. (1.549) and parameter 14.

Similarly, we now perform the downlink calculations:

$$\text{Satellite EIRP} = 17 - 4 + 23 = 36\,\text{dB} \tag{1.549b}$$

taking 4 dB for output backoff.

The noise power density is 208. Hence,

$$T = -208 + 228 = 20\,\text{dB}$$

$$\text{Earth } G/T = 57 - 20 = 37\,\text{dB} \tag{1.549c}$$

$$\text{Total loss} = -(205 + 2 + 4 + 1) = -212\,\text{dB} \tag{1.549d}$$

Note that earth's $G = 57$ dB implies that the diameter of the antenna is 7 meters.

The total downlink interference is

$$-(2 + 1 + 1 + 1) = -5\,\text{dB} \tag{1.549e}$$

The contribution due to k is

$$-10\log_{10} k = 228\,\text{dB} \tag{1.550}$$

Adding Eqs. (1.550a) through (1.550e), we get the downlink carrier-to-noise power density

$$(C/N_0)_D = 84\,\text{dB}$$

The downlink carrier-to-noise power is

$$(C/N)_D = 84 - 78 = 6\,\text{dB}$$

The total carrier-to-noise power density is

$$(C/N_0)_T = (10^{-9.1} + 10^{-8.4})^{-1}$$

$$= 83.21 \text{ dB}$$

Suppose that a binary FSK modulation is used and it is desired to have a probability of bit error of 10^{-6}. It is assumed that the channel is slow, flat Rayleigh fading. The required E_b/N_0 is 17 dB when rate $-1/2$ convolution coding and constraint length 5 with soft-decision Viterbi decoding is used. Therefore, the power-limited signaling rate is

$$R_b = 83.21 - 17 = 66.79 \text{ dB} = 4.8 \text{ Mbps}$$

There are also other models used for ionospheric scintillations. The fading intensity and phase are also modeled log normal [see Eq. (1.75)] and normal random processes. The third model is an approximation to Gaussian statistics and the probability density function of the intensity is given by Nakagami density [Eq. (1.91)], where $m = (S_4)^{-2}$ and S_4 is the *scintillation index*, see Eq. (1.522a). When $m = 1$, the Nakagami density approaches Rayleigh density; however, when m is large, Nakagami density is close to Rician density [see Eq. (1.135a)].

1.11 NUCLEAR EFFECTS ON PROPAGATION

1.11.1 Nuclear Propagation

In the event of a nuclear detonation, there will be several atmospheric effects including (1) a large amount of dust lofted high into the atmosphere, (2) large fires and fireballs, and (3) large amounts of release of chemical and physical species such as α, β, and γ rays and neutrons. In addition, there will be the release of a huge electromagnetic pulse (EMP), tidal waves, volcanic eruptions, shock waves, thunderstorms, and earthquakes (see Fig. 1.67). Readers are referred to a special issue of the *Proceedings of the IEEE* (Dec. 1967); Glasstone and Dolan (1977); The Effects of Nuclear Weapons; National Research Council (1975), Long-term Worldwide Effects of Multiple Nuclear Weapons Detonations; National Research Council (1985), The Effects on the Atmosphere of a Major-Nuclear Exchange; K. Ya Kondratyev, (1988); Climate Shocks, and Tsipis (1983). Arsenal: Understanding Weapons in the Nuclear Age. In this section, we will summarize nuclear detonations effects on RF signals. Nuclear bursts induces ionization, electromagnetic pulse and chemical particulates. Ionization creates electron density and changes the complex refraction index. The real part of the refraction index causes phase shift, Friday rotation, time delay and dispersion, and the imaginary part of the index attenuates the signal.

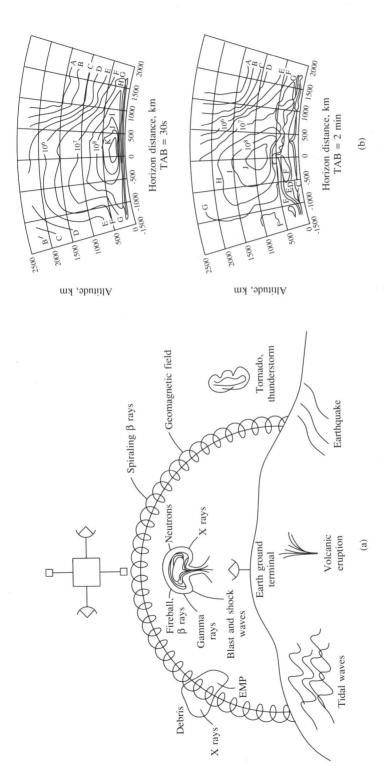

FIGURE 1.67 (a) Possible scenario for a nuclear burst. (b) Electron density contours (cm^{-3}) ($A = 1 \times 10^4$, $B = 3 \times 10^4$, $C = 1 \times 10^5$, etc.) (From R. W. Middlestead, R. E. LeLevier, and M. D. Smith, "Satellite Crosslink Communications Vulnerability in a Nuclear Environment," *IEEE J. Selected Areas Commun.*, February 1987. © 1987 IEEE.)

1.11.2 Atmospheric Effects

The radiation[s] from nuclear bursts is shown in Fig. 1.68. The amount of each radiation category depends on the amount of detonation and the height of burst. A detonation of 1 kiloton (kT) releases about 2.6×10^{31} electronvolts. A 10-kT bomb was detonated in Hiroshima. The present nuclear arsenal might be of 6 thousand megatons (6×10^6 kT). Fires from the explosion of a megaton bomb will create smoke into the troposphere for several days. The smoke will reduce the light level, causing intense cooling of the lower level and heating of the upper level of the atmosphere. Hundreds of teragrams (Tg) of dust and debris would be injected into the atmosphere from the surface detonations. A significant amount of dust particles would remain in the stratosphere and would reduce the sunlight at the earth's surface. The height of the dust cloud increases from the ground to 20 km within 3 minutes after the burst. In the event of a high-altitude burst of several nuclear bombs, each of several megatons, the height of the cloud could be more than 40 km. Clouds produced by nuclear explosions and by the fires can hold water in the order of 6000 Tg (1 tetragram = 10^{12} g). Most of the water would remain in the troposphere, and could cause severe signal degradation of the order of 60 dB or more.

The prediction of nuclear effects on the atmosphere has been performed by general circulation models (GCMs). The GCMs yield pressures, tempera-

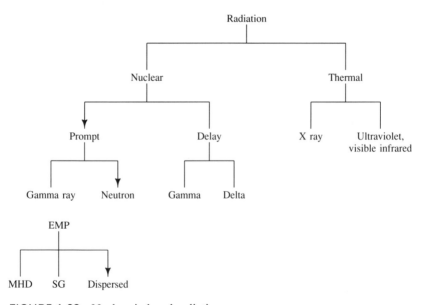

FIGURE 1.68   Nuclear-induced radiation.

---

[s] See also Chapter 8.

ture, wind, and moisture at different points and levels, along with climate dynamics. However, these models are inadequate for the prediction of thermal and optical effects. Smoke clouds reduce solar tansmission more than water clouds and dust clouds. It is estimated that surface temperature change will be $-14°C$ to $-42°C$ for a minimum of 8 days to a maximum of 50 days. The height of elevated temperature above the ground is estimated to be about 12 km. Fig. 1.67b shows results for a high-yield detonation at 600 km for time after burst (TAB) of 0.5 and 2 min.

Ozone, concentrated in the stratosphere, absorbs the extreme ultraviolet (UV) radiation from the sun and protects the earth from the dangerous effects from UV radiation. Nitrogen oxide (NO_2) releases from the nuclear explosion will destroy this ozone layer. It is estimated that UV solar radiation will increase three- to fourfold, and solar radiation thus will decrease 100 times for several months in the event of a 10-thousand-MT burst. It is known that there are more than 800 nuclear bomb testings, with a total energy yield of about 340 mT. Based on tests and simulations, it is predicted that stratospheric ozone layer can be reduced up to 90 percent. Less than 5 percent of the energy released from a nuclear explosion is in the form of gamma rays and X rays. These rays produce protons, and protons interact with the materials in the radial path, creating free electrons and positive ions. These electrons create spatial current densities. This current creates giant electric and magnetic fields extending several kilometers, and thus two fields travel together away from the point of detonation as a single *electromagnetic pulse* (EMP).

The EMP is a time-varying electromagnetic radiation that increases very rapidly. Its spectrum varies from very low frequencies to several magahertz. It is expected that its highest frequency will be below the EHF frequencies. The peak current may be 12 kA for a period of 2 μs and may reduce to 2 kA after 4 μs. For a 1-MT explosion at an altitude of 300 km, the EMP is roughly 10^{11} joules. The gamma rays interact with air molecules and atoms by Compton and photoeffect processes and produce an ionized region surrounding the burst known as the *deposition (source) region*. The deposition area takes the shape of pancakes of radius 800 km and a thickness of 50 km, with an electric field of 100 kV/m. For surface bursts the deposition region is primarily hemispherical. The electric field is very strong, but the radiated field decays with increasing distance from the deposition region. Electromagnetic signals are also generated at much later times because of the hydromagnetic movement of atmosphere and released debris. These EMPs are called *magnetohydrodynamic* (MHP) electromagnetic pulses. The MHP EMPs create magnetic bubbles of expanding size. These bubbles may deflect communication and radar signals. The electric fields are generated by the interaction of gamma and X rays with solid materials, including the satellite and ballistic missile surfaces. These effects are called *system-generated EMP*.

The radiation will penetrate the structure and will create free electrons inside the satellite systems. The electric field generated near the surface by direct interactions of ionizing radiation with the solid material in electronic systems can induce electric currents in components, cables, and ground wires and can cause failure in power systems. Some background material on radiation hardening is discussed in Chapter 8.

Some of the approaches to protecting the devices are the following: metal shields to prevent access of radiation, good grounding to divert large currents, and the provision of surge arrestors. The shield consists of continuous metal, such as soft iron, copper, or steel. Systems, including computers, employing transistors or semiconductors and capacitors are most susceptible to the EMP. The least susceptible equipment includes transformers, motors, battery-operated systems, and vacuum tubes. The reader is referred to Lee (1986) for a discussion of satellite systems applications.

1.11.3 Effects on RF Propagation

The number of electrons to be released from a nuclear detonation depends on the burst yield and height. It is estimated that electrons from a prompt radiation can be given by

$$N_e = 2.4 \times 10^{18} \frac{KW}{4\pi D^2} \rho \mu_m e^{-\mu_m M} \text{ cm}^3 \qquad (1.551)$$

where

$$\rho = \rho(h) = \rho_0 \exp[h/Hp] \ g \cdot \text{cm}^{-3}$$

ρ_0 = atmospheric density at higher air density at sea level

H_p = scale height

μ_m = mass absorption coefficient of the air

D = slant distance from the observation point to the explosion source

W = kilotons of bursts

K = 0.7 for X-rays

    = 0.003 for gamma rays

    = 0.01 for neutrons

M = *penetration mass*, the mass of air per unit area between the radiation source and the observation point.

$$M = 6.8 \times 10^5 \left(\frac{D}{H - H_0}\right) \rho_0 \left[\exp\left(\frac{-H_0}{4.3}\right) - \exp\left(\frac{-H}{4.3}\right)\right] g/\text{cm}^2 \qquad (1.552)$$

where

H_0 = altitude of explosion point

H = altitude of observation point

The *refractive index*[t] of the ionized medium is given by

$$n = (1 - N_e/10^4 f_c^2)^{1/2} \tag{1.553}$$

$$= x + iy \tag{1.554}$$

where

    x = real part of refractive index (ordinary)

    y = imaginary part of the refractive index

In order that the signal can penetrate the ionized medium, it is necessary that

$$f_c \geq 10^{-2} \sqrt{N_e} \sec i \tag{1.555}$$

where i is the angle of incidence.

The attenuation of the signal in the *ionized medium* is given by

$$A = 7.4 \times 10^4 \left(\frac{N_e v}{(4\pi^2 f_c^2 + v^2)} \right) \sec i \; \text{dB/dB/mile} \tag{1.556}$$

where

    N_e = electron density

    v = number of collisions per second that an electron makes with ions, molecules, or atoms

    f_c = carrier frequency

For a constellation of satellites at an altitude of 1100 km with a detonation of 1-MT burst at an altitude of 600 km, it is estimated that the electron density may be 10^{18} to 10^{20} electrons/cm^3. It may last for several hours. The total electron concentration N_T is given by

$$N_T = \int_{L_\rho} N_e(z) \, dz \tag{1.557}$$

where L_ρ is the path length.

There will be two types of disturbances. The ionization will create *static disturbances* such as absorption, noise (EMP), dispersion of time and frequency, phase shift, and time delay. The dynamic motion of electrons in the ionized medium will create *dynamic disturbances*, such as amplitude and phase scintillation, scattering, and time and frequency jitters. The antenna temperature will be increased by

$$T_{\Delta T} = 1000(1 - 10^{-k_a/10})$$

where

$$k_a = \int_0^{L_\rho} 1.16 \times 10^{19} \, N_e(z)/f_c^2 \, dz \tag{1.558}$$

---

[t] In free space, $n = 1$, $n > 1$ in troposphere, and $n < 1$ in ionosphere $n = \sin i$ yields Eq. (1.555).

TABLE 1.6 Parameters for a High-Altitude Megaton Burst

	Frequency		
	10 GHz	30 GHz	60 GHz
τ_0 (decorrelation time)	1 ms	4 ms	8 ms
f_0 (decorrelation frequency)	15 kHz	28 MHz	40 MHz
l_0 (decorrelation length)	40 m	1.5 m	3 m
Antenna loss	32 dB	22 dB	16 dB
Absorption	78 dB	5 dB	2 dB
Antenna temperature	10,000 K	9000 K	5000 K

The change in phase shift and time delay are given by

$$\theta_\Delta = 8.45 \times 10^{-6} \, N_T/f_c$$
$$T_\Delta = 1.34 \times 10^{-25} \, N_T/f_c^2$$

(1.559)

where N_T is given by Eq. (1.557). Some other parameters are derived in Middlestead, LeLevier, and Smith (1987). These parameters depend on the channel model, a most challenging area of research. The decorrelation time, frequency, length, antenna loss, absorption and antenna temperature for three frequencies are given in Table 1.6 for a megaton burst at a high altitude. For radar systems, the antenna loss and absorption will be double of the communication and navigation systems.

EXERCISES

1. *Flat-top Sampling.* The flat-top sampling method of holding time is shown in Fig. 1.69. Let $s(t)$ be a band-limited signal of bandwidth B (one-sided) and $T_s = 1/2B$. The filter to generate flat-topped pulses has a transfer function given by

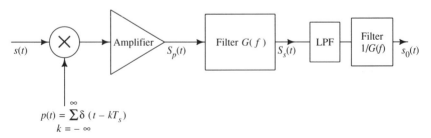

FIGURE 1.69   Flat-top sampling.

$$G(f) = (\sin \pi f \tau)/\pi f \tau, \qquad |f| < B$$
$$= 0, \qquad\qquad \text{elsewhere}$$

The amplifier gain is T_s/τ. The flat-top sampled signal is given by

$$S_p(t) = \sum_{k=-\infty}^{\infty} S(kT_s) r(t - kT_s)$$

where $r(t)$ is a rectangular pulse of duration τ seconds.

(a) Show that the amplified spectrum of $s(t)$ is

$$S_s(f) = \sum_{-\infty}^{\infty} \frac{\sin \pi \tau}{\pi \tau} S\left(\frac{f-k}{T_s}\right)$$

where $S(f)$ is the Fourier transform of $s(t)$.

(b) Show that the output of equalizer filter with the transfer function $1/G(f)$ is

$$s_0(t) = s(t) \qquad G(f) = (\sin \pi \tau)/\pi \tau$$

This process is also known as *zero-order-hold* approximation.

2. Let the transmitted signal be

$$x(t) = \sum_{k=-\infty}^{\infty} a_k g(t - kT) \cos 2\pi f_c t$$

where

$$a_k = 1 \text{ with probability } p$$
$$= -1 \text{ with probability } 1 - p$$

Show that the PSD of $x(t)$ is given by

$$S_x(f) = \left(\frac{2p(1-p)}{T}\right) [|G(f + f_c)|^2 + |G(f - f_c)|^2]$$
$$+ \left(\frac{(2p-1)^2}{T^2}\right) \left\{ \sum_{n=-\infty}^{\infty} \left|G\left(\frac{n}{T}\right)\right|^2 \left[\delta\left(\frac{f-n}{T+f_c}\right) + \delta\left(\frac{f-n}{T-f_c}\right)\right] \right\}$$

where $G(f)$ is the FT of the pulse $g(t)$.

3. Let

$$x(t) = a(t) \cos 2\pi f_c t + b(t) \sin 2\pi f_c t$$

where

$$a(t) = \sum a_k g(t - kT), \qquad b(t) = \sum b_k r(t - kT)$$

where $\{a_k\}$ and $\{b_k\}$ are uncorrelated and

$$E(a_k) = 0, \; E(a_k a_{k+m}) = A^2 \delta(m)$$
$$E(b_k) = 0, \; E(b_k b_{k+m}) = A^2 \delta(m)$$

Let

$$g(t) = r(t - T) = 1/\sqrt{2}, \qquad 0 \le t \le T$$
$$= 0, \qquad\qquad\qquad \text{elsewhere}$$

Show that the PSD of OQPSK $x(t)$ is $S_x(f) = \frac{1}{2}[G(f - f_c) + G(f + f_c)]$ where $G(f) = 2A^2T(\sin \pi 2Tf)/\pi 2Tf$.

4. Let

$$y(t) = \sum_{m=-\infty}^{\infty} X_m P(t - mT + \theta)$$

where X_m is a zero-mean stationary random process, and θ is a uniform random variable in $(0, T)$ and is independent of X_m. $P(t)$ is a pulse of duration T seconds. Show that the power spectral density of pulse amplitude modulation (PAM) signal $y(t)$ is

$$S_y(f) = \left(\frac{1}{T}\right)|P(f)|^2 S_x \exp(i2\pi fT)$$

where $P(f)$ is the FT of $P(t)$, and $S_x(\cdot)$ is the PSD of $\{X_m\}$ and is defined as

$$S_x \exp(i2\pi fT) = \sum_{k=-\infty}^{\infty} R_x(k) \exp(-ik2\pi fT) \qquad R_x(k) = E(X_m X_{m+k})$$

5. Let

$$x(t) = m(t) + n(t)$$
$$y(t) = x^2(t)$$
$$0 \le t \le T$$

$m(t)$ and $n(t)$ are error mean stationary uncorrelated processes and $n(t)$ is Gaussian with PSD $S_n(f)$ and variance σ_n^2. The PSD of $m(t)$ is $S_m(f)$ and its variance is σ_m^2. Show that the PSD of $y(t)$ is

$$S_y(f) = S_{mxm}(f) + 4S_m(f) * S_n(f) + 2\sigma_m^2 \sigma_n^2 \delta(f) + \sigma_n^4 \delta(f) + 2S_n(f) * S_n(f)$$

where

$$S_{mxm}(f) = FT[R_m(\tau) \cdot R_m(\tau)]$$
$$S_n(f) * S_n(f) = FT[R_n(\tau) \cdot R_n(\tau)]$$
$$S_m(f) * S_n(f) = FT[R_m(\tau) \cdot R_n(\tau)]$$

(Hint: $E[n(t)n(t + \tau)n(t)n(t + \tau)] = 2R_n(\tau) \cdot R_n(\tau) + \sigma_n^4$; see Mohanty, 1986, p. 21.)

6. Let us define the figure of merit of a demodulator by

$$d = \frac{\text{output signal-to-noise ratio}}{\text{input signal-to-noise ratio}}$$

The FM and AM signals are given by

$$S_{FM}(t) = A \cos(\omega_c t + \beta \sin \omega_m t)$$
$$S_{AM}(t) = A(1 + \cos \omega_m t) \cos \omega_c t$$

The input noise $n(t)$ at the demodulator is white Gaussian with variance $\eta/2$. Show that

$$\frac{d_{FM}}{d_{AM}} = \frac{9}{2\beta^2} = \frac{9}{2(B_{FM}/B_{AM} - 1)^2}$$

where $B_{FM} = 2(\beta + 1)f_m$ and $B_{AM} = 2f_m$.

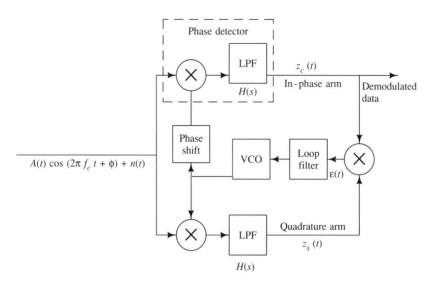

FIGURE 1.70   Costas loop.

7. A carrier tracking loop known as a *Costas loop* is given in Fig. 1.70. The received signal is

$$y(t) = A(t)\cos(2\pi f_c t + \phi) + n(t) \qquad 0 \le t \le T$$

where $A(t) = \pm 1$ and $n(t)$ is a zero-mean white Gaussian noise with variance $N_0/2$. Show that $\varepsilon(t)$ is independent of modulation $A(\theta)$ and carrier frequency f_c in the absence of noise. Derive one stochastic differential equation for the Costas loop. Note that the Costas loop can lock with the voltage-controlled oscillator (VCO) at two different angles relative to the phase of the received signal. $\varepsilon(t)$ is zero when the phase error is zero or 90°.

8. An FDM FM system has the following parameters: $B = 30$ MHz, $b = 3.1$ kHz, $f_\Delta =$ peak deviation $= 283$ kHz, $N = 600$ voice channels, $f_m = 2.54$ MHz, and $T = 293$ K.
   (a) Show that the noise power is -99.2 dBm.
   (b) If S/N is 69 dB, then the required carrier power is -48.1 dBm.
   (c) Calculate also the FM noise improvement factor.

9. Let a PM signal be given by

$$S(t) = A\cos[\omega_c t + \theta_0 + k_p m(t)]$$

where $m(t)$ is the message. Let the received signal be

$$y(t) = S(t) + n(t)$$

where $n(t)$ is a zero-mean white Gaussian noise with variance $N_0/2$. The demodulation process for a PM signal is shown in Fig. 1.71.
   (a) Show that the output signal-to-noise ratio of the demodulator is given by

FIGURE 1.71   Demodulation for a PM signal.

$$\left(\frac{S_0}{N_0}\right)_{\text{PM}} = \frac{A^2}{2}\left\{N_0\frac{k_p}{B_m}E[m^2(t)]\right\} = \frac{\Delta\omega_{\text{PM}}E(m^2(t))}{\left|\dfrac{dm}{dt}\right|^2_{\text{max}}}\left(\frac{S_0}{N_0}\right)_{\text{Baseband}}$$

when B_m is the message bandwidth and $\Delta\omega_{\text{PM}} = K_p\left.\dfrac{dm}{dt}\right|_{\text{max}}$.

(b) Show also that

$$\left(\frac{S_0}{N_0}\right)_{\text{FM}} = 3\left(\frac{\Delta\omega_{\text{FM}}}{\Delta\omega_{\text{PM}}}\right)\frac{\left|\dfrac{dm}{dt}\right|^2_{\text{max}}}{B_m^2|m(t)|^2_{\text{max}}}\left(\frac{S_0}{N_0}\right)_{\text{PM}}$$

$$\Delta\omega_{\text{FM}} = k_f|m(t)|_{\text{max}}$$

$$\Delta\omega_{\text{PM}} = k_p|dm/dt|_{\text{max}}$$

(c) Let

$$m(t) = \sin 2\pi f_m t \qquad f_m = B_m$$

Derive parts (a) and (b) above.

10. A channel is given by

$$H(f) = H_0(1 + g_1 f + g_2 f^2 + g_3 f^3 + g_4 f^4)x\exp[i(b_2 f^2 + b_3 f^3 + b_4 f^4)]$$

The input PM signal is given by

$$S(t) = A\cos[2\pi f_c t + \phi(t)]$$

Show that the output signal of the channel (Bell Telephone Laboratories, 1971, p. 500) is approximately given by

$$y(t) = AH_0\{[1 + c(t)]^2 + d^2(t)\}^{1/2}\cos[2\pi f_c t + \phi(t) + \phi_d(t)]$$

where

$$c(t) = g_1\phi'(t) + g_2[\phi'(t)]^2 + b_2\phi''(t) + 3(b_3 + g_1 b_2)\phi''(t)\phi'(t)$$
$$+ g_3[\phi'(t)]^3 - \phi'''(t) + (g_4 - b^2/2)\{-4\phi'''(t)\phi'(t) - 3[\phi''(t)]^2\}$$
$$+ (b_u + b_3 g_1 + b_2 g_2)\{[6\phi''(t)\phi'(t)]^2 - \phi''''(t)\}$$

$$d(t) = -g_2\phi''(t) + b_2[\phi'(t)]^2 + (b_3 + g_1 b_2)x[\phi'(t)]^3 - \phi'''(t) - 3g_3 g'(t)\phi''(t)$$
$$+ (b_4 b_3 g_1 + b_2 g_3)[-4\phi'''\phi' - 3\phi'']^2$$
$$+ (g_4 - b_2^2/2)(\phi'''' - 6\phi''[\phi'(t)]^2$$

$\phi_d(t) =$ the distorted noise (intermodulation)

$$= \arctan\left[\frac{d(t)}{1 + c(t)}\right]$$

$$\approx d(t) \text{ if } a(t), d(t) \ll 1$$

$$\phi'(t) = \frac{d}{dt}\phi(t), \; \phi''(t) = \frac{d^2}{dt^2}\phi(t), \ldots$$

Note that both amplitude and phase of the output are distorted. The amplitude distortion can be removed by passing the signal through a limiter. The output of the frequency discriminator will be proportional, in this case, to $\phi(t) + \phi_d(t)$, where $\phi_d(t)$ is known as the *intermodulation noise* distortion.

11. As an exercise in bandpass sampling, let $|X(f)|$, the amplitude spectrum of $x(t)$ be zero outside the band $|f - f_0| \le B/2$ and $|f + f_0| > B/2$. Show that

$$x(t) = \sum_{n=\infty}^{\infty} \text{sinc}(Bt - n)|z(n/B)| \cos[2\pi f_c(t - n/B)] + \arg[z(n/B)]$$

where $\text{sinc}\, x = (\sin \pi x)/\pi x$

$$z(t) = x(t) + i\hat{x}(t)$$

$$\hat{x}(t) = \text{Hilbert transform of } x(t)$$

$$\arg[z(t)] = \arctan(\hat{x}(t)/x(t))$$

$$|z(t)| = \text{amplitude of the analytic signal } x(t) + i\hat{x}(t)$$

12. For a PCM system, let the quantization noise be a uniform random variable. There are M number of binary bits per sample of a baseband signal having bandwidth W. Let B be the required bandwidth of the PCM signal. Show that the signal to quantization noise is

$$\text{SNR} = (3/2)2^{B/W}$$

Note that $M = \log_2 Q$, where Q is the number of quantization levels. $B = MW$ and $(S/N) = (3/2)Q^2$. If the signal-to n is 30 dB, find Q and M. Let 20 baseband signals be multiplexed to form an 8-bit PCM signal, with each baseband signal band-limited to 4 kHz. Show that the bandwidth of the multiplexed PCM system is $20 \times 2 \times 4 \times 8$ kHz and the pulse duration is 781 μs.

13. For an M-word binary PCM PSK signal, show that the signal-to-noise ratio is given by

$$\left(\frac{S_0}{N_0}\right) = \frac{2^{2M}}{1 + 4P_e c/2} = \frac{2^{2M}}{\left[1 + \left(\text{erfc}\sqrt{\frac{E_b}{\eta}}\right)^2 c\right]}$$

where $\eta/2$ is the PSD or the input noise, E_b is given by

$$E_b = S_i T_b = \frac{S_i T_s}{M} = \frac{S_i}{M}\left(\frac{1}{2f_m}\right) \qquad c = 2(2^{2M} - 1) \qquad Q = 2^M$$

where S_i is the input signal power and f_m is the bandwidth of the message.

14. Show that the signal-to-noise (granular) ratio for DPCM and DM are given by the following relation

$$\left(\frac{S_0}{N_g}\right)_{\text{DPCM}} = \frac{(2^q - 1)^2}{2q^3}\left(\frac{S_0}{N_g}\right)_{\text{DM}}$$

DPCM is superior to DM provided that the codeword length $q \geq 4$. N_g is the granulator noise power.

15. If the signal is a white Gaussian process with zero mean and unit variance, show that the signal to quantization noise power ratio is given by

$$\left(\frac{S_0}{N_g}\right) = Q^2\left\{1 + \left[\frac{\pi\sqrt{27}}{6} - 1\right][1 - (1 + q)]e^{-q}\right\}^{-1}$$

where Q is number of quantization levels $= 2^q$.

16. The received signal is a binary *amplitude-shift keying* (ASK) signal in the presence of white Gaussian noise $n(t)$, with variance $N_0/2$, and is given by

$$y(t) = A(t)\cos(2\pi f_c t + \theta) + n(t) \qquad 0 \leq t \leq T$$

where $A(t) = 0$ or A with equal probability and $f_c T = k$, an integer, and θ is the known phase. Show that the probability of a bit error is given by

$$P_e = \frac{1}{2}\text{erfc}\left(\sqrt{\frac{A^2 T_b}{8N_0}}\right)$$

17. The phase θ in Exercise 16 is a uniform random variable in $(-\pi, \pi)$. Show that the probability of a bit error is approximately given by

$$P_e = \tfrac{1}{2}[1 + (2\pi d)^{-1/2}]\exp(-d/2)$$

where $d = A^2 T/4N_0$.

18. A 16 *quadrature amplitude modulation* (QAM) signal is given by

$$x(t) = \sqrt{2}R_e\left[\exp(i\omega_c t)\sum_{m=-\infty}^{\infty} a_m p(t - mT)\right]$$

where $a_m = +a$ or $\pm 3a + i(\pm a$ or $\pm 3a)$ with equal probability. Show that the symbol error for the 16 QAM signal is

$$F_s = 3Q(\sqrt{\text{SNR}}) - 2.25Q^2(\sqrt{\text{SNR}})$$

where $\text{SNR} = a^2/\sigma^2$; σ^2 is the noise variance.

$$Q(x) = \frac{1}{\sqrt{2\pi}}\int_x^{\infty} e^{-t^2/2}\,dt = \tfrac{1}{2}[1 - \text{erf}(x/\sqrt{2})] = \tfrac{1}{2}\text{erfc}(x/\sqrt{2})$$

and $P(t)$ is a rectangular pulse of duration T seconds. Show also that the spectral efficiency is $2\log_2 N$ bits per second per hertz, where $N = 4$ (N being the number of levels).

19. A *continuous-phase PSK* (CPFSK) signal is received in the presence of a zero-mean white Gaussian noise $n(t)$ with variance $N_0/2$. The received signal is

$$y(t) = A\cos(2\pi f_c t + m\Delta\omega t + \theta) + n(t)$$
$$f_c T = 1 \qquad 0 \leq t \leq T$$

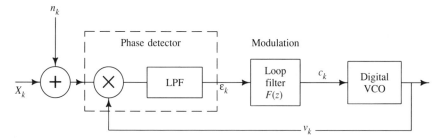

FIGURE 1.72 Digital phase-lock loop.

where $m = 1$ or -1 with equal probability. Show that the probability of a bit error is

$$P_e = \frac{1}{2}\mathrm{erf}\left\{\left[(1-b)\left(\frac{A^2T}{4N_0}\right)\right]^{1/2}\right\}$$

where $b = (\sin 2\,\Delta\omega T/2\,\Delta\omega T)$.

20. A *digital phase-lock loop* is given in Fig. 1.72. The transmitted signal is

$$X_k = A_c\cos(2\pi f_c kT + \theta_k)$$

The input of the discrete-time VCO is $v_k = A\cos(2\pi f_c kT + \phi_k)$, $\hat{\phi}_k$ is the estimate of the incoming signal phase ϕ_k, and f_v is the *natural frequency* of the VCO. The input to the VCO is given by

$$c_k = \phi_{k+1} - \phi_k$$

Let ε_k be the output of the phase detector and $\varepsilon_k = \theta_k - \phi_k$ and $F(z)$ is the z transform of the loop filter.

(a) If $\theta_k = \Omega_0 kT + \theta$, show that the *lock range* (hold-in range) of the phase-lock loop is given by

$$|\Omega_0| \le \frac{\pi}{T}|F(z)| \qquad z = 1$$

where θ is a constant and T is the sample interval.

(b) In the absence of noise, show that phase transfer function of the PLL is

$$\frac{\phi(z)}{\theta(z)} = \frac{F(z)}{F(z) + z - 1}$$

and the steady-state error is given by

$$e_{3S} = \lim_{z\to 1}\frac{\theta(z)(z-1)}{F(z) + z - 1}$$

where $\theta(z)$ and $\phi(z)$ are the z transforms of θ_k and ϕ_k.

21. (a) Let the received signal at the baseband be

$$r(t) = s(t)\exp[i\phi] + n_c(t) + in_s(t)$$

where $n_c(t)$ and $n_s(t)$ are independent Gaussian white noise with PSD $N_0/2$. Show that the variance of the phase error for high SNR is

$$\sigma^2 \phi = (2\varepsilon_s/N_0)^{-1}$$

where ε_s is the energy of the signal $s(t)$.

(b) Let the received signal baseband be

$$r(t) = s(t - \tau) + n_c(t) + in_s(t)$$

where $n_c(t)$ and $n_s(t)$ are independent white Gaussian noise with PSD $N_0/2$. Show that the variance of time delay error for high SNR is

$$\sigma_\tau^2 = (L/\pi^2 (2\varepsilon_s/N_0) \cdot f_N)^{-1}$$

where ε_s is the energy of the signal and f_N is defined in Eq. 1.520.

(c) The phase noise variance in a phase-lock loop is given by

$$\sigma_\phi^2 = \int_{-\infty}^{\infty} S_\psi(f) |1 - H(f)|^2 \, df$$

where $\psi(t)$ is the phase noise and $\phi = \psi(t) - \hat{\psi}(t)$, the error. Let

$$H(f) = \frac{1 + i(\sqrt{2}/\omega_n) 2\pi f}{1 + (\sqrt{2}\omega_n)\left(i2\pi f - \dfrac{i2\pi f}{\omega_n}^2\right)}$$

is the tracking loop transfer function. Let the PSD of the phase noise be

$$S_\psi(f) = \frac{S_F}{8\pi^3 f^3} + \frac{S_{WF}}{4\pi^2 f^2} + S_{WP}$$

Show that

$$\sigma_\phi^2 = \left(\frac{S_{WP}}{2\pi}\right)\omega_n + \frac{0.94 S_{WF}}{0.53 W_n} + \left(\frac{0.03 S_F}{0.53\omega_n}\right)^2$$

where ω_n is the loop natural frequency. [See Gagliardi (1978).]

22. Let $R(t)$ be the envelope of a signal. It is a Rayleigh process. N_{RS}, *the level crossing rate*, is the expected rate at which the envelope R crosses a level R_S in the positive direction. Let $\dot{R}(t) = d/dt[R(t)]$. The PDF of $\dot{R}(t)$ is given by

$$f(\dot{R}) = (2\pi f_d \sigma \sqrt{\pi})^{-1} \exp[-(\dot{R}/2\pi f_d \sigma)^2]$$

where $f_d = $ maximum Doppler frequency. Show that the level crossing rate

$$N_{RS} = \sqrt{2\pi f_d}(R_S/\sqrt{2\pi}) \exp(-R_S^2/2\sigma^2)$$

and the average value of the *fade duration* T_f is

$$E(T_f) = (\sqrt{2\pi}(R_S/\sqrt{2\sigma})f_D)^{-1}[\exp(R_S^2/2\sigma^2) - 1]$$

where $E(R^2) = 2\sigma^2$.

23. The received signal is

$$y(t) = R_e A \exp(i2\pi f_c t + \phi) + n(t)$$
$$+ I \exp(i2\pi f_c t + \phi_i + \theta) \qquad 0 \le t \le T$$

where $n(t)$ is the zero-mean complex noise with variance σ^2; ϕ and ϕ_i are indepen-

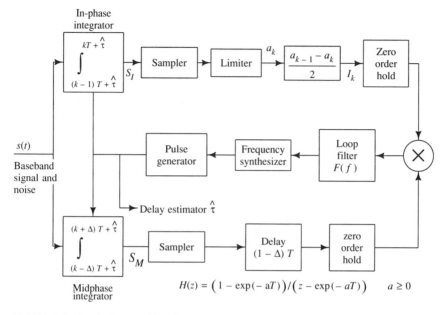

FIGURE 1.73   Delay tracking loop.

dent data symbols transmitted over desired and interfering carriers, and can take values of $\pm\pi/2$ with equal probability; θ is a uniformly random variable in $(0, 2\pi)$. Show that the probability of a bit error is

$$P_e = \frac{1}{2}\operatorname{erfc}\left(\frac{A}{\sigma\sqrt{2}}\right) + \frac{1}{\sqrt{\pi}}\exp\left(\frac{-A^2}{2\sigma^2}\right)\sum_{j=1}^{\infty} H_{2j-1}\left(\frac{A}{\sigma\sqrt{2}}\right)\left(\frac{1}{\sigma\sqrt{2}}\right)^{2j}\frac{I^{2j}}{2^{2j}(j!)^2}$$

where $H_{k+1}(x) = 2nH_k(x) - H_{k-1}(x)$, $k \geq 1$, $H_0(x) = 1$, and $x = A/\sigma\sqrt{2}$. In addition, $I^2/2$ is the power of the *cochannel interference*, and $H_k(n)$ is the *Hermite polynomial* of order k.

24. A delay tracking loop is shown in Fig. 1.73. The received signal is

$$y(t) = \sqrt{2E}\sum_{k=-\infty}^{\infty} a_k P(t - kT - \tau)\cos(2\pi f_c t + \theta) + n(t)$$

where $a_k = \pm 1$, E is the signal power, τ is the unknown time delay, T is the bit duration, θ is the known phase, and $n(t)$ is a zero-mean Gaussian process with variance $N_0/2$. Let $P(t)$ be a pulse with a duration of T seconds. Let

$$\varepsilon = \tau - \hat{\tau}$$

where $\hat{\tau}$ is the estimate of τ. Show that the mean-square time error for the linear loop is given by

$$\sigma_\varepsilon^2 = \frac{2\Delta TB_L}{2d[\operatorname{erf}(\sqrt{d})]}$$

where

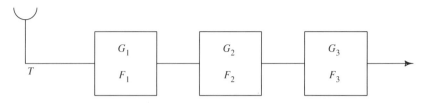

FIGURE 1.74   Earth station receiver with three amplifier/filters.

$$d = \frac{ET}{N_0} \qquad 0 < \Delta T < T/2$$

$$B_L = \int_{-\infty}^{\infty} \left| \frac{K'F(f)}{2\pi i f + k'F(f)} \right|^2 df$$

$k' = k\sqrt{2}\,\mathrm{erf}(\sqrt{d})$, k is the loop gain, and $F(f)$ is the loop transfer function.

25. Three devices are cascaded together in an earth station as shown in Fig. 1.74. G_i are the gains and F_i the noise figures of three devices, $i = 1, 2$, and 3. Show that the total noise figure is

$$F = F_1 + \frac{(F_2 - 1)}{G_1} + \frac{(F_3 - 1)}{G_1 G_2}$$

Let T be the environmental temperature, B the noise bandwidth, and $K =$ Boltzmann's constant. The equivalent temperature is given by

$$T_e = (F - 1)T$$

and the noise power is

$$N_0 = K T_e B \qquad K = -228.3 \text{ dBW}$$

Let $B = 36$ MHz, $T = 270$ K, $F_1 = F_2 = F_3 = 3$ dB, and $G_1 = G_2 = G_3 = 10$ dB. Show that the noise figure is 3.24 dB and the noise power is -125.3 dB.

26. The earth station and geosynchronous satellite parameters are given below:

Uplink	
Earth station transmitter power	2000 W
Backoff loss	3 dB
Other losses	4 dB
Antenna gain	64 dB
Atmospheric loss	0.6 dB
Transmit frequency	14 GHz
Satellite equivalent noise temperature	800 K

Downlink	
Satellite transmitter power	10 W
Satellite backoff loss	0.1 dB
Other losses and atmospheric loss	0.9 dB
Receive frequency	12 GHz
Earth station antenna gain	62 dB
Satellite antenna gain	30.8 dB
Earth station equivalent temperature	270 K

(a) Show that total carrier-to-noise ratio is 99.5 dB·Hz.

(b) The satellite system uses DQPSK signal. It is desired to have an error rate of 10^{-5} ($\varepsilon_b/N_0 = 10.3$ dB). Find the data rate and the minimum satellite channel bandwidth required to support the data rate.

(c) Assume that the data is convolutionally coded with code rate (see chapter 2) 1/2 constraint length 5. Assume that the coding gain is 3 dB. Find the corresponding data rate for $P_e = 10^{-5}$ and the minimum satellite channel bandwidth for this case.

(d) Let the voice channel bandwidth be 4 kHz. How many voice channels can be supported if the number of preamble bits for each channel is 25 percent of the data bits.

(e) Assuming an antenna efficiency of 0.55, find the diameter of the satellite antenna and the satellite antenna figure of merit G/T_S.

27. In an SCPC FDMA system, the number of accesses m is given by

$$m = \frac{(C/N_0)_{CH}^{-1}}{(C/N_0)_T^{-1}}$$

where

$(C/N_0)_{CH} = (C/N_0)$ required per channel

$(C/N_0)_T$ = composite carrier to noise ratio

$$= [(C/N_0)_u^{-1} + (C/N_0)_d^{-1} + (C/I)^{-1}]^{-1}$$

Let

$(C/N_0)_u = 105 + X$ dB·Hz

$(C/N_0)_d = 90 - 0.03X^2 + 0.07X$ dB·Hz

$(C/I) = 77.0 + 0.05X^2 - 0.5X$ dB·Hz

$\quad\quad = 77.0 + Z(X),\quad\quad$ dB·Hz

where X is the input backoff and Z is the output backoff.

$(C/N_0)_T = 10\log m + 10\log R + \text{(sum of margin)} + \varepsilon_b/N_0$

where ε_b/N_0 is the bit energy-to-noise ratio and R is the transmission rate.

(a) Show that optimum input backoff is 10.9 dB and $(C/N_0)_T = 83.4$ for $X = 10.9$ dB.

(b) Let the total margin available be 5 dB and data rate $R = 76$ Mbps at 10^{-5} bit error rate using binary noncoherent FSK. Find the $(C/N_0)_{CH}$ for each channel and number of accesses m.

(c) Let $b = B/M$, bandwidth for each channel. Let $B = 80$ MHz. Find the bandwidth b for each channel.

(d) Let the BFSK signal be convolutionally coded with rate 1/2, constraint length 5. Find $(C/N_0)_{CH}$ for each channel and m and b. Find the maximum number of PCM voice channels that can be carried in a TDMA frame if

(1) Frame length = 20 ms

(2) Burst rate = 90 mbps

(3) Number of stations = 10

(4) Number of reference bursts = 2

(5) Total traffic bursts = 20

(6) Length of carrier and clock recovery (CCR) sequence = 256 bits
(7) Length of unique word (UW) = 48 bits
(8) Length of order wire (OW) channel sequence = 256 bits
(9) Length of management channel = 128 bits
(10) Length of timing channel sequence = 200 bits
(11) Length of service channel sequence = 28 bits
(12) Length of guard time = 32 bits.

28. Let there be k number of users in a CDMA system and f_b and f_c be the data rate and chip rate, respectively, of the individual user. Each user transmits a BPSK signal. Assuming that the thermal noise is very small compared to the interference noise, show that the probability of bit error rate in CDMA DSBPSK system is

$$P_e = \frac{1}{2}\text{erfc}\left\{\left[2\frac{1}{(k-1)}\left(\frac{f_c}{f_b}\right)\right]^{1/2}\right\}$$

Note that f_c/f_b is called the *processing gain*. It is desirable for the processing gain to be much larger than $(k-1)/2$.

29. Let $f_c = 22\text{ GHz}$, elevation angle $= 60°$, and latitude $52°$. The rain rate that occurs for $p = 0.01$ percent of the time in an average year is 40 mm/h. The earth station above sea level $h_0 = 0.78$ km. Show that

$$h_R = 5.1\cos(1.06 \times 52°) = 2.916\text{ km}$$

$$L_S = \frac{2.916 - 0.78}{\sin 60°} = 2.136\text{ km}$$

$$L_G = L_S\cos 60° = 1.068\text{ km}$$

$$r_p = \frac{90}{90 + 4.3} = 0.95$$

$$A_{01} = 0.1338(40)^{1.062} \times 2.136 \times 0.95$$

$$= 11.3\text{ dB}$$

The attenuation due to rain for 0.01 percent of the time is 11.3 dB.

30. The model for binary digital communications with L diversity is shown in Fig. 1.75. The received signal in the kth channel is

$$y_k(t) = \alpha_k e^{i\phi_k}u_k(t) + n_k(t) \qquad k = 1, 2, 3, \ldots$$

where $u_k(t) = A\cos[\omega_c t + \phi_k(t)]$ (PSK)

$$= A\cos(2\pi f_k t + \theta)   \text{(FSK)}$$

α_k and ϕ_k are amplitude attenuation and phase shifts in the kth channel.

(a) Show that the probability of bit error for BPSK modulation is

$$P = \left(\frac{1-d}{2}\right)^L \sum_{k=0}^{L-1}\binom{L+k-1}{k}\frac{(1+d)^k}{2^k}$$

where

$$d^2 = \frac{E(\alpha_k^2)(A^2 T/2N_0)}{1 + E(\alpha_k^2)(A^2 T/2N_0)}$$

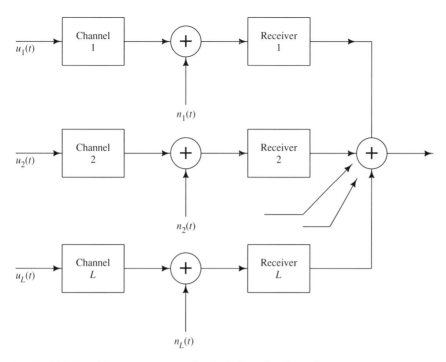

FIGURE 1.75   Binary communcation in L diversity channel.

where A is the amplitude of the kth signal and $N_0/2$ is the variance of the additive zero-mean white noise $n_k(t)$, $k = 1, \ldots, L$, and E is the expectation operator.

(b) Show that the probability of bit error for FSK modulation is

$$P = \left[(2^{2L-1}(L-1)!)^{-1} \left[E(\alpha_k^2) \left(\frac{A^2 T}{2N_0} \right) \right]^{-1} \left[\sum_{k=0}^{L-1} c_k (L + k - 1)! \, d^k \right] \right] d^L$$

where

$$c_k = \frac{1}{k!} \sum_{n=0}^{L-1-k} \binom{2L-1}{n}$$

where d is defined in part (a).

REFERENCES

A. Historical

Edelson, B. I., "Communication Satellites: The Experimental Years," *Acta Astronaut.*, vol. 2, no. 7–8, 1984.

Wheelon, A. D., "The Rocky Road to Communication Satellites," AIAA 24th Aerospace Sciences Meeting, January 1986.

B. General

Anderson, J. B., T. Aulin, and C. E. Sundberg, *Digital Phase Modulation*, New York: Plenum Press, 1986.

Beakley, G. W., "Overview of Commercial Satellite Communications," *IEEE Trans. Aerosp. and Electron. Syst.*, vol. AES-20, July 1984.

Bell Telephone Laboratories, *Transmission Systems for Communications*, Salem, N.C.: Western Electric Company, 1971.

Benedetto, S., E. Biglieri, and V. Castellani, *Digital Transmission Theory*, Englewood Cliffs, N.J.: Prentice-Hall, 1987.

Bogusch, R. L., *Digital Communications in Fading Channels*, Santa Barbara, Calif.: Mission Research Corporation, 1987.

Boithias, L., *Radio Wave Propagation*, London: North Oxford Academic, 1987.

CCIR, *Handbook on Satellite Communications*, Geneva, Switzerland: CCIR, 1985.

Carlson, A. B., *Communication Systems*, New York: McGraw-Hill, 1975.

Cooper, G. R., and C. D. McGillem, *Modern Communications and Spread Spectrum*, New York: McGraw-Hill, 1986.

Evans, B. G., Ed., *Communication Systems*, London: Peter Peregrinus, 1987.

Feher, K., Ed., *Advanced Digital Communications*, Englewood Cliffs, N.J.: Prentice-Hall, New York (1987)

Fontolliet, P., *Telecommunication Systems*, Dedham, Mass.: Artech House, 1986.

Freeman, R. L., *Radio System Design for Telecommunications*, New York: Wiley, 1987.

Fthenakis, E., *Manual of Satellite Communications*, New York: McGraw-Hill, 1984.

Gagliardi, R. M., *Communication Engineering*, New York: Wiley, 1988.

Gagliardi, R. M., *Satellite Communications*, New York: Van Nostrand Reinhold 1984.

Gardner, F. M., *Phaselock Techniques*, New York: Wiley, 1979.

Ha, T. T. *Digital Satellite Communication*, New York: Macmillan, 1986.

Haykin, S., *Digital Communications*, New York: Wiley, 1988.

Hershey, J. E., L. W. Pederson, H. M. Gates, and B. Bittner, "An Overview of EHF Communications for Normal and Stressed Environments," *NTC 83 Conf. Rec.*, pp. 6–11, December 1983.

Holmes, J. K., *Coherent Spread Spectrum Systems*, New York: Wiley, 1982.

Ippolito, L. J., *A Propagation Effects Handbook for Satellite Systems Design*, Washington, D.C.: NASA Headquarters, 1980.

Kadar, I., Ed., "Satellite Communications Systems," *AIAA Selected Reprint Series*, vol. 18, January 1976.

Kennedy, G., *Electronic Communication Systems*, New York: McGraw-Hill, 1985.

Killen, H. B., *Digital Communications with Fiber Optics and Satellite Applications*, Englewood Cliffs, NJ: Prentice-Hall, 1988.

Lee, E. A. and D. G. Masserschmit, *Digital Communication*, Boston: Kluwer Academic Publishers, 1988.

Lindsey, W. C., and C. M. Chie, "Performance Measures for Phase-Locked Loops—A Tutorial," *IEEE Trans. Commun. Technol.*, vol. COM-30, pp. 2224–2227, October 1982.

Lindsey, W. C., and M. K. Simon, *Telecommunication Systems Engineering*, Englewood Cliffs, N.J.: Prentice-Hall, 1973.

Long, M., *World Satellite Almanac*, Boise, Idaho: Comtek Publishing Company, 1985.

Martin, D. H., *Communication Satellites*, El Segundo, Calif.: The Aerospace Corporation, 1988.

Miya, K., Ed., *Satellite Communications Technology*, Tokyo: KDD Engineering and Consulting, Inc., 1982.

Mohanty, N. C., *Random Signals, Estimation, and Identification*, New York: Van Nostrand Reinhold, 1986.

Mohanty, N. C., *Signal Processing: Signals, Filtering, and Detection*, New York: Van Nostrand Reinhold, 1987.

Mural, G., and M. Bousquet, *Satellite Communication Systems*, New York: Wiley, 1986.

Omura, J. K., and B. K. Levitt, "Coded Error Probability Evaluation for Antijam Communication Systems," *IEEE Trans. Commun. Technol.*, vol. COM-30, pp. 896–903, 1982.

Panter, P. E., *Communication System Design*, New York: McGraw-Hill, 1972.

Papoulis, A., *Probability, Random Variables, and Stochastic Processes*, New York: McGraw-Hill, 1983.

Peebles, P. Z., Jr., *Digital Communication Systems*, Englewood Cliffs, N.J.: Prentice-Hall, 1987.

Pierce, J. R., and E. C. Posner, *Introduction to Communication Science and Systems*, New York: Plenum Press, 1980.

Pratt, T., and C. W. Bostian, *Satellite Communications*, New York: Wiley, 1986.

Prentiss, S., *Satellite Communication*, Blue Ridge Summit, Pa.: Tab Books, 1987.

Pritchard, W. L., "The History and Future of Commercial Satellite Communications," *IEEE Commun. Mag.*, vol. 22, May 1985.

Pritchard, W. L., and J. A. Sciulli, *Satellite Communication: Systems Engineering*, Englewood Cliffs, N.J.: Prentice-Hall, 1986.

Saveskie, P. N., *Radio Propagation Handbook*, Blue Ridge Summit, Pa.: Tab Books, 1980.

Schwartz, M., W. R. Bennett, and S. Stein, *Communication Systems and Techniques*, New York: McGraw-Hill, 1966.

Smith, D. R., *Digital Transmission Systems*, New York: Van Nostrand Reinhold, 1985.

Spilker, J. J., Jr., *Digital Communication by Satellite*, Englewood Cliffs, N.J.: Prentice-Hall, 1977.

Taub, H., and D. L. Schilling, *Principles of Communication Systems*, New York: McGraw-Hill, 1986.

Tomasi, W., *Advanced Electronic Communication Systems*, Englewood Cliffs, N.J.: Prentice-Hall, 1987.

Tzannes, N. S. *Communication and Radar Systems*, Englewood Cliffs, N.J.: Prentice-Hall, 1985.

Van Trees, H. L., Ed., *Satellite Communications*, New York: IEEE Press, 1979.

Viterbi, A. J. and J. K. Omura, *Principles of Digital Communication and Coding*, New York: McGraw-Hill, 1979.

Widrow, B. and S. D. Stearns, *Adaptive Signal Processing*, Englewood Cliffs, N.J.: Prentice-Hall, 1985.

Williams, R. A., *Communication Systems Analysis and Design*, Englewood Cliffs, N.J.: Prentice-Hall, 1987.

Wu, W. W., *Elements of Digital Satellite Communications*, Rockville, Md.: Computer Science Press, 1984.

Ziemer, R. E., and R. L. Peterson, *Digital Communications and Spread Spectrum Systems*, New York: Macmillan, 1985.

C. Military

Cummings, D., and C. G. Wildey, "Military Aeronautical Satellite Communications," *IEE Proc. (London)*, vol. 133, part F, no. 4, July 1986.

LaVean, G. E., "Interoperability in Defense Communications," *IEEE Trans. Commun. Technol.*, vol. COM-28, September 1980.

D. Intelsat

Astrain, S., "Satellite Communications: Intelsat and Global Patterns," *Space Commun. Broadcast.*, vol. 1, October 1983.

Welti, G. R., "Intelsat Architectures for the 1990s," Paper 3A.5, International Conference on Communications, June 1982.

E. Mobile

Bitzer, K. L., et al., "The Extension of the Inmarsat Coverage to the Polar Caps: A System Tradeoff," *Earth-Oriented Appli. Space Technol.*, vol 6, no. 2, 1986.

Sandrin, W. A., and D. W. Lipke, "A Satellite System for Aeronautical Data Communications," Paper 86-0602, AIAA 11th Communication Satellite Systems Conference, March 1986.

Yoshikawa, T., et al., "Link Configuration for Mobile Satellite Communication Systems," Paper 24.2, International Conference on Communications, June 1985.

F. Broadcasting

Cohen, H. D., "Spacecraft Technology for Broadcasting Satellites," *IEEE J. Selected Areas Commun.*, vol. 3, January 1985.
Dement, D. K., "United States Direct Broadcast Satellite System Development," *IEEE Commun. Mag.*, vol. 22, March 1984.
Reinhart, E. E., "An Introduction to the RARC '83 Plan for DBS Systems in the Western Hemisphere," *IEEE J. Selected Areas Commun.*, vol. 3, January 1985.
Siocos, C. A., "The International Telecommunication Union and Broadcasting from Satellites—A Brief Review," *IEEE J. Selected Areas Commun.*, vol. 3, January 1985.

G. Rural and Thin Routes

Abramson, N., "Satellite Data Networks for National Development," *Telecommun. Policy*, vol. 8, March 1984.
Briskman, R. D., "Domestic Satellite Services for Rural Areas," *IEEE Commun. Mag.*, vol. 22, March 1984.
Caccianani, E. R., W. B. Garner, and S. B. Salamoff, "The Emergence of Satellite Systems for Rural Communications," *Proc. IEEE*, vol. 72, November 1984.
Hudson, H. E., *New Directions in Satellite Communications: Challenges for North and South*, Dedham, Mass.: Artech House, 1985.

H. Advanced Concepts

Christopher, P. F., A. K. Kamal, and E. R. Edelman, "Space-Ground Tradeoffs for Millimeter Wave Satellite Communications," Paper 40.2, International Conference on Communications, June 1985.
Clopp, W., Jr., et al., "Geostationary Communications Platform Payload Concepts," Paper 86-0697, AIAA 11th Communication Satellite Systems Conference, March 1986.
Dreggers, T. F., and E. M. Hunter, "Geostationary Communications Platform Payload Concepts," Paper 86-0696, AIAA 11th Communication Satellite Systems Conference, March 1986.
Pelton, J. N. "Satellite Telenets: A Techno-Economic Assessment of Major Trends for the Future," *Proc. IEEE*, vol. 72, November 1984.
Teshrigi, T., and W. Chujo, "A Multiple-Access Link in an Inter-Satellite Data Relay System Using an On-Board Multibeam Antenna," Paper 24.7, International Conference on Communications, June 1985.

I. Applications

Block, C. H., "Satellite Linkages and Rural Development," in *New Directions in Satellite Communications: Challenges for North and South*, H. E. Hudson, Ed., Dedham, Mass.: Artech House, 1985.
Chakraborty, D., et al., "Wideband Digital Transmission Experiments in the Intelsat V System," *Comsat Tech. Rev.*, vol. 13, Fall 1983.

J. Ground Terminals

Barthle, R. C., and R. D. Briskman, "Trends in Design of Communications Satellite Earth Stations," *Microwave J.*, vol. 10, October 1967.
Bartholome, P., and C. D. Hughes, "Satellite Communications in Europe: The Earth-Segment Market," *ESA Bull.*, no. 44, November 1985.
Inoue, T., T. Saito, and K. Kagoshima, "30/20 GHz Band Small Earth Station for ISSDN Experiment," *IEEE Trans. Aerosp. Electron. Syst.*, vol. AES-17, November 1981.

McGann, W. E., "Ku-Band Satellite Digital Transmission Systems," *Internat. J. Satel. Commun.*, vol. 3, July–September 1985.

Parker, E. B., "Micro Earth Stations as Personal Computer Accessories," *Proc. IEEE*, vol. 72, November 1984.

K. Modulation and Multiple Access

Amitay, N., and L. J. Greenstein, "Multipath Outage Performance of Digital Radio Receivers Using Finite-Tap Adaptive Equalizers," *IEEE Trans. Commun. Technol.*, vol. COM-32, pp. 597–608, May 1984.

Arthurs, E., and H. Dym, "On the Optimum Detection of Digital Signals in the Presence of White Gaussian Noise—A Geometric Interpretation and a Study of Three Basic Data Transmission Systems," *IRE Trans. Commun. Syst.*, vol. CS-10, pp. 336–372, December 1962.

Aulin, T., and C. E. Sundberg, "Continuous Phase Modulation—Parts I and II," *IEEE Trans. Commun. Technol.*, vol. COM-29, pp. 196–225, March 1981.

Austin, M. C., M. U. Chang, D. F. Horwood, and R. A. Maslov, "QPSK, Staggered QPSK, and MSK—A Comparative Evaluation," *IEEE Trans. Commun. Technol.*, vol. COM-31, pp. 171–182, February 1983.

Bello, P. A., "Aeronautical Channel Characterization," *IEEE Trans. Commun. Technol.*, vol. COM-21, pp. 548–563, May 1973.

Berlekamp, E. R., "The Technology of Error-Correcting Codes," *Proc. IEEE*, vol. 68, pp. 563–593, May 1980.

Bhargava, V. K., *Digital Communications by Satellite*, New York: Wiley, 1981.

Blachman, N. M., "The Effects of Phase Error on DPSK Error Probability," *IEEE Trans. Commun. Technol.*, vol. COM-29, pp. 364–365, March 1981.

Bogusch, R. L., F. W. Guigliano, and D. L. Knepp, "Frequency-Selective Scintillation Effects and Decision Feedback Equalization in High Data-Rate Satellite Links," *Proc. IEEE*, vol. 71, pp. 754–767, June 1983.

Bogusch, R. L., F. W. Guigliano, D. L. Knepp, and A. H. Michelet, "Frequency Selective Propagation Effects on Spread-Spectrum Receiver Tracking," *Proc. IEEE*, vol. 69, pp. 787–796, July 1981.

Bussgand, J. J., and M. Leiter, "Error Rate Approximations for Differential Phase-Shift Keying," *IEEE Trans. Commun. Syst.*, vol. CS-12, pp. 18–27, March 1964.

Cain, J. B., G. C. Clark, Jr., and J. M. Geist, "Punctured Convolutional Codes of Rate $(n-1)/n$ and Simplified Maximum Likelihood Decoding," *IEEE Trans. Inform. Theory*, vol. IT-25, pp. 97–100, January 1979.

Campanella, S. J., and J. V. Carrington, "Satellite Communications Networks," *Proc. IEEE*, vol. 72, November 1984.

Chase, D., "Class of Algorithms for Decoding Block Codes with Channel Measurement Information," *IEEE Trans. Inform. Theory*, vol. IT-18, pp. 170–182, January 1972.

Chase, D., "A Combined Coding and Modulation Approach for Communication over Dispersive Channels," *IEEE Trans. Commun. Technol.*, vol. COM-21, pp. 159–174, March 1973.

Chase, D., "Digital Signal Design Concepts for a Time-Varying Rician Channel," *IEEE Trans. Commun. Technol.*, vol. COM-24, pp. 164–172, February 1976.

Ekanayake, N., and D. P. Taylor, "A Decision Feedback Receiver Structure for Bandlimited Nonlinear Channels," *IEEE Trans. Commun. Technol.*, vol. COM-29, pp. 539–548, May 1981.

Fang, R. J. F., "Quaternary Transmission Over Satellite Channels with Cascaded Nonlinear Elements and Adjacent Channel Interference," *IEEE Trans. Commun. Technol.*, vol. COM-29, pp. 567–581, May 1981.

Feher, K., *Digital Communications: Satellite/Earth Station Engineering*, Englewood Cliffs, N.J.: Prentice-Hall, 1983.

Feher, K., et al., Eds., "Special Issue on Digital Satellite Communications," *IEEE J. Selected Areas Commun.*, vol. 1, January 1983.

Forney, G. D., Jr., "The Viterbi Algorithm," *Proc. IEEE*, vol. 61, pp. 268–278, March 1973.

Foschini, G. J., and J. Salz, "Digital Communications Over Fading Radio Channels," *Bell Syst. Tech. J.*, vol. 62, pp. 429–456, February 1983.

Gardner, C. S., and J. A. Orr, "Fading Effects on the Performance of a Spread Spectrum Multiple Access Communication System," *IEEE Trans. Commun. Technol.*, vol. COM-27, pp. 143–149, January 1979.

Geraniotis, E. A., and M. B. Pursley, "Error-Probabilities for Slow Frequency-Hopped Spread-Spectrum Multiple Access Communications Over Fading Channels," *IEEE Trans. Commun. Technol.*, vol. COM-31, pp. 996–1009, May 1982.

Geraniotis, E. A., and M. B. Pursley, "Performance of Coherent Direct-Sequence Spread-Spectrum Communications Over Specular Multipath Fading Channels," *IEEE Trans. Commun. Technol.*, Vol. COM-33, pp. 502–508, June 1985.

Geraniotis, E. A., and M. B. Pursley, "Performance of Noncoherent Direct-Sequence Spread-Spectrum Communications Over Specular Multipath Fading Channels," *IEEE Trans. Commun. Technol.*, vol. COM-34, pp. 219–226, March 1986.

Gronemeyer, S. A. and A. L. McBride, "MSK and Offset QPSK Modulation," *IEEE Trans. Commun. Technol.*, vol. COM-24, pp. 809–819, August 1976.

Heller, J. A., and I. M. Jacobs, "Viterbi Decoding for Satellite and Space Communications," *IEEE Trans. Commun. Technol.*, vol. COM-19, pp. 835–848, October 1971.

Houston, S. W., "Modulation Techniques for Communication, Part I: Tone and Noise Jamming Performance of Spread Spectrum M-ary FSK and 2, 4-ary DPSK Waveforms," *Proc. Nat. Aerosp. Electron. Conf.*, pp. 51–58, June 1975.

Jones, J. J., "Multichannel FSK and DPSK Reception with Three-Component Multipath," *IEEE Trans. Commun. Technol.*, vol. COM-16, pp. 808–821, December 1968.

Jonnalgadda, K., and L. Schiff, "Improvements in Capacity of Analog Voice Multiplex Systems Carried by Satellite," *Proc. IEEE*, vol. 72, November 1984.

Larsen, K. J., "Short Convolutional Codes with Maximal Free Distance for Rates 1/2, 1/3, and 1/4," *IEEE Trans. Inform. Theory*, vol. IT-19, pp. 371–372, May 1973.

Ma, H. H., N. M. Shehadeh, and J. C. Vanelli, "Effects of Intersymbol Interference on a Rayleigh Fast-Fading Channel," *IEEE Trans. Commun. Technol.*, vol. COM-28, pp. 128–131, January 1980.

Modestino, J., and S. Mui, "Convolutional Code Performance in the Rician Fading Channel," *IEEE Trans. Commun. Technol.*, vol. COM-24, pp. 592–606, June 1976.

Monsen, P., "Theoretical and Measured Performance of a DFE Modem on a Fading Multipath Channel," *IEEE Trans. Commun. Technol.*, vol. COM-25, pp. 1144–1153, October 1977.

Monsen, P. "Fading Channel Communications," *IEEE Commun. Mag.*, pp. 16–25, vol. 18, January 1980.

Monsen, P., "MMSE Equalization of Interference on Fading Diversity Channels," *IEEE Trans. Commun. Technol.*, vol. COM-32, pp. 5–12, January 1984.

Natali, F. D., "Noise Performance of a Cross-Product AFC with Decision Feedback for DPSK Signals," *IEEE Trans. Commun. Technol.*, vol. COM-34, pp. 303–307, March 1986.

Oetting, J. D., "A Comparison of Modulation Techniques for Digital Radio," *IEEE Trans. Commun. Technol.*, vol. COM-27, pp. 1752–1762, December 1979.

Peterson, W. W., and E. J. Weldon, Jr., *Error Correcting Codes*, 2nd ed., Cambridge, Mass.: MIT Press, 1972.

Pursley, M. B., and W. E. Stark, "Performance of Reed-Solomon Coded Frequency-Hop Spread-Spectrum Communications in Partial-Band Interference," *IEEE Trans. Commun. Technol.*, vol. COM-33, pp. 767–774, August 1985.

Qureshi, S. U. H., "Adaptive Equalization," *Proc. IEEE*, vol. 73, pp. 1349–1387, September 1985.

Ramsey, J. L., "Realization of Optimum Interleavers," *IEEE Trans. Inform. Theory*, vol. IT-16, pp. 338–345, May 1970.

Rhodes, S. A., "Effect of Noisy Phase Reference on Coherent Detection of Offset-QPSK Signals," *IEEE Trans. Commun. Technol.*, vol. COM-22, pp. 1046–1055, August 1974.

Shimbo, O., R. J. Fang, and M. Celebiler, "Performance of M-ary PSK Systems in Gaussian Noise and Intersymbol Interference," *IEEE Trans. Inform. Theory*, vol. IT-19, pp. 44–58, January 1973.

Stark, W. E., "Coding for Frequency-Hopped Spread-Spectrum Communication with Partial-Band Interference—Part I: Capacity and Cutoff Rate," *IEEE Trans. Commun. Technol.*, vol. COM-33, pp. 1036–1044, October 1985.

Stark, W. E., "Coding for Frequency-Hopped Spread-Spectrum Communication with Partial-Band Interference—Part II: Coded Performance," *IEEE Trans. Commun. Technol.*, vol. COM-33, pp. 1045–1057, October 1985.

Stark, W. E., "Capacity and Cutoff Rate of Noncoherent FSK with Nonselective Rician Fading," *IEEE Trans. Commun. Technol.*, vol. COM-33, pp. 1153–1159, November 1985.

Stutzman, W. L., T. Pratt, D. M. Imrich, W. A. Scales, and C. W. Bostian, "Dispersion in the 10–30 GHz Frequency Range: Atmospheric Effects and Their Impact on Digital Satellite Communications," *IEEE Trans. Commun. Technol.*, vol. COM-34, pp. 307–310, March 1986.

Ungerboeck, G., "Channel Coding with Multilevel/Phase Signals," *IEEE Trans. Inform. Theory*, vol. IT-28, pp. 56–67, January 1982.

Viterbi, A. J., "Error Bounds for Convolutional Codes and an Asymptotically Optimum Decoding Algorithm," *IEEE Trans. Inform. Theory*, vol. IT-13, pp. 260–269, April 1967.

Viterbi, A. J., "Convolutional Codes and Their Performance in Communication Systems," *IEEE Trans. Commun. Technol.*, vol. COM-19, pp. 751–772, October 1971.

L. Multiple Access Techniques

Bertin, C., "PACKSATNET—An Alternative to Data Networks," *Proc. IEEE*, vol. 72, November 1984.

Brayer, K., "Packet Switching for Mobile Earth Stations via Low-Orbit Satellite Network," *Proc. IEEE*, vol. 72, November 1984.

Chang, J. F., "A Multibeam Packet Satellite Using Random Access Techniques," *IEEE Trans. Commun. Technol.*, vol. COM-31, October 1983.

Chang, J. F., and L. Y. Liu, "Demand Assigned Packet Broadcast Systems with buffers and Trailer Transmissions," Global Telecommunications Conference, November 1983.

Chethik, F., V. Gallagher, and C. Hoeber, "Waveform and Architecture Concepts for a High Efficiency TDMA Satcom System," Paper 86-0632, AIAA 11th Communication Satellite Systems Conference, March 1986.

Gopal, I., D. Coppersmith, and C. K. Wong, "Minimizing Packet Waiting Time in Multibeam Satellite System," *IEEE Trans. Commun. Technol.*, vol. COM-30, February 1982.

Gopal, I., and C. K. Wong, "Minimizing the Number of Switching in an SS/TDMA System," in *Satellite and Computer Communications*, J. L. Grange, Ed., Amsterdam: 1983; also in *IEEE Trans. Commun. Technol.*, vol. COM-33, June 1985.

Jabbari, B., "Cost-Effective Networking via Digital Satellite Communications," *Proc. IEEE*, vol. 72, November 1984.

Maral, G., M. Bousquet, and P. Wattier, "Performance of Fully Variable Demand Assignment SS/TDMA Satellite Systems," Paper 86-0630, AIAA 11th Communication Satellite Systems Conference, March 1986.

Matos, F., Ed., *Spectrum Management and Engineering*, New York: IEEE Press, 1985.

Muratani, T., "Satellite-Switched Time-Domain Multiple Access," *EASCON '74 Conv. Rec.*, October 1974.

Nuspl, P. O., "Synchronization Methods for TDMA," *Proc. IEEE*, vol. 65, March 1977.

Okinaka, H., Y. Yasuda, and Y. Hirata, "Intermodulation Interference—Minimum Frequency Assignment for Satellite SCPC Systems," *IEEE Trans. Commun. Technol.*, vol. COM-32, April 1984.

Pattini, F., and P. Porzio-Giusto, "Synchronization Technique for the On-Board Master Clock of a Regenerative TDMA Satellite Communications System," Paper 32.3, International Conference on Communications, June 1985.

Pennoni, G., "Bit and Burst Synchronization in Regenerative SS-TDMA Systems," Paper 32.8, International Conference on Communications, June 1985.

M. Miscellaneous Topics

Chakraborty, D., "Constraints in Ku-Band Continental Satellite Network Design," *IEEE Commun. Mag.*, vol. 24, August 1986.
Ramaswany, R., and P. Dhar, "An Algorithm for the Location of Satellite Ground Stations in the Design of Terrestrial/Satellite Computer Communication Networks," in *Satellite and Computer Communications*, J. L. Grange, Ed., Amsterdam: North-Holland, 1983.
Salmasi, A. B., and Y. Rahmat-Samii, "Beam Area Determination for Multiple-Beam Satellite Communication Applications," *IEEE Trans. Aerosp. Electron. Syst.*, vol. AES-19, May 1983.

N. Simulations

Ananasso, F., E. Biglieri, and E. Saggesse, "Counteracting High Noise Levels in a Satellite Link: A Computer Simulation Approach," Paper 48.5, International Conference on Communications, June 1985.
Aulin, T., and C. E., Sundberg, "CPM-The Effect of Filtering and Hard Limiting," IEEE National Telesystems Conference, November 1983.
Epstein, B. R. "A Program to Test Satellite Transponder for Spurious Signals," *RCA Rev.*, vol. 46, September 1985.
Kwatra, S. C., B. W. Maples, and G. A. Stevens, "Modeling of NASA's 20/30 GHz Satellite Communications System," Paper 35.7, International Conference on Communications, May 1984.
McKenzie, T., S. An, and J. Hsu, "A System to Demonstrate Integrated Approach of the Communications Link Modeling Programs of CLASS," Paper 48.4, International Conference on Communications, June 1985.
Meader, C. B., S. C. Kwatra, and G. H. Stevens, "Simulated Performance of the NASA 30/20 GHz Test Transponder Using Multi-H Phase Coded Modulation," Paper 86-0717, AIAA 11th Communication Satellite Systems Conference, March 1986.
Milstein, L., R. Pickholtz, and D. Schilling, "Comparison of Performance of Digital Modulation Techniques in the Presence of Adjacent Channel Interference," *IEEE Trans. Commun. Technol.*, vol. COM-30, August 1982.
Moens, C., and C. Kooter, "ECS-1 In-Orbit Measurements Programme and Results," *ESA J.*, vol. 8, No. 1 (1984).
Oka, I., "Intersymbol and CW Interference in QPSK, OQPSK and MSK Hard-Limiting Satellite Systems," *IEEE Trans. Aerosp. Electron. Syst.*, vol. AES-22, January 1986.

O. Network Control and Monitoring

Potukuchi, J. R., F. T. Assal, and R. C. Mott, "A Computer-Controlled Satellite Communications Monitoring System for TDMA," *Comsat Tech. Rev.*, vol. 14, Fall 1984.

P. Atmospheric Propagation

Aarons, J. "Construction of a Model of Equatorial Scintillation Intensity," *Radio Sci.*, vol. 20, pp. 397–402, May–June 1985.
Aarons, J., "Equatorial F-layer Irregularity Patches at Anomaly Latitudes," *J. Atmos. Terr. Phys.*, vol. 47, pp. 875–883, 1985.
CCIR, *Propagation in Ionized Media.* Geneva, Switzerland: CCIR, 1982.
Craine, R. K., "Comparative Evaluation of Several Rain Attenuation Prediction Models," *Radio Sci.*, vol. 20, July–August 1985.
Crane, R. K., "Prediction of Attenuation of Rain," *IEEE Trans. Commun. Technol.*, vol. COM-28, September 1980.

Fremouw, E. J., R. L. Leadabrand, R. C. Livingston, M. D. Cousins, C. L. Rino, B. C. Fair, and R. A. Long, "Early Results from the DNA Wideband Satellite Experiment—Complex-Signal Scintillation," *Radio Sci.*, vol. 13, pp. 167–187, January–February 1978.

Ippolito, L. J., *Radiowave Propagation in Satellite Communication*, New York: Van Nostrand Reinhold, 1986.

Ippolito, L. J., R. D. Kaul, and R. G. Wallace, *Propagation Effects Handbook for Satellite System Design*, 3rd ed., NASA Reference Publication 1082(03). Washington, D.C.: NASA, 1983.

Johnson, A., and J. Taagholt, "Ionospheric Effects on C^3I Satellite Communication Systems in Greenland," *Radio Sci.*, vol. 20, May–June 1985.

Rino, C. L., V. H. Gonzalez, and A. R. Hessing, "Coherence Bandwidth Loss in Transionospheric Radio Propagation," *Radio Sci.*, vol. 16, pp. 245–255, March–April 1981.

Rogers, D. V., "Propagation Considerations for Satellite Broadcasting at Frequencies above 10 GHz," *IEEE J. Selected Areas Commun.*, vol. 3, January 1985.

S. Stein, "Fading Channel Issues in Systems Engineering," *IEEE J. Selected Areas Commun.*, vol. 5, pp. 68–89, February 1987.

Q. Nuclear Effects

Arendt, P. R., and H. Soicher, "Effects of Arctic Nuclear Explosions on Satellite Radio Communication," *Proc. IEEE*, vol. 52, pp. 672–676, June 1964.

Baker, K. D., and J. C. Ulwick, "Measurements of Electron Density Structure in Striated Barium Clouds," *Geophys. Res. Lett.*, vol. 5, pp. 723–726, August 1978.

Boquist, W. P., and J. W. Snyder, "Conjugate Auroral Measurement from the 1962 U.S. High Altitude Nuclear Test Series," *Aurora and Airglow*, B. McCormac, ed., pp. 325–339, New York: Van Nostrand Reinhold, 1967.

Budden, K. G., *Radio Waves in the Ionosphere*, New York: Cambridge University Press, 1988.

Glasstone, S. and P. J. Dolan, *The Effects of Nuclear Weapons*, Washington, DC: U.S. Department of Defense 1977.

Ishimaru, A., *Wave Propagation and Scattering in Random Media*, Orlando, Fla.: Academic Press, 1978.

Kanellahos, D. P., "Response of the Ionosphere to the Passage of Acoustic Gravity Waves Generated by Low-Altitude Nuclear Explosions," *J. Geophys. Res.*, vol. 72, p. 4559, 1967.

Kelley, M. C., K. D. Baker, and J. C. Ulwick, "Late Time Barium Cloud Striations and Their Possible Relationship to Equatorial Spread F," *J. Geophy. Res.*, vol. 84, pp. 1898–1904, May 1, 1979.

Kondratyev, K. Ya, *Climate Shocks: Natural and Anthropogenic*, New York: Wiley, 1988.

Marshall, J., "PLACES—A Structured Ionospheric Plasma Experiment for Satellite System Effects Simulation," AIAA-82-0149, AIAA 20th Aerospace Sciences Meeting, January 11–14, 1982.

Middlestead, R. W., R. E. LeLevier, and M. D. Smith, "Satellite Crosslink Communications Vulnerability in a Nuclear Environment," *IEEE J. Selected Areas Commun.*, February 1987.

Moeneclay, M., "Comments on Tracking Performance of the Filter and Square Bit Synchronizer, *IEEE Trans. Commun. Technol.*, vol. COM-30, February 1982.

National Research Council, *Long-Term World Effects of Multiple Nuclear Weapons Detonation*, Washington, DC: National Academy of Sciences, 1975.

National Research Council, *The Effects on the Atmosphere of a Major Nuclear Exchange*, Washington, DC: National Academy Press, 1985.

"Nuclear Effects on RF Propagation, *Defense Electron*. pp. 228–232, October 1985.

Ratcliffe, J. A., "The Magneto-Ionic Theory and Its Applications to the Ionosphere," New York: Cambridge University Press, 1962.

Rufenach, C. L., "Ionospheric Scintillation by a Random Phase Screen: Spectral Approach," Radio Sci., Vol. 10, pp. 155–165, February 1975.

Rummler, W. D., "A New Selective Fading Model: Application to Propagation Data," *Bell Syst. Tech J.*, Vol. 54, pp. 1095–1125, July–August, 1979.

"Symposium on the Physical Effects of the High-Altitude Nuclear Explosion Above Johnston Island on 9 July 1962," *New Zeal. J. Geol. Geophys.* Special Nuclear Explosion Issue, vol. 5, December 1962.

Tatarskii, V. I., "The Effects of the Turbulent Atmosphere on Wave Propagation," translated by Israel Program for Scientific Translations, National Technical Information Service, Washington, D.C.: U.S. Department of Commerce, 1971.

Tsipis, K., *Arsenal: Understanding Weapons in the Nuclear Age*, New York: Simon and Schuster, 1983.

R. Launch Vehicles

Carey, J., "The Process of Launching Communications Satellites with the Shuttle: An Example Using Westar VI," Paper 84-0759, AIAA 10th Communication Satellite Systems Conference, March 1984.

Charhut, D. E., and J. E. Niesley, "Commercial Atlas/Centaur Program," *Earth Oriented Appli. Space Technol.*, vol. 4, 1984.

S. Reliability

Erdle, F. E., I. A. Feigenbaum, and J. W. Talcott, Jr., "Reliability Programs for Commercial Communication Satellites," *IEEE Trans. Rel.*, vol. R-32, August 1983.

Jesche, I. K., "Calculation of the Reliability Assessment," Paper 86-0711, AIAA 11th Communication Satellite Systems Conference, March 1986.

T. Communications Technology

"Advanced On-Board Processing," Global Telecommunications Conference, Session 6, December 1985.

"Advances in Satellite Antenna Technology," Global Telecommunication Conference, Session 47, November 1983.

Bagwell, J. W., "Technology Achievements and Projections for Communication Satellites of the Future," Paper 86-0649, AIAA 11th Communication Satellite Systems Conference, March 1986.

Bharj, J. S., and D. J. Flint, "Design Challenges Posed by Transponders for Direct Broadcast Satellites," *Space Commun. Broadcast.*, vol. 4, March 1986.

Chou, S., et al., "10-W Solid State Power Amplifier for C-Band TWTA Replacement," *Comsat Tech. Rev.*, vol. 14, Fall 1984.

Mizuno, H., and H. Kato, "30 GHz Band Low Noise Receiver for 30/20 GHz Single-Conversion Transponder," *IEEE J. Selected Areas Commun.*, vol. 1, September 1983.

U. Intersatellite Links (Crosslink), Laser Communication

Anzic, G., et al., "A Study of 60 GHz Intersatellite Link Applications," International Conference on Communications, June 1983.

Billerbeck, W. J., and W. E. Baker, "The Design of Reliable Power Systems for Communications Satellites," Paper 84-1134, AIAA/NASA Space Systems Technology Conference, June 1984.

Haugland, E. J., "High Efficiency Impact Diodes for 60 GHz Intersatellite Link Applications," Paper 84-0767, AIAA 10th Communication Satellite Systems Conference, March 1984.

"Intersatellite Link," International Conference on Communications, Session 70, June 1981.

Lee, K. S. H., *EMP Interaction Principles, Techniques, and Reference Data*, New York: Hemisphere Publishing Corporation, 1986.

Lee, Y. S., and R. E. Eaves, "Implementation Issues of Intersatellite Links for Future Intelsat Requirements," International Conference on Communications, ICC June 1983.

Ross, M., et al., "Space Optical Communications with the Nd : YAG Laser," *Proc. IEEE*, vol. 66, March 1978.

Takata, N., and S. Matsuda, "Development of New Solar Cells and Their Satellite Applications," *Space Commun. Broadcast.*, vol. 4, June 1986.

Welti, G. R., "Microwave Intersatellite Links for Communications Satellites," Paper 5E.4, International Conference on Communications, June 1982.

V. Policy and Economics

Astrain, S., "Telecommunications and the Economic Impact of Communications Satellites," *Acta Astronaut.*, vol. 8, November–December 1981.

W. Regulatory and Legal

Benko, M., and I. Damien, "United Nations: Current Developments in the Field of Space Law," *Space Commun. Broadcast.*, vol. 2, June 1984.

Stephens, L. C., "International DBS Regulations Brought Down to Earth," *IEEE J. Selected Areas Commun.*, vol. 3, January 1985.

X. Networks, Architecture, Protocol

Bertsekas, D. and R. Gallager, *Data Networks*, Englewood Cliffs, N.J.: Prentice-Hall, 1987.

Schwartz, M., *Telecommunication Networks*, Reading, Mass.: Addison-Wesley, 1987.

Stallings, W., "Computer Communications," New York: IEEE Press, 1987.

Tanenbaum, A. S., *Computer Networks*, Englewood Cliffs, N.J.: Prentice-Hall, 1988.

Y. Encryption and Data Security

Denning, D. E. R., *Cryptography and Data Security*, Reading, Mass.: Addison-Wesley, 1982.

Jueneman, R. R., "Electronic Document Authentication," *IEEE Network Mag.*, vol. 1, pp. 17–23, April 1987.

Leiss, E. L., *Principles of Data Security*, New York: Plenum Press, 1982.

Neawirth, L., "A Comparison of Our Key Distribution Methods," *Telecommunications*, July 1986.

Omura, J. K., "A Computer Dial Access System Based on Public Key Techniques," *IEEE Commun. Mag.*, vol. 25, pp. 73–79, July 1987.

Torrieri, D. J., *Principles of Secure Communications System*, Dedham, Mass.: Artech House, 1985.

P. VIJAY KUMAR

2

Error-Correcting Codes

An introduction to the theory of error-correcting codes as it applies to *communication channels* is presented here.[a] Such codes make transmission more reliable by using redundancy to combat errors introduced by the channel.

The chapter begins by using the band-limited additive white Gaussian noise (AWGN) channel as an example to demonstrate the energy savings that can be realized through error-correction techniques.

Codes employed in practice can be classified as belonging to either the class of (linear) block or convolutional codes. The treatment of linear block codes has emphasized those topics that do not require knowledge of the theory of finite fields. Thus, with the exception of Reed-Solomon codes, all codes discussed here such as the Hamming, BCH, Golay, Reed-Muller and Fire codes are binary codes. The discussion on convolutional codes features the Viterbi decoding algorithm.

Because of time and space limitations, no attempt was made to make this presentation complete. In particular, the emphasis here is on forward error correction and linear block codes, reflecting the author's interests in this area. Many excellent texts[1-5,8,9] and tutorial papers[13-15] on coding theory are currently available, for the reader who wishes to learn more on the subject.

---

[a] Thanks are due to Mrs. Amy Yung for the excellent job of word processing carried out on short notice, to Mr. Xiaowei Yin for producing the helpful figures appearing throughout the chapter, and to Mr. Serdar Boztas and Mr. Ara Patapoutian for helping proofread the manuscript. I would also like to thank my wife Sudha for evenings and weekends cheerfully given up.

2.1 CODING GAIN

The band-limited AWGN channel is used here as an example that shows how the use of block error correction can result in energy savings. The space communication channel is a good example of where one might expect to encounter an AWGN channel in practice. The gains actually obtainable in practice differ significantly from those shown by information theory to be achievable in principle. The practical considerations that are responsible for this difference in performance are discussed here. Some relevant material is also contained in Section 2.4.

The discussion in this section draws considerably upon the material contained in Chapter 1 of the textbook by Clark and Cain.[2]

2.1.1 The Channel

The physical channel under consideration here is bandwidth-limited and introduces zero-mean additive white Gaussian noise having two-sided power spectral density $N_0/2$ watts per hertz. As will be seen presently, practical considerations lead us to other derived channels, and in this section we discuss the capacity of the original and derived channels.

In Fig. 2.1, $s(t)$ denotes the transmitted signal, $n(t)$ represents the additive noise, and $r(t)$ is the received signal waveform. Consider transmission over the finite time interval $[0, T]$; that is, assume that

$$s(t) = 0 \qquad t \notin [0, T] \tag{2.1}$$

The signal is energy-limited, which means that the expected value of the signal energy is fixed; that is,

$$\varepsilon\left[\int_0^T s^2(t)\,dt\right] = E \tag{2.2}$$

Since a time-limited signal cannot at the same time be *strictly* band-limited, a looser definition of the notion of band-limitedness must be adopted. One possibility[7] is to assume that at most a fraction η_w of the energy of each sample waveform $s(t)$ lies outside the band $[-W, W]$; that is,

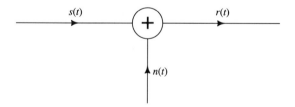

FIGURE 2.1 The additive white Gaussian noise waveform channel.

$$\int_{-W}^{W} |S(f)|^2 \, df \geq (1 - \eta_W)E \tag{2.3}$$

where $S(f)$ is the Fourier transform of $s(t)$. Let N be the dimension of the largest vector space contained in the set of all signals satisfying the above constraints. The value of N depends, of course, upon the particular interpretation attached to the bandwidth constraint. For the interpretation given above, with $\eta_W = 1/12$, it is shown[7] that

$$N \leq 2WT + 1 \tag{2.4}$$

In general, the rule-of-thumb estimate

$$N \approx 2WT \tag{2.5}$$

is most commonly used. Using a basis of orthonormal $\{\phi_i(t)|i = 1, 2, \ldots, N\}$ signals for this vector space, one can reduce the waveform channel to a collection of N identical discrete channels by defining

$$s_i = \int_0^T s(t)\phi_i(t) \, dt \tag{2.6}$$

$$n_i = \int_0^T n(t)\phi_i(t) \, dt \tag{2.7}$$

$$r_i = \int_0^T r(t)\phi_i(t) \, dt \tag{2.8}$$

The signal $s(t)$ can be recovered from the coefficients $\{s_i\}$ via

$$s(t) = \sum_{i=1}^{N} s_i \phi_i(t) \tag{2.9}$$

so there is no loss of information in passing from the continuous to the discrete. The N parallel discrete channels (Fig. 2.2) correspond to the equations

$$r_i = s_i + n_i \qquad 1 \leq i \leq N \tag{2.10}$$

It can be verified that the $\{n_i\}$'s are a collection of N independent Gaussian random variables, each having variance $N_0/2$ and zero mean. The energy constraint now takes the form

$$\varepsilon\left[\sum_{i=1}^{N} s_i^2\right] = E \tag{2.11}$$

The maximum amount of information C_{dc} that can be transmitted reliably (i.e., no errors) in a single use of each discrete channel is called the *capacity* of the discrete channel,[7] and is given by

$$C_{dc} = \frac{1}{2}\log_2\left(1 + \frac{E/N}{N_0/2}\right) \qquad \text{bits per channel use} \tag{2.12}$$

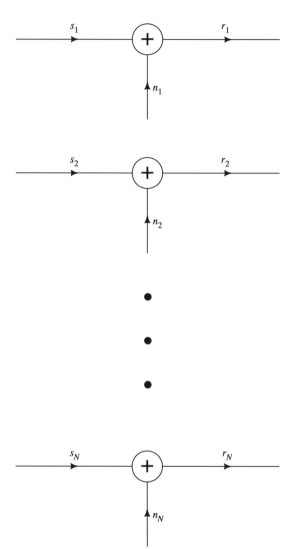

FIGURE 2.2   An equivalent discrete model for the channel.

The appearance of the fraction E/N in the capacity expression is due to the fact that maximum information transfer requires the signal energy to be spread equally amongst the N channels; that is,

$$\varepsilon[s_i^2] = E/N \triangleq E_{dc} \qquad 1 \le i \le N \tag{2.13}$$

Note, as an aside, that upon combining Eq. (2.12) with the approximation $N \approx 2WT$, one obtains Shannon's formula for the capacity C_W of the original waveform channel

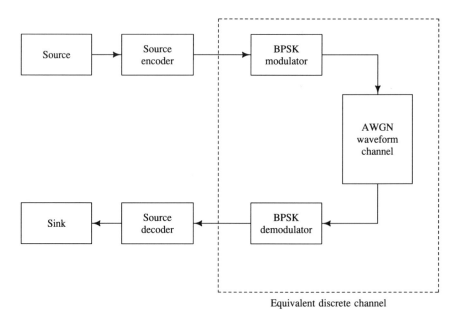

FIGURE 2.3   The equivalent discrete channel when the orthonormal signals employed correspond to binary phase-shift keying.

$$C_W = W \log_2\left(1 + \frac{P}{WN_0}\right) \qquad \text{bits per second} \qquad (2.14)$$

in which $P = E/T$ denotes the signal power.

As a practical and very common example of the conversion from the continuous to the discrete, consider a communication system (Fig. 2.3) employing binary phase-shift keying (BPSK) modulation. In this modulation scheme, the transmitted signal is either

$$\sqrt{\frac{2E_{dc}}{T/N}}\cos(\omega_0 t) \quad \text{or} \quad \sqrt{\frac{2E_{dc}}{T/N}}\cos(\omega_0 t + \pi) \qquad (2.15)$$

with successive phase changes occurring at time intervals of $\Delta = T/N$ seconds.

In this case, the orthonormal time functions can be taken to be given by

$$\phi_i(t) = \begin{cases} \sqrt{\dfrac{2}{\Delta}}\cos(\omega_0 t) & (i-1)\Delta \le t < i\Delta \\ 0 & \text{otherwise} \end{cases} \qquad 1 \le i \le N \qquad (2.16)$$

A typical function and its (single-sided) Fourier transform is shown in Fig. 2.4. Use of these orthonormal functions is equivalent to interpreting the bandwidth constraint as specifying that the only acceptable time functions are those that are linear combinations of the N orthonormal functions given above.

Thus, in this case, the N discrete channels correspond to the N sub-intervals into which the interval $[0, T]$ is subdivided. Also, in the block diagram of Fig. 2.3, the combination of the channel, the BPSK modulator, and demodulator together present to the source encoder, a discrete channel that can be used N times every T seconds.

A limitation imposed by the use of BPSK modulation is that in the equivalent set of N discrete channels, each s_i is restricted to take on the values $\pm\sqrt{E_{dc}}$. This reduces the maximum rate at which information transfer can take place. It can be shown that the capacity (C_{bs}) of the discrete channel with

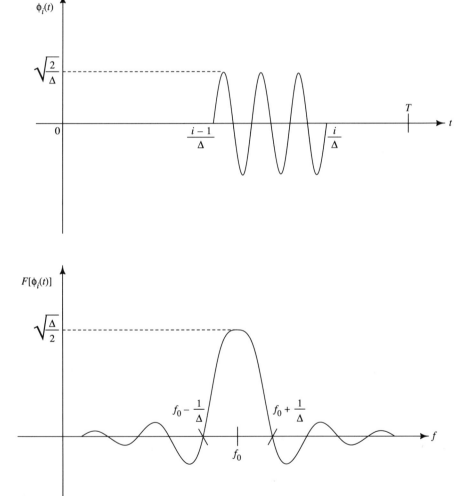

FIGURE 2.4   A typical orthonormal function and its (single-sided) Fourier transform.

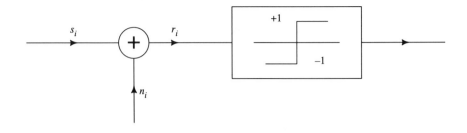

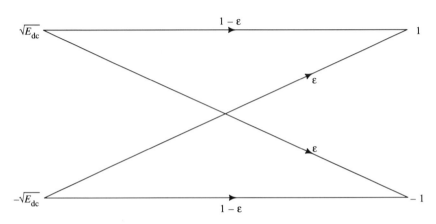

FIGURE 2.5   The binary symmetric channel resulting from hard-limiting the output of a discrete channel.

binary signaling has the lower bound[2]

$$C_{bs} > 1 - \log_2(1 + e^{-E_{dc}/N_0}) \qquad \text{bits/channel use} \qquad (2.17)$$

For the purposes of comparison, we will here treat Eq. (2.17) as a valid approximation to the capacity of the discrete channel.

A second simplification, often adopted in practice, is to quantize the received symbols r_i to two levels by using a threshold detector centered around zero (Fig. 2.5). The resulting channel, called the *binary symmetric channel* (BSC) has the simple equivalent model shown in the figure. The "crossover probability" ϵ is given by

$$\epsilon = \text{erfc}\left(\sqrt{\frac{2E_{dc}}{N_0}}\right) \qquad (2.18)$$

where erfc( ) denotes the complementary error function

$$\text{erfc}(y) = \int_y^\infty \frac{1}{\sqrt{2\pi}} e^{-x^2/2} \, dx \qquad (2.19)$$

Of course, this has the effect of further reducing the maximum possible rate of information transfer. The capacity C_{BSC} of the BSC is bounded below as follows:[2]

$$C_{BSC} > 1 - \log_2\{1 + 2\sqrt{\epsilon(1 - \epsilon)}\} \tag{2.20}$$

To obtain improved performance, quantization to four or eight levels is also commonly employed. The discrete memoryless channel (DMC) that results in the latter case is shown in Fig. 2.6. The capacity C_{DMC} of this channel satisfies the lower bound[2]

$$C_{DMC} > -\log_2\left\{\sum_{j=0}^{7}\left[\frac{1}{2}\sum_{k=0}^{1}[Pr(j|k)]^{1/2}\right]^2\right\} \tag{2.21}$$

and once again we will regard Eq. (2.21) as a valid approximation for the purposes of comparison.

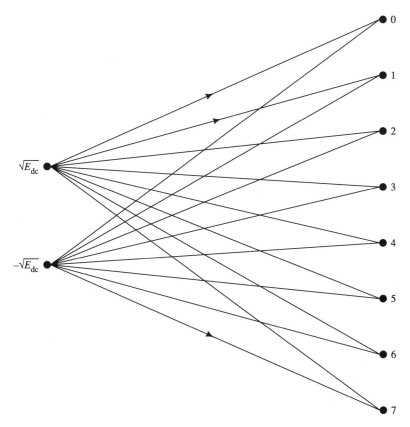

FIGURE 2.6   The discrete memoryless channel resulting from quantizing the output of a discrete channel to eight levels.

2.1.2 Uncoded System Performance

Consider a source having an alphabet of size M that puts out statistically independent symbols at discrete intervals of time. The *information*[6,32] conveyed by a single source symbol is a measure of the uncertainty dispelled from knowing that particular symbol. If all symbols are equally likely, the standard measure of information regards each symbol as providing $\log_2(M)$ bits of information. Thus, a binary ($M = 2$) source conveys 1 bit of information per binary symbol. If the rate at which the source puts out symbols is known, then the information rate of the source can also be expressed in bits per second.

From here onward it will be assumed that BPSK is the modulation scheme employed. Thus, as far as the source encoder and decoder are concerned, the channel appears as a discrete channel that can be used N times every T seconds. Furthermore, the input to the discrete channel is restricted to being $\pm\sqrt{E_{dc}}$.

When the output of the channel is detected using maximum-likelihood detection, it can be shown that to achieve a bit-error probability of 10^{-5}, a signal-to-noise energy ratio E_{dc}/N_0 of 9.6 dB is required.

2.1.3 Performance of Coded Systems

Error-correcting codes operate in general by introducing redundancy to combat errors introduced by the noise in the channel. Binary block codes (which are the codes under discussion in this section) partition the output of the binary source (or, source and source encoder combined) into blocks of k successive message symbols and then map each block onto a block of $n, n > k$, encoded symbols, which are then transmitted. Thus by introducing coding we have reduced the rate of information transfer from 1 bit per channel use to a fraction $R = (k/n)$ bits per channel use. The ratio R is called the *rate* of the block code.

A simple example of a block code is the *repetition code*, in which each source (message) symbol is transmitted n times across the channel. Thus, for this code, $k = 1$ and the rate $R = (1/n)$. At the receiver end, since there are now n received symbols upon which to base a decision regarding a single message symbol, the transmission has been made more reliable. However, such an analysis overlooks an important point.

To make a fair comparison of the performance of systems with and without coding, it must be assumed that the source has a fixed amount of energy E_b to expend *per bit of information transmitted*. Let $\sqrt{E_{dc}}$ denote, as before, the magnitude of the (binary) input to the discrete channel, either with or without coding. In the absence of coding, this is also the energy per bit of information transmitted, that is, $E_{dc} = E_b$. With coding, however, the energy available per transmission across the discrete channel is reduced to allow for the increased number (n) of transmissions across the channel for every k bits of information transmitted. Thus, in the presence of coding,

$$E_{dc} = RE_b = (k/n)E_b \qquad (2.22)$$

and the hope is that the benefits of using redundancy outweigh the decreased reliability with which *individual* symbols are received at the receiver. For medium to large values of the ratio E_b/N_0, this hope is realized. However, at sufficiently small signal-to-noise energy ratios, coding can actually cause a performance degradation.

Information theory tells us that, it is possible to transmit information at a rate R reliably across a channel only if R is less than the capacity C of the channel. Since the capacity of each of the three channels is a function of the code rate R, to determine the maximum possible rate at which information can be transmitted reliably across each of the three channels discussed above, one must find the largest value of R for which

$$R \leq C \qquad (2.23)$$

holds. Figure 2.7 plots[a] the maximum value of the rate R against the signal-to-noise energy required for the discrete channel with binary signaling, the 8-level DMC and the BSC. These plots show for instance that, given $(E_b/N_0) = 2.45$ dB, there exists in principle some coding scheme that allows error-free transmission at rates not exceeding $1/2$ across the discrete channel with binary signaling.

One would like to compare systems that do and do not employ coding by comparing the E_b/N_0 required in the two systems to achieve error-free transmission. However, it is impossible to achieve error-free transmissions in the absence of coding with finite values of (E_b/N_0). However, if one agrees to regard a bit-error rate of, say, 10^{-5} as corresponding to virtually error-free transmission, then such a comparison is possible. As noted earlier, using BPSK modulation, an E_b/N_0 ratio of 9.6 dB is required to achieve a bit-error rate of 10^{-5}. The plots show that if one is willing to suffer a 50 percent loss in transmission rate, the E_b/N_0 required for reliable transmission falls to 4.6, 2.6, and 2.45 dB or less[a] for the BSC, the DMC, and the discrete channel with binary signaling respectively. Thus, one speaks of the *coding gains* achievable in these cases as being equal to (or greater than) 5.0, 7.0, and 7.15 dB, respectively.

If one is willing to lower the rate of information even further, then still larger coding gains are available. The same is true if one lowers the bit-error

---

[a] The plots in Fig. 2.7 are reproduced with permission from the text by Clark and Cain.[2] These have been derived by replacing the capacities in Eq. (2.23) of the three derived channels with the corresponding lower bounds given in Eqs. (2.17), (2.20), and (2.21). As a result, the plots provide only an upper bound to the (E_b/N_0) ratio required to achieve reliable communication for a fixed rate R. When compared with exact expressions, the bounds in Eqs. (2.20) and (2.21) can differ by as much as 3 dB in the ratio (E_b/N_0) required. However, Clark and Cain point out that relative comparisons between the three curves accurately reflect differences observed in practice at an error rate of 10^{-5}.

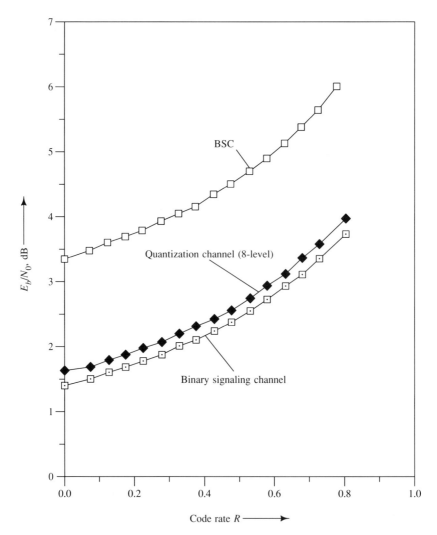

FIGURE 2.7   Plots showing upper bounds to the ratio required to communicate reliably at various rates R on each of the three derived channels. (From George C. Clark, Jr./J. Bibb Cain, *Error-Correction Coding for Digital Communications*, © 1981, p. 42. Reprinted by permission of Plenum Press, New York.)

rate at which transmission is declared reliable from, say, 10^{-5} to 10^{-8}. The plots also show that there is roughly a 2-dB loss arising from hard-limiting the received symbols.

However, it is not hard to show that it is possible to achieve capacity on any of these channels, only by using codes having a very large block length. No practical scheme has been devised to date to encode and decode such long codes. Thus the coding gains available from coded systems are smaller. Also, in view of the failure to use codes of large length, the transmission is no longer guaranteed to be reliable (i.e., there is a finite probability of bit error). In this case, the comparison with uncoded systems can be made directly on the basis of the ratio E_b/N_0 required to achieve a given bit-error rate. The plot in Fig. 2.8 shows the coding gains available with the Golay rate-(12/23) code having block length 23. At a bit-error probability equal to 10^{-5}, the available coding gain using hard decisions is 2.15 dB.

As presented here, coded systems accept a loss in rate of information transmission in exchange for either greater reliability in transmission or else an energy saving. Since the loss in data rate can be made up by expanding the bandwidth, an alternative viewpoint sees coding as acquiring greater reliability and energy savings at the expense of an expansion in bandwidth.

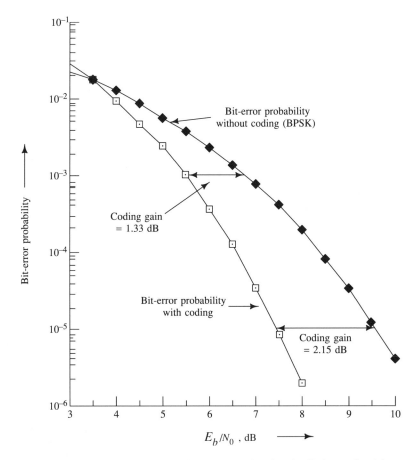

FIGURE 2.8   Bit-error probability versus E_b/N_0 for the Golay code with hard-decision coding.

TABLE 2.1 Comparison of Major Coding Techniques with BPSK or QPSK Modulation on a Gaussian Channel

Coding Technique	Coding Gain (dB) at 10^{-5}	Coding Gain (dB) at 10^{-8}	Data Rate Capability
Concatenated (RS and Viterbi)	6.5–7.5	8.5–9.5	Moderate
Sequential decoding (soft decisions)	6.0–7.0	8.0–9.0	Moderate
Concatenated (RS and biorthogonal)	5.0–7.0	7.0–9.0	Moderate
Block codes (soft decisions)	5.0–6.0	6.5–7.5	Moderate
Concatenated (RS and short block)	4.5–5.5	6.5–7.5	Very high
Viterbi decoding	4.0–5.5	5.0–6.5	High
Sequential decoding (hard decisions)	4.0–5.0	6.0–7.0	High
Block codes (hard decisions)	3.0–4.0	4.5–5.5	High
Block codes—threshold decoding	2.0–4.0	3.5–5.5	High
Convolutional codes— threshold decoding	1.5–3.0	2.5–4.0	Very high
Convolutional codes—table lookup decoding	1.0–2.0	1.5–2.5	High

Source: From George C. Clark, Jr./J. Bibb Cair, *Error-Correction Coding for Digital Communication*, © 1981. Reprinted by permission of Plenum Press, New York.

A listing of the kinds of coding gains currently available in practice is shown in Table 2.1. The speed (i.e., data rate, not to be confused with the rate of the code) at which a decoder of reasonable complexity, using present-day integrated-circuit technology, can operate is also shown. Although the bandwidth expansion factor for most options listed in the table lies typically in the range 1.25–4, the concatenated Reed–Solomon/biorthogonal code could require a larger bandwidth expansion (≈ 10).

2.2 INTRODUCTION TO CODES

2.2.1 Three Simple Coding Schemes

A discussion of three simple binary (block) coding schemes is presented below, at least two of which are very likely to suggest themselves to the novice faced with the problem of combating errors. It is assumed that the source alphabet is binary, and that the channel is the BSC introduced in the earlier section (Fig. 2.5).

In the first scheme (code 1), a *single parity bit* is appended to each string of six consecutive binary message symbols so as to ensure that the resulting

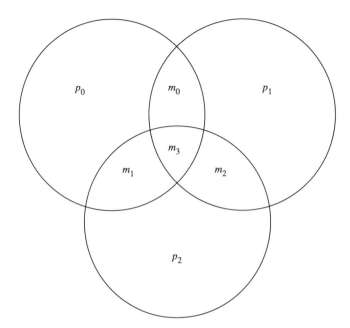

FIGURE 2.9   The rate $= (4/7)$ Hamming encoder. The parity symbols p_0, p_1, and p_2 must be chosen so as to ensure that each circle encloses an even number of 1's.

codeword (message plus parity) has even parity, i.e., to ensure that there are an even number of 1's in each codeword. Thus this code has rate $(6/7)$ and (block) length 7. By checking the parity of the received n-tuple $\mathbf{r}$, the receiver is able to detect, using this code, the presence of an odd number of errors.

The second scheme (code 2) extends this idea by partitioning the message sequence into blocks of four message symbols each, using *three* parity checks and a more sophisticated encoding/decoding algorithm. Let m_0, m_1, m_2, and m_3 denote the message symbols in a message block, with p_0, p_1, and p_2 denoting the parity bits to be appended. With respect to Fig. 2.9, the parity bits are chosen to ensure that the symbols within each circle have even parity. This code has rate $(4/7)$ and is an example of a family of codes known as *Hamming codes*, which will be discussed in greater depth in a later section. At the receiver end, decoding is performed simply by checking for parity within each circle. With this, it is easy to see that a single error in either the message or parity locations can always be corrected.[b]

In the final coding scheme (code 3), there is just a single message symbol and six parity symbols, each of which is set equal to the message symbol so that in effect, the message symbol is repeated seven times. Thus the only

---

[b] This viewpoint of a Hamming code has been taken from a seminar in Los Angeles presented by R. J. McEliece.

possible codewords are $(1111111)^t$ and $(0000000)^t$. This rate-$(1/7)$ code is a *repetition code*. The receiver uses majority logic to determine the transmitted codeword and thus will correctly decode provided that the number of errors is three or less.

In each of the codes above, let m_i, $i = 0, 1, \ldots, k - 1$ denote the message symbols with p_i, $i = 0, 1, \ldots, n - k - 1$ standing for the parity symbols. Then the linear equations defining the parity bits can be cast into a common matrix form

$$[1111111] \begin{bmatrix} m_0 \\ m_1 \\ m_2 \\ m_3 \\ m_4 \\ m_5 \\ p_0 \end{bmatrix} = 0 \tag{2.24}$$

for the single parity code; into the form

$$\begin{bmatrix} 1 & 1 & 0 & 1 & 1 & 0 & 0 \\ 1 & 0 & 1 & 1 & 0 & 1 & 0 \\ 0 & 1 & 1 & 1 & 0 & 0 & 1 \end{bmatrix} \begin{bmatrix} m_0 \\ m_1 \\ m_2 \\ m_3 \\ p_0 \\ p_1 \\ p_2 \end{bmatrix} = \begin{bmatrix} 0 \\ 0 \\ 0 \end{bmatrix} \tag{2.25}$$

for the $(4/7)$ Hamming code; and into the form

$$\begin{bmatrix} 1 & 1 & & & & & \\ 1 & 0 & 1 & & & & \\ 1 & 0 & 0 & 1 & & & \\ 1 & 0 & 0 & 0 & 1 & & \\ 1 & 0 & 0 & 0 & 0 & 1 & \\ 1 & 0 & 0 & 0 & 0 & 0 & 1 \end{bmatrix} \cdot \begin{bmatrix} m_0 \\ p_0 \\ p_1 \\ p_2 \\ p_3 \\ p_4 \\ p_5 \end{bmatrix} = \begin{bmatrix} 0 \\ 0 \\ 0 \\ 0 \\ 0 \\ 0 \\ 0 \end{bmatrix} \tag{2.26}$$

for the repetition code.

Thus in each case the code may be regarded as the null space of an appropriately constructed $(n - k) \times n$ matrix, called the *parity-check matrix*, where k is the number of message symbols and n is the length of the code. A little work shows that the rank of the parity-check matrices so constructed equals the number $(n - k)$ of parity bits, so that the codewords form a vector space of dimension k. Codes that have such a representation are called *linear (block) codes*. Although not all codes have this structure, the ones used most often in practice do belong to this class.

2.2.2 The Minimum-Distance Concept

Thus far, the error-detecting/error-correcting capability of each of the codes has been determined, by proposing a specific decoding algorithm for the code, rather than through an examination of its internal structure. As will be seen below, in many cases knowledge of just a single parameter will allow a good estimate to be obtained.

Implicit in the decoding schemes considered above was the fact that over a BSC, a lesser number of errors is a more likely event than a larger number of errors. For instance, if this was not true, it would be impossible to justify majority logic decoding for the repetition code. A second, more obvious observation is that, in each of the three codes above, the codewords form a subset of the set F^n of all binary n-tuples.

Taken together, these two observations suggest (1) that a binary code of length n can be regarded simply as any subset of the collection of all n-tuples and (2) that one should arrange n-tuples in the space F^n in such a way that all those n-tuples that differ in very few symbols from a codeword are placed close to the codeword (Fig. 2.10). Decoding can then be accomplished by searching

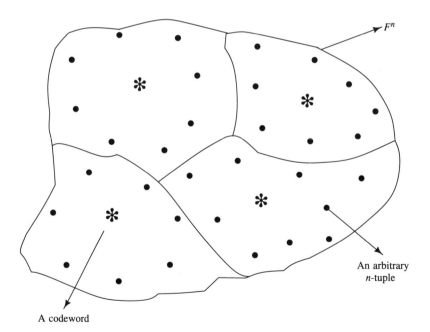

FIGURE 2.10   A binary code is simply a collection of M n-tuples. To decode a received n-tuple (possibly in error), one simply identifies the nearest (in Hamming distance) codeword. This partitions the space F_n into regions as shown.

for the codeword "nearest" to the received n-tuple. The error-correcting capability of a code can then be determined simply from the separation between codewords. A more precise formulation of these ideas is presented below.

Definition. The *Hamming weight* of a binary n-tuple is the number of non-zero symbols in the n-tuple. The *Hamming distance* between any two binary n-tuples $\mathbf{a}$ and $\mathbf{b}$ is the number of coordinates in which the two n-tuples differ; that is,

$$d(\mathbf{a}, \mathbf{b}) = \text{Hamming weight}(\mathbf{a} - \mathbf{b}) \tag{2.27}$$

To have all the attributes that one intuitively expects of a distance measure, it is necessary that the Hamming distance function satisfy the following axioms:

1. $d(\mathbf{a}, \mathbf{b}) \geq 0$, with equality holding if and only if $\mathbf{a} = \mathbf{b}$

2. $d(\mathbf{a}, \mathbf{b}) = d(\mathbf{b}, \mathbf{a})$

3. $d(\mathbf{a}, \mathbf{b}) \leq d(\mathbf{a}, \mathbf{c}) + d(\mathbf{c}, \mathbf{b})$ (triangle inequality)

The axioms are not hard to verify.

Assuming all codewords are equally likely to be transmitted, the algorithm that achieves the minimum probability of block error[4] is one that performs maximum-likelihood decoding (MLD)—that is, the algorithm that chooses the codeword $\mathbf{c}$ for which the conditional probability

$$p(\mathbf{r}|\mathbf{c}) \tag{2.28}$$

is a maximum, where $\mathbf{r}$ is the n-tuple received across the channel. Let $\mathbf{a}$ and $\mathbf{b}$ be two codewords having Hamming distances d_1 and d_2, $d_2 > d_1$, respectively from the received n-tuple $\mathbf{r}$. In the case of a BSC, the conditional probabilities $p(\mathbf{r}|\mathbf{a})$, $p(\mathbf{r}|\mathbf{b})$ can be determined directly from d_1 and d_2:

$$p(\mathbf{r}|\mathbf{a}) = (1 - \epsilon)^n(\epsilon/1 - \epsilon)^{d_1} \tag{2.29}$$

$$p(\mathbf{r}|\mathbf{b}) = (1 - \epsilon)^n(\epsilon/1 - \epsilon)^{d_2} \tag{2.30}$$

so that if the crossover probability ϵ satisfies (as is always the case) $\epsilon < 1/2$, we have $p(\mathbf{r}|\mathbf{a}) > p(\mathbf{r}|\mathbf{b})$ and thus we have justified the intuitive notion of decoding to the nearest codeword. This decoding procedure is known as *minimum-distance decoding*.

A measure of the error-correcting and/or error-detecting capability of a code can be obtained for a general code as follows:

Definition. The *minimum distance* of a code $C \subseteq F^n$ is the minimum distance between a pair of codewords; that is,

$$d_{\min} = \min\{d(\mathbf{a}, \mathbf{b})|\mathbf{a}, \mathbf{b} \in C, \mathbf{a} \neq \mathbf{b}\} \tag{2.31}$$

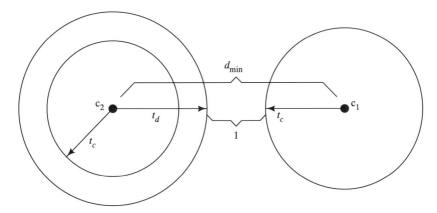

FIGURE 2.11   Illustration of the relation $d_{min} = t_c + t_d + 1$, relating the minimum distance of a code to its error-correcting/error-detecting capability.

Definition.   A (t_c, t_d) or error-correcting/error-detecting code, $t_d \geq t_c$, is one that can detect all combinations of t_d or fewer errors and in addition, correct all combinations of t errors, $t \leq t_c$.

It is quite possible that a given code is a (t_c, t_d) code for several pairs of values of t_c and t_d simultaneously; that is, t_c and t_d are not in general uniquely defined for a code.

THEOREM 1.   A binary code $C \subseteq F^n$ having minimum distance d_{min} is a t_c error-correcting, t_d error-detecting code, $t_d \geq t_c$, if and only if

$$t_d + t_c + 1 = d_{min} \tag{2.32}$$

Proof **("if" part).** (See Fig. 2.11.) Let t_d, t_c satisfy Eq. (2.32). Assume that no more than t_d errors have occurred. Assume the decoder carries out the following variation of the minimum distance decoding algorithm: if d is the distance to the nearest codeword **a** and $d \leq t_c$, declare **a** as having been transmitted. If $t_c < d \leq t_d$, declare an error as having been detected, but make no attempt to correct the error.

Case 1 $(d < t_c)$. The decoder can be mistaken only if, in fact, some other codeword was actually transmitted. But then, from the definition of d_{min} Eq. (2.32), and the triangle inequality, this is impossible.

Case 2 $(d > t_c, d \leq t_d)$. In this case, the decoder declares an uncorrectable error. A mistake could have been made only if there was a correctible error or else no error at all. But once again, by the triangle inequality and Eq. (2.32) neither is possible.

(only if part) Assume that Eq. (2.32) is violated; that is,

$$t_c + t_d + 1 > d_{min} \tag{2.33}$$

Let c_1 and c_2 be two codewords d_{min} apart. Assume c_1 is transmitted and that t_d errors occur in such a way that $d(\mathbf{r}, c_2) = d_{min} - t_d < t_c + 1$; that is, $d(\mathbf{r}, c_2) \leq t_c$. In this case, the decoder incorrectly decodes c_2.

Q.E.D.

It is shown in subsequent sections that the minimum distances of the three codes discussed above are 2, 3, and 7, respectively. Thus, by Theorem 1 above, the single parity-check code is guaranteed to detect all single-error patterns, the Hamming code can correct any single error and the repetition code can correct any pattern of three errors or less. Of course, as mentioned earlier, t_c, t_d are not uniquely defined for a given code, and the Hamming code can also be used to detect any pattern of two errors—i.e., used as a zero-error-correcting, double-error-detecting code.

The minimum distance of a code is handy as a single number that enables us to estimate the error-correcting/error-detecting capability of the code. We say estimate because it is often possible using a minimum distance decoder to correct/detect a larger number of errors than that guaranteed by Eq. (2.32). As a simple example of this note that whereas, by Theorem 1, the single parity-check code is *guaranteed* to detect only a single error, it can, as observed earlier, detect any error pattern involving an odd number of errors.

It turns out that for many codes, *bounded-distance decoding*—i.e., decoding only as many errors as guaranteed by Theorem 1—is considerably simpler than implementing a minimum-distance decoder for the code. The terminology *complete decoder* is often used to describe a minimum-distance decoder, so bounded-distance decoding is an example of *incomplete decoding*.

The codes described above belong to the rather special class of linear block codes. Other coding schemes are possible and are described below.

2.2.3 Convolutional Codes and ARQ Schemes

In a rate-(k/n) *tree code*, the input and output can once again be partitioned into blocks of lengths k and n, respectively. However, the difference lies in that, unlike the case of block codes, each output block can depend in general upon several input blocks. These codes are best described in terms of a tree code that describes the encoding map. At each node in the tree, there are 2^k branches corresponding to the 2^k possibilities for the current input block. The corresponding output block is in general a function not only of the current input block but also of some or all of the previous blocks. If the dependence on the message sequence is linear, we have a linear tree code. If the encoder is linear and, in addition, has finite memory and is time-invariant, we obtain a *convolutional code*, so called because the input and output can in this case be related via a convolutional equation.

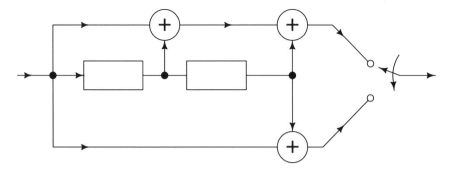

FIGURE 2.12   A convolutional encoder.

The encoder for a rate-(1/2) convolutional encoder is shown in Fig. 2.12. The corresponding tree is presented in Fig. 2.13. As indicated by the tree, we move along the higher branch at each node in the tree whenever the incoming symbol (a message block of length 1) is a 0 and take the lower branch otherwise. The branches are labeled with the corresponding output of the decoder. The lowercase letters appearing in the figure, relating each node in the tree to the contents of the shift register used to generate the code, can be ignored here.

Both block and tree codes are examples of *forward error correction* (FEC), in the sense that the decoder, having once processed the received *n*-tuple corresponding to a given message block, keeps constantly moving forward, without ever having to return to process a previous message symbol.

In *automatic repeat request* (ARQ) schemes, the emphasis is on the use of codes primarily for the purpose of error detection. In case an error is detected, the receiver puts out a request for retransmission along a feedback channel. Existing hybrid schemes combine some amount of forward error correction with error detection. We do not further discuss these codes in this chapter. A detailed discussion may be found in Lin and Costello.[3]

2.3 LINEAR BLOCK CODES

2.3.1 Systematic Codes

The previous section suggested that good codes might be found amongst the subspaces of F^n. Here some basic properties of such codes are studied.

Definition.   A *linear block code C* is a subspace of F^n.

Over the binary field, testing for a subspace is equivalent to testing for an (Abelian) subgroup and for this reason linear (block) codes are often also called *group codes.*

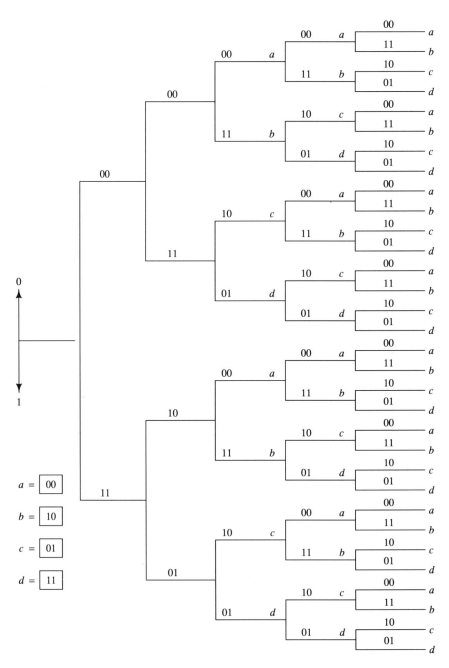

FIGURE 2.13 Tree code representation for encoder of Fig. 2.12. (From A. J. Viterbi and J. K. Omura, *Principles of Digital Communication and Coding*, © 1979. Reprinted by permission of McGraw-Hill Publishing Co., New York.)

The *dimension* of a linear code is its dimension as a subspace of F^n. The notation $[n, k, d]$ (sometimes simply $[n, k]$) is used to describe a linear block code of length n, dimension k, and minimum distance d. The corresponding notation for a general, quite possibly nonlinear, code of size M is (n, M, d) (or simply (n, M)).

Let $\mathbf{g}_0, \mathbf{g}_1, \ldots, \mathbf{g}_{k-1}$ be a basis for an $[n, k, d]$ block code C. Then every codeword $\mathbf{c}$ can be generated as a linear combination of these vectors, determined by the message symbols $m_0, m_1, \ldots, m_{k-1}$; that is,

$$\mathbf{c} = \sum_{i=0}^{k-1} m_i \mathbf{g}_i \tag{2.34}$$

This can be rewritten in matrix form:

$$[c_0\, c_1\, c_2 \ldots c_{n-1}] = [m_0\, m_1 \ldots m_{k-1}] \begin{bmatrix} \mathbf{g}_0^t \\ \mathbf{g}_1^t \\ \vdots \\ \mathbf{g}_{k-1}^t \end{bmatrix} \tag{2.35}$$

in which the $(k \times n)$ matrix G appearing on the right is called the *generator matrix* of the code. Unless otherwise indicated, every vector appearing in this chapter is to be regarded as a column vector. More generally, any $(k \times n)$ matrix having rank k whose rows span the code space can be regarded as a generator matrix for the code.

Let $C^\perp$ denote the vector space of all vectors in F^n that are orthogonal[c] to each vector in C; that is,

$$C^\perp = \{\mathbf{a} \in F^n | \mathbf{a}^t \mathbf{v} = 0, \qquad \mathbf{v} \in C\} \tag{2.36}$$

Then, $C^\perp$ has dimension $n - k$ and is called the *dual code*. Let H be any generator matrix for the dual code $C^\perp$. Then C is precisely the null space of H so that H is a valid parity-check matrix for the code. In general, for an $[n, k]$ code, the parity-check matrix is any $(n - k) \times n$ matrix of rank $(n - k)$ satisfying:

$$HG^t = [0] \tag{2.37}$$

where $[0]$ denotes an all-zero matrix.

E X A M P L E . The parity-check matrix of the single parity code (code 1) will be seen to be the generator matrix of the repetition code (code 3). Thus these codes are dual codes.

---

[c] The term "orthogonal" is applied here loosely since, strictly speaking, F^n is not a valid inner product space inasmuch as it is perfectly possible to find nonzero binary vectors that are orthogonal to themselves.

EXAMPLE. Consider the [6, 3] linear code (code 4) having the generator matrix

$$G = \begin{bmatrix} 0 & 0 & 0 & 1 & 1 & 1 \\ 1 & 1 & 0 & 1 & 0 & 1 \\ 0 & 1 & 1 & 0 & 1 & 1 \end{bmatrix} \qquad (2.38)$$

The matrix

$$H = \begin{bmatrix} 0 & 0 & 1 & 1 & 0 & 1 \\ 0 & 1 & 0 & 1 & 1 & 0 \\ 1 & 0 & 1 & 1 & 1 & 0 \end{bmatrix} \qquad (2.39)$$

has rank $(n - k) = 3$ and satisfies Eq. (2.37). It is thus a valid parity-check matrix for the code. Of course, H may also be regarded as a generator matrix for the dual code, which in this example also has parameters [6, 3].

Any linear code can be put into a simpler canonical form.

Definition.   Two codes C_1 and C_2 are said to be *equivalent* if they differ only in the order in which symbols appear within a codeword.

EXAMPLE. By interchanging the second and third symbols, we see that the two [4, 2] codes C_1 and C_2 below

$$C_1 = \{[0000]^t, [0011]^t, [1100]^t, [1111]^t\} \qquad (2.40)$$

and

$$C_2 = \{[0000]^t, [0101]^t, [1010]^t, [1111]^t\} \qquad (2.41)$$

are equivalent. Clearly over a *memoryless* BSC, two codes that are equivalent have exactly the same performance.

By interchanging columns of the generator matrix if necessary (this corresponds to replacing the given code with an equivalent one), one can always ensure that the first k columns of the generator matrix are linearly independent. Using elementary row reduction, the matrix can further be reduced to a form in which the first k columns form the $(k \times k)$ identity matrix. When the generator matrix is in this form, the message symbols appear explicitly as the first k symbols of each codeword.

EXAMPLE. By interchanging the first and last columns of the matrix G given earlier and then performing row reduction, the matrix

$$G_{sys} = \begin{bmatrix} 1 & 0 & 0 & 1 & 1 & 0 \\ 0 & 1 & 0 & 0 & 1 & 1 \\ 0 & 0 & 1 & 1 & 1 & 1 \end{bmatrix} \qquad (2.42)$$

is obtained.

Definition.   A linear block code is said to be *systematic* if its generator matrix G is of the form

$$G = [I_k P] \tag{2.43}$$

where I_k is the $(k \times k)$ identity matrix.

The discussion above leads to the next theorem.

THEOREM 2.   Every linear block code is equivalent to a systematic linear code.

When G is of the form in Eq. (2.43), a valid parity-check matrix is easily found; that is,

$$H = [P^t | I_{n-k}]. \tag{2.44}$$

E X A M P L E .   Continuing the example (code 4) of the $[6, 3]$ code, the corresponding parity-check matrix is given by

$$H_{sys} = \begin{bmatrix} 1 & 0 & 1 & 1 & 0 & 0 \\ 1 & 1 & 1 & 0 & 1 & 0 \\ 0 & 1 & 1 & 0 & 0 & 1 \end{bmatrix} \tag{2.45}$$

2.3.2 Minimum Distance of a Linear Code

For a linear code, computation of the minimum distance is made easier by the following theorem.

THEOREM 3.   The minimum distance of a linear block code C is the minimum Hamming weight of a (nonzero) codeword in C.

Proof. Let w_{min} denote the minimum Hamming weight in the code. Consider two codewords c_1, c_2 separated by Hamming distance d_{min}. Then, since $c_1 - c_2$ has Hamming weight d_{min} and is also a codeword,

$$w_{min} \leq d_{min} \tag{2.46}$$

Next, let c be a codeword of minimum Hamming weight in the code. Then, $d(o, c) = w_{min}$, and therefore

$$d_{min} \leq w_{min} \tag{2.47}$$

Q.E.D.

E X A M P L E .   The codewords in the single parity-check code (code 1) are precisely those that have an even Hamming weight. Thus for the code

$$d_{min} = w_{min} = 2 \tag{2.48}$$

Clearly,

$$d_{min} = w_{min} = 7 \qquad (2.49)$$

for the repetition code.

Sometimes it is easier to deduce the minimum distance of the block code by examining the columns of its parity-check matrix.

THEOREM 4.   Let s be the largest integer such that every set of s columns of the parity-check matrix H is linearly independent. Then $d_{min} = s + 1$.

Proof. Let $\mathbf{h}_1, \mathbf{h}_2, \dots, \mathbf{h}_n$ denote the n columns of the parity-check matrix. Let $\mathbf{c}$ be a codeword of Hamming weight w whose nonzero symbols have the coordinates $i_1, i_2, \dots, i_w$. Then the theorem follows from noting that

$$H\mathbf{c} = 0 \qquad (2.50)$$

implies that the corresponding columns $\mathbf{h}_{i_1}, \mathbf{h}_{i_2}, \dots, \mathbf{h}_{i_w}$ of the parity-check matrix are linearly dependent.

EXAMPLE
(a) Hamming code: The seven columns of the parity-check matrix for this code are all distinct. Further, it is possible to find three columns that are linearly dependent. Thus for this code $s = 2$, and therefore

$$d_{min} = 3 \qquad (2.51)$$

(b) The $[6, 3]$ code: Exactly the same argument shows that this code also has $d_{min} = 3$.

THEOREM 5 (Singleton Bound).   The minimum distance d_{min} of an $[n, k]$ linear code must satisfy the bound

$$d_{min} \leq n - k + 1 \qquad (2.52)$$

Proof. The parity-check matrix of the code has rank $n - k$ and thus the integer s in Theorem 4 above can be no larger than $n - k$.

Q.E.D.

Definition.   Codes whose parameters achieve the Singleton bound are called *maximum distance separables (MDS)*.

EXAMPLE.   Both the $[7, 6, 2]$ single parity code and its dual, the repetition code, are MDS codes.

It is always true that the dual of an MDS code is also necessarily an MDS code (see Exercise 4). The two codes cited above are representative of the only two possible classes of *binary* MDS codes. This is not true in the nonbinary case, and Reed-Solomon codes (to be discussed later) are an example of this. A comprehensive discussion of MDS codes can be found in MacWilliams and Sloane.[1]

2.3.3 Standard-Array Decoding

In the previous section the Hamming code was decoded by checking to see whether the received n-tuple $\mathbf{r}$ satisfied the three parity checks. An equivalent procedure that generalizes to any block code is to compute the matrix product $H\mathbf{r}$ at the receiver. The presence of a 1 in the resulting vector is an indication that that particular parity check has failed.

Definition.   The *syndrome* $\mathbf{s}$ associated with a received n-tuple $\mathbf{r}$ is the $(n-k)$-tuple given by

$$\mathbf{s} = H\mathbf{r} \tag{2.53}$$

Since the received n-tuple $\mathbf{r}$ is the sum

$$\mathbf{r} = \mathbf{c} + \mathbf{e} \tag{2.54}$$

of the transmitted codeword $\mathbf{c}$ and the error pattern $\mathbf{e}$ introduced by the channel, and each codeword $\mathbf{c}$ lies in the null space of the H matrix, it follows that

$$\mathbf{s} = H\mathbf{r} = H\mathbf{e} \tag{2.55}$$

Thus, the syndrome can also be regarded as being formed from the error vector.

As pointed out earlier, a linear code can be regarded as a subgroup of F^n, and therefore the code and its cosets partition the space F^n.

EXAMPLE (code 4).   In Fig. 2.14, all 64-tuples in F^6 are arranged in the form of an (8 × 8) array in which the elements of the top row are the eight codewords of the [6,3,3] code. Each row below the top row contains the elements of a coset of the code. The minimum-weight vector within each coset (called the *coset leader*) is listed in the first column of the matrix. When more than one element within each coset has the same minimum weight, an arbitrary choice is made. Elements within a coset are so arranged that the (i,j)th element in the array is the sum of the codeword associated with the column and the coset leader for that row.

An array that is constructed from a linear code in this fashion is called a *standard array* for the code. The following theorem shows that the syndrome uniquely identifies the coset to which the error pattern belongs.

Coset leader	Codewords / Elements of a coset							Syndrome
000000	100110	010011	001111	110101	101001	011100	111010	$(000)^t$
000001	100111	010010	001110	110100	101000	011101	111011	$(001)^t$
000010	100100	010001	001101	110111	101011	011110	111000	$(010)^t$
000100	100010	010111	001011	110001	101101	011000	111110	$(100)^t$
001000	101110	011011	000111	111101	100001	010100	110010	$(111)^t$
010000	110110	000011	011111	100101	111001	001100	101010	$(011)^t$
100000	000110	110011	101111	010101	001001	111100	011010	$(110)^t$
000101	100011	010110	001010	110000	101100	011001	111111	$(101)^t$

FIGURE 2.14 Standard array for the [6, 3, 3] code.

THEOREM 6.   There is a one-to-one correspondence between the cosets of the code and syndromes **s**. The correspondence is obtained by computing **s** = H**u** for any vector **u** in the coset.

Proof. To show that the map is well-defined, note that

$$H\mathbf{u} = H\mathbf{v} \Leftrightarrow H(\mathbf{u} - \mathbf{v}) = \mathbf{0}, \quad \Leftrightarrow \mathbf{u} \text{ and } \mathbf{v} \text{ belong to the same coset} \quad (2.56)$$

This also shows simultaneously that the map from coset to syndrome is one-to-one. Since the rank of the parity-check matrix equals $(n - k)$, its map is obviously one-to-one as well.

Q.E.D.

THEOREM 7.   In any coset of the code, there is at most one n-tuple whose Hamming weight is less than or equal to $[(d_{min} - 1)/2]$.[a]

Proof. Suppose there are two such n-tuples in some coset of the code. Then their difference yields a codeword of Hamming weight $w < d_{min}$, which is impossible.

Q.E.D.

Thus, for a linear block code, minimum distance decoding can be performed by first computing the syndrome from the received n-tuple and then determining the n-tuple in that coset having minimum Hamming weight. When more than one element within the coset has the same minimum weight, one makes an arbitrary choice. By subtracting this pattern from the received n-tuple, one obtains the codeword most likely to have been transmitted.

EXAMPLE (code 4).   Also shown in Fig. 2.14 is the syndrome associated with each coset of the code. Assume that the codeword **c** = $(010011)^t$ is transmitted and that the corresponding n-tuple **r** = $(110011)^t$ is received. The syndrome **s** is in this case given by

$$\mathbf{s} = \begin{bmatrix} 1 & 0 & 1 & 1 & 0 & 0 \\ 1 & 1 & 1 & 0 & 1 & 0 \\ 0 & 1 & 1 & 0 & 0 & 1 \end{bmatrix} \begin{bmatrix} 1 \\ 1 \\ 0 \\ 0 \\ 1 \\ 1 \end{bmatrix} = \begin{bmatrix} 1 \\ 1 \\ 0 \end{bmatrix} \qquad (2.57)$$

From the standard array in Fig. 2.14 we see that the corresponding coset leader is $(100000)^t$. Subtracting (or adding) this to the received vector **r** gives us the most likely codeword

$$\hat{\mathbf{c}} = (110011)^t + (100000)^t = (010011)^t \qquad (2.58)$$

---

[a] $[x]$ denotes the largest integer less than x.

Theorem 7 guarantees that for a t_c error-correcting, t_c error-detecting code (i.e., the code is guaranteed to detect no more errors than it can correct), all error patterns of t_c errors or less will appear as coset leaders. It does not guarantee, however, that all coset leaders will have a Hamming weight of t_c or less. The appearance of a coset leader of Hamming weight exceeding t_c is an indication that the code can sometimes correct a greater number of errors than that guaranteed by Theorem 1. This is precisely the difference between a bounded distance decoder and a complete decoder implemented using the standard array.

Standard-array decoding is a suitable means of decoding only when the redundancy $(n - k)$ of the block code is small and the mapping from syndrome to coset leader can be implemented using table lookup.

2.3.4 Probability of Error Computation

We assume transmission over a BSC having crossover probability $\epsilon < 1/2$. Clearly, an error pattern will be corrected by the minimum-distance decoding algorithm if and only if the actual error-pattern is indeed a coset leader. Thus, the standard array allows us to determine the *probability of block error* P_e (i.e., the probability that a codeword is incorrectly decoded) from

$$P_e = 1 - \text{(probability that the error pattern is a coset leader)} \qquad (2.59)$$

The *probability of undetected error* (for the case when the code is used solely for the purpose of error detection) can also be determined from the standard array. It is simply the probability that the error pattern corresponds to a codeword.

Definition.   The *weight distribution* of a block code is given by the collection of numbers $\{A_w | 0 \le w \le n\}$ in which

$$A_w = \{\text{number of codewords having Hamming weight } w\} \qquad (2.60)$$

In terms of the weight distribution $\{A_w\}$, the probability of undetected error P_u can be computed as follows:

$$P_u = \sum_{w=1}^{n} A_w \epsilon^w (1 - \epsilon)^{n-w} \qquad (2.61)$$

Of course, if the weight distribution of the coset leaders is also known for the code, then the probability of uncorrected error can also be computed in similar fashion.

E X A M P L E (Code 4).   Let the crossover probability of the BSC equal 10^{-5}. From the standard array of the $[6, 3]$ code in Fig. 2.14 we see that the code is capable of correcting all single errors and just one double-error pattern. Thus the probability of (block) error equals

$$P_e = 1 - \{(1 - \epsilon)^6[1 + 6(\epsilon/1 - \epsilon) + (\epsilon/1 - \epsilon)^2]\} \approx 14\epsilon^2$$
$$= 1.4 \times 10^{-9} \tag{2.62}$$

The weight distribution of the code (from the top row of the standard array) is given by

$$A_0 = 1, \quad A_1 = A_2 = 0, \quad A_3 = 4, \quad A_4 = 3, \quad A_5 = A_6 = 0 \tag{2.63}$$

The weight distribution of a code can be determined from that of its dual. This procedure can often simplify computation of the probability of undetected error.

Definition.   Let C be a linear block code of length n, having weight distribution $\{A_w | 0 \le w \le n\}$. The polynomial

$$A(z) = \sum_{w=0}^{n} A_w z^w \tag{2.64}$$

is called the *weight enumerator* of the code C.

In terms of the weight enumerator, the probability of undetected error P_u from Eq. (2.61) is given by

$$P_u = (1 - \epsilon)^n \left[A\left(\frac{\epsilon}{1 - \epsilon}\right) - 1 \right] \tag{2.65}$$

Let $B(z)$ denote the weight enumerator of the dual code. The identity

$$A(z) = 2^{-(n-k)}(1 + z)^n B\left(\frac{1 - z}{1 + z}\right) \tag{2.66}$$

was shown by MacWilliams[1] to hold for any linear code. Combining Eq. (2.65) with the MacWilliams identity, the undetected error probability P_u can be expressed completely in terms of the weight distribution of the dual code $C^{\perp}$; that is,

$$P_u = 2^{-(n-k)}B(1 - 2\epsilon) - (1 - \epsilon)^n \tag{2.67}$$

When the redundancy $(n - k)$ of the block code is smaller, Eq. (2.67) is easier to compute than Eq. (2.65). In the case of code 4, from Eq. (2.65), the probability of undetected error for this code is

$$P_u = 4\epsilon^3(1 - \epsilon)^3 + 3\epsilon^4(1 - \epsilon)^2 \approx 4\epsilon^3 = 4.0 \times 10^{-15} \tag{2.68}$$

The *bit-error probability* (bounded above by the probability of block error) can also be computed from the standard array, but we will not discuss this here. Instead we refer the reader to the text by Clark and Cain.[2]

2.3.5 Implementation

The circuits shown here are taken from Welch.[33] Figure 2.15 shows a circuit that can be used to generate a rate-(k/n) code when n and k are relatively small. The read-only memory (ROM) is an $(n \times k)$ array that stores the rows of the generator matrix along its columns with the leftmost column containing the first row of the generator matrix and so on. The AND gates and the modulo-2

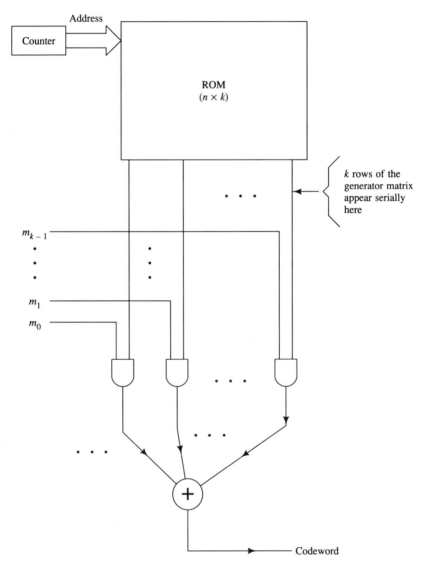

FIGURE 2.15   An encoder for a rate-(k/n) linear block code.

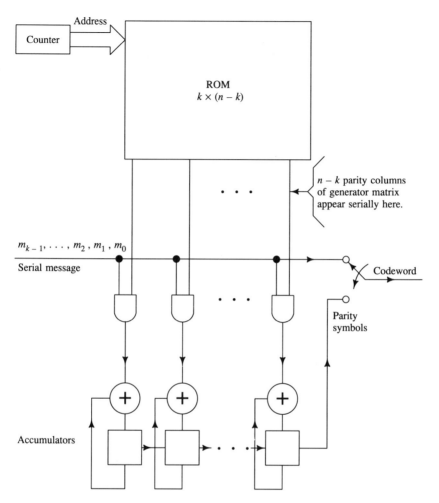

FIGURE 2.16   A systematic encoder for a (k/n) linear block code.

adder are seen to implement multiplication of the generator matrix by the message vector. The code symbols appear serially at the output of the adder as the counter steps through the n columns of the generator matrix.

When the code is systematic, and $(n - k)$ and k are small, the circuit in Fig. 2.16 can be used to encode the code. This time the $k \times (n - k)$ ROM stores only those *columns* of the generator matrix corresponding to the locations of the parity-check bits. After the counter has stepped through the k rows of the generator matrix, the $(n - k)$ parity bits have been computed and stored in the $(n - k)$-bit shift register at the bottom, and are then serially shifted out.

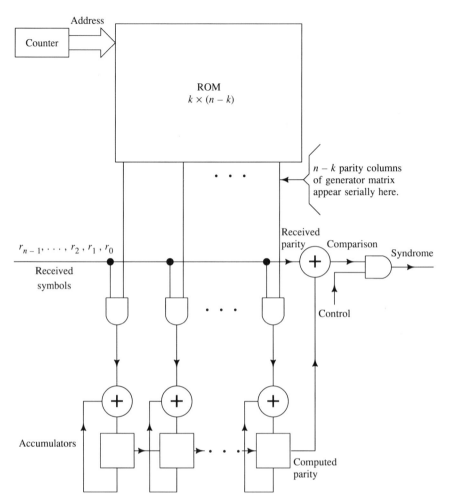

FIGURE 2.17   Circuit for syndrome computation when systematic encoding has been used.

Essentially the same circuit (Fig. 2.17) can be used to compute the syndrome, since computation of the syndrome is equivalent to recomputing parity from the received "message" bits and then comparing the result to the "parity" bits received. The modulo-2 adder at the extreme right performs this comparison, and the syndrome bits appear serially at the output.

From the point of view of implementation, the final step of determining the error pattern from the syndrome is the most complex. When the redundancy $(n - k)$ of the code is sufficiently small, this goal is easily accomplished by using a ROM to carry out the table lookup suggested by the standard array for the code. Most linear block codes are decoded using a decoding algorithm specific to the particular code family.

2.3.6 Hamming Codes

The parity-check matrix of the $[7,4,3]$ Hamming code given earlier in Eq. (2.25) has the feature that every nonzero 3-tuple appears precisely once as a column in the matrix. This property characterizes the family of Hamming codes.

Definition. A *Hamming code of redundancy* (number of parity bits) $= r$ is a linear block code having parameters $[2^r - 1, 2^r - 1 - r, 3]$ whose parity-check matrix has the property that every nonzero r-tuple appears precisely once as a column vector.

Note that by this definition it follows that any two Hamming codes having the same length are equivalent.

> E X A M P L E . In the case of the $[7,4,3]$ Hamming code discussed in the previous sections (code 2), note that the proof (based on Theorem 4) showing the $[7,4]$ code to have minimum distance $d_{min} = 3$ carries over in the general case. Identification of the error location from the syndrome is made simple by the fact (true of any single error-correcting code) that the syndrome vector is precisely the column in the parity-check matrix that corresponds to the error location. It turns out that the Hamming code is cyclic, thus allowing the decoding process to be further simplified (Section 2.7.5).

An examination of the standard array of the Hamming code will reveal that the (nonzero) coset leaders represent precisely the set of all n-tuples having weight 1 or less. Codes having this property are called *perfect* codes because they can be shown to have the largest size given the required minimum distance. These codes are treated in greater detail in the next section. Note that as a consequence, there is no distinction to be made between bounded-distance and minimum-distance decoding in the case of the Hamming code.

2.4 PERFORMANCE OF BOUNDED-DISTANCE DECODING

As mentioned in Section 2.1, reliable transmission across a BSC can be achieved only through the use of codes having very large block length. Conversely, there exist[6] long codes that achieve reliable transmission provided that the rate of transmission of information does not exceed the capacity of the channel. However, this statement does not hold if the decoder is restricted to bounded-distance decoding. In this case, one is forced to settle for a lesser rate of information transmission. This is clear from the bounds discussed below. A good discussion of these bounds may be found in Refs. 1 and 32. The material presented below draws considerably from the latter source.

2.4.1 Three Simple Bounds

Let $M_{\max}(n, d_{\min})$ denote the maximum possible size of a binary code having length n and minimum distance $d_{\min}$.

THEOREM 7 (Hamming Bound)

$$M_{\max}(n, d_{\min}) \leq \frac{2^n}{\sum_{i=0}^{t} \binom{n}{i}} \tag{2.69}$$

where $t = (d_{\min} - 1)/2$ is the guaranteed error-correcting capability of the code.

Proof. The proof follows from noting that for the code to have minimum distance $d_{\min}$, it must be that the spheres of radius t surrounding each codeword are nonintersecting.

Q.E.D.

Although not presented here, the Hamming bound has a straightforward extension to the case when the alphabet is nonbinary. A code which achieves the Hamming bound (either binary or nonbinary) is called a *perfect code* since this code has the largest possible size for the given length and minimum distance. It is clear from the bound that a perfect code can only exist when $d_{\min}$ is odd.

Every Hamming code (for any value of the parameter r) is a perfect code. Hamming codes can be generalized to the case of nonbinary alphabets when their parameters take on a slightly different form. These codes are also perfect. The bound in Theorem 7 can be used to narrow down the search for a perfect code. In the late 1940s, Golay observing that the parameters $[23, 12, 7]$ (with $M = 2^k$) satisfied the bound with equality, succeeded in identifying a linear binary code, known as the (binary) Golay code, having precisely these parameters.

These are, in fact, the only two nontrivial examples of a perfect binary code that exist. Perfect codes are rare even for nonbinary alphabets, and the only other known example of a perfect code is a ternary double-error-correcting code of length 11 also found by Golay. In fact, it is known that if a nontrivial perfect code over any finite field exists, it must have the same parameters as either the Hamming or Golay codes.[1]

THEOREM 8 (Gilbert-Varshamov Bound)

$$M_{\max}(n, d_{\min}) \geq \frac{2^n}{\sum_{i=0}^{d_{\min}-2} \binom{n}{i}} \tag{2.70}$$

Proof. The existence of a *linear* code whose size M equals or exceeds the right-hand side of Eq. (2.70) will now be established.

Consider attempting to construct the parity-check matrix of a code of longest length n possible having minimum distance d_{min} and redundancy r (thus the dimension of the code equals $k = n - r$) in the following way: Without loss of generality, the first r columns of H are picked to be the first r columns of the $(r \times r)$ identity matrix. Additional columns are then added to these, taking care to ensure that at any stage, the matrix formed by the columns selected thus far defines an $[n', n' - r, d_{min}]$ code.

After picking n' columns in this fashion, suppose that we find that the inequality

$$1 + \binom{n'}{1} + \binom{n'}{2} + \cdots + \binom{n'}{d_{min} - 2} < 2^r \tag{2.71}$$

is satisfied. It is thus implied that there exists at least one r-tuple in F^r that cannot be expressed as the linear combination of $(d_{min} - 2)$ or fewer columns drawn from the matrix. Thus, at this stage, at least one additional column can be added to the matrix. This process can be continued until a length n is reached for which

$$\sum_{i=0}^{d_{min} - 2} \binom{n}{i} \geq 2^r \tag{2.72}$$

But, then, setting $k = n - r$ and $M = 2^k$, we see that at this point the code defined by the resulting parity-check matrix has size M satisfying the relation

$$M \geq \frac{2^n}{\displaystyle\sum_{i=0}^{d_{min} - 2} \binom{n}{i}} \tag{2.73}$$

Q.E.D.

The Plotkin bound is based on the reasoning that the minimum distance of a block code of size M cannot exceed the average distance between codewords. When the *fractional minimum distance* $D = d_{min}/n$ of a code exceeds $1/2$, this argument can be translated into a bound on the size of the code. To extend the applicability of the Plotkin bound to other cases as well, the following argument is used: Let C be a code of length n and minimum distance d_{min} having maximum size $M_{max}(n, d_{min})$ and let C be partitioned into two subsets (cosets) such that the codewords within each subset have the same last symbol. At least one of these subsets must have size equal to or exceeding $M_{max}(n, d_{min})/2$. By dropping the last symbol of each codeword in the larger subset, we obtain an $(n - 1, M', d_{min})$ code with $M' \geq M_{max}(n, d_{min})/2$. Thus

$$M_{max}(n, d_{min}) \leq 2M_{max}(n - 1, d_{min}) \tag{2.74}$$

Repeated use of this argument gives

$$M_{max}(n, d_{min}) \leq 2^m M_{max}(n - m, d_{min}) \tag{2.75}$$

When m is such that the fraction $d_{min}/(n - m)$ exceeds 1/2, the Plotkin bound can then be applied. The final bound that results is given in the next theorem.

THEOREM 9 (Plotkin Bound).   Given n and d_{min}, let m be an integer for which $d_{min}/(n - m) > 1/2$. Then

$$M_{max}(n, d_{min}) \leq \frac{2^m d_{min}/(n - m)}{d_{min}/(n - m) - \frac{1}{2}} \qquad (2.76)$$

A table of the best-known bounds on minimum distance for binary, linear codes of given length and dimension is given in Verhoef.[27]

2.4.2 Asymptotic Bounds

To obtain reliable performance using a block code having parameters (M, n, d_{min}) and a decoder that is restricted to bounded-distance decoding, one needs to use long codes whose fractional minimum distance D exceeds twice the crossover probability ϵ of the BSC. Assuming all codewords are transmitted with equal probability, the maximum rate at which it is possible to transmit information using a block code of length n and minimum distance d_{min} is given by

$$R_{max}(n, d_{min}) = \frac{\log_2 [M_{max}(n, d_{min})]}{n} \qquad \text{bits/channel symbol} \qquad (2.77)$$

Our interest lies in determining for a fixed value of fractional minimum distance $D = d_{min}/n$ and very large block length n, how much information can be reliably transmitted per channel symbol. Since the sequence (one for each value of D in the interval $[0, 1]$) is not guaranteed to approach a limit as $n \to \infty$, we define

$$\bar{R}(D) = \limsup_{n \to \infty} R_{max}(n, nD) \qquad (2.78)$$

and

$$R(D) = \liminf_{n \to \infty} R_{max}(n, nD) \qquad (2.79)$$

Using Stirling's approximation[1] to the factorial, the Hamming and Gilbert-Varshamov bounds can be shown to imply

$$\bar{R}(D) \leq 1 - H\left(\frac{D}{2}, 1 - \frac{D}{2}\right) \qquad 0 \leq D \leq 1 \quad \text{(Hamming)} \qquad (2.80)$$

$$R(D) \geq 1 - H(D, 1 - D) \qquad 0 \leq D \leq 1 \quad \text{(Gilbert-Varshamov)} \quad (2.81)$$

where for x, $0 \leq x \leq 1$,

$$H(x, 1 - x) = x \log\left(\frac{1}{x}\right) + (1 - x) \log\left(\frac{1}{1 - x}\right) \qquad (2.82)$$

(with logarithms to base 2) is the entropy function of information theory.
    The following asymptotic version of the Plotkin bound

$$\bar{R}(D) \leq 1 - 2D \qquad 0 \leq D \leq \frac{1}{2} \tag{2.83}$$

can be derived using Theorem 9 in conjunction with the smallest integer m satisfying

$$\frac{d_{\min}}{n - m} > \frac{1}{2} \tag{2.84}$$

These bounds are plotted in Fig. 2.18. To date, the Gilbert-Varshamov bound

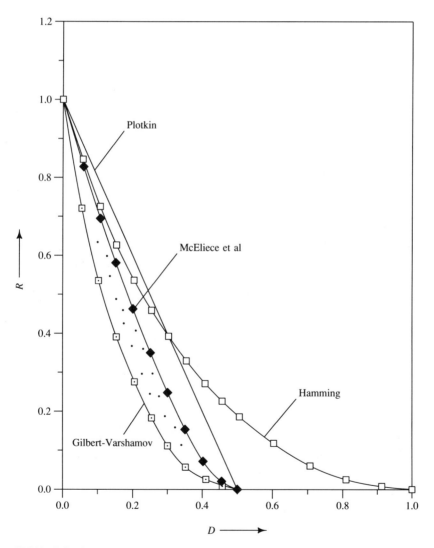

FIGURE 2.18 Asymptotic bounds on the best binary codes. Dots indicate region where the best long codes lie.

remains the best known asymptotic lower bound. The best known upper bound, however, is due to McEliece, Rodemich, Rumsey, and Welch[1] (not discussed here) and is also shown in the plot. These two bounds define the region in which the best long codes are likely to be found. There exist binary linear block codes that asymptotically meet the Gilbert-Varshamov bound.[1]

These bounds are easily extended to the nonbinary case. A recent method of code construction based on the theory of algebraic curves has shown that, when the size of the underlying field is an even power of a prime and exceeds 49, it is possible to construct long codes that *improve* upon the Gilbert bound.[29]

By setting $D = 2\epsilon$, the plots in Fig. 2.18 can be interpreted as indicating the maximum rate at which information transfer can be accomplished across a BSC having crossover probability ϵ using bounded distance decoding. The Hamming bound is, at the same time, a plot of the capacity of the channel! This is based just upon an examination of Eq. (2.80), and the true capacity of a BSC; note that Eq. (2.20) is an approximation. The discrepancy between this bound and the McEliece et al. bound clearly shows that one cannot achieve capacity if the decoder is restricted to bounded-distance decoding.

2.5 REED-MULLER CODES

2.5.1 Definition and Dimension

Reed-Muller codes are named after Reed[22] and Muller[23] and are best understood when the codewords are viewed as Boolean functions. The description below follows Ref. 33, which in turn is based to a considerable extent on the original paper by Reed.

Definition.   Any function defined on F^m, taking on values either 0 or 1, is called a *Boolean function*.

It follows that there are a total of 2^{2^m} Boolean functions in m variables. It is customary to represent a Boolean function in the form of a truth table. The truth table for a Boolean function in $m = 4$ variables is shown in Fig. 2.19. By spelling out the values of the function in some specified order, a one-to-one correspondence between Boolean functions and vectors of length 2^m can be set up. The order that we choose here is the one that would be determined by a binary counter, with the first element of a vector specifying the value of the associated Boolean function when all variables are equal to zero. Thus, the vector

$$(1110 \ 0001 \ 1110 \ 0001)^t \tag{2.85}$$

is associated with the Boolean function given by the truth table in Fig. 2.19.

Using Lagrange interpolation, a polynomial expression for a Boolean

x_1x_2 \ x_3x_4	00	01	11	10
00	1	1	0	1
01	0	0	1	0*
11	0	0	1	0
10	1	1	0	1

FIGURE 2.19   A Boolean function in four variables. Asterisk indicates presence of an error.

function can be derived:

$$f(x_1, x_2, \ldots, x_m) = \sum_{(a_1 a_2 \ldots a_m)^t \in F^m} f(a_1 a_2 \ldots a_m) \prod_{i=1}^{m} (x_i + a_i + 1) \qquad (2.86)$$

This equation shows that every Boolean function can be expressed in the form

$$f(x_1, x_2, \ldots, x_m) = c_0 + \sum_{i=1}^{m} c_i x_i + \sum_{\substack{i,j=1 \\ j>i}}^{m} c_{ij} x_i x_j + \cdots + c_{123\ldots m} x_1 x_2 \ldots x_m \qquad (2.87)$$

where addition is carried out modulo-2 and each coefficient $c_0, c_1, \ldots, c_{12\ldots m}$ is either 0 or 1. Conversely every such expression represents a Boolean function in m variables. Since there are a total of

$$1 + \binom{m}{1} + \binom{m}{2} + \cdots + \binom{m}{m} = 2^m \qquad (2.88)$$

coefficients, the total number of distinct expressions of the form in Eq. (2.87) cannot exceed 2^{2^m}. Since this is precisely the number of Boolean functions, the monomials $x_{i_1} x_{i_2} \ldots x_{i_r}$, $0 \le i \le r$, together with the all-1 function (the constant coefficient), form a basis for the vector space of all Boolean functions in m variables. The expansion of a Boolean function in terms of these basis functions is often called the *Reed-Muller canonical expansion* for the function. The *degree* of a monomial term $x_{i_1} x_{i_2} \ldots x_{i_r}$ is defined to equal r, and the degree of a Boolean function is set equal to the degree of highest-degree monomial appearing (with nonzero coefficient) in its canonical expansion.

Definition.   The rth-order Reed-Muller (R-M) code $R(r, m)$ in m variables is the collection of all Boolean functions in m variables whose degree does not exceed r.

Thus the length of the code equals 2^m and its dimension k is given by

$$k = 1 + \binom{m}{1} + \binom{m}{2} + \cdots + \binom{m}{r} \tag{2.89}$$

E X A M P L E . $R(2, 4)$ encompasses all Boolean quadratics in four variables. The length of the code is 16 and the dimension is given by

$$k = 1 + 4 + 6 = 11 \tag{2.90}$$

Every codeword is of the form

$$f(x_1 x_2 x_3 x_4) = m_0 + \sum_{i=1}^{4} m_i x_i + \sum_{\substack{i,j=1 \\ j>i}}^{4} m_{ij} x_i x_j \tag{2.91}$$

for some binary choice of the coefficients. As usual, the coefficients m_0, m_i, m_{ij}, etc., of the basis functions for the code are regarded as containing the message to be transmitted. The truth table in Fig. 2.19 represents the Boolean function

$$f(x_1 x_2 x_3 x_4) = 1 + x_2 + x_3 x_4, \tag{2.92}$$

and the associated vector in Eq. (2.85) is thus an element of the $R(2, 4)$ code.

By using the vectors associated with the basis functions it is possible to form a generator matrix for the code. However, in our description of the code here, we do not make use of the generator matrix. Figure 2.20 shows how with the aid of a binary counter, the encoder for the RM code $R(2, 4)$ may be implemented.

As the theorem below shows, the dual of an RM code is also an RM code.

THEOREM 10.   The codes $R(r, m)$ and $R(m - r - 1, m)$ are dual codes.

Proof. Computing the inner product of the vectors associated with a pair of Boolean functions is equivalent to forming the product function and then summing over the values of the function as the arguments run through all possibilities. The product function in this case has degree $m - 1$ whereas the summation yields the coefficient of the degree-m monomial, which in this case equals zero. Thus the codewords in the two codes are orthogonal to each other. To see that these codes are dual, we note that the sum of the dimensions of the two codes adds up to the length of the code:

$$\sum_{i=0}^{r} \binom{m}{i} + \sum_{i=0}^{m-r-1} \binom{m}{i} = \sum_{i=0}^{r} \binom{m}{i} + \sum_{i=r+1}^{m} \binom{m}{i} = 2^m \tag{2.93}$$

Q.E.D.

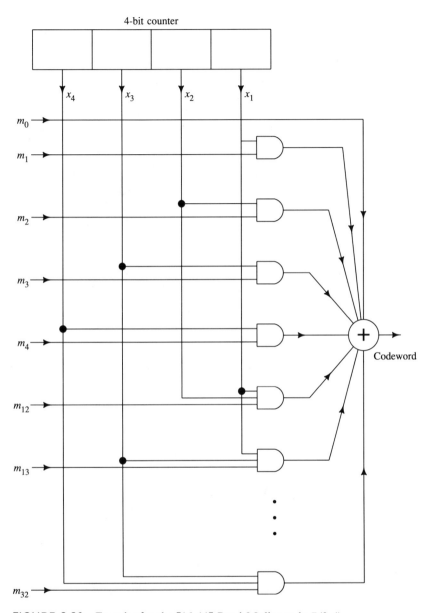

FIGURE 2.20   Encoder for the [16, 11] Reed-Muller code $R(2, 4)$.

2.5.2 Decoding and Minimum Distance

First, consider the problem of determining the coefficient of the highest-degree monomial $x_1 x_2 x_3 \ldots x_r$ in the RM canonical expansion of a Boolean function $f(x_1, x_2, \ldots, x_m)$, given only the truth table for the function.

This can be accomplished simply by summing over all possible values of the function—that is, forming the sum of all the entries in the truth table. The result is the desired coefficient. To see why this is so, consider the truth table of the function to be the modulo-2 sum of the truth tables of the individual monomial functions that sum up to give the function $f(\cdot)$. When one sums over all the entries of the truth table of a typical monomial of the form $x_{i_1} x_{i_2} \ldots x_{i_j}$, $j \le m$, one obtains a *real* sum of 2^{m-j}, since the function takes on the value 1 only when all the variables $x_{i_1}, x_{i_2}, \ldots, x_{i_j}$ equal 1 regardless of the values of the remaining variables. When reduced modulo-2, this yields the value 0 except in the case when $j = m$, in which case the real sum and its modulo-2 reduction both equal 1.

The decoding of the Reed-Muller code $R(r, m)$ can be accomplished by making repeated use of this idea. The decoding proceeds by first determining the coefficients of the degree-r terms of the code polynomial $f(x_1, x_2, \ldots, x_m)$. This is described below in greater detail. Then the effect of these is stripped away from the received polynomial $r(x_1, x_2, \ldots, x_m)$, which can now be regarded as corresponding to a codeword of degree $r - 1$ or less, belonging to $R(r - 1, m)$. This process is then repeated until all coefficients have been determined.

To determine the coefficients of a typical, single, degree-r term $x_1 x_2 \ldots x_r$, assign arbitrary values to the remaining $m - r$ variables. The code function f with this assignment can now be regarded as a degree-r function in only the r variables $x_1, x_2, \ldots, x_r$. Then one sums over all possible values of the received function $r(x_1, x_2, \ldots, x_m)$. In the absence of any errors in that portion of the truth table (in all m variables) corresponding to the arbitrary assignment for the variables $x_{r+1}, \ldots, x_m$, the sum will yield the desired coefficient as explained above. In the presence of errors, there are 2^{m-r} possible assignments for the remaining variables, each of which can be used to provide an independent estimate (there is no overlap in the segments of the truth table over which summation is carried out) of the desired coefficient.

Thus if the number of errors in the received function's truth table is $< 2^{m-r-1}$, by using majority logic one can decode this coefficient correctly. Similarly, one proceeds to determine the remaining degree-r coefficients.

This decoding strategy also allows us to determine the minimum distance of the code as follows: Since we are able to correct $(2^{m-r-1} - 1)$ errors or less it must be that

$$d_{\min} \ge 2^{m-r} - 1 \tag{2.94}$$

But, for $r < m$, the coefficient of $x_1 x_2 \ldots x_m$ is zero for any code polynomial $f(x_1, x_2, \ldots, x_m)$. Thus the values of the function sum to zero, and each codeword must have even weight. Thus, by Theorem 3, $d_{\min}$ must be even, so Eq.

(2.94) can be tightened to

$$d_{\min} \geq 2^{m-r} \tag{2.95}$$

On the other hand, the monomial function $x_1 x_2 \ldots x_r$ (which certainly is a codeword in $R(r, m)$) has Hamming weight 2^{m-r} and therefore, by Theorem 3 again,

$$d_{\min} = 2^{m-r} \tag{2.96}$$

The minimum distance can also be determined directly by viewing the Reed-Muller code as being constructed from two RM codes of lesser degree and length and applying the theory of the $|u|u + v|$ construction presented in Section 2.6 (see also Ref. 1).

E X A M P L E .   Consider the code $R(2, 4)$. This code has minimum distance $2^{m-r} = 4$ and can therefore correct all single-error patterns. Let the codeword

$$f(x_1, x_2, x_3, x_4) = 1 + x_2 + x_3 x_4 \tag{2.97}$$

be transmitted. Let the corresponding received vector $\mathbf{r}$ be given by

$$r = (1110 \ \ 0011 \ \ 1110 \ \ 0001)^t \tag{2.98}$$

that is, the channel has inserted a single error in the coordinate of the codeword corresponding to the variable assignment $(x_1, x_2, x_3, x_4) = (0110)$ (indicated with an asterisk in the truth table of Fig. 2.19).

We begin by decoding the coefficients of the degree-2 monomials. To determine the coefficient of $x_1 x_2$ we assign, in turn, arbitrary values to x_3 and x_4 and then sum over all the entries in the truth table consistent with this assignment. Setting $x_3 = x_4 = 0$, yields a modulo-2 sum equal to 1. Other assignments yield, in turn, the values 0, 0, and 0. Thus using majority logic, the coefficient m_{12} has been declared equal to zero. Proceeding in this fashion, one finds the coefficients of the monomial terms $x_1 x_3, x_1 x_4, x_2 x_3, x_2 x_4$, and $x_3 x_4$ to equal 0, 0, 0, 0, and 1, respectively. Thus the degree-2 portion of the codeword has been (correctly) decoded to be $x_3 x_4$. The truth table of this degree-2 position is now subtracted, entry be entry, from the truth table associated with the received vector. The decoder then proceeds to decode the degree-1 terms using the modified truth table. This stage of the decoding process will decode $m_1 = m_3 = m_4 = 0$ and $m_2 = 1$. After the effect of the degree-1 decoded part (that is, x_2 has been subtracted from the truth table), it remains to determine the value of the constant coefficient.

At this stage, each entry in the remaining truth table represents an independent estimate of this coefficient and, using majority logic as before, decoding can be completed yielding $m_0 = 1$ in this case.

RM codes are an important example of a class of codes that can be decoded using majority logic. Structure-wise, these codes fall into the category of codes constructed using the theory of finite geometries. A good discussion of these classes of codes may be found in Refs. 1, 3, and 8. By deleting the symbol in a single fixed coordinate of a RM code, one obtains a code that is equivalent to a cyclic code. This can be used to simplify both the encoding as well as the decoding of these codes.[1]

2.5.3 First-Order Reed-Muller Codes

This class of RM codes is important not only for their large minimum distance, but also because a relatively simple soft-decision decoding algorithm for these codes is available.

Consider transmission using binary signaling (see Section 2.1) over the unquantized discrete channel derived from the AWGN waveform channel. The transmitted sequence $s(x_1, x_2, \ldots, x_m)$, corresponding to the codeword generated by message block $\{m_i/i = 0, 1, \ldots, m_m\}$ in the first-order RM code $R(1, m)$ is given by:

$$s(x_1, x_2, \ldots, x_m) = \sqrt{E_{dc}}(-1)^{m_0 + \sum_{i=1}^m m_i x_i} \qquad (x_1, x_2, \ldots, x_m)^t \in F^m \qquad (2.99)$$

The inner product of any two of these (message symbols m_i and m_i') signals is given by

$$E_{dc}(-1)^{m_0 + m_0'} \sum_{(x_1, x_2, \ldots, x_m)^t \in F^m} (-1)^{\sum_{i=1}^m x_i(m_i + m_i')}$$

$$= \begin{cases} 0 \text{ if some } m_i \neq m_i', & 1 \le i \le m \\ 2^m E_{dc} \text{ if } m_i = m_i', \text{ for all } i, & 0 \le i \le m \\ -2^m E_{dc} \text{ if } m_i = m_i', \text{ for all } i, & 1 \le i \le m, m_0 \neq m_0' \end{cases} \qquad (2.100)$$

Thus we have the following theorem.

THEOREM 11. The transmitted sequences in a first-order RM code constitute a biorthogonal set of sequences of length 2^m.

Note that as a consequence of this theorem, it is a relatively simple matter to determine the Euclidean distance between any two transmitted sequences, thus making easier the job of estimating the probability of block error P_e. A closed-form expression for P_e over an AWGN channel for such sequences can be found in Ref. 7. Over such channels, when the received sequence

$$\{r(x_1, x_2, \ldots, x_m)/(x_1, x_2, \ldots, x_m)^t \in F^m\} \qquad (2.101)$$

is not quantized, MLD is reduced to a determination of the signal closest in Euclidean distance to the received sequence or, equivalently, the message sequence $\{m_i\}$ for which the correlation

$$(-1)^{m_0} \sum_{(x_1, x_2, \ldots, x_m)^t \in F^m} (-1)^{\sum_{i=1}^m x_i m_i} r(x_1, x_2, \ldots, x_m) \qquad (2.102)$$

is a maximum. However, computation of the inner product of the received and transmitted vectors is equivalent to computing the Hadamard (Walsh) transform of the received vector except that, along with each transform coefficient, one must compute the negative of that coefficient as well. As a result, the decoding strategy can be described as follows:

Step 1

Compute the transform coefficients $\hat{r}(m_1, m_2, \ldots, m_m)$ defined according to

$$\hat{r}(m_1, m_2, \ldots, m_m) = \sum_{(x_1, x_2, \ldots, x_m)^t \in F^m} r(x_1, x_2, \ldots, x_m)(-1)^{\sum_{i=1}^m x_i m_i}$$

$$(m_1, m_2, \ldots, m_m)^t \in F^m \qquad (2.103)$$

Step 2

Find the coefficient having maximum magnitude. At this stage, one has determined the most likely values of the message symbols $m_1, m_2, \ldots, m_m$. It remains to decode the message symbol m_0.

Step 3

If the transform coefficient having maximum magnitude is positive, declare $m_0 = 0$; otherwise, say $m_0 = 1$.

Use of a fast Hadamard transform algorithm [analogous to the fast Fourier transform (FFT)] allows Step 1 of the decoding algorithm to be speeded up. This is illustrated for the example below.

E X A M P L E . We consider the RM code $R(1, 3)$ with $E_{dc} = 1$. Let the message sequence correspond to the codeword $1 + x_3$ so that the sequence

$$\mathbf{s} = (-1 \ 1 \ -1 \ 1 \ -1 \ 1 \ -1 \ 1)^t \qquad (2.104)$$

is transmitted. Suppose

$$\mathbf{r} = (0 \ 2 \ 1 \ 1 \ -1 \ 2 \ -1 \ 1)^t \qquad (2.105)$$

is received.

Step 1

The Hadamard transform $\hat{\mathbf{r}}$ of $\mathbf{r}$ can be computed from the matrix equation:

$$
\begin{bmatrix} \hat{r}(000) \\ \hat{r}(001) \\ \hat{r}(010) \\ \hat{r}(011) \\ \hat{r}(100) \\ \hat{r}(101) \\ \hat{r}(110) \\ \hat{r}(111) \end{bmatrix} = \begin{bmatrix} 1 & 1 & 1 & 1 & 1 & 1 & 1 & 1 \\ 1 & -1 & 1 & -1 & 1 & -1 & 1 & -1 \\ 1 & 1 & -1 & -1 & 1 & 1 & -1 & -1 \\ 1 & -1 & -1 & 1 & 1 & -1 & -1 & 1 \\ 1 & 1 & 1 & 1 & -1 & -1 & -1 & -1 \\ 1 & -1 & 1 & -1 & -1 & 1 & -1 & 1 \\ 1 & 1 & -1 & -1 & -1 & -1 & 1 & 1 \\ 1 & -1 & -1 & 1 & -1 & 1 & 1 & -1 \end{bmatrix} \begin{bmatrix} r(000) \\ r(001) \\ r(010) \\ r(011) \\ r(100) \\ r(101) \\ r(110) \\ r(111) \end{bmatrix} \qquad (2.106)
$$

Computation of this transform can be speeded up by factoring this matrix into the product of 3 (m in the general case) more "elementary" matrices:

$$
\begin{bmatrix}
1 & 1 & 1 & 1 & 1 & 1 & 1 & 1 \\
1 & -1 & 1 & -1 & 1 & -1 & 1 & -1 \\
1 & 1 & -1 & -1 & 1 & 1 & -1 & -1 \\
1 & -1 & -1 & 1 & 1 & -1 & -1 & 1 \\
1 & 1 & 1 & 1 & -1 & -1 & -1 & -1 \\
1 & -1 & 1 & -1 & -1 & 1 & -1 & 1 \\
1 & 1 & -1 & -1 & -1 & -1 & 1 & 1 \\
1 & -1 & -1 & 1 & -1 & 1 & 1 & -1
\end{bmatrix}
$$

$$
=
\begin{bmatrix}
1 & 1 & & & & & & \\
& & 1 & 1 & & & & \\
& & & & 1 & 1 & & \\
& & & & & & 1 & 1 \\
1 & -1 & & & & & & \\
& & 1 & -1 & & & & \\
& & & & 1 & -1 & & \\
& & & & & & 1 & -1
\end{bmatrix}
$$

$$
\times
\begin{bmatrix}
1 & 1 & & & & & & \\
& & 1 & 1 & & & & \\
& & & & 1 & 1 & & \\
& & & & & & 1 & 1 \\
1 & -1 & & & & & & \\
& & 1 & -1 & & & & \\
& & & & 1 & -1 & & \\
& & & & & & 1 & -1
\end{bmatrix}
$$

$$
\times
\begin{bmatrix}
1 & 1 & & & & & & \\
& & 1 & 1 & & & & \\
& & & & 1 & 1 & & \\
& & & & & & 1 & 1 \\
1 & -1 & & & & & & \\
& & 1 & -1 & & & & \\
& & & & 1 & -1 & & \\
& & & & & & 1 & -1
\end{bmatrix}
$$

The initial, intermediate and final vectors that result from this three-stage matrix multiplication are reproduced here:

$$\mathbf{r} = \begin{bmatrix} 0 \\ 2 \\ 1 \\ 1 \\ -1 \\ 2 \\ -1 \\ 1 \end{bmatrix} \Rightarrow \begin{bmatrix} 2 \\ 2 \\ 1 \\ 0 \\ -2 \\ 0 \\ -3 \\ -2 \end{bmatrix} \Rightarrow \begin{bmatrix} 4 \\ 1 \\ -2 \\ -5 \\ 0 \\ 1 \\ -2 \\ -1 \end{bmatrix} \Rightarrow \begin{bmatrix} 5 \\ -7 \\ 1 \\ -3 \\ 3 \\ 3 \\ -1 \\ -1 \end{bmatrix} = \hat{\mathbf{r}}$$
(2.107)

Step 2

From the computed transform, we see that $\hat{r}(001)$ has the largest magnitude. Thus at this point we have decoded the message symbols m_1, m_2, m_3 as being equal to 0, 0, and 1, respectively.

Step 3

Since the sign of the largest transform coefficient is negative, we declare $m_0 = 1$.

Thus the (correctly) decoded codeword is $1 + x_3$. The performance curves (with hard and soft decisions) for the Reed-Muller code $R(1, 5)$ used on the Mariner '69 expedition are shown in Fig. 2.21.

2.6 NEW CODES FROM OLD

Some methods of modifying linear codes to obtain codes having slightly different parameters are described below. These methods are especially useful when the parameters of a known good code do not match the requirements and have also recently been used in the construction of tables of the best-known minimum distance bounds for binary linear codes[27] of given length and minimum distance. In all the methods described below, we use the notation $[\mathbf{n}, \mathbf{k}, \mathbf{d}]$ and $[\mathbf{n'}, \mathbf{k'}, \mathbf{d'}]$ to describe the parameters of the original and modified code respectively. For more details, the reader is referred to the text by MacWilliams and Sloane.[1]

Extending the Code

In this method, the length of the code is extended from n to $n + 1$ by adding an overall parity check that ensures that all codewords in the new code have even weight. Thus, if the minimum distance of the original code was odd, for the new code it will have increased by 1. The dimension, of course, remains the same because no new codewords are added.

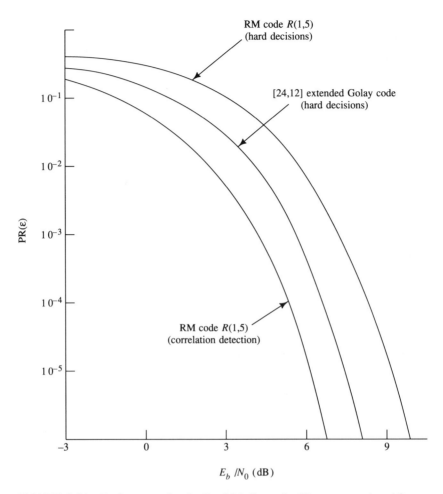

FIGURE 2.21   Performance for the Reed-Muller code. (Curves reproduced from S. W. Golomb, ed., *Digital Communications with Space Applications*, © 1964. Reprinted by permission of Prentice Hall, Inc., Englewood Cliffs, N.J.)

E X A M P L E .   By extending the $[7, 4, 3]$ Hamming code (code 2), one obtains an $[8, 4, 4]$ code. Thus the extended codes can detect double errors as well. The parity-check matrix for the new code can be derived by adding a column of all 0's followed by a row of all 1's as follows:

$$\begin{bmatrix} 1 & 1 & 1 & 1 & 1 & 1 & 1 & 1 \\ 1 & 1 & 0 & 1 & 1 & 0 & 0 & 0 \\ 1 & 0 & 1 & 1 & 0 & 1 & 0 & 0 \\ 0 & 1 & 1 & 1 & 0 & 0 & 1 & 0 \end{bmatrix} \qquad (2.108)$$

Puncturing the Code

By puncturing a code, we simply mean that the symbol in some specific coordinate is dropped from all the codewords. In terms of the generator matrix of the code, this technique simply corresponds to deleting the column corresponding to the specific coordinate. Unless the original code was a trivial code, the new code will have the same dimension. Usually, the minimum distance drops by 1.

> E X A M P L E .   Consider the first-order RM code $R(1,4)$ having parameters $[16, 5, 8]$. The generator matrix for this code as well as that for the punctured code are shown in Fig. 2.22. Note that by adding the top and bottom rows of G' one obtains a code vector having Hamming weight equal to 7. Thus the new code has parameters $[15, 5, 7]$. Moreover, by permuting columns as shown in the Fig. 2.22, it turns out that one obtains the generator matrix of a cyclic code (these codes are discussed in Section 2.7).

This phenomenon is true in general: by puncturing the RM code $R(r, m)$, one obtains a code equivalent to a cyclic code having parameters $[n - 1, k, d - 1]$.

Expurgating the Code

By this, we mean throwing away some codewords to obtain a smaller code. The most common example of this technique is to drop all codewords in the code having odd Hamming weight, leaving behind an even-weight subcode.

The new code can be shown to be a subgroup of the original group. Moreover, this subgroup has precisely one coset—namely, the set of codewords in the original code that had odd Hamming weight. Thus the new code is linear (a subgroup) and has dimension $k - 1$. If the original code had $d_{\min}$ odd, then the new code has minimum distance $d_{\min} + 1$; otherwise, the minimum distance remains unchanged.

The Hamming weight of the sum of two n-tuples of odd weight is necessarily even. Thus the generator matrix of any linear code can through elementary row reduction techniques, always be put in the form in which only one row has odd Hamming weight.

This is illustrated in Fig. 2.22 for the cyclic $[15, 5, 7]$ code derived by puncturing the first-order RM code $R(1, 4)$. Notice that only the top row of the generator matrix for the code has odd Hamming weight. By deleting this row, we obtain a cyclic code [with generator matrix G' (equivalent)] having parameters $[15, 4, 8]$. It can be shown that for this code, all the codewords are cyclic shifts of each other. Such a linear cyclic code is called a *maximum-length*

$$G = \begin{pmatrix} 1 & 1 & 1 & 1 & 1 & 1 & 1 & 1 & 1 & 1 & 1 & 1 & 1 & 1 & 1 & 1 \\ 0 & 0 & 0 & 0 & 0 & 0 & 0 & 0 & 1 & 1 & 1 & 1 & 1 & 1 & 1 & 1 \\ 0 & 0 & 0 & 1 & 1 & 1 & 1 & 0 & 0 & 0 & 0 & 1 & 1 & 1 & 1 \\ 0 & 0 & 1 & 1 & 0 & 0 & 1 & 1 & 0 & 0 & 1 & 1 & 0 & 0 & 1 & 1 \\ 0 & 1 & 0 & 1 & 0 & 1 & 0 & 1 & 0 & 1 & 0 & 1 & 0 & 1 & 0 & 1 \end{pmatrix}$$

Column $\Downarrow$ Puncture
deleted

$$G' = \begin{pmatrix} 1 & 1 & 1 & 1 & 1 & 1 & 1 & 1 & 1 & 1 & 1 & 1 & 1 & 1 & 1 \\ 0 & 0 & 0 & 0 & 0 & 0 & 0 & 1 & 1 & 1 & 1 & 1 & 1 & 1 & 1 \\ 0 & 0 & 0 & 1 & 1 & 1 & 1 & 0 & 0 & 0 & 0 & 1 & 1 & 1 & 1 \\ 0 & 1 & 1 & 0 & 0 & 1 & 1 & 0 & 0 & 1 & 1 & 0 & 0 & 1 & 1 \\ 1 & 0 & 1 & 0 & 1 & 0 & 1 & 0 & 1 & 0 & 1 & 0 & 1 & 0 & 1 \end{pmatrix}$$

$\Downarrow$ Permute
columns

$$G' \text{ (equivalent)} = \begin{pmatrix} 1 & 1 & 1 & 1 & 1 & 1 & 1 & 1 & 1 & 1 & 1 & 1 & 1 & 1 & 1 \\ 0 & 0 & 0 & 1 & 1 & 1 & 1 & 0 & 1 & 0 & 1 & 1 & 0 & 0 & 1 \\ 0 & 0 & 1 & 1 & 1 & 1 & 0 & 1 & 0 & 1 & 1 & 0 & 0 & 1 & 0 \\ 0 & 1 & 1 & 1 & 1 & 0 & 1 & 0 & 1 & 1 & 0 & 0 & 1 & 0 & 0 \\ 1 & 1 & 1 & 1 & 0 & 1 & 0 & 1 & 1 & 0 & 0 & 1 & 0 & 0 & 0 \end{pmatrix}$$

Argument $\Uparrow$ $\Downarrow$ Expurgate

$$G_{ml} = \begin{pmatrix} 0 & 0 & 0 & 1 & 1 & 1 & 1 & 0 & 1 & 0 & 1 & 1 & 0 & 0 & 1 \\ 0 & 0 & 1 & 1 & 1 & 1 & 0 & 1 & 0 & 1 & 1 & 0 & 0 & 1 & 0 \\ 0 & 1 & 1 & 1 & 1 & 0 & 1 & 0 & 1 & 1 & 0 & 0 & 1 & 0 & 0 \\ 1 & 1 & 1 & 1 & 0 & 1 & 0 & 1 & 1 & 0 & 0 & 1 & 0 & 0 & 0 \end{pmatrix}$$

FIGURE 2.22   The generator matrices of the codes obtained through successive modification of the RM code $R(1,4)$.

(ml) linear-feedback shift register code (ml code for short; sometimes also referred to as a *simplex code*) and is discussed further in Section 2.7.5.

Once again this example generalizes, and one always obtains an ml code by expurgating in this fashion the punctured cyclic first-order RM code.

Augmenting the Code by Adding Codewords

The most important example is when one adds the all-one row to the generator matrix of the original code (assuming that the all-one n-tuple is not already a codeword in the original code). This augmentation has the effect of adding to the code the complement of every codeword. The code length remains the same, and since the codesize is doubled, the dimension is increased by 1. The minimum distance of the modified code is given by

$$d' = \min(d, n - d_{max}) \tag{2.109}$$

E X A M P L E . By augmenting the $[15, 4, 8]$ ml code with the all-one codeword to the generator matrix, we recover the punctured, cyclic first-order RM code (Fig. 2.22).

Shortening the Code

In this method, one first picks a coordinate and throws away all codewords having a 1 in that coordinate. The next step is to delete that coordinate from each of the remaining codewords.

It is easy to show, using arguments similar to that made in the above discussion on expurgating the code, that the resulting code is still linear, with length $n - 1$ and dimension $k - 1$. The minimum distance either remains unchanged or increases. Using elementary row reduction techniques, the generator matrix of any linear code can always be put into a form in which all the entries in the column corresponding to the coordinate in question are zero except for one. By deleting this row and then discarding the resulting all-zero column in the generator matrix, one obtains the generator matrix for the shortened code.

E X A M P L E . By deleting the second coordinate in the generator matrix of the $[7, 4, 3]$ Hamming code (code 2) as shown below, we derive the $[6, 3, 3]$ code (code 4).

$$\begin{bmatrix} 1 & 0 & 0 & 0 & 1 & 1 & 0 \\ 0 & 1 & 0 & 0 & 1 & 0 & 1 \\ 0 & 0 & 1 & 0 & 0 & 1 & 1 \\ 0 & 0 & 0 & 1 & 1 & 1 & 1 \end{bmatrix} \Rightarrow \begin{bmatrix} 1 & 0 & 0 & 1 & 1 & 0 \\ 0 & 1 & 0 & 0 & 1 & 1 \\ 0 & 0 & 1 & 1 & 1 & 1 \end{bmatrix} \tag{2.110}$$

The $|u|u + v|$ Construction

Given two linear codes C_1 and C_2 having the same length n but having dimensions k_1 and k_2 and minimum distances d_1 and d_2, an obvious means of combining the two codes to obtain a code having twice the length of the original code is to declare as codewords all $2n$-tuples of the form $\{|u|v|/u \in C_1, v \in C_2\}$. The new code has dimension $k_1 + k_2$. However, with this construction technique, the minimum distance of the new code is given by:

$$d_{min} = \min(d_1, d_2) \tag{2.111}$$

A better approach is to let the new code consist of all $2n$-tuples of the form $|u|u + v|$. With this method, it can be shown that the minimum distance now improves to

$$d' = \min(2d_1, d_2). \tag{2.112}$$

This result can be used to determine in a recursive fashion the minimum distance of the Reed-Muller code.[1]

2.7 CYCLIC CODES

Most block codes employed in practice are either cyclic, or else very closely related to this class of codes. Good treatments of this subject can be found in a number of textbooks.[1,3,8,9]

2.7.1 Structure

Definition.   A linear block code is said to be a *cyclic code* if, in addition to being linear, every cyclic shift of a codeword is also a codeword.

For example, if $\mathbf{a} = (a_0, a_1, a_2, \ldots, a_{n-1})^t$ belongs to a cyclic code of length n, then the $n - 1$ cyclic shifts of the codeword

$$(a_{n-i}, a_{n-i+1}, \ldots, a_{n-i-1})^t   \quad 1 \le i \le n - 1 \tag{2.113}$$

must also be codewords. In polynomial terms, if $a(x)$ is the polynomial associated with the codeword $\mathbf{a}$, that is, if

$$a(x) = a_0 + a_1 x + a_2 x^2 + \cdots + a_{n-1} x^{n-1} \tag{2.114}$$

then every polynomial of the form

$$x^i a(x) \bmod (x^n - 1) \tag{2.115}$$

is also a code polynomial.

Let C be an $[n, k]$ cyclic code. One can then establish the following:[1]

1. The set of all code polynomials contains a unique polynomial of minimum degree, called the *generating polynomial* $g(x)$.

2. $g(x)$ has degree $n - k$ and every code polynomial $c(x)$ can be expressed as a multiple $a(x)g(x)$ with $\deg[a(x)] \le k - 1$.

3. $g(x)$ divides $(x^n - 1)$; note that this implies as a consequence that x does not divide $g(x)$.

Conversely, given any polynomial $g(x)$ of degree r dividing $(x^n - 1)$, for some n, the set of all multiples $a(x)g(x)$, $\deg[a(x)] \le n - r - 1$, forms a cyclic code C with parameters $[n, k = n - r]$. The polynomial $g(x)$ is said to *generate* the code. An algebraic description of a cyclic code would describe a cyclic code as the ideal in the ring of polynomials over F modulo $(x^n - 1)$ generated by $g(x)$.

EXAMPLE.   Let $g(x) = x^6 + x^5 + x^4 + x^3 + 1$. It can be verified that $g(x)|x^{15} + 1$. Thus, the set of all multiples of $g(x)$ of degree ≤ 8 forms a $[15, 9]$ code.

Given an $[n, k]$ cyclic code generated by the polynomial $g(x)$, the set of polynomials

$$g(x), xg(x), x^2g(x), x^3g(x), \ldots, x^{k-1}g(x) \tag{2.116}$$

form a basis for the code generated by $g(x)$ and can be used to construct a generator matrix for the code. However, the resulting code is not systematic and we therefore do not pursue this any further.

It is not hard to see that the dual of a cyclic code must also be cyclic. Given a polynomial $f(x)$ of degree d, the polynomial $f_r(x)$ given by:

$$f_r(x) = x^d f(x^{-1}) \tag{2.117}$$

is for obvious reasons, called the *reciprocal* of $f(x)$. If the original cyclic code is generated by $g(x)$, then it turns out that the dual is generated by the polynomial

$$h_r(x) = \text{reciprocal}[h(x)] \tag{2.118}$$

where

$$h(x) = (x^n - 1)/g(x) \tag{2.119}$$

Therefore the parity-check equations can be expressed in terms of the coefficients of the polynomial $h(x)$; for this reason, $h(x)$ is called the parity-check *polynomial* of the code. In the example above, the dual code is generated by the polynomial

$$h_r(x) = \text{reciprocal}\left(\frac{x^{15} + 1}{x^6 + x^5 + x^4 + x^3 + 1}\right) = 1 + x + x^4 + x^5 + x^6 + x^9 \tag{2.120}$$

2.7.2 Systematic Encoding

Systematic encoding can be accomplished as follows: Given a message polynomial $m(x)$—that is, a polynomial of degree $\leq k - 1$—whose coefficients represent the information symbols to be encoded, multiply $m(x)$ by x^{n-k} and then compute the remainder $q(x)$, obtained upon division of this polynomial by $g(x)$. The sum polynomial

$$m(x)x^{n-k} + q(x) \tag{2.121}$$

is a multiple of $g(x)$, and the information symbols now appear explicitly as the coefficients of the powers $n - k + 1, n - k + 2, \ldots, n - 1$.

EXAMPLE. Consider the $[15, 9]$ cyclic code generated by the polynomial

$$g(x) = x^6 + x^5 + x^4 + x^3 + 1 \tag{2.122}$$

Let

$$m(x) = x^5 + x + 1 \tag{2.123}$$

be the message polynomial; that is,

$$(m_0 m_1 m_2 m_3 m_4 \ldots m_8) = (110001000) \tag{2.124}$$

To derive the code polynomial, first multiply $m(x)$ by x^6 and divide by $g(x)$. This leaves $x^3 + 1$ as the remainder. Adding the remainder to the product $m(x)x^6$, we obtain the code polynomial $x^{11} + x^7 + x^6 + x^3 + 1$.

Let the polynomials $q_i(x)$, $i = 0, 1, \ldots, k - 1$ be the remainders resulting from division of the polynomials x^{n-k+i}, $0 \leq i \leq k - 1$, by $g(x)$. Then the k polynomials

$$b_i(x) = x^{n-k+i} + q_i(x) \tag{2.125}$$

form a basis for the systematic code. The corresponding generator and parity-check matrices[e] are shown below in which the entry q_{ij}, $0 \leq i \leq k - 1$, $0 \leq j \leq n - k - 1$, denotes the coefficient of x^j in the polynomial $q_i(x)$.

$$G = \begin{bmatrix} q_{00} & q_{01} & \cdots & q_{0,n-k-1} & 1 & 0 & \cdots & 0 \\ q_{10} & q_{11} & \cdots & q_{1,n-k-1} & 0 & 1 & \cdots & 0 \\ \vdots & \vdots & & \vdots & \vdots & \vdots & & \vdots \\ q_{k-1,0} & q_{k-1,1} & \cdots & q_{k-1,n-k-1} & 0 & 0 & \cdots & 1 \\ & & P & & & & I_k & \end{bmatrix} \tag{2.126}$$

$$H = [I_{n-k} P^t] \tag{2.127}$$

EXAMPLE. The generator matrix obtained in this fashion for the $[15, 9]$ cyclic code with generator polynomial $g(x) = x^6 + x^5 + x^4 + x^3 + 1$ is shown below:

$$G = \begin{bmatrix} 1 & 0 & 0 & 1 & 1 & 1 & 1 & & & & & & & & \\ 1 & 1 & 0 & 1 & 0 & 0 & & 1 & & & & & & & \\ 0 & 1 & 1 & 0 & 1 & 0 & & & 1 & & & & & & \\ 0 & 0 & 1 & 1 & 0 & 1 & & & & 1 & & & & & \\ 1 & 0 & 0 & 0 & 0 & 1 & & & & & 1 & & & & \\ 1 & 1 & 0 & 1 & 1 & 1 & & & & & & 1 & & & \\ 1 & 1 & 1 & 1 & 0 & 0 & & & & & & & 1 & & \\ 0 & 1 & 1 & 1 & 1 & 0 & & & & & & & & 1 & \\ 0 & 0 & 1 & 1 & 1 & 1 & & & & & & & & & 1 \end{bmatrix} \tag{2.128}$$

There is no simple means for determining the minimum distance of a general cyclic code. Estimates are available for specific subfamilies of cyclic codes such as the Hamming codes (see Section 2.7.5), the Golay code (Section

---

[e] The identity matrix is positioned at the opposite end from that given in the definition of Section 2.3.1, but this clearly is a minor detail.

2.7.7), ml codes (Section 2.6), Bose-Chaudhuri-Hocquenghem (BCH) and Reed-Solomon (RS) codes (Sections 2.7.9 and 2.7.10), and quadratic-residue codes (discussed elsewhere[1]).

2.7.3 Syndrome Computation

As noted earlier in Section 2.3.5, the syndrome can be determined by recomputing parity from the received version of the message symbols and then forming the modulo-2 sum with the received parity bits. With systematic encoding, this amounts to reducing the received polynomial

$$r(x) = r_0 + r_1 x + \cdots + r_{n-1} x^{n-1} \qquad (2.129)$$

modulo the generating polynomial $g(x)$.

The polynomial

$$s(x) = r(x) \bmod [g(x)] \qquad (2.130)$$

that results from this reduction is called the *syndrome polynomial*. Having obtained the syndrome, decoding can then be accomplished using table lookup to determine the coset leader in the standard array. The size of the table required to be stored by the decoder can be substantially reduced by making use of the cyclic nature of the code (see discussion below on "error trapping"). When the redundancy $(n - k)$ of the code is large, making table lookup impractical, one is forced to use other decoding techniques specific to the particular subclass of cyclic codes as discussed in the subsections to follow.

2.7.4 Implementation

Consider the 4-bit shift register shown in Fig. 2.23. Let the initial contents a_0, a_1, a_2, a_3 of the shift register be regarded as the coefficients of a polynomial of degree 3 or less with the coefficient a_3 of x^3 being on the extreme right. Loosely speaking, the shift register initially "contains" the polynomial $a(x)$. If the shift register is then clocked once, the contents change to the polynomial

$$xa(x) \bmod x^4 + x + 1 \qquad (2.131)$$

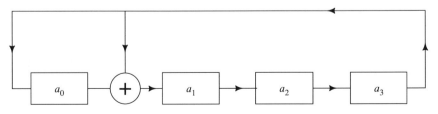

FIGURE 2.23   Shift register whose initial contents are the coefficients of the polynomial $a(x) = a_0 + a_1 x + a_2 x^2 + a_3 x^3$. In succeeding clock intervals, the contents will be $xa(x)$, $x^2 a(x)$, ..., modulo the "feedback" polynomial $x^4 + x + 1$.

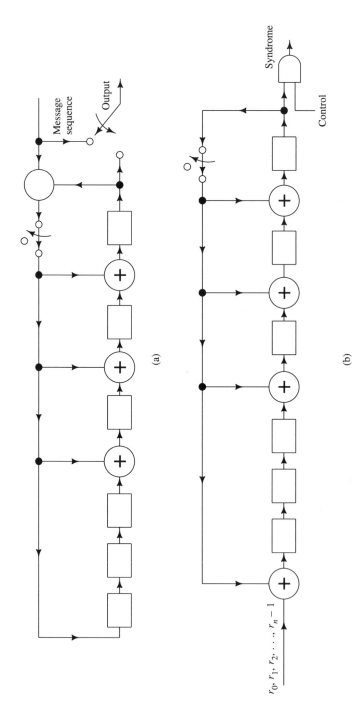

FIGURE 2.24 Encoder and syndrome generation circuit for the systematic cyclic code having generator polynomial $x^6 + x^5 + x^4 + x^3 + 1$.

because of the locations of the feedback taps which are "wired" according to the polynomial $x^4 + x + 1$. Thus each time we clock the shift-register, we are in effect, implementing multiplication of its contents by x and reducing the result modulo $x^4 + x + 1$.

E X A M P L E .   Let the initial polynomial stored in the shift-register of Fig. 2.23 be x^3. After clocking, the new contents will be

$$x(x^3) \bmod x^4 + x + 1 = x + 1 \qquad (2.132)$$

Such circuitry (see, for example, Ref. 3) can be used to encode a cyclic code. Figure 2.24a shows the encoding circuit in the case of the [15, 9] cyclic code discussed earlier. First one wires the shift register according to the generator polynomial of the cyclic code

$$g(x) = x^6 + x^5 + x^4 + x^3 + 1 \qquad (2.133)$$

The next step is to feed in the message bits into the shift register from the right [with the message symbol m_{k-1} (m_8 in this case) entering first] and then clocking the register k nine times. Note that feeding in the message symbols in this fashion is equivalent to premultiplying the message polynomial by $x^{n-k}(x^6)$. At the same time, the message bits are also transmitted across the channel. At the $(k + 1)$th (10th in this case) time instant, the shift register contains the parity bits that are now ready for transmission. Thus, the final step is to shift out to the channel the contents of the shift register after making sure that the feedback loop has been disconnected.

The same circuit can be used to compute the syndrome as well, the only difference being that this time the received symbols are entered into the shift register from the left (see Fig. 2.24b).

2.7.5 Cyclic Hamming and Maximum-Length Codes

Every Hamming code is equivalent to a cyclic code. This fact can be shown as follows:

Definition.   A polynomial is said to be *irreducible* if it cannot be expressed as the product of two polynomials of lesser degree.

Fact.   Every binary irreducible polynomial of degree r divides $x^{2^{r-1}} - 1$.

Definition.   An irreducible polynomial $p(x)$ of degree r is said to be *primitive* if the smallest n for which $p(x)|(x^n - 1)$ is $n = 2^{r-1}$.

Fact.   There exist binary primitive polynomials of all degrees $r, r \geq 1$.

Consider the $[2^{r-1}, 2^r - 1 - r]$ cyclic code generated by a primitive polynomial $p(x)$ of degree r. Each code polynomial is some multiple of $p(x)$. Using the fact that $p(x)$ is primitive, it can be shown that it is impossible for a code polynomial to have Hamming weight ≤ 2. On the other hand, by the Hamming bound, the minimum weight cannot exceed 3 simply because the code with parameters $[2^{r-1}, 2^r - 1 - r, 3]$ is perfect. Thus the minimum weight of the code equals 3, and the code has the same parameters as the Hamming code. But then, any linear code with these parameters is equivalent to the Hamming code simply from noting that the parity-check matrix of such a code must contain all nonzero r-tuples with no column repeats.

E X A M P L E .   The polynomial $p(x) = x^3 + x + 1$ is primitive and therefore divides $x^7 + 1$. The generator matrix G of the $[7, 4]$ systematic code generated by $p(x)$ is shown below along with the associated parity-check matrix H. From an examination of H it is clear that $p(x)$ generates a Hamming code.

$$G = \begin{bmatrix} 1 & 1 & 0 & 1 & 0 & 0 & 0 \\ 0 & 1 & 1 & 0 & 1 & 0 & 0 \\ 1 & 1 & 1 & 0 & 0 & 1 & 0 \\ 1 & 0 & 1 & 0 & 0 & 0 & 1 \end{bmatrix} \tag{2.134}$$

$$H = \begin{bmatrix} 1 & 0 & 0 & 1 & 0 & 1 & 1 \\ 0 & 1 & 0 & 1 & 1 & 1 & 0 \\ 0 & 0 & 1 & 0 & 1 & 1 & 1 \end{bmatrix} \tag{2.135}$$

Next, consider the dual of the cyclic Hamming code of length $2^r - 1$ generated by the primitive polynomial $p(x)$ of degree r. This code is cyclic, has length $2^r - 1$ and dimension r. But it turns out this can only happen if all the codewords are cyclic shifts of one another, which means that this code is a maximum-length (ml) code (see Section 2.6). The parity-check matrix of the cyclic Hamming code serves as a generator matrix for this code. Since all nonzero r-tuples appear precisely once in this matrix, it follows from counting that each row contains 2^{r-1} ones and $2^{r-1} - 1$ zeros. Thus the minimum distance of the code equals 2^{r-1}.

Since the reciprocal of a primitive polynomial can be shown to also be primitive, it follows that ml codes of length $2^r - 1$, $r \geq 2$, can be defined as the $[2^r - 1, r, 2^{r-1}]$ cyclic codes generated by generator polynomials $g(x)$ of the form

$$g(x) = \frac{x^{2^r - 1} - 1}{p(x)} \tag{2.136}$$

where $p(x)$ is some primitive polynomial having degree r. Since any two Hamming codes of the same length are equivalent, it follows that the same is true of ml codes.

Maximum-length codes are better known in the literature as maximal-length, linear-feedback, shift-register sequences (*m*-sequences for short) and in this form, have been extensively studied.[1, 24, 37, 38] Let $(a_0, a_1, a_2, \ldots, a_{n-1})^t$ and $(a_{n-\tau}, a_{n-\tau+1}, \ldots, a_{n-1}, a_0, \ldots, a_{n-\tau-1})^t$ and $(n = 2^r - 1$ for some $r)$ be two code-words belonging to an ml code (note that any two nonzero codewords can be so represented). Then the inner product (correlation) of the sequences corresponding to these codewords, transmitted across the channel is given by

$$E_{dc} \sum_{i=0}^{n-1} (-1)^{a_i + a_{n-r+i}} \tag{2.137}$$

and can be shown, using distance properties of the code, to equal

$$\begin{cases} -E_{dc}, & \text{for } \tau, 1 \leq \tau \leq n - 1 \\ +nE_{dc}, & \text{for } \tau = 0 \end{cases} \tag{2.138}$$

This has led to a variety of applications in spread-spectrum communications[37] and elsewhere.

2.7.6 Error Trapping

A simple decoding technique can be used in conjunction with systematic cyclic codes. Using this method, if t is the error-correcting capability of the code, the code is guaranteed to correct all patterns of t or fewer errors *provided that the errors are confined to $n - k$ consecutive locations.* In interpreting the word "consecutive," one should regard the coordinates of the code as being arranged along a circle with the first coordinate immediately following the last.

To see how the decoding is simplified, first consider the case in which all errors are confined to the parity-bit locations; that is, the received symbols r_0, r_1, ..., r_{n-k-1} are the only symbols possibly in error. We assume of course, that the number of errors does not exceed the error-correcting capability $t = [(d_{min} - 1)/2]$ of the code. In this case it is easy to see that the syndrome computed from the received symbols (as described earlier) is itself the error pattern. Thus no error correction of the message symbols is required in this case. The receiver can recognize this situation by testing the weight of the syndrome ($\leq t$) computed.

This technique can be extended to the case when the errors are confined to any set of $(n - k)$ consecutive received symbols. To see this, let $r(x)$ be the received polynomial and $s(x)$ the syndrome polynomial computed from

$$s(x) = r(x) \bmod g(x) \tag{2.139}$$

Note next that

$$x^i s(x) \bmod g(x) = x^i r(x) \bmod g(x) \qquad 1 \leq i \leq n - 1 \tag{2.140}$$

that is, the syndrome corresponding to the ith cyclic shift of the received vector can be obtained simply by clocking the shift register containing the syndrome

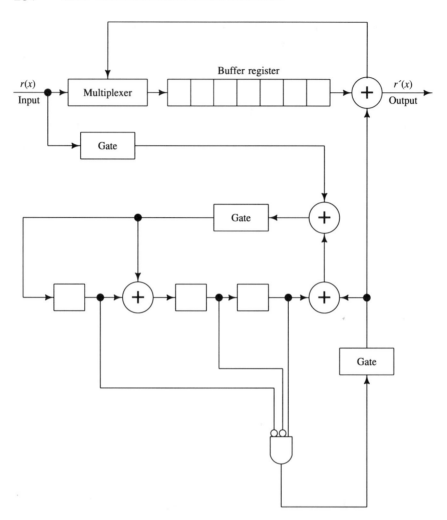

FIGURE 2.25   Error-trapping decoding circuit for the $[7, 4, 3]$ cyclic Hamming code generated by $g(x) = 1 + x + x^3$. (From Shu Lin/Daniel J. Costello, Jr., *Error Control Coding: Fundamentals and Applications*, © 1983. Reprinted by permission of Prentice Hall, Inc., Englewood Cliffs, N.J.)

(see Fig. 2.24b) i times. Thus by clocking the shift register the correct number of times, the errors can be "trapped" in the syndrome register.

The same observation can also be used to reduce the memory requirements of a table lookup decoder. The circuit in Fig. 2.25 essentially uses error trapping to decode the Hamming code. Since the Hamming code can correct single errors only, clearly error-trapping decoding is equivalent to complete decoding for this code. This decoder can also be regarded as a table lookup decoder that gets away with storing just a single pair of entries in its truth table, by making use of the fact that all the syndrome polynomials are cyclic shifts of a single polynomials $s(x)$ of the form $x^i s(x) \bmod g(x)$.

A good discussion of this decoding technique may be found in Refs. 1 and 3.

2.7.7 The Golay Code

This $[23, 12, 7]$ linear binary cyclic code was discussed earlier, in connection with the Hamming bound, as an example of a perfect code. The generator polynomial for these codes can be taken to be either the polynomial

$$g(x) = x^{11} + x^9 + x^7 + x^6 + x^5 + x + 1 \tag{2.141}$$

or else its reciprocal. Encoding and syndrome computation in the case of the Golay code can be performed as with any other cyclic code. Error-trapping can be used to decode the code when the errors all lie in a string of $n - k = 11$ successive bits. Modification[1,3] of the error-trapping technique allow for the correction of all possible patterns of three errors or less. Other algorithms also exist.[10,28] Performance curves for the Golay code are shown in Figs. 2.8 and 2.21.

2.7.8 Shortened Cyclic Codes

As described in Section 2.6, shortened cyclic codes (the resultant code is no longer cyclic) can be used whenever the parameters of a cyclic code having good distance properties do not match the requirements of the application. The advantage of shortening a cyclic code is that one retains all the simplicity in terms of encoding and syndrome computation associated with a cyclic code. The same methods of encoding/syndrome computation are applicable because one can choose to drop the high-order message bits (i.e., the message symbols $m_{k-1}, m_{k-2}, \ldots$) in that order to achieve shortening, and then perform encoding and syndrome computation by pretending that these symbols are always equal to zero and further, that, no errors ever occur in these locations.

2.7.9 Bose-Chaudhuri-Hocquenghem Codes

The Bose-Chaudhuri-Hocquenghem (BCH) codes form an important subclass of cyclic codes. For lengths up to several thousand, BCH codes are among the best-known binary block codes.[1] It will be assumed in this section and the next that the reader is familiar with the theory of finite fields. A good introduction to the subject can be found in Refs. 1 and 9 and other textbooks on coding theory.

Given an integer n, let m be the smallest integer such that $n | 2^m - 1$. Then, the finite field of 2^m elements $GF(2^m)$, contains an element α which is a primitive nth root of unity. Let C be a cyclic code of length n. Since the generator polynomial $g(x)$ of C divides $x^n - 1$, its roots lie among the powers α^i, $0 \leq i \leq n - 1$, of α.

Definition.   Given an integer n, let m and α be as defined above. A binary cyclic code of length n is said to be a *BCH code of designed distance* δ if the generator polynomial $g(x)$ [coefficients in $GF(2)$] of the code, is for some integer b $0 \leq b \leq n - 1$, the least-degree polynomial having among its roots the consecutive powers $\{\alpha^{b+i} | 0 \leq i \leq \delta - 2\}$ of α.

The BCH code is said to be *primitive* if n is of the form $2^m - 1$ for some integer m, $m \geq 2$ and *narrow-sense* if $b = 1$.
  A polynomial $c(x)$ of degree $\leq n - 1$ is a code polynomial iff $g(x)|c(x)$; that is, iff

$$c(\alpha^i) = 0 \qquad b \leq i \leq b + \delta - 2. \tag{2.142}$$

This condition can be used to derive a form of parity-check equation for the code:

$$\begin{bmatrix} 1 & \alpha^b & \alpha^{2b} & \cdots & \alpha^{(n-1)b} \\ 1 & \alpha^{2b} & \alpha^{4b} & \cdots & \alpha^{2(n-1)b} \\ \vdots & \vdots & \vdots & & \vdots \\ 1 & \alpha^{(b+\delta-2)} & \alpha^{2(b+\delta-2)} & \cdots & \alpha^{(n-1)(b+\delta-2)} \end{bmatrix} \begin{bmatrix} c_0 \\ c_1 \\ c_2 \\ \vdots \\ c_{n-1} \end{bmatrix} = \begin{bmatrix} 0 \\ 0 \\ \vdots \\ 0 \end{bmatrix} \tag{2.143}$$

The parity-check matrix H_{ef} occurring in this equation is different from that encountered earlier in two respects: (1) Its entries belong to the *extension field* $GF(2^m)$ and (2) the number of rows of the matrix is not necessarily equal to the redundancy $(n - k)$ of the code. Nevertheless, a binary vector $(c_0 c_1 c_2 \ldots c_{n-1})^t$ is a codeword iff it satisfies Eq. (2.143).
  It can be verified that any set of $(\delta - 1)$ columns picked from the matrix H above form a Vandermonde matrix and are thus linearly independent. Thus, by Theorem 4, the code has a minimum distance

$$d_{min} \geq \delta \tag{2.144}$$

which justifies the terminology used in the definition. Note that the above bound holds for the minimum distance of any cyclic code of the same length n, whose generator polynomial $g(x)$ is a multiple of that of the BCH code. The bound in Eq. (2.144) is therefore called the BCH bound.[1] Other, stronger results have recently been derived.[30,31]
  Given a sequence $\{a_k\}$ (having symbols belonging to some field) satisfying the linear recursion (over the same field)

$$a_k = \sum_{i=1}^{r} c_i a_{k-i} \tag{2.145}$$

the polynomial

$$c(x) = 1 - \sum_{i=1}^{r} c_i x^i \tag{2.146}$$

is called the *characteristic polynomial*[24] of the sequence. Let $i_1, i_2, \ldots, i_r$ be the error locations in the n-tuple $\mathbf{r}$ received across a BSC. We assume channel encoding was performed using a BCH code of length n and designed distance δ. Then the polynomial

$$\sigma(x) = \sum_{j=1}^{r} (x - \alpha_{i_j}) \tag{2.147}$$

is called the *error-locator polynomial*.

Let $\mathbf{s}$ be the syndrome vector associated with the received vector $\mathbf{r}$; that is,

$$\begin{bmatrix} s_1 \\ s_2 \\ \vdots \\ s_{\delta-1} \end{bmatrix} = \mathbf{s} = H\mathbf{r} \tag{2.148}$$

In the number of errors t inserted by the channel does not exceed the designed error-correcting capability of the code C, that is, if

$$t \le \left[\frac{\delta - 1}{2} \right] \tag{2.149}$$

it turns out that the smallest-degree linear recursion satisfied by the syndrome "sequence" $\{s_1, s_2, \ldots, s_{\delta-1}\}$ has $\sigma(x)$ as the associated characteristic polynomial.

Thus a BCH code can be decoded as follows:

1. Compute the syndrome $\mathbf{s}$.

2. Determine the error-locator polynomial using continued-fraction expansion[35] or some other equivalent technique.

3. Factor $\sigma(x)$ to determine the error locations.

Berlekamp[9] was the first to provide a bounded-distance decoding algorithm for BCH codes. Subsequently, Massey[34] showed Berlekamp's algorithm to be equivalent to shift-register synthesis. The connection between Berlekamp's algorithm and continued-fraction expansion was made explicit in the paper by Welch and Scholtz.[35] A table of primitive BCH codes of length ≤ 255 and their minimum distances is contained in Chapter 9 of Ref. 1 (see also Appendix A of Ref. 2 and Appendix C of Ref. 3).

The bit-error probability (P_{be}) using bounded-distance decoding of 5, 10, and 15 error-correcting BCH codes of block length 127 as a function of the crossover probability ϵ of the BSC is shown in Fig. 2.26. Also shown alongside is the performance of the $[127, 112, 6]$ even-weight subcode (see the subsection on expurgating the code in Section 2.6) of a $[127, 113, 5]$ BCH code used on some Intelsat satellite channels.[5]

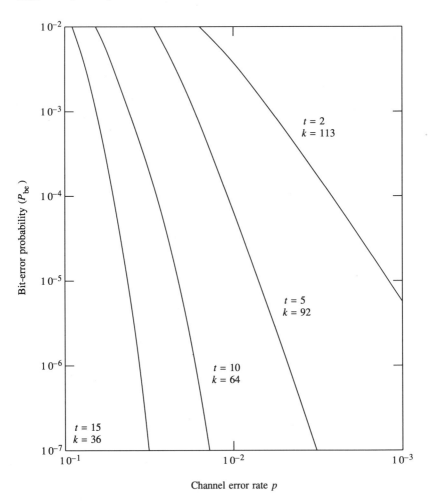

FIGURE 2.26   Bit-error probability versus channel error-rate performance of several block-length-127 BCH codes. (From Odenwalden, Error-Correcting Codes, in T. C. Bartee, *Data Communication, Networks and Systems,* © 1985. Reprinted by permission of Howard W. Sams & Co., Inc., Indianapolis.)

2.7.10 Reed-Solomon Codes

Reed-Solomon (RS) codes are the only nonbinary codes that will be encountered in this chapter. Although RS codes can be defined over other fields as well, most practical applications involve only codes over finite fields $GF(2^m)$—that is, having characteristic 2. The theory of cyclic codes carries over almost unchanged from that in the binary case. Thus cyclic codes over $GF(2^m)$ having length n correspond to ideals in the ring R_n of polynomials (over $GF(2^m)$ $\mod(x^n - 1)$), exactly as in the binary case. Every such ideal is principal and the monic generator of the ideal is the generator polynomial $g(x)$ of the code. Once again, for an $[n, k]$ code, $g(x)$ has degree $n - k$.

Except for the difference in the field in which code symbols lie, the definition of RS codes is very nearly identical to that of BCH codes. Given an integer n, let m and α be as in the definition of a BCH code. A Reed-Solomon code of length n and designed distance δ is the cyclic code generated by the polynomial

$$g(x) = \prod_{i=b}^{b+\delta-2} (x - \alpha^i) \tag{2.150}$$

Again, the code is said to be primitive if n is of the form $2^m - 1$ and is called narrow-sense if $b = 1$. Note that for these codes, the size of the code alphabet equals or exceeds the length of the code. It follows that the code is precisely the null space of the parity-check matrix

$$H = \begin{bmatrix} 1 & \alpha^b & \alpha^{2b} & \cdots & \alpha^{(n-1)b} \\ 1 & \alpha^{2b} & \alpha^{4b} & \cdots & \alpha^{2(n-1)b} \\ \vdots & \vdots & \vdots & \cdots & \vdots \\ 1 & \alpha^{(b+\delta-2)} & \alpha^{2(b+\delta-2)} & \cdots & \alpha^{(n-1)(b+\delta-2)} \end{bmatrix} \tag{2.151}$$

The same argument as in the case of BCH codes show that even for these codes

$$d_{min} \geq \delta \tag{2.152}$$

However, unlike the case of BCH codes, the entries of the parity-check matrix and code symbols lie in the same field. As a result, the redundancy of the code equals the rank of the parity-check matrix, which in this case equals $\delta - 1$. The Singleton bound (Theorem 5), which applies unchanged to linear codes over any finite field, now shows that

$$d_{min} = \delta \tag{2.153}$$

Thus these codes are MDS. It can be shown that primitive RS codes of length $2^m - 1$ can be extended to lengths 2^m and $2^m + 1$ while still preserving this MDS property.[1]

An RS code may be regarded as the parent code of the BCH code sharing the same parity-check matrix since, by selecting only the binary n-tuples belonging to the RS code, one obtains the BCH code. Also, RS codes can be decoded using the same Berlekamp/shift register synthesis/continued fraction algorithm described earlier for BCH codes.

Note that when RS codes are used in conjunction with BPSK modulation, each symbol of the encoded output [an element of $GF(2^m)$] is mapped onto a block of m binary symbols and then transmitted across the channel.

There are three properties of RS codes that makes them attractive in practical applications: (1) The codes are MDS, which means that they have the largest possible size for a given length and minimum distance. (2) The MDS property also means that any k columns of the generator matrix for the code are linearly independent. Thus a codeword can be completely recovered if *any* k symbols are correctly known. This makes them very attractive for error correction over channels that erase symbols (known as erasure channels). By using appropriate soft-decision information for the discrete channel discussed in Section 2.1, one can often cause a channel error to be replaced by an erasure. Erasures are easier to correct because, roughly speaking, they correspond to errors whose locations are already known. (3) Since the code symbols belong to a higher alphabet, a burst of b binary errors introduced by the channel appears to the RS decoder as a burst of length approximately given by b/m. Thus RS codes are very effective in combating error bursts (see the next section for other techniques used).

These codes do, however, suffer from two disadvantages. First, there are no easy techniques to decode the code using soft-decision data received over the channel. Second, the mapping from output code symbol to a binary m-tuple prior to transmission across the channel, makes these codes vulnerable to random binary errors introduced by the channel.

Both these disadvantages can be offset by the use of RS codes in a concatenated coding scheme. In such a scheme the output of the RS encoder after being mapped onto a binary m-tuple is presented to a second, inner, binary code. This code can be chosen to either take advantage of the unquantized output of the channel (popular choices here include first-order RM codes as well as convolutional codes) or else, short-length binary block codes to achieve high speed, random error-correction.

Unfortunately it is not possible to treat this subject in any greater depth here. Instead, the reader is referred to the more detailed discussion on the merits/demerits of different possibilities in Chapter 8 of the text by Clark and Cain[3] as well as Refs. 13 and 14. Table 2.1 in Section 2.1.3 lists the coding gains typically available using such schemes.

Examples where RS codes have found practical implementation include a deep-space standard, the Joint Tactical Information Distribution System and Compact Discs.[14] Figure 2.27 shows the performance curve for an extended RS code of length 256 (that is, $m = 8$). Figure 2.28a is a block diagram of a concatenated scheme, and Fig. 2.28b shows a performance curve involving a [255, 223] RS outer code and a rate-(1/2), $K = 7$ convolutional code (discussed in Section 2.9). The output of the RS encoder is interleaved before it is fed to the inner code. This coding scheme has been adapted as the telemetry channel coding standard by the Consultative Committee for Space Data Systems.[14]

2.8 BURST-ERROR CORRECTION

Examples of channels producing errors in bursts include fading channels as well as channels in which interference is present that can be blanked. This section draws considerably from the material in Refs. 3 and 5.

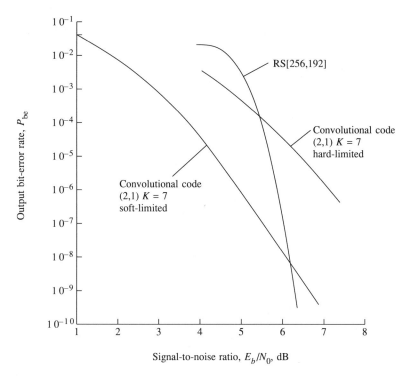

FIGURE 2.27   Performance of a [256, 192] RS code as well as of the rate-(1/2), $K = 7$ convolutional code. (From E. R. Berlekamp, R. E. Peile, S. P. Pope, "The Application of Error Control to Communication," *IEEE Communications Magazine*, vol. 25, April 1987. © 1987 IEEE.)

2.8.1 Rieger Bound

Definition.   If the errors are confined to b consecutive locations $\{i, i + 1, \ldots, \overline{i + b - 1}\}$ within the codeword, and if in addition, the ith as well as the $(i + b - 1)$th bits are in error for some $0 \leq i \leq n - b$, an *error burst of length* b is said to have occurred.

Definition.   A *b-burst error-correcting code* is a code that can correct all error bursts of length b or less.

Less redundancy is required in general to correct bursts of length b as opposed to correcting b random errors. Intuitively, this is because there is less uncertainty with regard to the error pattern. Some measure can be obtained as follows:[33]

Consider the problem of correcting three random errors in an n-tuple of length 31. The Hamming bound tells us that, for a linear code, the redundancy

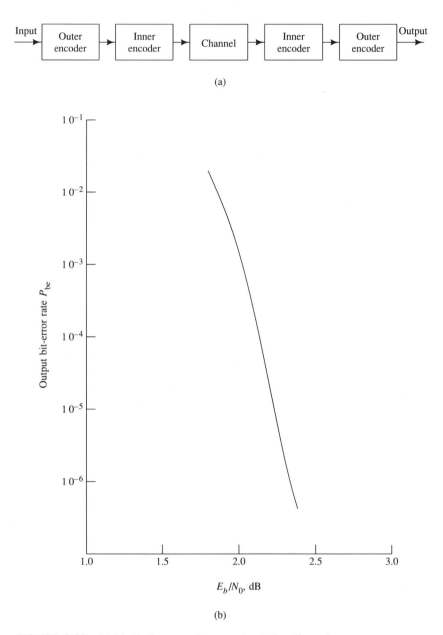

(a)

E_b/N_0, dB

(b)

FIGURE 2.28   (a) Block diagram of a concatenated coding scheme.
(b) Performance of a concatenated coding system using the RS [255, 223] and
convolutional codes (2, 1) $K = 7$. (From E. R. Berlekamp, R. E. Peile, S. P. Pope,
"The Application of Error Control to Communication," *IEEE Communications
Magazine*, vol. 25, April 1987. © 1987 IEEE.)

required is

$$n - k \geq \log_2\left[1 + \binom{31}{1} + \binom{31}{2} + \binom{31}{3}\right] \geq 15 \tag{2.154}$$

The corresponding sphere-packing bound for a four-burst error-correcting code yields

$$n - k \geq \log_2(1 + 31 + 30 + 29 \times 2 + 28 \times 4) \geq 7 \tag{2.155}$$

The theorem below relates the burst-error correcting capability of a linear code to its redundancy. Note that the minimum distance of a code is not very useful for estimating burst-error correcting capability.

THEOREM 13 (Rieger Bound).   If a linear code of redundancy $r = n - k$ and burst-error correcting capability b exists, then it must be that

$$b \geq \frac{n - k}{2} \tag{2.156}$$

Proof. First, note that no code word can be a burst of length $\leq 2b$. If such a codeword $\mathbf{c}$ does exist, then it is possible to construct two burst-error patterns $\mathbf{b}_1 + \mathbf{b}_2$ such that $\mathbf{c} + \mathbf{b}_1 = 0 + \mathbf{b}_2$; that is, the burst causes the received symbols for two distinct codewords to be identical. It is clear that in such cases the decoder will sometimes decode incorrectly.

Second, we claim that if a linear code has the property that no codeword can be regarded as a burst (of 1's) of length $2b$ or less, then $2b \leq n - k$. To prove the claim, consider all burst-error patterns in which the bursts are confined to the first $2b$ coordinates. Clearly, these must lie in distinct cosets of the standard array; otherwise, one would be able to construct a codeword that is a burst of length $\leq 2b$.

Combining these two observations, the bound is obtained.

2.8.2 Fire Codes

A necessary and sufficient condition for a *cyclic* code to correct all bursts of length b or less is that for all i and j, $0 \leq i, j \leq n - 1$, $i \neq j$, and all burst error polynomials $b_1(x)$ and $b_2(x)$, the generating polynomial $g(x)$ of the cyclic code must have the property that

$$g(x) \text{ does not divide } x^i b_1(x) + x^j b_2(x) \tag{2.157}$$

Fire codes are an example of such a cyclic code and are constructed as follows: Given any irreducible polynomial $p(x)$, the period (or order) of $p(x)$ is defined to be the smallest exponent ρ for which

$$p(x) | x^\rho + 1 \tag{2.158}$$

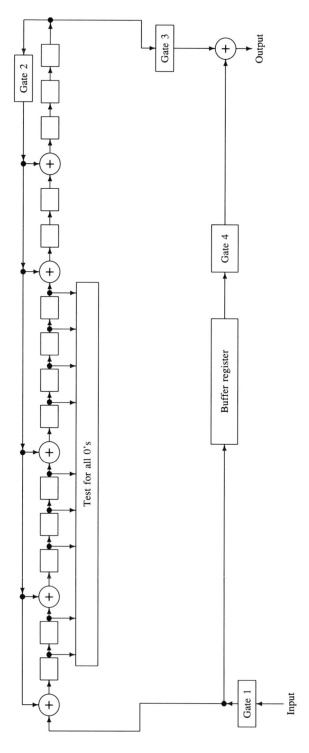

FIGURE 2.29   Error-trapping decoder for the [279, 265] Fire code. (From Shu Lin/Daniel J. Costello, Jr., *Error Control Coding: Fundamentals and Applications*, © 1983. Reprinted by permission of Prentice Hall, Inc., Englewood Cliffs, N.J.)

To construct a b-burst-error correction Fire code, choose $p(x)$ of degree $m \geq b$ such that its period ρ does not divide $2b - 1$. Then the cyclic code generated by the polynomial

$$g(x) = p(x)(x^{2b-1} + 1) \tag{2.159}$$

is a Fire code of length

$$n = lcm(\rho, 2b - 1) \tag{2.160}$$

and can correct all bursts of length b or less. This can be shown[3] using the divisibility criterion laid out above.

Fire codes also have the property (as do all cyclic codes) that they can also correct all end-around bursts of length b as well.

Decoding a burst-error-correcting cyclic code is very simply accomplished by using error-trapping decoding (see Section 2.7.6).

E X A M P L E .[3]   Let $p(x) = 1 + x^2 + x^5$. This polynomial has order = $31 = \rho$. Let $l = 5$. The Fire code generated by

$$g(x) = (x^9 + 1)(1 + x^2 + x^5) = 1 + x^2 + x^5 + x^9 + x^{11} + x^{14} \tag{2.161}$$

has length $n = lcm(9, 31) = 279$ and dimension 265. All bursts of length 5 or less can be corrected (see Fig. 2.29).

2.8.3 Interleaving

Interleaving of the encoded symbols prior to transmission across the channel is an alternative technique that can be employed to combat bursts of errors. In *block interleaving*,[5] the (channel) encoded data are written into an array of size $(N \times B)$ column by column, starting at the left. After the array is full, the data are then read out serially from the array, this time in a row-by-row fashion, starting at the top. At the receiver end, deinterleaving is performed by reversing the roles of row and column in the read-write sequence. Figure 2.30 shows an implementation in which two buffers of size $(N \times B)$ are used at both the transmitter and receiver end.

When encoding block codes, N usually is simply the block length n of the code. The parameter B is called the *depth of interleaving*. If the channel introduces an error burst of length B or less, then we see that after deinterleaving, the decoder sees at most one erroneous symbol per codeword. If a burst of length $B\lambda$ occurs, then each codeword contains at most a burst of λ errors. Thus by interleaving a block code having a b burst-error-correcting capability and interleaving the code to depth B, we obtain a code that can correctly decode any burst of length Bb.

If both burst- and random-error correction are desired, then one can employ a code that corrects t random errors and then interleave the code. Note that if soft-decision decoding is employed, the deinterleaver must

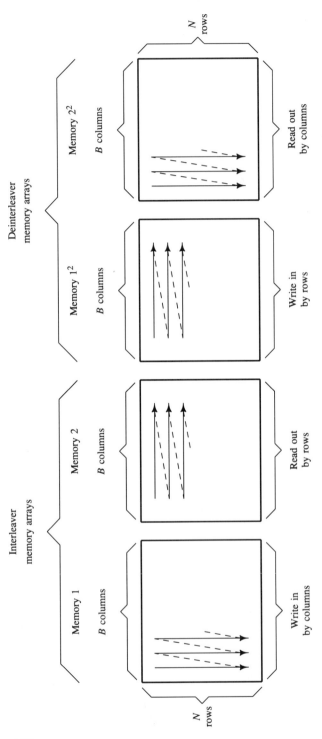

FIGURE 2.30   Block interleaving. (From Odenwald, Error-Correcting Codes, in T. C. Bartee, *Data Communication, Networks and Systems,*
© 1985. Reprinted by permission of Howard W. Sams & Co., Inc. Indianapolis.)

Note:

1. When Memory 1 is filled, write in Memory 2 and read from Memory 1.
2. Each deinterleaver memory element, in general, stores a soft-decision number.

deinterleave *soft-decision data*. The total memory requirement when two arrays are used (as in Fig. 2.30) both at the transmitter and receiver thus equals $2(Q + 1)nB$ bits, where Q is the number of levels of quantization used at the receiver. Clearly, interleaving also introduces a delay proportional to the size nB used for interleaving. Also, additional synchronization is required to determine the beginning and end of each array.

Other techniques for combating burst errors include helical interleaving, interleaved convolutional codes, RS codes with and without interleaving, concatenated codes, and cyclic redundancy check (CRC) codes. A discussion of these may be found in Refs. 2–5, 13, 14, and 36 (see also Section 2.7.10 and Exercise 10).

2.9 CONVOLUTIONAL CODES

A brief introduction to the subject is given below. A more in-depth treatment may be found in Refs. 3 and 4.

2.9.1 Introduction

The encoders for two convolutional encoders are shown in Fig. 2.31. In the first, two output symbols are produced for each incoming message symbol; in the second, the input presents itself in blocks of two message bits at a time, and there are three output symbols for each message symbol block. The convolutional codes are said to have *rates* (1/2) and (2/3), respectively.

Let $u_i(l)$, $v_j(l)$ denote the lth, $l = 0, 1, 2, \ldots$, symbols of the ith, $1 \leq i \leq k$, and jth, $1 \leq j \leq n$, output sequences $\{u_i\}$ and $\{v_i\}$ respectively. The input/output relation can be put in the form of convolutional equations:

$$v_1(l) = u(l) + u(l - 1) + u(l - 2) \tag{2.162}$$

$$v_2(l) = u(l) + u(l - 2) \tag{2.163}$$

and

$$v_1(l) = u_1(l) + u_1(l - 1) + u_2(l - 1) \tag{2.164}$$

$$v_2(l) = u_1(l) + u_2(l - 1) + u_2(l) \tag{2.165}$$

$$v_3(l) = u_1(l) + u_1(l - 1) \tag{2.166}$$

for the rate (1/2) and (2/3) codes, respectively.

Defining the formal power series $u_i(D)$ and $v_i(D)$

$$u_i(D) = \sum_{l=0}^{\infty} u_i(l)D^l \qquad 1 \leq i \leq k \tag{2.167}$$

$$v_i(D) = \sum_{l=0}^{\infty} v_i(l)D^l \qquad 1 \leq j \leq n \tag{2.168}$$

corresponding to the input and output sequences, we can express the relation

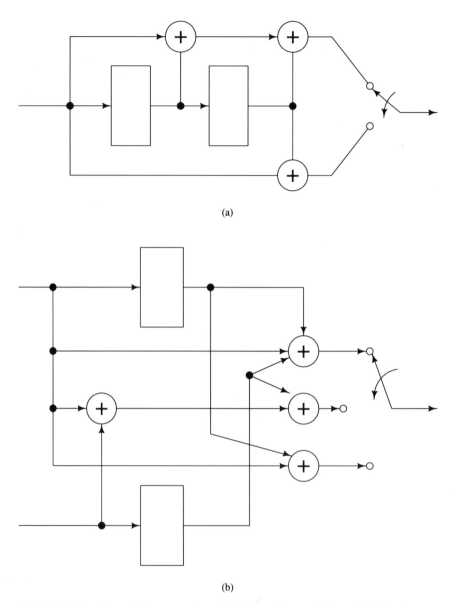

(a)

(b)

FIGURE 2.31   Convolution encoders with (a) $K = 3$, rate $(1/2)$ and (b) $K = 2$, rate $(2/3)$. (From A. J. Viterbi and J. K. Omura, *Principles of Digital Communication and Coding*, © 1979. Reprinted by permission of McGraw-Hill Publishing Co., New York.)

between input and output power series, via a matrix $G(D)$ (called the *transfer function matrix*) of polynomials whose (i, j)th entry is the polynomial $g_{ij}(D)$ showing the dependence of the jth output sequence $v_j(D)$ upon the ith input sequence $u_i(D)$.

Thus, for the general-rate (k/n) convolutional encoder, the input and output power series are related via

$$[v_1(D)v_2(D)\ldots v_n(D)]$$

$$= [u_1(D)u_2(D)\ldots u_k(D)] \begin{bmatrix} g_{11}(D) & g_{12}(D) & \cdots & g_{1n}(D) \\ g_{21}(D) & \cdots & \cdots & g_{2n}(D) \\ g_{k1}(D) & \cdots & \cdots & g_{kn}(D) \end{bmatrix} \qquad (2.169)$$

The transfer function matrices $G(D)$ for the two examples are as follows:

$$G(D) = [1 + D + D^2, 1 + D^2] \qquad \text{rate } (1/2) \qquad (2.170)$$

$$G(D) = \begin{bmatrix} 1 + D & 1 & 1 + D \\ D & 1 + D & 0 \end{bmatrix} \qquad \text{rate } (2/3) \qquad (2.171)$$

The general convolutional equation expressing the dependence of the jth output on the input is given by

$$v_j(l) = \sum_{i=1}^{k} \sum_{m=0}^{K-1} u_i(l - m)g_{ij}(m) \qquad 1 \le j \le n \qquad (2.172)$$

in which we have assumed that the maximum degree of any of the polynomials $g_{ij}(D)$ does not exceed $K - 1$. The integer product kK has the interpretation of being the largest number of input bits that can effect a single output symbol and is called the *constraint length* of the code. Of course, when $k = 1$, K itself is the constraint length.

When the input message sequence is of infinite length, one can view the output as being generated by a semi-infinite generator matrix. The elements of this matrix are $(k \times n)$ submatrices with the (i, j)th submatrix showing the dependence of the jth output block on the ith.

If the message sequence is of finite length, then the encoded output is terminated by inserting a tail of $(K - 1)$ blocks, each block containing k zeros, at the end of the sequence of message blocks. Transmission is then terminated after the last message block has cleared the shift register. Thus if the message sequence contains B blocks, then there are $B + K - 1$ blocks in the encoded output so that the rate (defined as for block codes) of the convolutional code is actually given by

$$\text{Rate} = \frac{Bk}{[B + (k - 1)]n} \qquad (2.173)$$

However, B usually is large compared to K, so this is very nearly equal to (k/n).

Consider the rate-$(1/2)$, constraint length $K = 3$ encoder whose transfer function matrix $G(D)$ is given by

$$G(D) = [1 + D^2, 1 + D] \tag{2.174}$$

Let the input sequence to this encoder have power series

$$u(D) = 1 + D + D^2 + \cdots \tag{2.175}$$

Since $(1 + D)u(D) = 1$, one can also write

$$u(D) = \frac{1}{1 + D} \tag{2.176}$$

The output sequence can then be seen to be given by

$$[v_1(D)v_2(D)] = u(D)[1 + D^2, 1 + D]$$
$$= [1 + D, 1] \tag{2.177}$$

Thus the resulting pair of (infinite-length) output sequences contain only three 1's. This means that three channel errors can render the output sequence indistinguishable from the corresponding all-zero message sequence [corresponding to $u(D) = 0$], leading to an infinite number of decoded information bit errors.

Such convolution codes—i.e., codes for which a finite number of channel errors can result in an infinite number of decoded information bit errors—are called *catastrophic codes*. It was shown by Massey and Sain[21] that a necessary and sufficient condition for a rate $(1/n)$ convolutional code not to be catastrophic is that

$$\gcd[g_1(D), g_2(D), \ldots, g_n(D)] = D^l \qquad \text{some } l \geq 0 \tag{2.178}$$

In contrast, for the example code above

$$\gcd(1 + D^2, 1 + D) = 1 + D \tag{2.179}$$

A similar condition on the elements of the transfer function matrix can be stated for general rate (k/n) convolutional codes.[3]

2.9.2 State and Trellis Diagrams

The distance properties of a convolutional code are easier to determine when the code is viewed as being generated by a finite-state machine. This viewpoint also leads to an efficient decoding algorithm for these codes. The state diagrams for the rate-$(1/2)$, $K = 3$ example code is shown in Fig. 2.32. To develop a better feeling for the relation between input and output sequences, it is convenient to expand and redraw the state diagram in the form of a trellis, which allows one to differentiate between the same state at different instants of time. Figure 2.33 shows the trellis corresponding to the rate-1/2, constraint-length-3 example code. From this it can be seen that the output corresponding to the all-zero input sequence is the all-zero sequence. The output corresponding to the input 10111 ... is 11 10 00 01 10.... The figure also shows how the

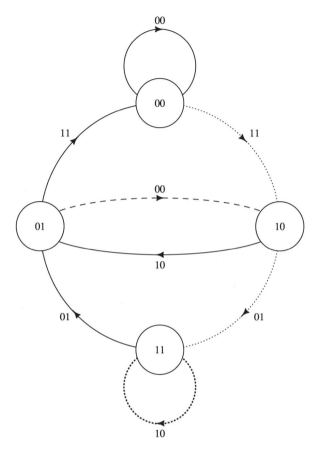

FIGURE 2.32   State diagram for the rate-$(1/2)$, $k = 3$ code. The solid line indicates a zero input and the dashed line a 1. (From A. J. Viterbi and J. K. Omura, *Principles of Digital Communication and Coding*, © 1979. Reprinted by permission of McGraw-Hill Publishing Co., New York.)

encoder is returned to the all-zero state and the code terminated by the addition of $K - 1 = 2$ zeros. Since the discussion to follow is largely in terms of the trellis, some convenient terminology will now be introduced.

Each state in the trellis diagram is called a *node*. The states in the trellis diagram corresponding to the lth time instant (starting from time $l = 0$) will be referred to as the nodes at the lth *node level*. Each node within a node level is identified by the contents of the encoding shift register—i.e., the two past message symbols that most recently entered the shift register in the case of the code in Fig. 2.31a. The transitions between nodes belonging to successive node levels are called *branches*. Each branch in the trellis is labeled with the output corresponding to that particular branch. Those branches corresponding to the zero input symbol are shown using a solid line; the broken line is used to

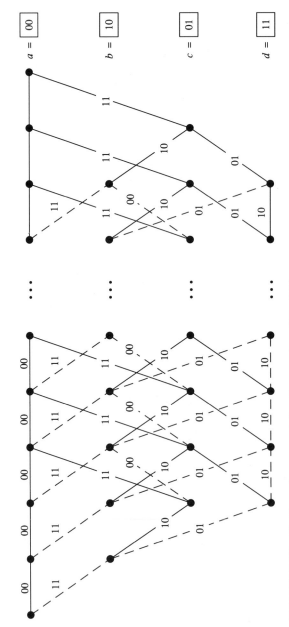

FIGURE 2.33   Trellis code representation for encoder of Fig. 2.31.

indicate those identifying the message symbol 1. A sequence of branches in the trellis is called a *path* and is thus associated with a message as well as an output sequence.

2.9.3 Distance Distribution of the Code

A commonly used measure of distance for convolutional codes is the *minimum free distance*. Since a convolutional code is also a linear code, the minimum distance between any two codewords of infinite length (i.e., between any two paths in the trellis) is also the minimum weight of a path in the trellis that does not correspond to the all-zero message sequence. Except in the case of catastrophic codes, the minimum weight path will be a path that deviates from the all-zero path only for a finite number of message symbols.

In the example code, the minimum-weight path can be seen to be the path that has only one message symbol equal to 1—as, for example, the input sequence 10000000..., which produces the output 11 10 11 00 00... of weight 5. Thus the minimum free distance (or simply the free distance) equals 5 in this case.

Using generating-function techniques,[4] information relevant to determining the probability of error for these codes can be obtained. Let $A_{00}(L, D, I)$, $A_{01}(L, D, I)$, $A_{10}(L, D, I)$ and $A_{11}(L, D, I)$ represent power series providing distance and length information concerning paths leading from the all-zero path to the state identified by the subscripts, without returning in between to visit the 00 state. Thus the coefficient of $L^a D^b I^c$ in the polynomial $A_{10}(L, D, I)$ identifies the number of paths of length a originating from node 00 at level 0 and terminating at node 10 at node level a, having Hamming distance b from the all-zero path of the same length and whose associated message sequence has Hamming weight c. A modified state diagram is shown in Fig. 2.34, obtained by splitting the all-zero state in the state diagram path into two and labeling branches with monomials of the form $L^a D^b I^c$ whenever the weight of the output (input) for that branch has weight $b(c)$ and length a (a is thus 1 for each branch corresponding to a single message symbol).

The two states that result from splitting the all-zero state, (labeled I for initial and F for final) are associated with the generating functions $A_I(L, D, I)$ and $A_F(L, D, I)$. Clearly $A_I(L, D, I) = 1$. From the modified state diagram, the generating functions associated with the various states can be seen to satisfy the following linear equations:

$$\begin{bmatrix} A_{01}(L, D, I) \\ A_{10}(L, D, I) \\ A_{11}(L, D, I) \end{bmatrix} = L \begin{bmatrix} 0 & D & D \\ I & 0 & 0 \\ 0 & DI & DI \end{bmatrix} \begin{bmatrix} A_{01}(L, D, I) \\ A_{10}(L, D, I) \\ A_{11}(L, D, I) \end{bmatrix} + \begin{bmatrix} 0 \\ LD^2I \\ 0 \end{bmatrix} \qquad (2.180)$$

$$A_F(L, D, I) = LD^2 A_{01}(L, D, I) \qquad (2.181)$$

This equation can be solved for $A_F(L, D, I)$ and the result is given by

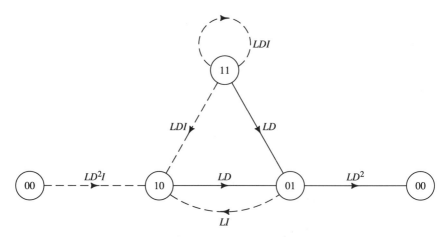

FIGURE 2.34   Modified state diagram for the encoder of Fig. 2.31a. (From A. J. Viterbi and J. K. Omura, *Principles of Digital Communication and Coding,* © 1979. Reprinted by permission of McGraw-Hill Publishing Co., New York.)

$$A_F(L, D, I) = \frac{L^3 D^5 I}{1 - LDI(1 + L)} \tag{2.182}$$

In the power series $A_F(L, D, I)$, the minimum free distance appears as the least power of D having a nonzero coefficient, which in the example equals 5.

An alternative method of obtaining the generating function is to use a formula developed by Mason.[3,4]

2.9.4 The Viterbi Decoding Algorithm

It will be assumed that the message sequence has been terminated by the addition of $(K - 1)$ blocks of zeros at the tail. The trellis for the rate-$(1/2)$ constraint-length-3 code would then appear as shown in Fig. 2.33. An algorithm, called the Viterbi algorithm,[4,11] will now be described. It performs maximum-likelihood sequence detection; i.e., it identifies the most likely path in the trellis, given the received symbols. The algorithm is described, as it applies to the decoding of the example code. To begin with, it is assumed that the channel is the memoryless BSC.

Let $\{u(i)\}$, $\{v_1(i)\}$, $\{v_2(i)\}$, $\{r_1(i)\}$, and $\{r_2(i)\}$ denote the message and the encoded and received sequences, respectively. First, each branch of the trellis is labeled with the Hamming distance (referred to as the metric) of the encoded output corresponding to that branch, from the corresponding received symbols. The decoder assigns the metric 0 to the 00 node at level 0. From the 00 node, the encoder can transit to either the 00 or 10 node at node level 1. It computes the metric for these nodes by adding the metric for node 00 at level 00 to the metric of the branch leading to that state from the 00 node. The decoder also stores at each node the sequence of information bits leading up to that node. The decoder then moves onto the second node level and computes metrics for nodes at this level in a similar fashion, by adding the metric of the previous node to the metric associated with the branch leading from the previous node to the present node. Once again, the sequence of information bits leading up to that node are stored.

Things are slightly different at node level 3. Here there are two branches leading up to each node. For each node at this node level, the decoder computes two metrics corresponding to the paths leading up to that node; i.e., it forms the sum of the metric of the node from which the branch originated as well as the branch metric. It then compares the two paths on this basis and chooses the path having the lesser metric (distance). The chosen path at this node level is called the *survivor*. When two competing paths have identical metric, an arbitrary choice is made.

Thus at the third node level, at each node the decoder retains only the message sequence corresponding to the survivor and the associated metric. This procedure is repeated at each successive node level until the decoder reaches the tail of the trellis, where the number of nodes at the final node levels shrinks from four to two to one. Thus, just as in the case of the initial node levels, toward the end the comparison and selection of a survivor is made unnecessary because there is only one path leading to each node. At the final node level, there is only one node and, therefore, only a single survivor, which is declared to be the decoded sequence. An example of this decoding algorithm is presented in the next section.

In general, maximum-likelihood decoding amounts to examining all possible encoded sequences of length $(B + K - 1)$ blocks and choosing the one closest to the received sequence. Thus it would appear that the number of sequences needed to be examined grows exponentially with increase in the length B of the transmitted sequence. However, the Viterbi algorithm described above recognizes that it is not necessary to examine all these paths. It avoids an exhaustive search by proceeding sequentially and retaining only the survivors at each node level. This technique can be justified as follows: Let us assume that the most likely path found by exhaustive search turned out to have a prefix of l blocks, which at the lth node level was not a survivor. Then by replacing that prefix by the survivor corresponding to that node at the lth node level reached by the prefix, one obtains a second path whose minimum distance is smaller than that of the most likely path, in contradiction to our earlier assumption.

In the case of a general rate $(1/n)$ constraint-length-K encoder, there would be 2^{K-1} states, and therefore 2^{K-1} survivors, at each node level. Except for this difference, the decoding would proceed in exactly the same manner.

For more general rates of the form (k/n) and constraint length K, the number of states would equal $2^{k(K-1)}$ with 2^k paths emerging from and leading to each node. Thus in determining the survivor, one would choose from among 2^k paths rather than two, as in the case of rate-$1/n$ codes. Apart from this, the decoding procedure remains the same.

The decoding algorithm described above is a special case of the algorithm for arbitrary memoryless channels. On such channels, maximum-likelihood sequence docoding for the example code amounts to choosing that encoded sequence $\{v_1(i), v_2(i)/i = 0, 1, 2, \ldots, N - 1\}$ $(N = B + K - 1)$ for which the conditional probability (or conditional probability density function in the continuous case)

$$p(\{r_1(i), r_2(i)|0 \le i \le N - 1\}|\{v_1(i), v_2(i)|0 \le i \le N - 1\}) \qquad (2.183)$$

is a maximum. In this case, by taking logarithms (to any convenient base) a convenient metric is obtained:

$$\ln[p(\{r_1(i), r_2(i)\}|\{v_1(i), v_2(i)\})] = \sum_{i=0}^{N-1} \ln[p(r_1(i), r_2(i)|v_1(i), v_2(i))] \qquad (2.184)$$

This result once again allows us to think of the metric of a path in the trellis as being the sum of the metrics of the individual branches forming that path. Note, however, that here the decoded sequence is the sequence having the *largest* metric. It can be shown that for a BSC, the metric is directly proportional to the *negative* of the Hamming distance of the received symbol segment from the encoded sequence sgement. Hence, for this channel, the survivor is the one whose Hamming distance from the received sequence is the least.

In the case of the discrete channel resulting from the use of BPSK signaling over the additive white Gaussian noise waveform channel (Section 2.1), the transmitted symbols are either $\pm\sqrt{E_{dc}}$.

Let us assume that the convolutional code used is the same as that in Fig. 2.31. Let $\{r_1(i), r_2(i)\}$ denote the received sequence as before, and let $\{s_1(i), s_2(i)\}$ denote the transmitted sequence corresponding to a particular code sequence. Since

$$p[r_1, r_2(i)|s_1(i), s_2(i)]$$

$$= \frac{1}{\sqrt{2\pi N_0/2}} \exp\left\{-\frac{1}{N_0}[(r_1(i) - s_1(i))^2 + (r_2(i) - s_2(i))^2]\right\} \qquad (2.185)$$

taking natural logarithms we find that

$$\ln[p(r_1(i), r_2(i)|s_1(i), s_2(i))] \text{ is proportional to } [r_1(i)s_1(i) + r_2(i)s_2(i)] \qquad (2.186)$$

Thus, one can simply use the inner product of the received and transmitted signals as the metric in this case.

2.9.5. Practical Implementation Aspects

Although in the above description we have assumed that the decoder output is terminated by inserting a tail of $(K - 1)$ blocks of zeros into the message sequence, this approach is not carried out in practice because either the decoding delay would be unacceptable or else the loss in rate would not be

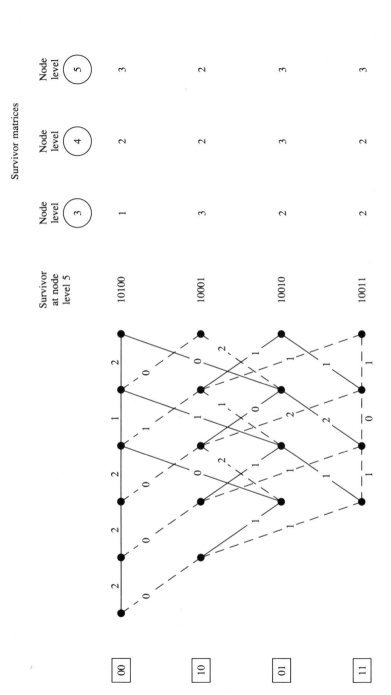

FIGURE 2.35   Illustrating the Viterbi decoding algorithm.

307

negligible. In practice,[5] it is found that the survivors at the lth node level all have the same prefix up to node level $l - 4K$ or $l - 5K$. This suggests that after reaching the lth node level, there is negligible loss in performance in decoding the $(l - 5K)$th information bit(s).

Block synchronization is simpler in the case of convolutional codes than in the case of block codes, inasmuch as there are n positions to search through and n for a convolution code is small compared to the block length of a block code.

The path memory and metric requirements grow exponentially [there are $2^{k(K-1)}$ survivors that need to be stored along with their associated metrics and message sequences] with increase in the constraint length. Thus Viterbi decoding becomes impractical for large constraint lengths. For larger constraint lengths, one uses sequential decoding.[3,4,7] A major advantage convolutional codes enjoy over block codes is their ability to process the unquantized output of the channel.

Good convolutional codes are identified usually by means of a computer search. Tables of the convolutional codes having largest free distance for short constraint lengths may be found in Refs. 3 and 5. A rate-(1/2), $K = 7$ code is used in many commercial satellite applications.[5,14] Figure 2.35 illustrates the Viterbi decoding algorithm as it applies to the rate (1/2), $K = 3$ (noncatastrophic) code discussed previously. Each branch in the trellis is labeled with the Hamming distance of the associated output sequence from the appropriate segment of the received sequence shown. Also shown in the trellis are the metrics of the survivors at node levels 3, 4, and 5, as well as the surviving message sequences at node level 5. In case of a tie at any node, the upper branch has always been selected. Arrows identify the surviving path at each node.

Figure 2.27 shows the performance of a rate-(1/2), $K = 7$ convolution code with hard and soft decisions.

2.9.6 Probability of Bit Error

An upper bound for the bit-error probability of the Viterbi decoder that makes use of the generating function $A_F(L, D, I)$ and is simple to evaluate, can be derived as follows:[11] Let a message sequence of length Bk be divided into B blocks and encoded into an output sequence of length $[B + (K - 1)] n$, as described earlier. Since convolutional codes are linear, the error probability, as in the case of block codes, is the same no matter which code sequence is transmitted. Without loss of generality, therefore, it will be assumed that the all-zero sequence is the transmitted sequence. The following definitions will be used in the derivation:

- $P_{be,i}$ = probability that the ith message bit $(0 \leq i \leq kB - 1)$ is incorrectly decoded.
- P_{be} = the average bit error probability.
- S denotes a survivor.

- IP denotes an incorrect path (i.e., a path different from the all-zero path) through the trellis that originates and terminates at the all-zero state at node levels 0 and $(K + B - 1)$, respectively.
- W_{IP} is the Hamming weight of the message sequence underlying the incorrect path IP;
- D_j a divergent path segment that starts at the all-zero node at some node level j, $0 \le j \le B - 1$ and terminates at some later node level l, $l > j$, *without having visited the all-zero state at some intermediate node level i, j < i < l.*
- W_{D_j} is the Hamming weight of the segment of the message sequence associated with the divergent path segment D_j.
- E_j is the event that the segment of the received sequence that overlaps in time with the path segment D_j is closer to it than to the corresponding all-zero path.

With these definitions, we have

$$P_{be} = \frac{1}{kB} \sum_{i=0}^{kB-1} P_{be,i}$$

$$= \frac{1}{kB} \sum_{i=0}^{kB-1} \sum_{\substack{\text{all incorrect} \\ \text{paths IP}}} P_r(IP = S) P_r(i\text{th message bit is in error} \,|\, IP = S)$$

$$= \frac{1}{kB} \sum_{\substack{\text{all incorrect} \\ \text{paths IP}}} P_r(IP = S) W_{IP}$$

$$= \frac{1}{kB} \sum_{j=0}^{B-1} \sum_{\substack{\text{all divergent path} \\ \text{segments } D_j}} \sum_{\substack{\text{all incorrect paths} \\ \text{IP through } D_j}} P_r(IP = S) W_{D_j}$$

$$\le \frac{1}{kB} \sum_{j=0}^{B-1} \sum_{\text{all } D_j} P_r(E_{D_j}) W_{D_j}$$

$$= \frac{1}{k} \sum_{\text{all } D_j} P_r(E_{D_j}) W_{D_j} \qquad (2.187)$$

The right-hand side can now be estimated using the generating function $A_F(L, D, I)$.

Case 1 (BSC). Let a divergent path segment D_j be fixed and define $W = W_{D_j}$. Then

$$P_r(E_{D_j}) = \sum_{i=(W+1)/2}^{W} \binom{W}{i} \epsilon^i (1 - \epsilon)^{W-i} \qquad \text{if } W \text{ is odd}$$

$$= \sum_{i=(W/2)+1}^{W} \binom{W}{i} \epsilon^i (1 - \epsilon)^{W-i}$$

$$+ \frac{1}{2} \binom{W}{W/2} \epsilon^{W/2} (1 - \epsilon)^{W/2} \qquad \text{if } W \text{ is even} \qquad (2.188)$$

In either case,

$$P_r(E_{D_j}) < \sum_{i=(W+1)/2}^{W} \binom{W}{i} \epsilon^i (1-\epsilon)^{W-i}$$

$$< (1-\epsilon)^W \left(\frac{\epsilon}{1-\epsilon}\right)^{W/2} \sum_{i=(W+1)/2}^{W} \binom{W}{i}$$

$$< (1-\epsilon)^W \left(\frac{\epsilon}{1-\epsilon}\right)^{W/2} 2^W$$

$$< (\sqrt{2\epsilon(1-\epsilon)})^W \tag{2.189}$$

A comparison of Eqs. (2.187) and (2.189) and the definition of the generating function $A_F(L, D, I)$ now shows that the right hand side of Eq. (2.187) can now be upper bounded by

$$P_{be} < \frac{1}{k} \frac{\partial A_F(L, D, I)}{\partial I}\bigg|_{\substack{D=\sqrt{2\epsilon(1-\epsilon)}, \\ L=1, I=1}} \tag{2.190}$$

Case 2 (discrete channel derived from the AWGN waveform channel). Let D_j be fixed and set

$$W_{D_j} = W \tag{2.191}$$

as before. In this case,

$$P_r(E_{D_j}) = P_r\left\{ \sum_{i=1}^{W} (x_i + \sqrt{E_{dc}})^2 < \sum_{i=1}^{W} (x_i - \sqrt{E_{dc}})^2 \right\} \tag{2.192}$$

where the $\{x_i\}$ are independent, identically distributed Gaussian random variables having mean $\sqrt{E_{dc}}$ and variance $N_0/2$.

Thus,

$$P_r(E_{D_j}) = P_r\left(\sum_{i=1}^{W} x_i < 0 \right) \tag{2.193}$$

The sum $y = \sum x_i$ is Gaussian with mean $W\sqrt{E_{dc}}$ and variance $W(N_0/2)$; therefore,

$$P_r(y < 0) = \text{erfc}\left(\sqrt{\frac{2WE_{dc}}{N_0}} \right)$$

$$< \frac{1}{2} e^{-WE_{dc}/N_0}$$

$$< (e^{-E_{dc}/N_0})^W \tag{2.194}$$

In this case we have

$$P_{be} < \frac{1}{k} \frac{\partial A_F(L, D, I)}{\partial I}\bigg|_{\substack{D=\exp(-E_{dc}/N_0), \\ L=1, I=1}} \tag{2.195}$$

EXERCISES

1. It is desired to employ a code that can correct three errors and detect four errors as a single-error-correcting, x-error-detecting code. Determine x.

2. Determine the minimum distances of the codes C_1 and C_2 whose generator and parity-check matrices are given respectively by

$$G = \begin{bmatrix} 0 & 0 & 1 & 1 & 1 & 0 & 1 \\ 1 & 1 & 1 & 0 & 1 & 0 & 0 \end{bmatrix}$$

and

$$H = \begin{bmatrix} 1 & 0 & 0 & 0 & 1 \\ 0 & 1 & 0 & 0 & 1 \\ 0 & 0 & 1 & 0 & 1 \\ 0 & 0 & 0 & 1 & 0 \end{bmatrix}$$

3. Determine the probability of uncorrected error for the $[8, 4, 4]$ extended Hamming code whose parity-check matrix H is

$$H = \begin{bmatrix} 1 & 1 & 1 & 1 & 1 & 1 & 1 & 1 \\ 0 & 0 & 0 & 1 & 1 & 1 & 1 & 0 \\ 0 & 1 & 1 & 0 & 0 & 1 & 1 & 0 \\ 1 & 0 & 1 & 0 & 1 & 0 & 1 & 0 \end{bmatrix}$$

when the standard array is used to decode the code. Assume a BSC with crossover probability ϵ.

4. Prove that the dual of a binary MDS code is also MDS.

5. A message sequence is encoded using the binary RM code $R(2, 4)$ and then transmitted across a BSC. Determine, using majority-logic decoding, the message symbols corresponding to the received 16-tuple:

$$r = (0011 \ 0110 \ 0111 \ 1010)^t$$

$$\uparrow \qquad\qquad\quad \uparrow$$
$$r(0000) \qquad\quad r(1111)$$

The mapping from truth table to vector is performed as described in the text.

6. Prove that the $(2^m \times 2^m)$ Hadamard transform matrix for every integer $m \geq 2$ can be decomposed into the product of m matrices as shown in Section 2.5.3 for the case $m = 3$.

7. Let C^* be the code obtained by puncturing the $[n, k, d \geq 2]$ linear code C in the first coordinate. How are the duals of the two codes related?

8. Let $f(x_1, x_2, \ldots, x_m)$ be a code polynomial belonging to the RM code $R(r, m)$. By expressing $f(\cdot)$ in the form

$$f(x_1 x_2 \ldots x_m) = x_1 g(x_2 x_3 \ldots x_m) + h(x_2 x_3 \ldots x_m)$$

show that $R(r, m)$ can be viewed as being constructed out of two RM codes of length 2^{m-1} using the $|u|u + v|$ construction of Section 2.6.

9. A binary cyclic code has length 15 and generating polynomial

$$g(x) = 1 + x + x^2 + x^3 + x^4$$

(a) Determine the code polynomial resulting from the systematic encoding of the message polynomial

$$m(x) = x^2 + x^3 + x^4$$

(b) Determine the syndrome polynomial $s(x)$ corresponding to the received polynomial

$$r(x) = x^{10} + x^6 + 1$$

10. Let $g(x)$ be the generator polynomial of an $[n, k]$ cyclic code. Interleaving this code to depth λ using block interleaving results in an $[\lambda n, \lambda k]$ linear code. Show that the interleaved code is cyclic with generator polynomial $g(x^\lambda)$. (From Lin and Costello.[3])

11. Draw the state and trellis diagrams corresponding to the rate-2/3, $K = 2$, convolutional encoder shown in Fig. 2.31b.

12. Given the rate-1/2, $K = 3$, convolutional code of Fig. 2.31a of the text, suppose the code is used on a BSC and that the received sequence for the first eight branches is

    00 01 10 00 00 00 10 01

Trace the decisions on a trellis diagram labeling the survivor's Hamming distance metric at each node level. If a tie occurs in the metrics required for a decision, always choose the upper (lower) path.

REFERENCES

1. F. J. MacWilliams and N. J. A. Sloane, *The Theory of Error-Correcting Codes*, Amsterdam: North-Holland, 1977.
2. G. C. Clark, Jr., and J. B. Cain, *Error-Correction Coding for Digital Communications*, New York: Plenum Press, 1981.
3. S. Lin and D. J. Costello, Jr., *Error Control Coding: Fundamentals and Applications*, Englewood Cliffs, N.J.: Prentice-Hall, 1983.
4. A. J. Viterbi and J. K. Omura, *Principles of Digital Communication and Coding*, New York: McGraw-Hill, 1979.
5. T. C. Bartee, *Data Communication, Networks and Systems*, Indianapolis: Howard W. Sams, 1985.
6. R. G. Gallager, *Information Theory and Reliable Communication*, New York: Wiley, 1968.
7. J. M. Wozencraft and I. M. Jacobs, *Principles of Communication Engineering*, New York: Wiley, 1965.
8. W. W. Peterson and E. J. Weldon, Jr., *Error-Correcting Codes*, 2nd ed., Cambridge, Mass.: MIT Press, 1972.
9. E. R. Berlekamp, *Algebraic Coding Theory*, New York: McGraw-Hill, 1968.
10. V. Pless, "Decoding the Golay Codes," preprint.
11. A. J. Viterbi, "Convolutional Codes and Their Performance in Communication Systems," *IEEE Trans. Commun. Technol.*, vol. COM-19, pp. 751–772, October 1971.
12. J. A. Heller and I. M. Jacobs, "Viterbi Decoding for Satellite and Space Communication," *IEEE Trans. Commun. Technol.*, vol. COM-19, pp. 835–848, October 1971.
13. E. R. Berlekamp, "The Technology of Error-Correcting Codes," *Proc. IEEE*, vol. 68, pp. 564–593, May 1980.
14. E. R. Berlekamp, R. E. Peile, and S. P. Pope, "The Application of Error Control to Communications," *IEEE Communi. Mag.*, vol. 25, pp. 44–57, April 1987.

15. V. K. Bhargava, "Forward Error Correction Schemes for Digital Communications." *IEEE Communi. Mag.*, vol. 21, pp. 11–19, January 1983.

16. D. Chase, "A Class of Algorithms for Decoding Block Codes with Channel Measurement Information," *IEEE Trans. Inform. Theory*, vol. IT-18, pp. 170–182, January 1972.

17. J. M. Wozencraft and B. Reiffen, *Sequential Decoding*, Cambridge, Mass.: MIT Press, 1961.

18. R. M. Farro, "A Heuristic Discussion of Probabilistic Decoding," *IEEE Trans. Inform. Theory*, vol. IT-9, pp. 64–74, April 1963.

19. K. Zigangirov, "Some Sequential Decoding Procedures," *Publ. Peredachi Inf.*, vol. 2, pp. 13–25, 1966.

20. F. Jelnick, "A Fast Sequential Decoding Algorithm Using a Stack," *IBM J. Res. Develop.*, vol. 13, pp. 675–685, November 1969.

21. J. L. Massey and M. K. Sain, "Inverses of Linear Sequential Circuits," *IEEE Trans. Comput.*, vol. C-17, pp. 330–337, April 1968.

22. I. S. Reed, "A Class of Multiple-error-correcting Codes and the Decoding Scheme," *IRE Trans. Inform. Theory*, vol. 4, pp. 38–49, 1954.

23. D. E. Muller, "Application of Bolean Algebra to Switching Circuit Design and to Error Detection," *IRE Trans. Electron. Comput.*, vol. 3 pp. 6–12, 1954.

24. S. W. Golomb, *Shift Register Sequences*, San Francisco: 1967; Holden-Day, revised edition, Laguna Hills, Calif.: Aegean Park Press, 1982.

25. S. W. Golomb, ed., *Digital Communications with Space Applications*, Englewood Cliffs, N.J.: Prentice-Hall, 1964.

26. E. R. Berlekamp, ed., *Key Papers in the Development of Coding Theory*, New York: IEEE Press, 1974.

27. T. Verhoeff, "An Updated Table of Minimum-Distance Bounds for Binary Linear Codes," *IEEE Trans. Inform. Theory*, vol. IT-33, September 1987.

28. I. S. Reed, T. K. Truong, Xiaowei Yin, and J. K. Holmes, "A Simplified Procedure for Correcting Three Errors in a [23, 12] Golay Code," preprint.

29. M. A. Tsfasman, "Goppa Codes That Are Better Than the Varshamov-Gilbert Bound," *Proble. Inform. Transmiss.*, vol. 18, pp. 163–165, 1982.

30. T. Schaub, "A Linear Complexity Approach to Cyclic Codes," Ph.D. Dissertation, Swiss Federal Institute of Technology, Zurich, 1988.

31. J. H. vanLint and R. M. Wilson, "On the Minimum Distance of Cyclic Codes," *IEEE Trans. Inform. Theory*, vol. IT-32, pp. 23–40, January 1986.

32. S. W. Golomb, R. E. Peile, and R. A. Scholtz, *Notes on Information Theory*, to be published.

33. L. R. Welch, Unpublished Lecture Notes, University of Southern California.

34. J. L. Massey, "Shift-Register Synthesis and BCH Decoding," *IEEE Trans. Inform. Theory*, vol. IT-15, pp. 122–127, 1969.

35. L. R. Welch and R. A. Scholtz, "Continued Fractions and Berlekamp's Algorithm," *IEEE Trans. Inform. Theory*, vol. IT-25, pp. 19–27, January 1979.

36. E. R. Berlekamp, "Helical Interleaving," in *Workshop Record for University of Southern California–Naval Research Office Workshop on Research Trends in Military Communications*, pp. 169–183, May 1983.

37. M. K. Simon, J. K. Omura, R. A. Scholtz, and B. K. Levitt, *Spread Spectrum Communications*, vol. 1, Rockville, Md.: Computer Science Press, 1985.

38. D. V. Sarwate and M. B. Pursley, "Cross Correlation Properties of Pseudorandom and Related Sequences," *Proce. IEEE*, vol. 68, pp. 593–619, May 1980.

R. B. DYBDAL

3

. . . .

SHF/EHF Antennas and Natural Propagation Limitations

Antennas designed for operation at superhigh and extremely high frequencies (SHF/EHF) are an important part of present and future satellite communication systems. Present communication systems capitalize on their well-developed technology base, and future systems will incorporate and extend this technology into designs that have more flexibility and performance. A significant advantage of SHF/EHF antennas is their ability to generate high-gain, narrow-beamwidth antenna patterns from antennas with modest physical size and weight, important parameters for practical spacecraft applications. The importance of this advantage is seen in developing efficient system designs, limiting the coverage of antenna systems to isolated portions of the earth's field of view, and the ability to cancel interference that is located close to desired system users. These features will be discussed in greater depth throughout this chapter.

Similarly, a second advantage of SHF/EHF satellite communication systems is the existence of relatively wide spectral allocations in the SHF/EHF range.[1] These allocations permit high-data-rate communications with relatively benign propagation limitations. The natural atmospheric medium in the SHF/EHF range does indeed have absorption bands and sensitivity to inclement weather conditions that do not occur at lower microwave frequencies. However, the SHF/EHF regime is isolated from ionospheric limitations and a relatively high noise level at ultrahigh frequencies (UHF) and is significantly less affected by atmospheric propagation limitations experienced at higher optical frequencies. The atmospheric propagation limitations in the SHF/EHF range will also be discussed in greater depth in this chapter.

314

The present development effort in the SHF/EHF spectrum results from the ability to transfer high-data-rate communications, a well-demonstrated technology base, the capability to achieve good angular resolution from antennas of modest physical size, and relatively favorable propagation conditions. This chapter will review these features starting with a communication system overview, antenna techniques for satellite payloads, and a review of SHF/EHF propagation limitations.

3.1 COMMUNICATION SYSTEM DESIGN

Communication system design and achievable performance inherently involve an examination of the link equations. The performance of antennas in communication systems is critical: they must provide adequate coverage and sensitivity to achieve the desired link performance. In addition, the present interest in negating interference and maintaining isolation between different geographic areas that simultaneously use the same frequency band results in requirements that minimize the sidelobe response from the antennas in regions beyond the desired coverage area.

The sensitivity of a communication system depends in part on the gain of the antennas used in the link. The gain of an antenna is defined in a manner analogous to the gain of an amplifier, which is referenced to an ideal "straight wire" whose output is identical to its input. Since an antenna response must quantify the variations in sensitivity with angular location, the reference for antennas is an ideal antenna having uniform sensitivity in all directions, which is referred to as an *isotropic antenna*. An antenna must also be connected to electronics, so its terminal characteristics also affect its sensitivity. Thus, significant mismatch or ohmic loss in the antenna terminals results in loss in system sensitivity. The maximum gain G of an antenna is related to its aperture area A by

$$G = \eta \left(\frac{4\pi A}{\lambda^2} \right) \tag{3.1}$$

where η is the antenna efficiency and λ is the operating wavelength, defined as $\lambda = c/f$, where c is the speed of light and f is the operating frequency. This equation can also be interpreted by its electrical size as measured in wavelengths. Since SHF/EHF wavelengths are smaller than the lower-frequency microwave values, SHF/EHF antennas are more compact than their microwave counterparts.

The antenna efficiency is comprised of several different factors that reduce system sensitivity. One factor is the uniformity with which the aperture area is used. The aperture distribution and its antenna pattern are related by a transform; in particular, for a planar aperture surface the fields in the aperture plane are related to the antenna pattern by a Fourier transform. Like any transform relation, the maximum resolution (for an antenna, its beamwidth) is achieved by a uniform weighting (uniform aperture illumination); such weighting also produces the highest sensitivity (maximum antenna gain). However, the uniform aperture is accompanied by high sidelobes in the pattern. When the system application requires lower sidelobes, the aperture amplitude distribution is weighted toward the center; there is a corresponding reduction in antenna gain compared with a uniformly weighted aperture. This resulting loss is referred to as aperture taper loss, one of the factors affecting antenna efficiency. Similarly, maximum sensitivity is achieved by uniform phasing, and aperture phase perturbations reduce sensitivity.

In some cases, the aperture distribution is unavoidably modified by the structure of the antenna. For example, a reflector antenna has a feed system located at the focal point that distorts the fields radiated by the reflector surface and creates what is referred to as blockage losses. Again for reflector antennas, some of the radio-frequency (RF) energy processed by the feed is not utilized by the reflector. Such losses are referred to as spillover losses since the energy is not focused by the reflector surface. The antenna is designed for a particular polarization, which specifies the orientation of the electric field. Practical antenna designs unavoidably contain undesired polarization components that reduce the antenna efficiency. Finally, ohmic loss and impedance mismatch at the antenna terminals reduce the antenna efficiency. These individual factors comprising the antenna efficiency can be evaluated for a specific antenna design. For system planning purposes, however, an antenna efficiency representative of practical designs is often used. An antenna efficiency of 55 percent is a commonly assumed value, and the corresponding antenna diameter for circular apertures shown in Fig. 3.1 can be used in gain calculations.

The coverage of the antenna depends on the antenna beamwidth. This coverage is particularly important for satellite antennas since it defines the portion of the earth that can be serviced by the satellite antenna. The projection of the antenna pattern on the earth's surface and the resulting contours of constant system sensitivity are commonly referred to as the antenna "footprints." The beamwidth of the antenna can also be interpreted by the transform relation between the aperture field and the antenna pattern, and like antenna gain, depends on the electrical size in wavelengths. Thus, the half-power beamwidth, θ_{HP}, which is the common reference level, is given by

$$\theta_{HP} = \frac{K\lambda}{D} \tag{3.2}$$

Again, the electrical size of the antenna, D/λ, arises. As the electrical size of an antenna increases, its gain increases and its beamwidth decreases. The value of the constant K depends on the amplitude taper used by the antenna. While amplitude tapering reduces antenna sidelobes, it also reduces the antenna gain and broadens the antenna beamwidth. A typical value for K in practical reflector designs that is commonly used in system estimates is 70, measured in degrees.

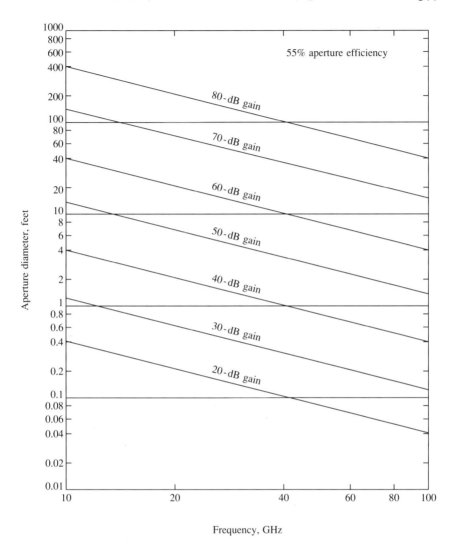

FIGURE 3.1   Antenna diameter versus frequency. (From R. B. Dybdal, "Millimeter Wave Antenna Technology," *IEEE J. Selected Areas Commun.*, vol. SAC-1, September 1983. © 1983 IEEE.)

The shifting theorems used in transforms also apply to determining the beam maximum position. When an antenna aperture is uniformly in phase, the beam maximum lies on the axis. When the beam is scanned off-axis, a linear phase gradient is required and direction of the beam maximum is normal to the phase gradient.

Measurements of a reflector antenna in Fig. 3.2 illustrate typical pattern characteristics. This 6-in.-diameter parabolic reflector, measured at 92 GHz, has a highly tapered aperture distribution; the diagonal feed horn illuminates

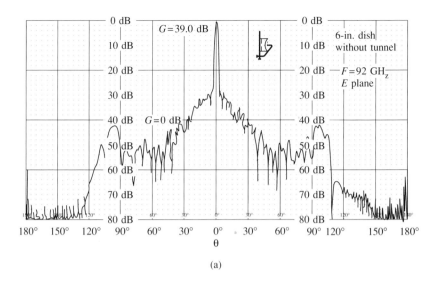

(a)

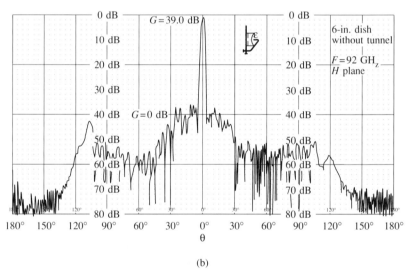

(b)

FIGURE 3.2   Measured reflector antenna patterns. (a) E = plane pattern.
(b) H = plane pattern. (From R. B. Dybdal, "Millimeter Wave Antenna
Technology," *IEEE J. Selected Areas Commun.*, vol. SAC-1, September 1983.
© 1983 IEEE.)

the reflector edges 25 dB lower than the central portion. The antenna has
a narrow main beam with a 39-dB gain level corresponding to a 37 percent
efficiency. Some efficiency loss results from the strong tapering used in the
aperture distribution. Ohmic loss in the waveguide section connecting the
antenna feed is not insignificant; a straight waveguide has an attenuation of
1 dB per foot at these frequencies. This antenna has low sidelobe levels near

the main beam that result from the aperture amplitude taper; however, the nulls in this region are not deep and extend over a relatively wide angular region. This effect results from the blockage of the feed and connecting waveguide. At wider angles from the main beam, 100°, spillover lobes occur in the pattern resulting from energy from the feed that is not intercepted by the reflector. The measured antenna patterns have an 80-dB dynamic range, which is not unusual for high-gain antennas. The earlier discussion described antenna radiation by its aperture distribution mechanism, which not only is a convenient way to describe antenna operation but also is the dominant mechanism in the main beam region used in communication systems. This measured pattern illustrates some of the second-order effects that contribute to the overall antenna pattern; detailed analysis techniques[2] exist to project their effect on antenna performance.

The link equation relates the power transfer from the transmitting antenna to the receiving antenna. The RF power radiated from an antenna propagates as an outward spreading spherical wave. The power density P_d is given by

$$P_d = \frac{P_t G_t}{4\pi R^2} \tag{3.3}$$

where G_t is the gain of the antenna and R is the range to the point of observation. The angular dependence of the antenna pattern is included within the antenna gain value. Typically, the antenna is located at the origin of a spherical coordinate system, and the pattern variation is specified by the two angles θ and ϕ. The gain value in Eq. (3.1) is the maximum gain of the antenna, and the gain in other directions is included as $G(\theta, \phi)$, which is the antenna pattern. The product of the transmitted power P_t and the transmitter antenna gain G_t is referred to as the EIRP (effective isotropic radiated power), which is the measure of the transmitter effectiveness.

The power received by an antenna is the product of the incident power density and the effective aperture A_e, which, by definition, is related to the receiving antenna gain as

$$A_e = \left(\frac{\lambda^2}{4\pi}\right) G \tag{3.4}$$

Thus the received power is related to the transmitted power by

$$P_r = \frac{(\text{EIRP})G_r}{(4\pi)^2 (R/\lambda)^2} \tag{3.5}$$

This relationship is referred to as the *Friis transmission formula*.[3] In cases where the link has other losses such as rain attenuation, discussed later, the received power is reduced by an additional loss factor L.

The received signal must also compete with noise components. The noise power at the receiver equals

$$P_N = kT_s B \tag{3.6}$$

where k is Boltzmann's constant $(-198.6 \text{ dBm/Hz} \cdot \text{K}$ is convenient in link calculations), T_s is the total system noise temperature, and B is the receiver bandwidth. The total system noise temperature is the sum of noise contributions from the antenna (T_a) and noise contributions from the receiver electronics (T_r). The antenna noise temperature arises from noise generated by ohmic losses and external noise received by the antenna. An uplink antenna on a satellite basically views the earth's surface, which radiates an average noise temperature of about 300 K. The noise temperature of a ground terminal antenna depends on the sky temperature, which is increased by propagation loss as described later.

The signal-to-noise ratio (SNR) is the quotient of the received signal power and the noise power, given by

$$\text{SNR} = \frac{(\text{EIRP})G_r}{(4\pi)^2 (R/\lambda)^2 k T_s B} \tag{3.7}$$

The performance of digital communication systems is typically stated in terms of bit error rate (BER), which is related to the E_b/N_0 (energy per bit/noise spectral density) value achieved in the link. The energy per bit equals the received power multiplied by the bit time T_b. Thus

$$\frac{E_b}{N_0} = \frac{(\text{EIRP})G_r T_b}{(4\pi)^2 (R/\lambda)^2 k T_s} \tag{3.8}$$

Communication system performance stated either in terms of SNR or E_b/N_0 depends on the EIRP of the transmitter and the G/T ratio of the receiver,[a] its measure of effectiveness. Communication systems are typically required to achieve a minimum performance stated in terms of SNR or a minimum BER corresponding to a required E_b/N_0. The actual system performance in excess of these minimum requirements is the "margin" achieved by the system design.

3.2 SHF/EHF ANTENNA TECHNOLOGY

SHF/EHF communication systems benefit from a well-developed technology base in antennas, and further development remains active. Early satellites necessarily used exceedingly simple antennas that could be isolated from the RF electronics in the communication payload. Present satellite antennas are increasingly integrated into the overall payload design. These antennas are required to route information to and from various geographic regions, process signals from desired user segments, and reject unwanted interference. The ability to perform these functions within practical size, weight, and

---

[a] The G/T ratio is the antenna gain divided by the system noise temperature, where G is expressed in decibels and T in kelvins.

power limitations results in the present development of SHF/EHF antenna technology.

An earlier survey paper[4] has described EHF antenna technology for communication systems. Most technology at the lower microwave frequencies has been demonstrated in the EHF region. EHF antenna hardware must be produced with greater precision for EHF applications than the lower microwave frequency technology, but such techniques have been amply demonstrated. The antenna technology for terminal segments typically involves straightforward designs. For example, many ground terminals use reflector antennas that can be manufactured with a high degree of precision. The popularity of reflector antennas stems from their simplicity and corresponding cost-effectiveness, an important issue with the trend toward more numerous terminal segments. Present and future spaceborne antennas tend to be more complex and more integrated, with the RF electronics within the payload, as the requirements for satellite systems become more demanding. These requirements include increased data throughput, greater flexibility in servicing terminal user segments, and increased demands for electronic survivability in negating undesired interference. These trends become particularly apparent in three technologies for payload antennas that will be discussed in greater detail. The first technology is multiple-beam antennas that service different regions within the earth's field of view. The second technology is adaptive antenna designs that can separate interference from desired user signals. The third technology is active aperture designs for satellite downlinks that sequentially service different user segments distributed within the earth's field of view.

3.2.1 Multiple-Beam Antennas

Multiple-beam antennas typically consist of an aperture whose overall size dictates the minimum beamwidth of an individual beam and a multiplicity of feed elements that result in multiple independent beams having different pointing directions. Typically, multiple-beam antennas are based on an optical design—i.e., a reflector or lens antenna—and the multiple beams arise from a cluster of feed elements located in the focal region of the antenna. These designs follow geometric optics principles; i.e., the on-axis beam results from a feed element centered in the focal region and off-axis beams result from feed elements displaced from the center of the focal region.

Multiple-beam antennas may be used in several different ways. A general block diagram of a multiple-beam antenna system for a satellite payload[5] is shown in Fig. 3.3. The uplink antenna has K beams, which are received and routed to N beams in the downlink antenna. The "message router" may take a variety of forms. A simple example is a switching network that maps the uplink beams into their downlink counterparts. The switching network may have the flexibility to combine individual beams to increase coverage to user ground segments. Channelization based on the frequency plan that divides

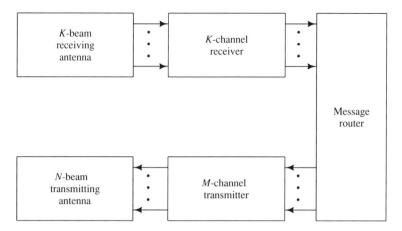

FIGURE 3.3   Multiple-beam antenna system. (From R. B. Dybdal, "Multiple Beam Communication Satellite Antenna Systems," *Proc. IEEE Internat. Conf. Commun.*, Minneapolis Minn., June 17–19, 1974. © 1974 IEEE.)

the allocated spectrum into portions assigned to individual user communities can also be included within the message router. The message router may also contain demodulation/remodulation equipment that is capable of reformatting the uplink information for distribution on the downlink. Finally, the message router can also contain adaptive processing elements to separate interference from desired signals and to use efficiently the available transmitter power for transmitting desired information rather than interference power.

The differences between domestic and military communications requirements become apparent as multiple-beam antenna technology is applied. Domestic communication satellites service fixed user communities, and emphasize the efficient use of the available frequency allocations. In practice, individual beams are combined in a fixed manner to service specific user communities; a typical example is combining a number of beam ports together to generate a footprint that conforms to a specified geographic region. Synthesizing specific coverage areas from a collection of multiple beams tends to maximize the gain within the coverage area and to increase the isolation between different geographic regions. Several different regions can be serviced simultaneously by the same antenna aperture, and beam patterns with sufficient separation can use the same frequency band simultaneously. Thus the principal interest in multiple-beam antenna technology for domestic communications is efficient coverage to specified regions, reusing the frequency in different regions to increase the overall data throughput and isolating the coverage to desired geographic areas.

Multiple-beam antenna systems also appeal to military satellite communication users but for different reasons. Military users experience variation in their coverage needs; e.g., communications to naval elements vary with fleet movements. Changing the coverage of a multiple-beam antenna requires flexibility in the switching and combining of the individual beams to vary the coverage. Electronic survivability is extremely important to military users, so that the capability to steer nulls automatically in the direction of interference through adaptive processing has high interest for military users. When spread spectrum modulation is used to reduce interference, the satellite transponder can also demodulate and remodulate to remove interference power in the satellite uplink. Although the demodulation/remodulation increases the complexity of the satellite, removing interference power prior to rebroadcasting uses the downlink transmitter for the desired information rather than interference. Thus, the military communications users have a principal interest in multiple-beam antennas for the flexibility in varying the antenna coverage to accommodate changing communication needs and for adaptively combining the individual beams to reduce interference.

The design of a multiple-beam system involves both the antenna hardware per se and the processing contained within the message router. The antenna hardware generally favors optical designs with clusters of feed horns in the focal region to generate the individual beams. Antenna sidelobe control of the individual beams is important to maintain isolation between beam positions and to control interference beyond the desired coverage area. The optics of the antenna must be designed so that the off-axis beam patterns are not degraded by phase perturbations. The beams must be closely spaced so that the gain in regions between the individual beams is not excessively low, and a trade-off between low sidelobe patterns and high beam crossover levels results from physical interference between feed horns in the focal region. Generally, the beam centers of the individual beams are placed at the vertices of equilateral triangles to maximize the coverage.

The overall complexity of a multiple-beam antenna system depends in part on the number of beams in its design. For geosynchronous satellites, the number of beams to achieve full earth coverage versus the beamwidth of each beam is shown in Fig. 3.4, and the corresponding minimum footprint size occurring at the subsatellite point is shown in Fig. 3.5 as a function of the antenna beamwidth. Minimizing the beamwidth has many attractive advantages—e.g., increasing the G/T or EIRP, tighter conformance to a specified geographic region, and improved interference rejection. Although the overall physical aperture size may not be excessive at these frequencies, reducing the beamwidth is also accompanied by a dramatic increase in the number of beam positions for full earth coverage. Although a large number of beam positions can be generated by the antenna hardware, the processing complexity of a large number of beams in practical satellite designs may become prohibitive. When narrow beamwidths are required, they may be restricted to a limited portion of the earth referred to as "spot coverage."

The processing of multiple-beam systems takes place within the message router, which can have a variety of forms. The simplest satellite transponder

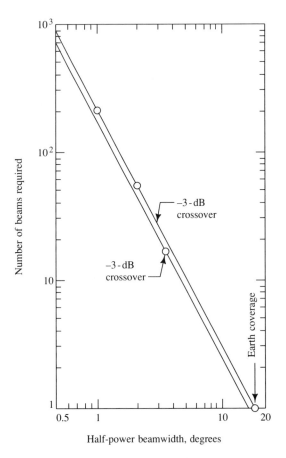

Half-power beamwidth, degrees

FIGURE 3.4   Number of beams for earth coverage. (From R. B. Dybdal, "Multiple Beam Communication Satellite Antenna Systems," *Proc. IEEE Internat. Conf. Commun.*, Minneapolis, Minn., June 17–19, 1974. © 1974 IEEE.)

is a frequency-translating repeater, whose message router consists of switch matrices that map uplink beam positions into their downlink counterparts, frequency translation between the uplink frequencies and downlink frequencies, and suitable amplification that drives the downlink transmitters. Channelization can also be incorporated into the transponder to implement the frequency plan that divides the allocated spectrum into smaller portions assigned to the user segments. The switching matrices are generally referred to as beam-forming networks that select and/or combine the static beams formed by the antenna hardware. The beam-forming network complexity increases with the number of available beams, the flexibility in the routing, and redundancy. The uplink beamformers can be performed at RF prior to the receiver electronics with the penalty of degrading the uplink G/T by the switching losses. Alternatively, the beam forming can be performed after RF preamplification and/or down-conversion to intermediate frequencies (IF) to avoid degrading the G/T and achieve a smaller, lighter beam-forming network at the penalty of an increased amount of RF electronics for the additional preamplifiers and/or down-converters. The downlink beam-forming networks route a limited number of transmitters, which are more bulky and power-consuming than the receiving electronics.

The Intelsat VI satellite[6] illustrates the use of multiple-beam antennas. Its multiple-beam antennas are offset-reflector designs to avoid the blockage degradation of the feed system. Multiple beams are combined to isolate the antenna coverage to desired geographic locations, as illustrated in Fig. 3.6. The antennas are designed for orthogonal polarization to increase the data rate, and significant design attention to minimizing cross polarization was made to achieve isolation between simultaneously used frequency bands. The overall transponder in Fig. 3.7 illustrates the routing and channelization for this system.

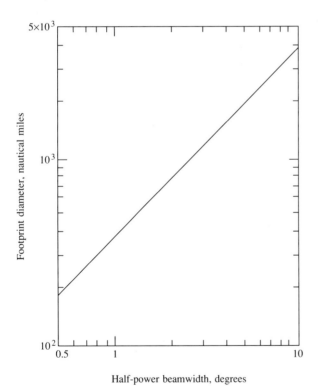

FIGURE 3.5   Minimum footprint size. (From R. B. Dybdal, "Multiple Beam Communication Satellite Antenna Systems," *Proc. IEEE Internat. Conf. Commun.*, Minneapolis, Minn., June 17–19, 1974. © 1974 IEEE.)

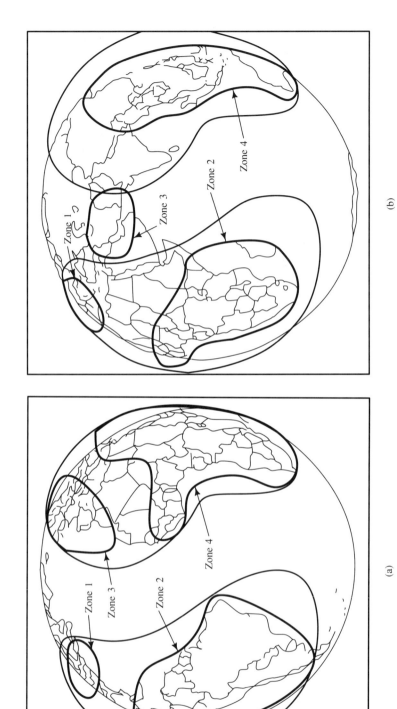

FIGURE 3.6 Intelsat VI footprints generated by multiple-beam combining. (a) Atlantic Ocean. (b) Indian Ocean. (From Donald H. Martin, *Communication Satellites 1958–1988*, The Aerospace Corp., 1988. Used with permission of The Aerospace Corporation.)

(a)

(b)

The Intersat VI satellite design illustrates the attractive features multiple-beam antennas afford the domestic community. The military community has the additional requirements of maintaining communications in the presence of interference, somewhat more flexibility in beam positioning to accommodate changing coverage conditions, and somewhat less demand on efficient use of the allocated spectrum. The DSCS III satellite transponder in Fig. 3.8 illustrates a smaller amount of channelization in its design. The multiple-beam antennas use a waveguide lens design, with 61 beam positions on the uplink and two 19-beam antennas on the downlink. These multiple-beam antennas use a cascaded switch network to combine the beam ports into a single input port for the receiver or a single output port for the transmitter. Thus, the multiple-beam antennas are used to vary the coverage of a single, simultaneous beam that is used by the transponder; the coverage can be varied from complete earth coverage, which uses all beam positions, to the coverage provided by a single beam.

The trend toward increased electronic survivability for military communication satellites also leads towards a different transponder design. The regenerative repeater demodulates the uplink signal, reformats it, and re-modulates it for the downlink format. This repeater also provides the necessary translation between the uplink and downlink frequencies and the necessary amplification to drive the downlink transmitters. The modulation uses a spread spectrum format to reduce the effectiveness of interference. Although the benefits of spread spectrum can be realized at either the satellite transponder or at the ground terminal, the advantage of despreading on the transponder is that the transmitter is not required to broadcast the interference. Thus the regenerative repeater design can maintain its limited transmitter power for desired signals and avoid loss in transponder performance by broadcasting interference power. The FLTSATCOM EHF package in Fig. 3.9 employs a regenerative transponder package to provide initial operational experience with regenerative repeaters and 44-GHz uplink and 20-GHz downlink frequency operation. This experimental package uses earth coverage horns and narrower spot-beam antennas; multiple-beam technology will be used in future system designs.

The principal development issues for multiple-beam antennas involve the multiple-beam hardware and to a greater extent the processing equipment used. The antenna hardware issues include how to generate the required number of beams with acceptable pattern performance, the isolation between beam positions, and the trade-off between beam crossover levels and the sidelobe levels of individual beams. The multiple-beam processing referred to as the message router can have a variety of forms as described. Efficient beam-forming networks at RF or IF, channelization, demodulation/remodulation, and in the future adaptive interference cancellation are included within multiple-beam antenna system designs to meet future mission performance requirements.

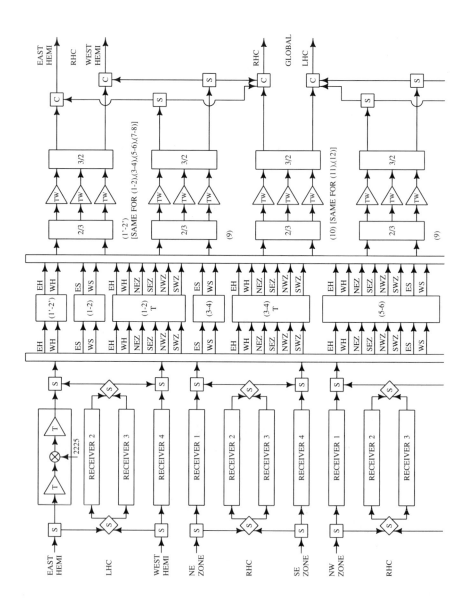

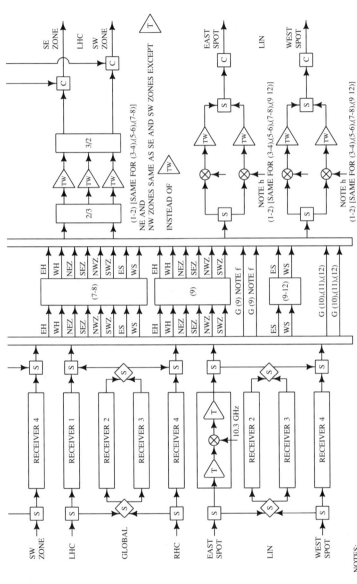

NOTES:

a. The numbers in parentheses are channel numbers. The multiple numbers indicate channel bandwidths > 36 MHz, e.g., (1-2) is a channel occupying the spectrum used by Channels 1 and 2 on intelsat IV.   b. (1'-2') occupies the spectrum just below (1-2).   c. Each switch matrix can form any one-to-one combination of inputs and outputs. d. The switch matrices marked "T" can be used for SS/TDMA.   e. LHC/RHC = Left/right-hand circular polarization; LIN = Linear polarization.   f. Channel 9 may be used on both EH and WH or on the co-polarized global beam; it may also be used in all 4 zones or on the co-polarized beam.   g. Spot beam antennas have 11/14 GHz diplexers (not shown).   h. 7.25 GHz for (1-2), (3-4), (5-6); 7.50 GHz for (7-8), (9-12).   i. Combiners after transmitters also have inputs from unillustrated transmitters.

FIGURE 3.7 Intelsat VI communication system. (From Donald H. Martin, *Communication Satellites 1958–1988*, The Aerospace Corp., 1988. Used with permission of The Aerospace Corporation.)

329

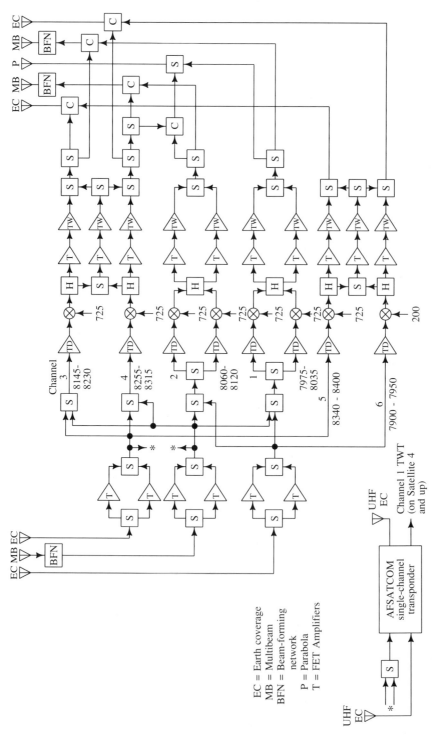

EC = Earth coverage
MB = Multibeam
BFN = Beam-forming
     network
P = Parabola
T = FET Amplifiers

FIGURE 3.8   DSCS III communication subsystem. (From Donald H. Martin, *Communication Satellites 1958–1988*, The Aerospace Corp., 1988. Used with permission of The Aerospace Corporation.)

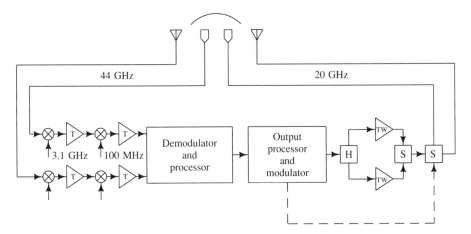

FIGURE 3.9 FLTSATCOM EHF regenerative repeater transponder. (From Donald H. Martin, *Communication Satellites 1958–1988*, The Aerospace Corp., 1988. Used with permission of The Aerospace Corporation.)

3.2.2 Adaptive Interference Cancellation

Adaptive antennas reduce interference in receiving antenna systems by combining different antenna elements using closed-loop circuitry. Two benefits exist for satellite systems: (1) the signal-to-interference power ratio is increased at the input to the satellite transponder, and (2) the limited transmitter power is used for desired signals rather than interference. Adaptive cancellation systems for future satellites thus improve the overall electronic survivability of the system and effectively use the available transmitter power.

Three techniques can reduce uplink interference signals: (1) spread spectrum modulation, (2) passive control of antenna sidelobes, and (3) adaptive interference cancellation. Spread spectrum modulation is a well-developed technology, but its benefit is ultimately limited by the bandwidth of the allocated spectrum. Passive antenna sidelobe control has inherent design limitations but it can be very effective in reducing interference that is widely separated from coverage areas containing the desired users. Adaptive antennas inherently require additional circuitry that increases the complexity of the satellite and consumes valuable prime power; however, adaptive antennas can reduce interference that is close to desired user terminals. The challenge for practical system designs is to combine these three techniques in a way that minimizes the overall satellite complexity, development risk, weight, and power while achieving the required interference protection.

Adaptive antenna techniques, because of their additional circuitry, are required in those communications satellite systems that cannot achieve adequate interference protection from spread spectrum or passive sidelobe control. Interference protection requirements are generally described by scenarios that specify the interference levels, spectral distributions, and number and geographic locations of interfering sources. Adaptive processing is generally required to satisfy scenarios that have high-level interference close to desired signal users. Systems with high data rate requirements and correspondingly limited, spread spectrum benefits are particularly susceptible to interference close to desired users. When interference is close to desired users, very narrow antenna beams are required for effective interference reduction.

The ability to cancel interference close to desired signals depends on the antenna beamwidth, the tolerable G/T loss for desired users, and the location of the interference and desired signals. Example calculations illustrate this point in Fig. 3.10. These calculations are based on the pattern of a reflector antenna. The interference is aligned with the first null next to the main beam. The separation between the interference source and the desired source is calculated under two conditions: (1) that the desired signal has adequate G/T from the antenna at its half-power point, and (2) that the desired signal has adequate G/T from the antenna when it is 10 dB lower than the peak gain. The minimum separation between the desired signal and the interference source is reduced as the antenna beamwidth narrows. Similarly, the separation is smaller when 10 dB of G/T can be sacrificed rather than 3 dB. Finally, the minimum separation occurs at the subsatellite point and increases along with the footprint dimensions as the beam moves toward the limb of the earth; values for a 20° user elevation angle are indicated. This simple example quantifies the effect of the half-power beamwidth of the antenna, which in turn specifies the overall antenna size, the advantage of maximizing the uplink G/T so that some G/T can be sacrificed in canceling interference, and the loss in resolution that occurs for users at low elevation angles.

The precise requirements for the uplink antenna design depend on the coverage of desired users, the minimum G/T needed for link closure, the details of the interference scenario, and the necessary bandwidth and allocated spectrum. The coverage of the desired users defines the beamwidth of the uplink antenna design or the number of multiple beams required to service that area. The minimum G/T required for link closure dictates not only a minimum antenna beamwidth but also the amount of pattern loss that can be tolerated in adaptive processing. The interference scenario defines the required protection that can be allocated between spread spectrum benefits, passive sidelobe rejection, and adaptive cancellation requirements. The required bandwidth to support the user data rate(s) and the allocated bandwidth define the spread spectrum protection; in addition, effective interference cancellation over the allocated bandwidth also imposes requirements on the adaptive cancellation design.

In operation, the desired signals must be distinguished from the interference. Spread spectrum modulation used by desired signals and not by interference sources is a distinguishing feature. The adaptive circuitry is generally adjusted by correlation circuitry in a closed-loop manner. Correlation techniques using the protected spread spectrum code can be a very

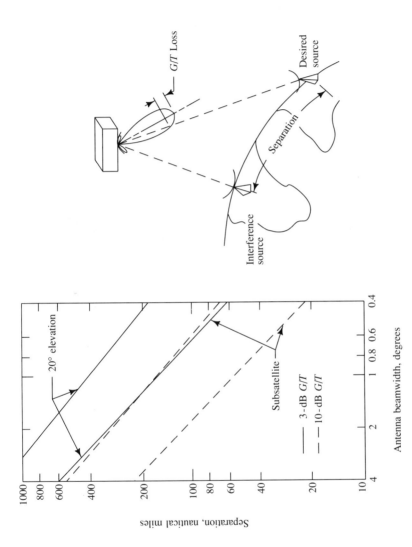

Antenna beamwidth, degrees

FIGURE 3.10 Resolution between interference and desired signals. (From J. T. Mayhan, A. J. Simmons, and W. C. Cummings, "Wideband Adaptive Antenna Nulling Using Tapped Delay Lines," *IEEE Trans. Antennas Propagat.*, vol. AP-29, November 1981. © 1981 IEEE.)

powerful means of distinguishing desired signals from interference and are used by the adaptive algorithms that govern the circuitry adjustment. These adaptive algorithms typically are based on minimum mean-square error criteria, and are described in detail elsewhere.[7] When a spread spectrum is used, the interference power must be significantly stronger than the desired signals; if this were not the case, the spread spectrum benefit by itself would provide adequate interference protection. Thus, the power differences between the interference and the desired signals can also be used as a means of distinguishing interference from desired signals.

The adaptive algorithms controlling the circuitry are configured to meet an optimumization criterion. One commonly used criterion is based on maximizing the SINR (signal-to-noise plus interference ratio). This criterion defines the steady-state performance of the adaptive system design after the circuitry reaches its final values. The amount of time the circuitry requires to reach its steady-state value after the initiation of the interference is referred to as the convergence time. This time depends on the algorithms used to adjust the adaptive circuitry, the relative difference between the interference and desired signal power levels, and the circuitry used in the system design. A third parameter, referred to as the "degrees of freedom" of the adaptive system design, represents the number of parallel interference cancellation circuits. The degrees of freedom in an adaptive system design roughly relates to the number of interference sources that can be adaptively cancelled. Each adaptive circuit is capable of generating a null in the direction of the interference. In some cases the interference may be sufficiently reduced by the passive sidelobes of the antenna, and adaptive cancellation is not required. In other cases the interference sources may be sufficiently close that one null cancels more than one interferer. However, wide-bandwidth jammers may require cancellation from more than one circuit, expending more than one degree of freedom to cancel the interference.

In recent years, candidate antenna designs for adaptive uplink systems have tended to be divided into two broad categories. One category is multiple-beam technology for which the adaptive circuitry combines the individual beams to generate nulls in the interference direction(s). The second category is thinned-array designs, which typically use a set of overlapping earth coverage horns separated sufficiently to resolve interference and desired signals. Although the choice between these two technology categories depends on the overall system requirements, the generic differences between their designs can be described.

Adaptive processing techniques applied to multiple-beam antennas dynamically combine the collection of individual received beams to generate a pattern that provides coverage to desired users and nulls in the direction of interference sources. A typical application would service desired users lying within a required coverage region and cancel a collection of interference sources whose global distribution is described in the scenario. Interference sources that are widely separated from the desired coverage area can have

significant protection from the sidelobes of the multiple beams. The combination of sidelobe attenuation and spread spectrum can achieve adequate
protection for interference that is closer to the required coverage area. Thus,
adaptive cancellation is required to provide interference protection only in
the vicinity of the desired coverage area. In this case, only a limited number
of beams are used in the adaptive design, and the adaptive processing circuitry
can be switched to different locations as the position of the desired coverage
area changes.

A thinned-array design typically consists of a number of earth coverage
horn antennas that are separated from one another. The separation between
adjacent elements in the thinned array is determined from the minimum
angular separation between interference and desired signal sources, and the
elements can be fairly widely spaced to resolve closely spaced interference.
The individual thinned-array elements are combined by the adaptive circuitry
to generate nulls in the direction of the interference.

These two systems differ in both their level of performance and the types
of interference scenarios that can be satisfied. Multiple-beam adaptive designs
can effectively use sidelobe control techniques to suppress a large number of
interference sources that are widely separated from the desired coverage and
to concentrate the adaptive circuit resources on interference that is close to
desired users. Its ability to resolve closely spaced interference depends on
the beamwidth of an individual beam. When the scenario contains closely
spaced interference sources, the required beamwidth may result in an excessive
number of beams to achieve full earth coverage and the overall design becomes
impractical. In this case the adaptive system might use a limited number of
narrow beams that can be mechanically steered to the desired coverage area.
The multiple-beam coverage thus is restricted to a limited spot size rather
than full earth coverage,[8] as shown in Fig. 3.11.

Thinned-array designs can increase the resolution of closely spaced interference with sufficiently separated array elements, whereas increased resolution
of interference for multiple-beam designs inherently increases the weight
and complexity. However, thinned-array designs have several built-in disadvantages compared with multiple-beam designs. The pattern of a thinned
array does not have inherent sidelobe protection, so this design is very
susceptible to interference sources widely separated from desired signals.
Although such interference can be nulled, additional complexity in the necessary circuitry results. A thinned array basically has less G/T than a multiple-
beam array, so sharper nulls must be achieved; i.e., less G/T can be sacrificed
to cancel interference. The separation between antenna elements limits the
bandwidth over which effective cancellation can be achieved. A narrow-
bandwidth interferer can be cancelled by a single amplitude and phase adjustment. However, as the bandwidth of the interference increases, the null position
of the thinned array changes, frequency sensitivity common in array designs.
Accommodating wide-bandwidth interference signals increases the complexity
of the adaptive electronics, an effect that the common phase center of a single

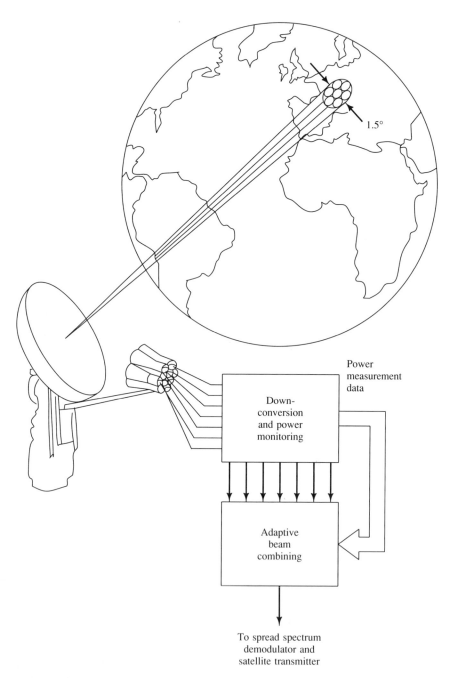

FIGURE 3.11   Adaptive system concept for spot coverage. (From J. T. Mayhan, A. J. Simmons, and W. C. Cummings, "Wideband Adaptive Antenna Nulling Using Tapped Delay Lines," *IEEE Trans. Antennas Propagat.*, vol. AP-29, November 1981. © 1981 IEEE.)

336

multiple-beam aperture avoids. Furthermore, when a thinned array design is required to provide coverage over an extended area, the generation of a null in the direction of the interference unavoidably produces additional periodically spaced nulls that can reduce the required coverage performance. These additional nulls are called grating nulls.[8]

The principal development issues for adaptive antennas depend on the required interference protection. Adaptive interference cancellation, spread spectrum modulation, and antenna sidelobe control should be combined to meet the interference protection requirements in a way that minimizes the overall system weight, prime power, complexity, and development risk. The choice between multiple-beam and thinned-array designs generally depends on the interference scenario.[9-13]

The ability to achieve effective interference cancellation over the required bandwidth depends on the inherent frequency sensitivity, or dispersion, of the antenna system; further efforts are required to understand the limitations of the antenna system design and to develop antenna hardware that minimizes the loss in adaptive system performance. Adaptive systems based on multiple-beam designs have relatively little dispersion since each beam has a common phase center; their dispersion properties depend on second-order effects that can be addressed by diffraction analyses.[14] Thinned-array designs have a bandwidth limitation caused by the phase center separation of the array elements. Standard complex weighting circuitry cannot compensate for this inherent dispersion over a wide bandwidth, but variable true time delay circuitry, when available, can produce the required compensation. The algorithms used to adjust the cancellation circuitry also require development; manufacturing suitable space-qualified hardware, simplifying the hardware complexity, and developing algorithms that can withstand the scrutiny of investigation by interferers over the long operational life of the satellite.

3.2.3 Active Aperture Antennas

Active aperture antennas are small-phased arrays used in the satellite downlinks as shown in Fig. 3.12. The array design consists of a number of waveguide elements typically contiguous to one another. Each element contains an active amplifier module connected to the radiating waveguide element that is preceded by a phase shifter. Active amplifiers are also used within the cascaded arrangement of power dividers between the input from the satellite transponder and the individual radiating elements.

Active aperture antennas have several innate advantages. The antenna coverage can be rapidly moved to different regions of the earth by resetting the phase shifters. Recent development of GaAs field-effect transistor (FET) power amplifiers has been dramatic, and space-qualified devices are available. At X-band frequencies, sufficient device output power exists so that only a limited number of devices can be combined[15] to replace TWT amplifiers; however, at 20 GHz, the device power outputs are lower and a larger number

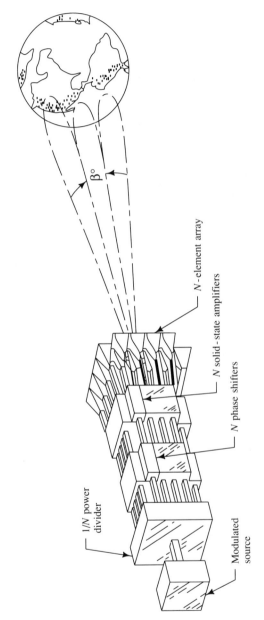

FIGURE 3.12 Active aperture concept.

of devices must be combined to compete with TWT performance. The active aperture design is an effective power-combining technique, and the inherent high isolation between adjacent waveguide elements results in graceful degradation in the event of device failure.[16] The ability to increase the down-link ERP by increasing the number of active devices in the active aperture is another advantage compared with the development risk involved in increasing the power output of existing TWT amplifiers.

The overall size of the active aperture like any antenna depends on its beamwidth. The maximum efficiency of the amplifiers in the active aperture results when the amplifiers are operated near saturation. Thus, the active aperture has a uniform amplitude distribution, and for a square array with overall size $D \times D$, the beamwidth is given by

$$\theta_{HP} = \frac{50\lambda}{D} \tag{3.9}$$

The aperture is subdivided into the individual waveguide apertures. The number of waveguide apertures comprising the active aperture, the dimensions of each waveguide, and the power output of the individual amplifier modules determine the EIRP generated by the active aperture design. The EIRP at the beam peak is expressed by

$$EIRP = N^2 P_e G_e \eta$$

where N is the number of elements, P_e is the average power in each element, G_e is the waveguide element gain, and η is the antenna efficiency. The antenna efficiency has loss components for the aperture efficiency of each waveguide element, pattern loss over the earth's field of view, the effects of amplitude and phase matching in the active devices, power delivered to grating lobes, ohmic loss, and so forth.

The waveguide element size in active apertures can exceed one wavelength, and the possibility of secondary lobes, called grating lobes, exists. The aperture is phased to produce the desired main beam direction and each waveguide element adds in phase in this direction. Grating lobes arise in other directions when the elements also combine in phase, a possibility when the element aperture spacing exceeds one wavelength. Generally, the grating lobes should be kept off the earth's surface. Systems that sequentially service user segments that are globally distributed typically use acquisition codes with a beam-hopping pattern; reception of the grating lobe could confuse the acquisition process. For geosynchronous satellites, the restriction to keep grating lobes off the earth's surface results in a maximum element separation of 3.4 wavelengths. However, if the element is that large, a significant pattern loss at the edge of the earth also results; if the maximum pattern loss at the edge of the earth is 1 dB, the maximum element size is 1.78 wavelengths. Thus, several factors dictate the bounds on the element size in active aperture designs.

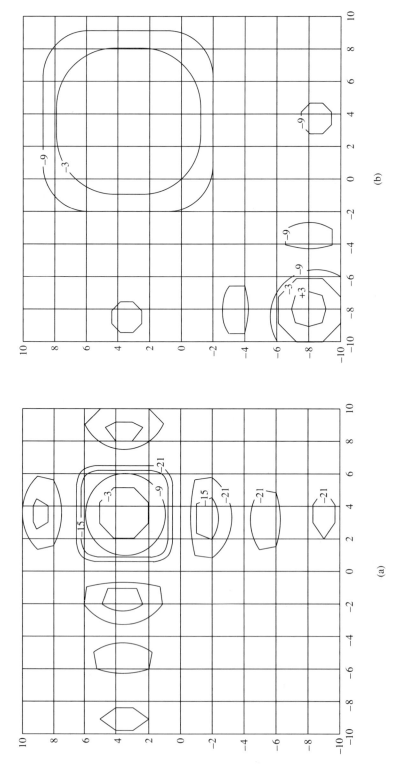

FIGURE 3.13 Active aperture patterns with and without beam broadening. (a) Minimum beamwidth. (b) Broadened beamwidth. (From G. M. Shaw and R. B. Dybdal, "Beam Broadening for Active Aperture Antenna," **IEEE** Antennas and Propagation Symposium, 1989. © 1989 IEEE.)

340

The actual design of an active aperture antenna involves several conflicts. A minimum number of elements is desired to minimize the system complexity, but the required output of each amplifier module with the waveguide element is increased and the antenna efficiency is reduced with an increased element size. Small elements result in a large number of active aperture elements that are tightly spaced; the tight spacing together with the relatively low power efficiency of the GaAs FET amplifiers exacerbates the thermal control problem. A large number of elements also increases the command requirements to the individual phase shifters at each element and increases the complexity of the combining circuitry between the input to the downlink and the individual active aperture elements.

The principal development issues for active aperture technology are separated into several categories. Although GaAs FET devices with usable power levels are available, improvements in their efficiency and gain per device can greatly reduce the prime power consumption. Similarly, improved phase shifters can reduce prime power requirements, and techniques to command the phase shifters require further development. Recently, a new technique[17] has been developed to increase the coverage area from its minimum value by rephasing the active aperture; such techniques increase the flexibility of active aperture designs. Patterns for a diagonal scan that accentuates the grating lobe problem are shown for the unbroadened and broadened beamwidths in Fig. 3.13. The overall power consumption also depends on how the gain is partitioned between the input and the radiating elements; some redundancy is also required to achieve the required lifetime. The thermal design of active apertures needed to dissipate the heat resulting from the device inefficiency must be developed for space operation. Finally, techniques for on-orbit diagnostics must be developed to determine the cause of reduced ERP results; the selection of other data, such as thermal measurements, for telemetry to the ground can aid in the diagnostic process.

3.3 NATURAL SHF/EHF PROPAGATION LIMITATIONS

SHF/EHF satellite communications experience some natural propagation limitations principally caused by molecular absorption and inclement weather. The frequency allocations for satellite uplinks and downlinks have been selected to correspond to the broad "windows" between the absorption frequencies. Frequency assignments within the strong 60-GHz absorption band have been allocated to satellite crosslink systems to provide protection from ground-based interference.[18] The sensitivity to inclement weather results from absorptive scattering from hydrometric particles; the principal concern is attenuation from rain. Systems that simultaneously use orthogonal polarization to increase the data rate must maintain isolation between the independent data streams; rain depolarization that reduces the isolation is particularly important in such systems. These propagation limitations are described in turn.

3.3.1 Molecular Absorption

The molecular absorption for frequencies less than 1000 GHz is caused principally by the absorption resonance of oxygen and water vapor.[19-21] The principal absorption frequencies of interest for SHF/EHF satellite communications are the 22- and 183-GHz absorption lines for water vapor and the 60-GHz line structure and the 118-GHz line for oxygen, as shown in Fig. 3.14. The water vapor absorption line at 22 GHz is relatively weak and

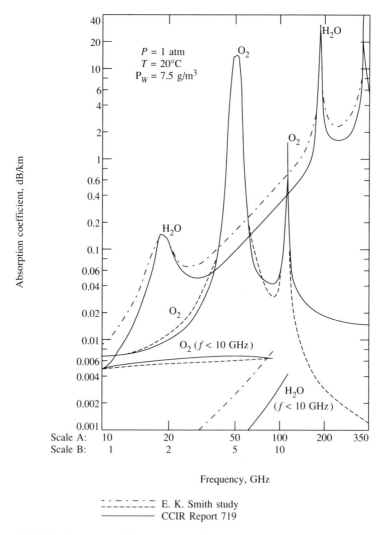

FIGURE 3.14   Specific attenuation for oxygen and water vapor absorption. (From E. K. Smith, "Centimeter and Millimeter Wave Attenuation and Brightness Temperature Due to Atmospheric Oxygen and Water Vapor," *Radio Sci.*, vol. 17, November–December 1982, © 1981 *Radio Science*.)

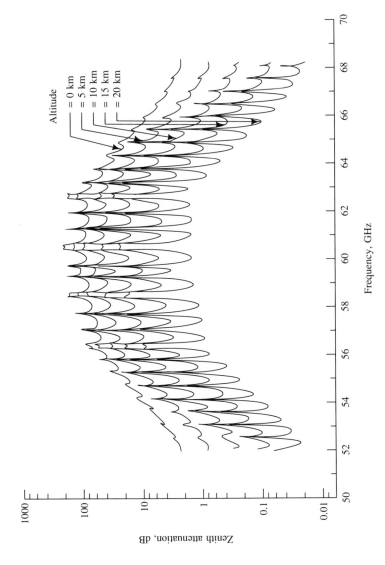

FIGURE 3.15 Variation of 60-GHz oxygen absorption with altitude. (From R. B. Dybdal and F. I. Shimabukuro, "Electronic Vulnerability of 60 GHz Crosslinks," *MILCOM 84 Conf. Rec.*, IEEE Military Communications Conference, Los Angeles, Calif., October 21–24, 1984. © 1984 IEEE.)

depends on the relative humidity along the path profile. The water vapor lines result from electric dipole transitions. The absorption spectra at 60 GHz contains a large number of resonant lines, which are associated with the spin states of the magnetic dipole moment of oxygen. This absorption spectra at 60 GHz persists over a wide frequency range. At sea level, the spectra is pressure-broadened so that a broad continuum of high-level absorption results. At high altitudes, both less oxygen is present and the effects of pressure-broadening decrease, and the individual resonant lines become distinct.

Computer programs have been developed to project the absorption characteristics over a wide range of frequencies. One of the most widely used programs has been developed by Liebe.[22] His programs cover the frequency range from 1 to 1000 GHz and permit the user to select the pressure, temperature, and humidity profiles as input data. At very high altitudes the earth's magnetic field interacts, producing an anisotropic medium, a factor treated in a second program that becomes important only near the resonance frequencies and at very high altitudes.

An example of the oxygen absorption spectra at 60 GHz is shown in Fig. 3.15, and was calculated using Liebe's program. This set of spectra illustrates the broad attenuation at sea level provided by oxygen absorption, and the decrease of the absorption at high altitudes to a set of discrete resonances. This example indicates the zenith attenuation values—i.e., the attenuation that would be experienced looking straight up into empty space. The program uses a standard atmospheric model and treats the atmosphere as a sequence of spherical shells. Thus, when absorption for paths other than zenith is required, the zenith attenuation can be multiplied by $\csc \epsilon$, where ϵ is the elevation angle. This approximation is valid for elevation angles greater than $10°$. The program can also be used to calculate more precise path geometries.

3.3.2 Hydrometeor Attenuation and Depolarization

SHF/EHF satellite communications are also adversely affected by inclement weather, specifically weather resulting from absorptive scattering by hydrometeors. Hydrometeors are the nonvaporous water particles in the atmosphere, and include raindrops, ice crystals, hail, sleet, snow, clouds, and fog, which have a definite particle size. The most significant effects typically result from raindrops, because their size is larger than the water within clouds and fog. Liquid water has a much higher index of refraction than frozen water, so the absorption from ice crystals, snow, and dry hail is generally small.

The importance of hydrometeor attenuation and depolarization at SHF/EHF frequencies has resulted in extensive research during recent years. A significant experimental effort has resulted in a large data base, and statistical modeling efforts to predict the expected long-term outages at different geographical locations have been made. The expected long-term outages are generally used to determine appropriate margin requirements for communication links.

The initial measurements of rain attenuation resulted in the observation that the attenuation per unit distance along the path followed an aR^b relationship, where R is the rain rate and a and b are numerical coefficients that depend on the operating frequency and drop-size distributions. More recent examination[23] has developed the theoretical basis of this relationship from the absorptive scattering from raindrops and has obtained values of the numerical coefficients a and b, which are regression fits for a variety of drop-size distributions matched with experimental data. The results are presented in terms of attenuation per path length (dB/km), sometimes referred to as the specific attenuation, and rain rate (mm/h) for frequencies between 1 and 1000 GHz.

The actual rain attenuation depends on the specific attenuation of the frequency of operation, the rain rate, and the extent of the rain along the path. Additionally, the occurrence of rain must also be determined. Much recent research has been devoted to the statistical characterization of rain and its corresponding attenuation of EHF signals. A significant variation in these parameters can be anticipated, and the statistics inherently are long-term, multiyear values. Although such statistics have application to defining the margin requirements for a particular site that is to be operated for many years, inherent shortcomings exist. For example, determining the probability of link availability of a given site on next Tuesday afternoon has the same inherent shortcomings as weather forecasting. These statistical models are only stable when long-term (multiyear) periods are used.

Commonly used approaches[24−28] typically divide the earth into different climatic zones as shown in Fig. 3.16, assign rain rate statistics in each zone as shown in Fig. 3.17, and adjust the 0° isotherm height as a function of latitude such as in Fig. 3.18. At high altitudes, liquid water changes to ice, at a point referred to as the 0° isotherm height, above which little attenuation occurs. These models individually have some variation, and debates on portions of the models, such as rain rate statistics[29] as well as the complete models,[30] have been conducted and will continue in the future. When these models are applied, the basic tasks are to calculate the specific attenuation rate for the frequency used by the system, decide on an appropriate path length model and statistical distribution of the rain rate. The references provide guidance and comparisons with measurements that can be examined.

The prediction of rain attenuation involves determining the absorptive scattering by rain, the path extent of rain, and the statistical occurrence of rain. This process becomes progressively less precise as it evolves. The absorptive scattering by a given raindrop can be calculated with great precision, the distribution of drop sizes has some variation, the path length distribution of rain has still more variation, and the occurrence of rain can be highly variable.

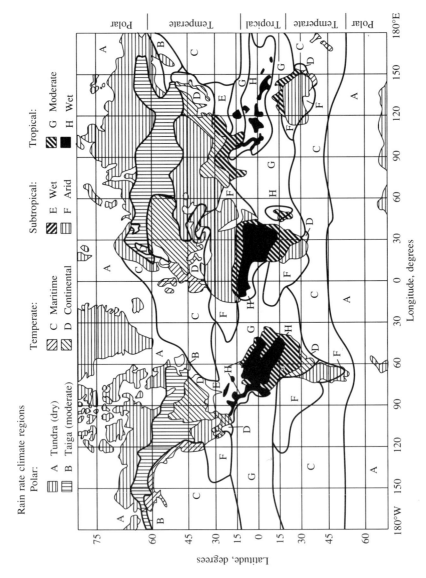

FIGURE 3.16 Example of global rain model. (From **R. K. Crane,** "Prediction of Attenuation by Rain," *IEEE Trans. Commun. Technol.,* vol. COM-28, September 1980. © 1980 IEEE.)

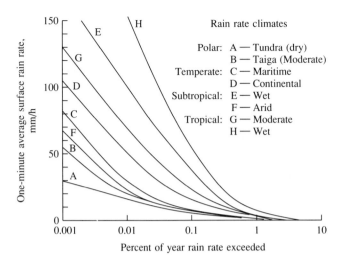

FIGURE 3.17 Example of rain rate statistics. (From R. K. Crane, "Prediction of Attenuation by Rain," *IEEE Trans. Commun. Technol.*, vol. COM-28, September 1980. © 1980 IEEE.)

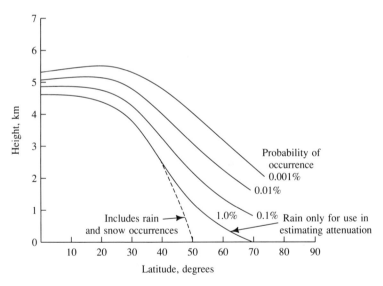

FIGURE 3.18 Example of 0° isotherm height statistics. (From R. K. Crane, "Prediction of Attenuation by Rain," *IEEE Trans. Commun. Technol.*, vol. COM-28, September 1980. © 1980 IEEE.)

The interest in this statistical projection of rain attenuation arises from the practical problem of determining an appropriate margin for rain outage. A basic decision must be reached prior to examining the statistical projections of rain attenuation. The availability of the system is a key parameter. The statistical projections result in cumulative values of attenuation; e.g., the attenuation is less than X dB for Y percent of the time. If the link is to be available Y percent of the time, then the rain margin is X dB. This procedure works well for reasonable values of the availability; e.g., 95 percent of the time. For very high availability requirements principally resulting from severe thunderstorms, the statistics themselves are not stable, and the required margins are sufficiently excessive to exceed the bounds of practical system design and affordable costs. Three courses of action are recommended in this case: (1) Reexamine the availability requirements and accept a lower availability corresponding to a reasonable margin; (2) if the availability is truly required, examine the possibility of site diversity,[31] which uses two ground terminals sufficiently separated to achieve a low mutual probability of simultaneously having high attenuation; and (3) reduce the data rate during periods of high attenuation to maintain adequate link performance.

The measurement of the total path attenuation can be performed by two different techniques, which have their individual advantages and limitations.[32] A number of measurements have been performed using actual satellite links. This technique has the advantages of providing attenuation measurements up to the dynamic range limitations of the link and the coherency to measure depolarization and dispersion. However, such measurements are inherently costly and at a particular site, data can only be obtained for a single elevation angle for a given geosynchronous satellite. The results of 19- and 28-GHz measurements using the COMSTAR satellite have been discussed in detail elsewhere.[33]

The second measurement technique uses a radiometric receiver to measure the variations in the sky temperature and infer the path attenuation from the measured sky temperature values. The radiometer is significantly less expensive than satellite measurements and, unlike a satellite link, measurements of different elevation angles can be made. However, the dynamic range of the attenuation measurements is limited and, since these measurements are not coherent, depolarization characteristics and dispersion measurements cannot be made. When systems with reasonable availability requirements are being planned, radiometric measurements performed over several years can yield an accurate projection of the statistics of rain attenuation at the specific site in a cost-effective manner.

The radiometric measurements will be described further because a second factor in link degradation is inherently related to this measurement technique. When attenuation occurs, noise is also produced. Loss in waveguide components produce a well-known contribution to the system noise temperature; loss in the propagation path produces similar contributions to the total system temperature. When a radiometric receiver is pointed toward the region of the

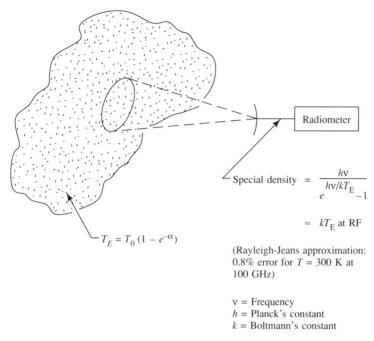

$$\text{Special density} = \frac{h\nu}{e^{h\nu/kT_E} - 1}$$

$$\approx kT_E \text{ at RF}$$

(Rayleigh-Jeans approximation:
0.8% error for $T = 300$ K at
100 GHz)

ν = Frequency
h = Planck's constant
k = Boltmann's constant

$T_E = T_0 (1 - e^{-\alpha})$

FIGURE 3.19   Radiometric measurement.

path containing loss as shown in Fig. 3.19, noise is received as described by the Rayleigh-Jeans approximation to Planck's law. The emission temperature for the loss is shown in Fig. 3.20 for three different values of the empirical constant T_0. The value of 290 K is the normal room temperature, 270 K is a typical path average value, and 250 K is typical in heavy rainfall and at higher EHF where energy is scattered out of the beam. This uncertainty in the value of the empirical value of T_0 results in the dynamic range limitations for the radiometric technique.

This same increase in the total system temperature also degrades the G/T of receiving ground terminal antennas. When rain is present, not only does the link suffer from attenuation in the propagation path, the system noise temperature is also increased from the thermal-noise radiation resulting from the attenuation. Low-noise receiver technology has made significant recent process; however, the thermal contributions from rain ultimately determine the effectiveness with which this technology can be exploited.

The radiometric receiver has been described for making attenuation measurements; moreover, its use with an operational ground terminal may have several benefits. A simple, low-cost way to incorporate a radiometer into a terminal that uses a frequency-hopped spread spectrum has been developed, as shown in Fig. 3-21.[34] The radiometer would provide a real-time indication of link availability, a means of transmitter power control which is particularly important for systems that use frequency division multiple access, sensing

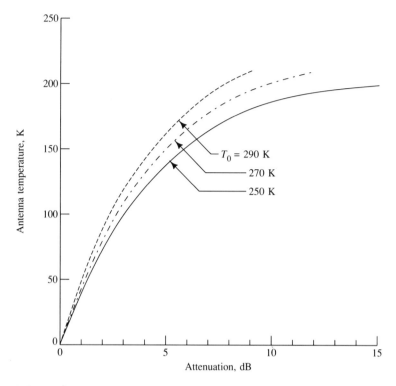

$T_0 = 290$ K

270 K

250 K

FIGURE 3.20   Attenuation versus emission temperature.

interference, and a diagnostic capability that complements the normal built-in test equipment (BITE).

Although attenuation caused by hydrometeors limits the performance of all EHF systems, those systems that simultaneously use orthogonal polarization become limited by another factor. Raindrops are not ideally spherical, so the polarization of the incident field is modified by the presence of rain. When rain is present in the propagation path, the scattering from the nonspherical raindrops leads to differential amplitude and phase components that depend on the incident polarization. The resulting polarization after the wave traverses the rain region differs from the incident polarization. In practice, the change in polarization is not significant enough to result in serious polarization efficiency losses; however, when two orthogonal polarizations are used, the resulting cross polarization reduces the isolation between channels.

The overall polarization isolation depends on the polarization purity of both the transmitting and receiving antennas[35] and the depolarization introduced by the propagation medium. The cross polarization induced by rain is commonly related to the rain attenuation.[36,37] The importance of depolarization depends on the margin for rain attenuation as well; when

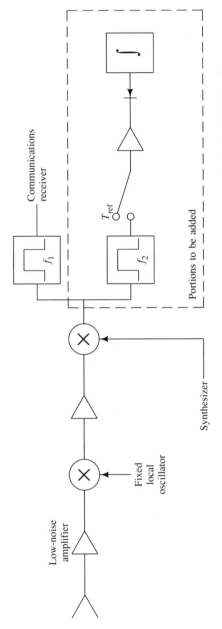

FIGURE 3.21 Radiometer integrated into frequency-hopped ground terminal. (From R. B. Dybdal, "Radiometer Integrated with Communication Terminal," *Proc. 1984 IEEE Antennas Propagat. Symp.*, Boston, Mass., June 25–29, 1984. © 1984 IEEE.)

the system uses a small rain margin, the corresponding depolarization is small. Thus, significant cross polarization is accompanied by significant rain attenuation.

The depolarization caused by rain depends on the frequency, the tilt angle, the incident polarization, and the rain attenuation. The polarization of the incident field is an important factor. Since the rain scattering produces both differential amplitude and phase components that lead to the loss in isolation, circular polarization has a greater sensitivity to loss in isolation than linear. Thus, the empirical relations between depolarization and attenuation distinguish linear and circular polarizations, and systems such as Intelsat VI that use simultaneous orthogonal polarizations use linear polarization. Measured data[33] have a significant spread in depolarization at low rain attenuation rates; this spread results from scattering from ice crystals that also contribute to depolarization.

This discussion has concentrated on the weather sensitivity within the propagation medium. An additional weather-sensitive factor in the link arises when radomes are used to protect the ground terminal antenna. When wet, the radome decreases in efficiency, and further depolarization may occur. When the water layer has a uniform thickness, its effect is similar to adding an additional dielectric layer to the radome surface, and hence the radome reflection loss increases. Similarly, when the water beads and flows in an irregular pattern referred to as rivulet flow, depolarization results. Specialized coatings for radomes have been developed to reduce radome sensitivity to rain; however, the effectiveness of these coatings can decrease with aging. Recent measurements[38,39] have been made with radome material that has been coated. Further development of radomes and their coatings can be anticipated to continue in future years.

REFERENCES

1. Joint Special Issue on the 1979 World Administrative Radio Conference (WARC-79), *IEEE Trans. Commun.*, vol. COM-29, August 1981.
2. S. H. Lee and R. C. Rudduck, "Aperture Integration and GTD Techniques Used in the NEC Reflector Code," *IEEE Trans. Antennas Propagat.*, vol. AP-33, p. 189, February 1985.
3. H. T., Friis, "A Note on a Simple Transmission Formula," *Proc. IRE*, pp. 254–256, May 1946.
4. R. B. Dybdal, "Millimeter Wave Antenna Technology," *IEEE Journal Selected Areas Commun.*, vol. SAC-1, pp. 633–644, September 1983.
5. R. B. Dybdal, "Multiple Beam Communication Satellite Antenna Systems," *Proc. IEEE Internat. Conf. Commun.*, Minneapolis, Minn., June 17–19, 1974, pp. 27D-1–5.
6. D. H. Martin, "Communication Satellites, 1958–1988," Los Angeles, Calif.: The Aerospace Corp., 1988.
7. R. T. Compton, Jr., *Adaptive Antennas: Concepts and Performance*, Englewood Cliffs, N.J.: Prentice-Hall, 1988.
8. K. M. SooHoo and R. B. Dybdal, "Resolution Performance of Adaptive Multiple Beam Antennas," paper for MILCOM 89.
9. J. T. Mayhan, "Nulling Limitations for a Multiple-Beam Antenna," *IEEE Trans. Antennas Propagat.*, vol. AP-24, pp. 769–779, November 1976.

10. J. T. Mayhan, "Some Techniques for Evaluating the Bandwidth Characteristics of Adaptive Nulling Antennas," *IEEE Trans. Antennas Propagat.*, vol. AP-27, May 1979.

11. J. T. Mayhan, A. J. Simmons, and W. C. Cummings, "Wideband Adaptive Antenna Nulling Using Tapped Delay Lines," *IEEE Trans. Antennas Propagat.*, vol. AP-29, November 1981.

12. J. T. Mayhan, "Thinned Array Configurations for Use with Satellite-Based Adaptive Antennas," *IEEE Trans. Antennas Propagat.*, vol. AP-28, November 1980.

13. J. T. Mayhan, "Area Coverage Adaptive Nulling From Geosynchronous Satellites: Phased Arrays Versus Multiple Beam Antennas," *IEEE Trans. Antennas Propagat.*, vol. AP-34, pp. 410–419, March 1986.

14. R. H. Ott and R. B. Dybdal, "The Effects of Reflector Antenna Diffraction on the Interference Cancellation Performance of Coherent Sidelobe Cancellers," *IEEE Trans. Antennas Propagat.*, vol. AP-34, pp. 432–439, March 1986.

15. K. Chang and C. Sun, "Millimeter-Wave Power-Combining Techniques," *IEEE Trans. Microwave Theory Tech.*, vol. MTT-31, pp. 91–107, February 1983.

16. Z. Galani, J. L. Lampen, and S. J. Temple, "Single Frequency Analysis of Radial and Planar Amplifier Combiner Circuits," *IEEE Trans. Microwave Theory Tech.*, vol. MTT-29, pp. 642–654, July 1981.

17. G. M. Shaw and R. B. Dybdal, "Beam Broadening for Active Aperture Antennas," IEEE Antennas and Propagation Symposium, 1989.

18. R. B. Dybdal and F. I. Shimabukuro, "Electronic Vulnerability of 60 GHz Crosslinks," *MILCOM 84 Conf. Rec.*, IEEE Military Communications Conference, Los Angeles, Calif., October 21–24, 1984.

19. E. E. Reber, R. L. Mitchell, and C. J. Carter, "Attenuation of the 5 mm Wavelength Band in a Variable Atmosphere," *IEEE Trans. Antennas Propagat.*, vol. AP-18, pp. 472–490, July 1970.

20. P. W. Rosenkrantz, "Shape of the 5 mm Oxygen Band in the Atmosphere," *IEEE Trans. Antennas and Propagat.*, vol. AP-23, pp. 498–506, 1975.

21. E. K. Smith, "Centimeter and Millimeter Wave Attenuation and Brightness Temperature Due to Atmospheric Oxygen and Water Vapor," *Radio Sci.*, vol. 17, pp. 1455–1464, November–December 1982.

22. H. J. Liebe, "Modeling Attenuation and Phase of Radio Waves in Air at Frequencies Below 1000 GHz," *Radio Sci.*, vol. 16, pp. 1183–1199, November–December 1981.

23. R. L. Olsen, D. V. Rodgers, and D. B. Hodge, "The aR^b Relation in the Calculation of Rain Attenuation," *IEEE Trans. Antennas Propagat.*, vol. AP-26, pp. 318–329. March 1978.

24. R. K. Crane, "Prediction of the Effects of Rain on Satellite Communication Systems," *Proc. IEEE*, vol. 65, pp. 456–474, March 1977.

25. R. K. Crane, "Prediction of Attenuation by Rain," *IEEE Trans. Commun. Technol.*, vol. COM-28, pp. 1717–1733, September 1980.

26. R. K. Crane, "A Two-Component Rain Model for the Prediction of Attenuation Statistics," *Radio Sci.*, vol. 17, pp. 1371–1387, November–December 1982.

27. R. R. Persinger, W. L. Stuzman, R. E. Castle, Jr., and C. W. Bostian, "Millimeter Wave Attenuation Prediction Using a Piecewise Uniform Rain Rate Model," *IEEE Trans. Antennas Propagat.*, vol. AP-28, pp. 149–153, March 1980.

28. W. L. Stutzman and W. K. Dishman, "A Simple Model for the Estimation of Rain-Induced Attenuation Along Earth-Space Paths at Millimeter Wavelengths," *Radio Sci.*, vol. 17, pp. 1465–1476, November–December 1982.

29. R. K. Crane, "Evaluation of Global and CCIR Models for Estimation of Rain Rate Statistics," *Radio Sci.*, vol. 20, pp. 865–879, July–August 1985.

30. R. K. Crane, "Comparative Evaluation of Several Rain Attenuation Prediction Models," *Radio Sci.*, vol. 20, pp. 843–863, July–August 1985.

31. D. B. Hodge, "An Improved Model for Diversity Gain on Earth-Space Propagation Paths," *Radio Sci.*, vol. 17, pp. 1393–1399, November–December 1982.

32. D. C. Hogg and T. S. Chu, "The Role of Rain in Satellite Communications," *Proc. IEEE*, vol. 63, pp. 1308–1331, September 1975.

33. D. C. Cox and H. W. Arnold, "Results from the 19 and 28 GHz COMSTAR Satellite Propagation Experiments at Crawford Hill, *Proc. IEEE*, vol. 70, pp. 458–488, May 1982.

34. R. B. Dybdal, "Radiometer Integrated with Communication Terminal," *Proc. 1984 IEEE Antennas Propagat. Symp.*, Boston, Mass., pp. 645–647, June 25–29, 1984.

35. S. I. Ghobrial, "Cross Polarization in Satellite and Earth Station Antennas, *Proc. IEEE*, vol. 65, pp. 378–387, March 1977.

36. T. S. Chu, "A Semi-Empirical Formula for Microwave Depolarization Versus Rain Attenuation on Earth-Space Paths," *IEEE Trans. Commun.*, vol. COM-30, pp. 2550–2554, December 1982.

37. D. C. Cox, "Depolarization of Radio Waves by Atmospheric Hydrometeors in Earth-Space Paths: A Review," *Radio Sci.*, vol. 16, pp. 781–812, September–October 1981.

38. F. J. Dietrich and D. B. West, "An Experimental Radome Panel Evaluation," *IEEE Trans. Antennas Propagat.*, vol. AP-36, pp. 1566–1570, November 1988.

39. C. E. Hendrix, J. E. McNalley, and R. A. Monzingo, "Depolarization and Attenuation Effects of Radomes at 20 GHz," *IEEE Trans. Antennas Propagat.*, vol. AP-37, pp. 320–328, March 1989.

S. HOVANESSIAN
and
NIRODE MOHANTY

4

· ·

Radar Systems

This chapter provides the basic principles and technology involved in the design and development of modern radar systems. Various types of radars— including those with low, medium, and high *pulse-repetition frequencies (PRF)*—together with their governing principles are discussed, and several numerical examples are given. During the past few years, important advances have been made in digital radar signal processing methods. Applications of these methods are presented in this chapter. The discussion of radar and signal processing principles is followed by the derivation of the radar signal-to-noise ratio equation and its application in the design of radar systems. Another area of current radar interest, electronic scan antennas, are also discussed in this chapter. The final section includes methods of calculation of the probability of detection of sinusoidal signals in noise and the computation of threshold settings and false alarm rates.

Radar is an acronym, coined in 1942 by the U.S. Navy, for radio detection and ranging. It is an electromagnetic device for detecting distant objects by sending known signals at them and analyzing the echoes. Radar signals have frequencies from the microwave region to the infrared and optical regions of the electromagnetic spectrum. Pulse-modulated magnetrons, klystrons, traveling-wave-tube amplifiers (TWTAs), or crossed-field amplifiers are used as transmitters, and the first stage of the receiver is a diode mixer. The antenna in the radar system is usually a parabolic-reflector type.

Pulse radars for measuring radar-target range have been in operation for more than 40 years. These radars measure range (distance between the radar and the target) by transmitting a pulse and measuring the elapsed time between the pulse and its return from the target. The elapsed time between transmitted and received pulses is proportional to the radar-target range. These radars, also known as low-PRF radars, have a PRF of anywhere between 30 and 4000 pulses per second (pps).

Doppler radars, on the other hand, use the Doppler (frequency) principle to measure radar-target range *rate*, rather than the range itself, as was the case in the low-PRF radars. The range rate is proportional to the frequency shift between the transmitted and the received signals. The radar-target range in Doppler radars is measured using specialized techniques such as frequency modulation of the carrier frequency or multiple pulse-repetition frequencies. Also known as high-PRF radars, Doppler radars have a characteristically high PRF, usually of the order of 65,000 to 300,000 pps. Doppler radars are more complex to mechanize, but they are characterized by good discrimination between moving and nonmoving (e.g., ground) target returns. These radars are capable of detecting targets that are masked by ground clutter, which consists of those radar signals that are reflected from the undesired ground scatterers.

In recent years the desirable features of low-PRF pulse radars (accurate range) and high-PRF pulse Doppler radars (moving-target indication) have been combined in medium-PRF radars. These radars use time to calculate radar-target range and use Doppler (frequency) information to indicate a moving target or calculate radar-target range rate. Medium-PRF radars use PRFs in the range of 10,000 to 30,000 pps.

4.1 PULSE RADARS

Pulse radars measure range (distance between the radar and the target) by transmitting a pulse and timing the returned pulse from the target, as shown in Fig. 4.1. Since electromagnetic energy travels at the speed of light c,

$$t = \frac{2R}{c} \quad \text{or} \quad R = \frac{ct}{2} \tag{4.1}$$

where R is the radar-target range, t is the time between transmit and target return signals, and $c = 3 \times 10^8$ m/s. The angular target information (azimuth and elevation of the target with respect to the radar) is obtained from the antenna position data at the time of target detection. With the values of these angles and range R, it is possible to establish the position of the target with respect to the radar.

In the pulse radar of Fig. 4.1, the maximum measurable range R_{max} is a function of T, the interpulse period, and is given by

$$R_{max} = \frac{cT}{2} \tag{4.2}$$

Beyond this range, the elapsed time between transmitted pulse and target return will include multiples of interpulse period T. For this reason, one

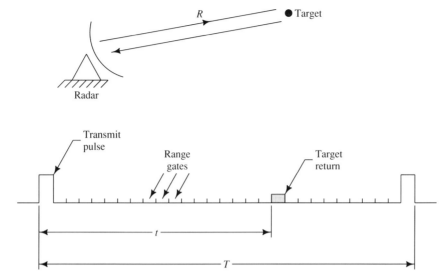

FIGURE 4.1   Pulse radar range measurement.

cannot determine correctly the value of elapsed time between transmitted and received pulses, making the measured range ambiguous.

Note that the transmitted pulse of Fig. 4.1 is generated by pulsing a pure sine wave, as illustrated in Fig. 4.2. The frequency of the sine wave f_0 is usually in the thousands of megahertz; for example, in the X band it can range from 8 GHz to 12 GHz.

The elapsed time t between transmit and receive pulses is measured by placing "range gates" all along the receive time, as shown in Fig. 4.1. The consecutive range gates open for the duration of a single pulse starting with each transmitted pulse. The presence of a signal in these gates corresponds to

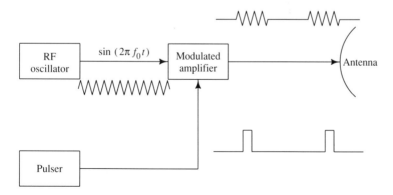

FIGURE 4.2   Generation of transmitted pulses from a sine wave.

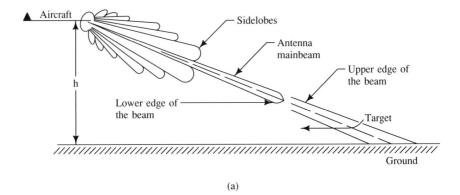

(a)

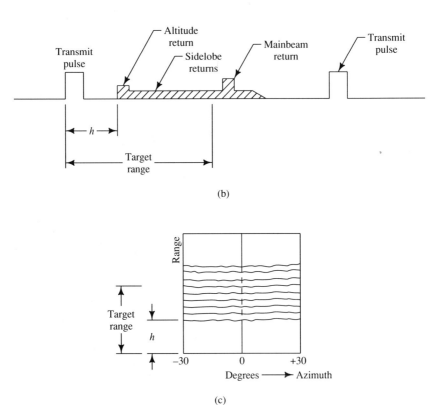

(b)

(c)

FIGURE 4.3   Pulse radar detection of ground targets.

the elapsed time t, which is used in Eq. (4.1) to compute the radar-target range. *In radars using a single antenna for transmit and receive functions, the receiver is turned off when the transmitter is on; that is, target returned pulses cannot be accepted during transmitter on-times.*

4.1.1 Discussion of Pulse Radars

In airborne applications, pulse radars encounter difficulty in detecting targets under "lookdown" conditions. Figure 4.3 presents the encounter geometry and the detection process under these conditions. Figure 4.3a shows the radar-target geometry, while Fig. 4.3b and 4.3c show the return pulse measurement and the range-azimuth display, respectively. From Fig. 4.3b note that ground returns emanate from ranges equal to aircraft altitude to ranges exceeding the slant range to the ground along the main beam. Most of these returns are due to antenna sidelobes. Nevertheless, they interfere with the detection process, making it difficult to detect targets under lookdown conditions. The radar display characteristics are shown in Fig. 4.3c. This figure depicts a cluttered display resulting from ground returns.

The desire for reliable lookdown detection capability together with the requirements of increased target detection range led radar engineers to the design of Doppler radars.

4.2 PULSE DOPPLER RADARS

To increase the maximum detection range of a pulse radar, one needs to increase both the transmitted power per pulse and the interpulse period that determines the unambiguous radar-target range. Although the interpulse period can be increased, there are hardware limitations to increasing the transmitted power, especially in airborne radars where severe weight and volume constraints exist.

To increase the amount of transmitted energy and thereby the detection range, as well as to detect moving targets in the presence of ground clutter, pulse Doppler radars were invented. These radars operate on the Doppler principle first observed by Christian Doppler early in the nineteenth century. Figure 4.4 illustrates this principle, which is based primarily on the frequency change caused by motion. Figure 4.4a shows the Doppler effect on a non-moving target, which results in *no* frequency change between transmit and receive waveforms. Figure 4.4b shows a closing target ($\dot{R} < 0$), which results in an increased frequency as compared to the transmitted frequency. Figure 4c illustrates the conditions for opening targets ($\dot{R} > 0$), where a decrease of frequency will be observed.

Using the Doppler effect rather than timing transmit-receive pulses allows one to use a very high PRF to increase the transmitted energy, as shown in

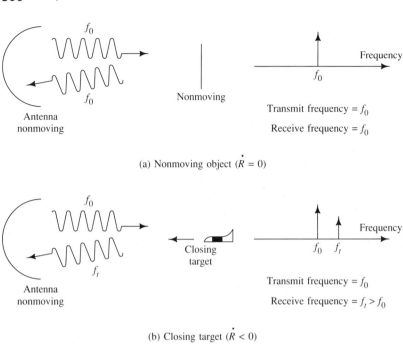

(a) Nonmoving object ($\dot{R} = 0$)

(b) Closing target ($\dot{R} < 0$)

(c) Opening target ($\dot{R} > 0$)

FIGURE 4.4   Doppler effect on targets.

Fig. 4.5. This high PRF, however, eliminates the capability of the radar to obtain a radar-target range by timing transmit and target return pulses. High-PRF radars use the Doppler principle, which shows range rate $\dot{R}$ as a function of transmit-receive frequency shift, given by the equation

$$\Delta f = \frac{-2\dot{R}}{\lambda} \tag{4.3}$$

where Δf is the difference between transmitted frequency and the target return frequency and λ is the wavelength. The negative sign of Eq. (4.3) is used to show that approaching targets (decreasing range or negative range rate

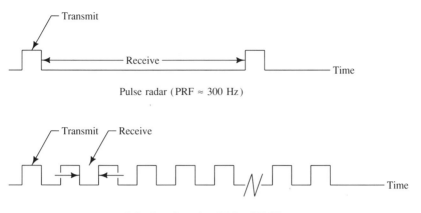

FIGURE 4.5   Pulse and pulse Doppler radars.

$\dot{R}$) produce positive frequency shifts Δf. Doppler radars are mechanized to measure $\dot{R}$ from the measured frequency shift Δf. Special techniques are used in these radars to measure radar target range R.

4.2.1 Range Measurement

The increased number of pulses in high-PRF radars destroys the ability of the radar to determine radar-target range by simple time-discrimination ranging. To obtain radar-target range in high-PRF radars, one may apply either frequency modulation of the carrier frequency or multiple-PRF ranging methods. Frequency-modulation ranging methods are based on frequency shifts of the returned signal with respect to the transmitted signals. This frequency shift is translated into a time delay that corresponds to the radar-target range. Multiple-PRF ranging methods, on the other hand, utilize time as the basis of measurements—similar to low-PRF radars. The operating principles of the two ranging schemes are described below.

Frequency-Modulation Ranging

With this method, the transmitted frequency is increased and decreased periodically. The frequency difference between the transmitted and received signals contains information from which the values of range R and range rate $\dot{R}$ can be derived. Figure 4.6 gives the relationship of the transmitted and received signal in a *linear* frequency-modulation ranging scheme. Note that the received signal is delayed because of radar-to-target-to-radar round-trip distance and frequency-shifted due to the radar-target closing rate. The applicable equations are as follows:

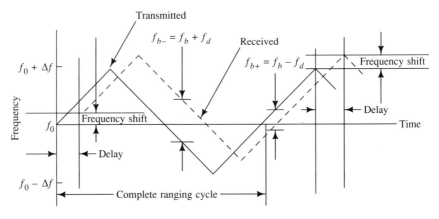

FIGURE 4.6   Linear modulation to obtain range and range values

$$f_{b+} = \frac{2R\dot{f}}{c} + \frac{2\dot{R}}{\lambda} \triangleq f_b + f_d$$

$$f_{b-} = \frac{2R\dot{f}}{c} - \frac{2\dot{R}}{\lambda} \triangleq f_b - f_d$$

(4.4)

where $\dot{f}$ is the modulation slope and R is the radar-target range.

Adding the above equations, one obtains the expression for radar-target range R

$$R = \frac{c(f_{b+} + f_{b-})}{4\dot{f}}$$

(4.5)

Subtracting Eqs. (4.4), one obtains range rate $\dot{R}$

$$\dot{R} = -\frac{\lambda(f_{b-} - f_{b+})}{4}$$

(4.6)

The frequency differences f_{b-} and f_{b+} of Eqs. (4.5) and (4.6) can be measured as the difference between target return frequency and transmitted frequency. Note that one complete ranging cycle should occur during the target illumination period to solve Eqs. (4.5) and (4.6) for R and $\dot{R}$ values. The target illumination period, depending on the antenna sweep rate and the antenna beamwidth, is usually of the order of 10 to 50 ms.

EXAMPLE.   Consider an airborne radar using a transmit frequency of $f_0 = 9000$ MHz with the frequency-modulation ranging method of Fig. 4.6. The complete ranging cycle is 20 ms and $\Delta f = 20$ kHz. For an f_{b-} of 6 kHz and an f_{b+} of 2 kHz, obtain the values of target radar range R and range range $\dot{R}$.

The slope $\dot{f}$ can be calculated as

$$f = \frac{\Delta f}{(T/4)} = \frac{20 \times 10^3}{(20 \times 10^{-3}/4)} = 4 \times 10^6 \text{ Hz/s}$$

where, from Fig. 4.6, $T/4$ is the corresponding time period for a Δf frequency change. The wavelength λ is equal to

$$\lambda = \frac{c}{f_0} = \frac{3 \times 10^{10}}{9 \times 10^9} = 3.333 \text{ cm}$$

Using Eqs. (4.5) and (4.6), together with the above values, we obtain

$$R = \frac{3 \times 10^{10}(2 \times 10^3 + 6 \times 10^3)}{4 \times 4 \times 10^6} = 1.5 \times 10^7 \text{ cm}$$

$$= 150 \text{ km}$$

$$\dot{R} = -\frac{3.333\,(6 \times 10^3 - 2 \times 10^3)}{4} = -3.333 \times 10^3 \text{ cm/s}$$

$$= -33.33 \text{ m/s}$$

Multiple-PRF Ranging

In radars using time discrimination to calculate range, the pulse-repetition frequency PRF determines the maximum unambiguous range that can be achieved by the radar. Figure 4.7 shows radar returns for two targets, which are widely separated in range, being detected by a pulse radar. If the target returns were not labeled, there would be no way of telling which transmitted pulse caused which target return. To overcome this difficulty, the method of

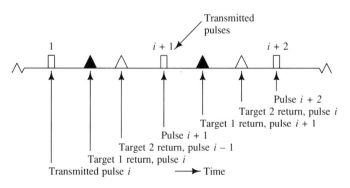

FIGURE 4.7   Transmit and receiver pulses

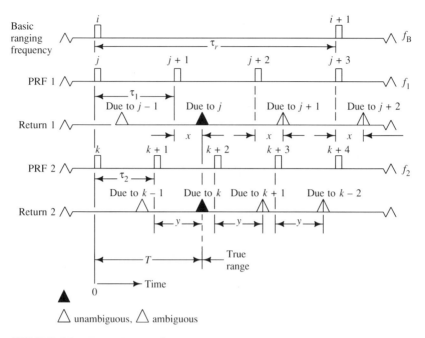

FIGURE 4.8   Two PRF ranging system

multiple-PRF ranging has been used successfully. The principles involved in multiple-PRF methods are illustrated in Fig. 4.8, where the basic ranging frequency with an interpulse period of τ_r and the two PRFs of periods $\tau_r/3$ and $\tau_r/4$ are shown. Target returns of each PRF are illustrated by triangles. Note that if timed from the beginning of each PRF transmission, only the return time T corresponding to the true target range will match for both PRFs.

In actual mechanization, during a single target illumination period, PRF 1 is transmitted first. The values of x in Fig. 4.8, representing the time between the last transmitted pulse and target return, are obtained. Following this, PRF 2 is transmitted and the value of y is obtained. With the values of x and y, the radar-target range can be determined. Thus, in the multiple-PRF mechanization, a great number of target return pulses are utilized in measurements of x and y, as compared to using the basic PRF where only a single returned pulse would be utilized.

Theoretically, a large number of PRFs can be used to approach the unambiguous range of the basic PRF of a multiple-PRF ranging system. However, in practice, two or three PRFs are sufficient to resolve most amgibuous range problems.

To construct an effective two-PRF ranging system[2] we may start by selecting a basic PRF, f_B, and an integer N (see Fig. 4.8). The two ranging PRFs can then be calculated, for example, by

$$f_1 = Nf_B$$

$$f_2 = (N + 1)f_B$$

(4.7)

Note that the multipliers N and $N + 1$ are relatively prime numbers in that they do not share a common factor.

Using the notation of Fig. 4.8, we can write the expression for the time elapsed between transmission and the true target return. Designating this time by T, we can express it in terms of parameters of PRFs 1 and 2 as follows:

$$T = (n_1 + x)\tau_1 = (n_1 + x)\frac{1}{f_1}$$

(4.8a)

$$= (n_2 + y)\tau_2 = (n_2 + y)\frac{1}{f_2}$$

(4.8b)

where n_1 and n_2 are integers representing the number of elapsed pulses between transmission and target return for PRFs 1 and 2, respectively, x and y are the corresponding fractions. Note that the radar system outputs only the values x and y. These values are usually in terms of range cell numbers (or range bins).

Given the PRFs f_1 and f_2 and the values of x and y from the radar, and using equations derived from Eq. (4.8), it is possible to calculate T of Eq. (4.8) and subsequently the radar-target range from $R = cT/2$. By using multiple-PRF ranging, in effect we increase the unambiguous range from

$$R_{ua_1} = \frac{c\tau_1}{2} = \frac{c}{2f_1}$$

$$R_{ua_2} = \frac{c\tau_2}{2} = \frac{c}{2f_2}$$

(4.9)

corresponding to multiple PRFs f_1 and f_2 of Fig. 4.8 to

$$R_{ua} = \frac{c\tau_r}{2} = \frac{c}{2f_B}$$

(4.10)

corresponding to the basic PRF f_B. The advantage of multiple-PRF ranging is the increased number of returned pulses and the resulting improvement in signal-to-noise ratio.

Additionally, the multiple-PRF method of range calculation allows us to use a high-PRF radar and calculate range in addition to range rate values, or moving target indication, which are available from Doppler information.

EXAMPLE. As an example of multiple-PRF ranging, consider a radar with a maximum unambiguous range of approximately 100 nautical miles (nmi). For $N = 79$, calculate the two PRFs f_1 and f_2, the corresponding unambiguous ranges, and the values of x and y that will be obtained from the radar if the target is at a range of 53 nmi.

The basic ranging PRF f_B is calculated from

$$f_B = \frac{c}{2R} = \frac{161,875}{2 \times 100} \approx 810 \text{ Hz}$$

where 161,875 is the velocity of light in nautical miles. From Eq. (4.7), the two PRFs are

$$f_1 = (79)(810) = 63.990 \text{ kHz}$$

$$f_2 = (80)(810) = 64.800 \text{ kHz}$$

The corresponding unambiguous ranges, from Eq. (4.9), are

$$R_{ua_1} = \frac{c}{2f_1} = \frac{161,875}{2 \times 63.990 \times 10^3} = 1.2648 \text{ nmi}$$

$$R_{ua_2} = \frac{c}{2f_2} = \frac{161,875}{2 \times 64.800 \times 10^3} = 1.2490 \text{ nmi}$$

For a radar target range of 53 nmi, the elapsed time between the transmit and receive pulse will be

$$T = \frac{2 \times 53}{161,875} = 654.8 \times 10^{-6} \text{ s}$$

Using this, we can calculate $(n_1 + x)$ and $(n_2 + y)$ from Eqs. (4.8a) and (4.8b) as follows:

$$n_1 + x = Tf_1 = 41.902$$

$$n_2 + y = Tf_2 = 42.432$$

The values of x and y will be the decimal fraction of the above numbers; that is,

$$x = 0.902$$

$$y = 0.432$$

These values correspond to the number of the "range gates" where the target return pulse is timed divided by the total number of range gates. Note that by using multiple PRFs, the unambiguous range increased to 100 nmi from about 1.25 nmi, given above.

4.2.2 Detection in Clutter and Ground Returns

In addition to the higher number of transmitted pulses, which results in longer detection ranges, Doppler radars obtain very accurate range rate $\dot{R}$ values as their primary measurement. This characteristic is exploited in separating ground clutter (nonmoving targets) from actual moving-target returns. This feature is very desirable, especially in airborne radars, where ground clutter interferes with actual target detection (see Fig. 4.3). Figure 4.9 shows an

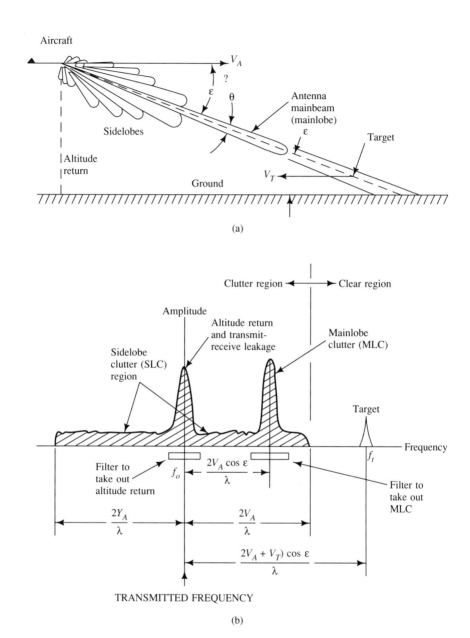

FIGURE 4.9 Amplitude-frequency characteristics of a Doppler radar. (a) Encounter geometry. (b) Amplitude-frequency diagram.

aircraft/target encounter geometry, together with the corresponding ampli-
tude/frequency diagram of the returned target signal using a Doppler radar.
Note from this figure that the relative velocity, with respect to the aircraft, of
stationary objects (ground) are smaller than those of moving targets. Thus,
all ground clutter return of Fig. 4.9b occurs at velocities below those of the
aircraft. In Fig. 4.9b, the mainlobe clutter return, which results from the
interception of the antenna mainbeam and the ground, is considerably larger
than other returns. The altitude return at frequency f_0 also accounts for
considerable power return. In actual radar mechanizations, mainlobe clutter
and altitude returns are usually taken out by placing filters in proper locations
(see Fig. 4.9b).

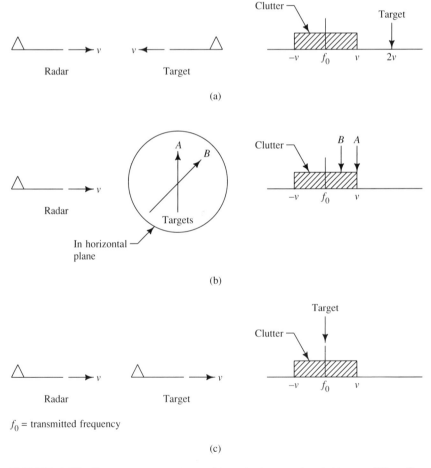

FIGURE 4.10   Encounter geometry and target return under clutter conditions for
a Doppler radar. (a) Head-on (nose). (b) Beam. (c) Tail.

Since the Doppler shift is proportional to the velocity, the ground return (clutter region) can be separated from the regions where target returns are expected. Figure 4.10 summarizes target detection capability of Doppler radars under several encounter geometry conditions. These figures are shown for "lookdown" geometries, where the radar mainbeam intercepts the ground. Under "lookup" conditions, where the radar beam does not intercept the ground, ground clutter is less and targets can be detected more easily.

4.3 COMPARISON OF PULSE RADARS AND PULSE DOPPLER RADARS

The general background of pulse radars and pulse Doppler radars, presented above, allows the comparison shown in Table 4.1. As developed in the preceding discussion, note that Doppler radars exploit the Doppler effect, resulting in range rate as the primary measured parameter. On the other hand, pulse radars use time discrimination between transmitted and received signals, resulting in range as the primary measured parameter. The ground clutter discrimination is accomplished more efficiently in high-PRF Doppler radars. The average transmitted power of high-PRF radars is considerably higher than that of the corresponding low PRF radars under the same peak power restriction.

The advantages of high-PRF Doppler radars (as described above) are obtained at the expense of more complex mechanization than is required by pulse radars. One area of difficulty is the frequency stability of the transmitted signal and the subsequent signal processing mechanization to maintain the frequency shift achieved between transmit and receive frequencies. This signal processing should be accomplished during the frequency acquisition.

Under the lookdown clutter conditions of Fig. 4.10, head-on closing targets (Fig. 4.10a) can be detected easily because their frequency shifts are beyond the frequency shift of the clutter region. The Doppler radar, however, will have difficulty detecting beam targets, as in Fig. 4.10b, because these targets are in the mainlobe clutter frequency region. The target return in this

TABLE 4.1 Comparison of High-PRF Doppler Radar and Low-PRF Pulse Radar

Item	High-PRF Doppler Radar	Low-PRF Pulse Radar
Average power	High	Low
Range rate measurement	Primary	Secondary
Range measurement	Secondary	Primary
Range determination	Sophisticated	Simple
General system characteristics	Complex	Simple
Detection in clutter	Excellent	Fair

case will be competing with the stationary ground ahead of the aircraft. The tail geometries of Fig. 4.10c will also produce difficult-to-detect targets under lookdown conditions. With a closing rate of zero ($\dot{R} = 0$), the targets will appear at the transmitted frequency f_0. The target return in this case is competing with the ground return directly below the aircraft (altitude return) and the transmitter noise.

Pulse radars, on the other hand, will have difficulty detecting targets under all lookdown conditions such as those illustrated in Fig. 4.10. These radars, however, show a better performance for coaltitude beam and tail targets (Figs. 4.10b and 4.10c) than the Doppler radar, which loses the target completely in clutter. In addition, pulse radars are used to measure range to ground and ground mapping. For these reasons, modern radars often are built with *both* pulse and pulse Doppler capabilities.

Under tactical conditions the Doppler mode is used for long-range detection and tracking of incoming targets at all altitudes, whereas the pulse mode is used for the coaltitude or lookup beam and tail geometries similar to those of Figs. 4.10b and 4.10c.

4.3.1 Frequency-Domain Representations

The general frequency-time characteristics of various radars are shown in Figs. 4.11a–4.11d. In the case of continuous-wave (CW) radar of Figure 4.11a, the time-domain representation shows a continuous sine wave, $\sin 2\pi f_0 t$, representing the transmitted waveform. In the frequency domain, the transmitted frequency is represented by a single carrier frequency f_0. The coherent pulse radar time domain illustrated in Fig. 4.11b shows a train of pulses that are cut from a continuous sine wave. Note that the phase relationship between pulses is maintained. The length of a given pulse is labeled t, and the period between consecutive pulses is labeled T. The reciprocal of this pulse period ($1/T$) is the PRF, f_P. The frequency-domain representation of the pulse train of Fig. 4.11b will consist of several distinct frequency lines at a "distance" f_P apart, as shown. Figure 4.11c gives a similar diagram to Fig. 4.11b, but this time for a high-PRF radar with a larger f_P. Here the "distance" between frequency lines is greater, as shown in the corresponding frequency diagram.

Figure 4.11a describes a low-PRF, noncoherent pulse radar in time and frequency diagrams. Note that the phase relationship of the transmitted waveform is *not* maintained. Frequency representation, therefore, is not in distinct lines as given in the coherent case of Figs. 4.11b and 4.11c. The noncoherent waveform of Fig. 4.11a represents, in effect, a "smearing" of frequencies over the frequency region rather than distinct frequency lines. Since high- and medium-PRF radars use frequency separation in the measurement of range and/or range rate, the coherency of the transmitted waveform is essential to their operation. In low-PRF radars, where timing is used for radar-target range calculation, the returned target pulses can be integrated

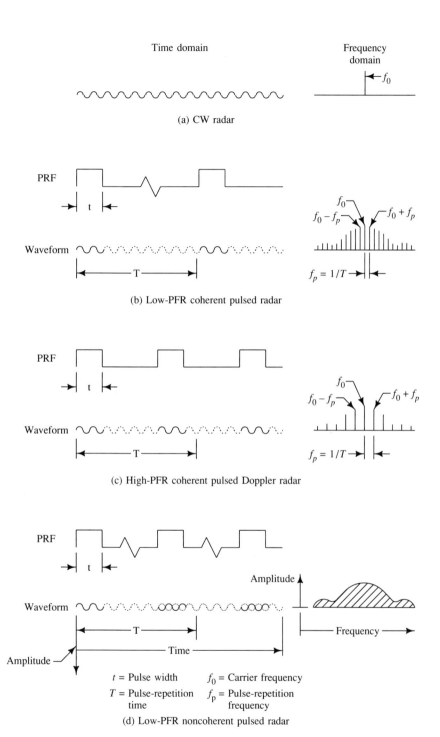

FIGURE 4.11 Types of radar transmissions.

371

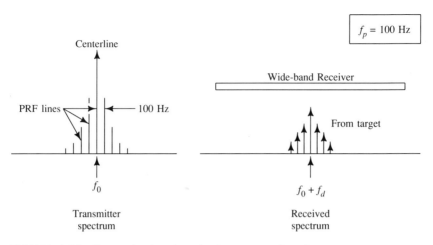

FIGURE 4.12   Transmitted and received spectrum of a coherent low-PRF radar.

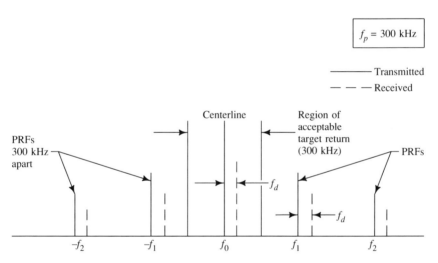

FIGURE 4.13   Transmitted and received spectrum of a high-PRF radar.

noncoherently (postdetection integration); therefore, coherency of the transmitted waveform is not essential to the operation of low-PRF radars.

Figures 4.12 and 4.13 show expanded versions of Figs. 4.11b and 4.11c for a low-PRF radar with a PRF of 100 Hz and a high-PRF radar with a PRF of 300 kHz. Note that in low-PRF radar, target return signals over the entire frequency band are processed in the receiver. In the high-PRF radar of Fig. 4.11c, only the target return due to the centerline frequency at f_0 is processed, and the rest of the returns are filtered out.

The frequency-domain representation of Fig. 4.11c under "lookdown" conditions where ground clutter is present is shown in Figure 4.11d. This figure is drawn to point out the fact that all the PRF lines also receive their own clutter and target returns, although they are filtered out before entering the receiver. The approximate frequency difference of various returns at X-band frequencies is also shown in Figure 4.11.

4.4 MEDIUM-PRF RADARS

Medium-PRF radars are constructed utilizing both time discrimination for ranging and frequency discrimination for moving-target indication or range rate calculation. Medium-PRF radars usually are built in the PRF range of 10 to 30 kHz.[3] In the following discussion, coherency of the transmitted signal is assumed.

Consider a medium-PRF radar with a PRF of 15 kHz. This PRF is selected so as to put the *mainlobe clutter* (MLC) return of Fig. 4.14 at the PRF line. The frequency spectrum of this medium-PRF radar is shown in Fig. 4.15, as is the target return, which is repeated for each PRF line. Note also that the target return competes with sidelobe clutter for detection when the radar is at low altitude and with system noise when the radar is at high altitude. Thus, in order to increase target detection capability in medium-PRF radars, it is crucial to reduce sidelobe clutter levels. These levels can be effectively reduced by reducing antenna sidelobes.

Limiting the area of detection to the frequency range between the centerline and the first PRF line, we may place detection filters as shown in Fig. 4.15. An examination of Fig. 4.15 reveals frequency ambiguity much like the

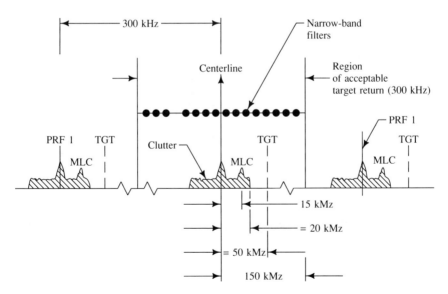

FIGURE 4.14 Transmitted and received spectrum of a high-PRF pulse Doppler radar including clutter return.

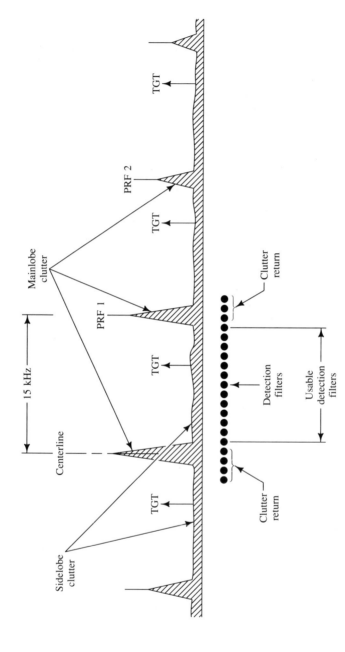

FIGURE 4.15 Medium-PRF radar frequency spectrum in presence of ground clutter.

range ambiguity discussed earlier. We cannot be sure whether the target return shown in the "usable detection area" is due to the centerline or one or more PRFs before the centerline. To resolve this frequency ambiguity, one may utilize PRF switching similar in principle to the previously discussed multiple-PRF ranging method.

Consider the case in which we use PRFs of 15 and 12 kHz. The frequency spectrum from these PRFs is illustrated in Fig. 4.16. The location of 1-kHz detection filters is shown by dots. A target with a Doppler shift of 22 kHz can be seen on the eleventh filter with the 15-kHz PRF and on the fourteenth filter with the 12-kHz PRF. Knowing the location of these filters and having the PRF values of 15 and 12, one can resolve the frequency ambiguity and

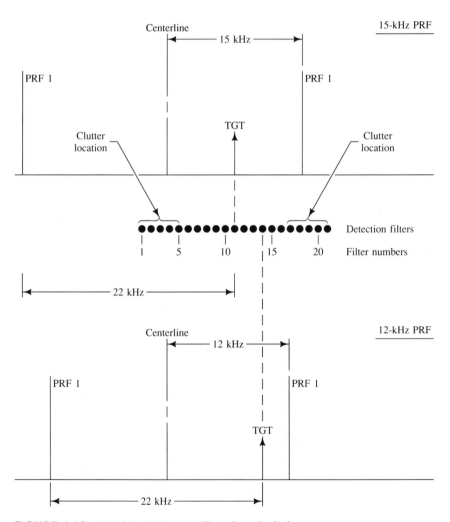

FIGURE 4.16   Multiple-PRF target Doppler calculation.

calculate radar-target range rate. The PRF selection to resolve Doppler shift ambiguity is made on the basis of usable detection area. This selection also includes a consideration of the location of clutter and the fact that targets in the clutter filters will not be detected. It is sometimes necessary to go to many PRFs to ensure that targets will be detected, in the usable detection filters of Fig. 4.15, in at least enough of these PRFs to permit resolution of true velocity.

In the medium-PRF radars, the radar-target range is usually calculated by using the multiple-PRF ranging methods described previously. The PRFs used for multiple ranging are usually not the same as those used for frequency ambiguity resolution, since the criteria of selection are different. In these radars, two sets of PRF may be used—one for range calculation and one for range rate calculation.

The selection of PRFs in a medium-PRF radar can also be accomplished through a computer simulation of several sets of candidate PRFs. In this method, several sets of range and Doppler PRFs are selected, and each set is tested by a complete simulation for clear range and velocity regime. Targets are detected in the clear regime and are not detected in the "blind zones." Figure 4.17 is a range-Doppler plot showing typical simulation outputs. This figure shows "blind zones" as well as clear regions. Range blind zones occur when target return pulses are eclipsed by transmitted pulses; that is, target return pulses arrive at transmitter on-times. Doppler blind zones occur when Doppler frequency shifts of the target returns coincide with mainlobe clutter or altitude returns.

The above-described method can be used when independent measurements of range R and range rate $\dot{R}$ are desired. Note that range rate $\dot{R}$ can also be obtained by differentiating range R; this $\dot{R}$ information can be used in the selection of Doppler PRFs and the resolution of Doppler ambiguity.

EXAMPLE.   As an example of a medium-PRF radar, let us assume a basic frequency resolution PRF of 15 kHz. Assume that filters are placed every 1000 Hz and that the mainlobe clutter and altitude return take up a 3000-Hz region (three filters). To ensure that targets will not fall in the clutter filters, we may select two additional PRFs 3 kHz apart—i.e., 12 and 18 kHz (see Fig. 4.18). From this figure we note that MLC is displaced 15 kHz from the centerline and the PRF lines. Also note that a target with a Doppler shift of 30 kHz will fall in the clutter filters of the 15-kHz PRF. This target, however, will be in the clear region of the 12 and 18 kHz PRFs. Note that the location of MLC is known by the data processing computer, and the input from corresponding clutter filters can be ignored.

To calculate the Doppler shift of the target, we proceed as follows: In the 12-kHz PRF, the target will fall in filter number 6; in the 18-kHz PRF, it will fall in filter number 12 (see Fig. 4.18). To obtain the correct

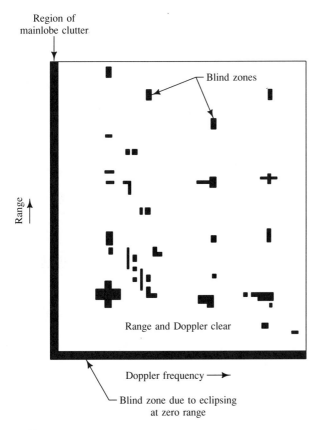

Region of
mainlobe clutter

Blind zones

Range

Doppler frequency →

Range and Doppler clear

Blind zone due to eclipsing
at zero range

FIGURE 4.17   Selection of PRFs using clear range and Doppler regions.

target Doppler shift, we add the corresponding PRF increments to the filter numbers and seek a match between resulting values; that is,

PRF	Doppler Shift
12 kHz	$6 + 12 + 12 = 30$
18 kHz	$12 + 18 = 30$

Thus the ambiguity is resolved; the correct target Doppler shift is 30 kHz.

Ranging PRFs are selected by using the criteria described in Eq. (4.7). For example, taking the basic PRF ($f_B = 810$ Hz) for the 100-nmi unambiguous range and selecting $N = 17$, we obtain

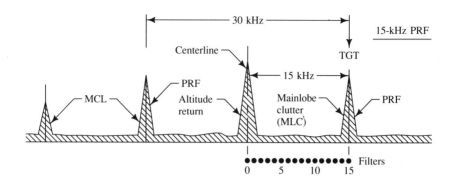

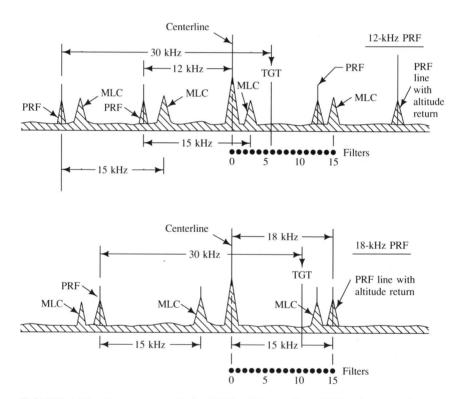

FIGURE 4.18 Frequency resolution PRFs of the medium-PRF radar example.

$$f_1 = (17)(810) = 13.770 \text{ kHz}$$

$$f_2 = (18)(810) = 14.580 \text{ kHz}$$

Using these PRFs, assume that we have measured an $x = 0.508$ and a $y = 0.832$ [Eq. (4.8)] from the radar. We wish to calculate the corresponding radar-target range.

The solution can be obtained by matching the time (distance) between transmit and receive pulses from each of the two PRFs [Eq. (4.8)]. The unambiguous range (equivalent distance between two consecutive pulses) can be computed for each PRF as follows:

$$R_{ua_1} = \frac{c}{2f_1} = \frac{3 \times 10^5}{2 \times 13770} = 10.893 \text{ km}$$

$$R_{ua_2} = \frac{c}{2f_2} = \frac{3 \times 10^5}{2 \times 14580} = 10.288 \text{ km}$$

Using this and the values of x and y, we can write the radar-target range

Range $= (0.508)(10.893) + n(10.893)$ PRF 1

Range $= (0.832)(10.288) + n'(10.288)$ PRF 2

when n and n' are integers. Computing range for values of n and $n' = 1$, 2, ..., using two decimal places, we obtain

$R = 16.43, 27.32, 38.21, 49.11, 60.00, 70.89, 81.78, 92.68, \ldots$ PRF 1

$R = 18.88, 29.14, 39.42, 49.71, 60.00, 70.29, 80.58, 90.86, \ldots$ PRF 2

We note a match between range values calculated for PRFs 1 and 2 at 60 km; therefore, 60 km represents the measured radar-target range.

Thus two ranging PRFs, together with frequency resolution PRFs, will result in radar-target range and also *moving-target indication* or radar-target range rate.

4.5 THE AMBIGUITY FUNCTION

Let the transmitted signal be

$$S(t) = u(t)e^{i2\pi f_c t} \quad 0 \le t \le T$$

$$u(t) = a(t)\exp[i\phi(t)]$$

(4.11a)

where $u(t)$ is the complex envelope and f_c the carrier frequency. Let the returned signal be

$$r(t) = u(t + \tau)e^{i2\pi(f_c - f_d)(t+\tau)}$$

(4.11b)

where τ is the time delay and f_d the Doppler shift. The output of the matched filter is

$$Y = \int_0^T r^*(t)s(t)\,dt$$

$$= e^{-i2\pi(f_c-f_d)\tau}\int_0^T u(t)u^*(t+\tau)e^{-i2\pi f_d t}\,dt \tag{4.11c}$$

The envelope of the output is

$$|Y| = |X(\tau,f_d)| \tag{4.11d}$$

where

$$X(\tau,f_d) = \int_0^T u(t)u^*(t+\tau)e^{-i2\pi f_d t}\,dt \tag{4.12}$$

is called the *ambiguity function*. The ambiguity function is a function of τ_1 (the delay) and f_d (the Doppler shift). This function is also known as time-frequency correlation function and matched filter response function. This gives the distribution of range and Doppler resolution. When $f_d = 0$, the ambiguity function yield the time autocorrelation function of the complex envelope. At ranges and radial velocities such that $|X(\tau,f_d)|$ is about the same as $|X(0,0)|$, targets are indistinguishable to the radar. The *resolution* of a radar is the ability to discriminate between multiple targets with similar range and velocity. The width of the peak $\tau = 0$ and $f_d = 0$ determines the resolution of the waveform. Two targets with two different time delays can be resolved if the separation is greater than the delay resolution of the waveform. The ambiguity function $X(\theta,\phi)$ can be related to the gain function of the antenna at the angle θ, ϕ.

Ambiguity functions exhibit the following properties:

1. $|X(\tau,f_d)| \le X(0,0) = 2E$ \hfill (4.13a)

where E is the energy of the signal.

2. $X(\tau,f_d) = \displaystyle\int_{-\infty}^{\infty} U(f+f_d)U^*(f)e^{-i2\pi f_d t}\,df$ \hfill (4.13b)

where

$$U(f) = \int_{-\infty}^{\infty} u(t)e^{-i2\pi ft}\,dt \tag{4.13c}$$

3. $X(\tau,0) = \displaystyle\int_{-\infty}^{\infty} u(t)u^*(t+\tau)\,d\tau \qquad$ autocorrelation of $u(t)$

\hfill (4.13d)

4. $X(0,f_d) = \displaystyle\int_{-\infty}^{\infty} U(f+f_d)U^*(f)\,df \qquad$ autocorrelation of $U(f)$

\hfill (4.13e)

5. $\displaystyle\int_{-\infty}^{\infty}\int |X(\tau,f_d)|^2\,d\tau\,df_d = 4E^2$ \hfill (4.13f)

where E is the transmitted energy.

Let us define

$$\alpha = 2\pi \left[\frac{1}{2E \int_{-\infty}^{\infty} (t)^2 |u(t)|^2 \, dt} \right]^{1/2} \tag{4.14a}$$

$$\beta = 2\pi \left[\frac{1}{2E \int_{-\infty}^{\infty} (f)^2 |U(f)|^2 \, df} \right]^{1/2} \tag{4.14b}$$

$$T_r = \frac{1}{4E^2 \int_{-\infty}^{\infty} |X(\tau, 0)|^2 \, d\tau} \tag{4.14c}$$

$$F_r = \frac{1}{4E^2 \int_{-\infty}^{\infty} |X(0, f_d)|^2 \, df_d} \tag{4.14d}$$

α and β are called the *effective or root-mean-square (RMS) duration* and *RMS bandwidth* when the signal has zero mean time and zero mean frequency. The mean time and mean frequency are defined by

$$\bar{t} = \int_{-\infty}^{\infty} t |u(t)|^2 \, dt \tag{4.15a}$$

$$\bar{f} = \int_{-\infty}^{\infty} f |U(f)|^2 \, df \tag{4.15b}$$

T_r and F_r are the *time resolution* and *frequency resolution* constants, respectively. $1/T_r$ and $1/F_r$ measure the total *occupied frequency span* and *time span* of the signal. If

$$u(t) = 1/\sqrt{T} \qquad |t| \leq T/2$$
$$= 0 \qquad \text{elsewhere} \tag{4.16a}$$

then the ambiguity function is

$$X(\tau, f_d) = \frac{\exp[i\pi f_d \tau](1 - |\tau|/T) \sin \pi f_d(T - |\tau|)}{\pi f_d(T - |\tau|)} \qquad |\tau| < T$$
$$= 0 \qquad \text{elsewhere} \tag{4.16b}$$

We note that time resolution can be found from the ambiguity function with $f_d = 0$ and Doppler resolution with $\tau = 0$. Small range resolution can be obtained by using wide-bandwidth signals. As seen from the properties of the ambiguity function, the Doppler resolution increases with increasing coherent integration time.

4.6 ADAPTIVE TEMPORAL FILTERING

Clutter and other noise sources may yield erroneous range and Doppler estimates if the signal returned from the target is weak. We present here an adaptive noise cancellation method to improve the performance of the signal-to-noise ratio. Let us denote the estimator of a parameter as

$$\hat{d}_k = \sum_{j=0}^{n-1} w_j x_{k-j} \tag{4.17a}$$

where $x_k, x_{k-1}, \ldots, x_{k-n}$ are observed values, x_k is the current observed value, and w_j, $j = 0, 1, 2, \ldots, n-1$ are the unknown weights. Let us denote

$$J = E[(d_k - \hat{d}_k)]^2 \tag{4.17b}$$

$$R_{xx}(m) = E(x_k x_{k-m}) \quad \text{and} \quad R_{xd}(m) = E[x_{k-m} d_k] \qquad m = 0, \ldots, n-1$$

$$E(d_k^2) = R_{dd}(0) \tag{4.17c}$$

where $E(\cdot)$ is the ensemble average operator. We assume that $\{x_k\}$ is a wide-sense stationary process. It can be shown that

$$J = R_{dd}(0) + \sum_{l=0}^{n-1} \sum_{m=0}^{n-1} W_l W_m R_{xx}(l - m) - 2 \sum_{e=0}^{n-1} W_l R_{xd}(l)$$

using Eqs. (4.17a), (4.17b), and (4.17c). Differentiating J with respect to W_l, we get

$$\sum_{l=0}^{n-1} W_l R_{xx}(k - l) = R_{xd}(k) \tag{4.17d}$$

The optimum weight that minimizes the mean square error ξ is obtained by taking z transform of Eq. (4.17d). Hence

$$W^o(z) = \frac{S_{xd}(z)}{S_{xx}(z)} \tag{4.17e}$$

where

$$W^o(z) = \sum_j W_j z^{-j}$$

$$S_{xd}(z) = \sum_j R_{xd}(j) z^{-j}$$

$$S_{xx}(z) = \sum_j R_{xx}(j) z^{-j}$$

are z transforms of the weights and correlation functions. $W^o(z)$ is the Weiner filter in z domain.

In order to cancel the noise, in 1985 Widrow proposed an adaptive noise-canceling scheme, which is shown in Fig. 4.19. The signal source is distorted by clutter noise $\{c_k\}$ and thermal noise $\{u_k\}$. Clutter noise is observed in another sensor noise in addition to thermal noise. It is assumed that all noises are stationary.

Referring to Fig. 4.19, we write

$$d(t) = \int h_1(t - \tau)[S(\tau) + c(\tau)] \, d\tau + u(t) \tag{4.17f}$$

where $s(t)$, $c(t)$, and $u(t)$ are signal, clutter noise, and additive noise. This

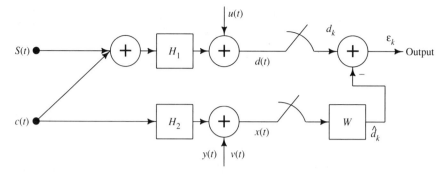

FIGURE 4.19   Adaptive clutter canceler.

channel (source) is called the *primary source*. The clutter is observed in another sensor in the presence of noise $v(t)$. The second channel or source is called the *reference source*. All inputs $s(t)$, $c(t)$, $u(t)$, and $v(t)$ are stationary and uncorrelated. The input to the adaptive filter W is denoted by $x(t)$.

$$d(t) = m(t) + n(t) + u(t)$$
$$x(t) = y(t) + v(t)$$

(4.17g)

where

$$m(t) = \int (h_1(t - \tau)s(\tau)\,d\tau$$

$$n(t) = \int h_1(t - \tau)c(\tau)\,d\tau$$

$$y(t) = \int h_2(t - \tau)c(\tau)\,d\tau$$

The power spectral density of x_k in the z domain is

$$S_{xx}(z) = S_{nn}(z) + S_{vv}(z)$$
$$= |H(z)|^2 S_{cc}(z) + S_{vv}(z)$$

(4.17h)

The cross correlation of x_k and d_k is

$$E[x_k d_{k+m}] = E[(y_k + v_k)(n_{k+m} + u_{k+m} + S_{k+m})]$$
$$= E[y_k n_{k+m}]$$

The cross correlations of other terms are zero. Therefore, the cross-power spectral density is

$$S_{xd}(z) = S_{yn}(z) = H_1(z)H_2(z^{-1})S_{cc}(z)$$

(4.17i)

Substitution of Eqs. (4.17g) and (4.17h) in Eq. (4.17e) yields

$$W^o(z) = S_{cc}(z)H_2(z^{-1}) \cdot \frac{H_1(z)}{[S_{vv}(z) + S_{cc}(z)|H_2(z)|^2]} \tag{4.18a}$$

The filter is independent of primary source signal $s(t)$ and noise $u(t)$.
Assume that

$$S_{cc}(z) = N_c/2 \qquad\qquad H_1(z) = S(z^{-1})z^{-1}$$

$$S_{uu}(z) = N_0/2 \qquad\qquad H_2(z) = S(z^{-1})z^{-2} \tag{4.18b}$$

$$S_{vv}(z) = N_1/2 = 1$$

where $S(z)$ is the z transform of the signal $s(k)$, $z = \exp(i2\pi f T)$. Substitution of Eq. (18b) in Eq. (18a) gives

$$W^o(z) = \frac{|S(z)|^2 z}{1 + |S(z)|^2} \tag{4.18c}$$

Note that

$$\varepsilon_k = m_k + n_k + u_k + \hat{d}_k$$

$$= m_k + n_k^o$$

where

$$m_k = \int h_1(\tau)s(kT - \tau)\,d\tau$$

$$n_k = \int h_1(\tau)c(kT - \tau)\,d\tau$$

$$u_k = u(kT)$$

$$\hat{d}_k = \int \omega(\tau)([y(KT - \tau)] + [v(KT - \tau)])\,d\tau$$

n_k^o = total output noise

$$= u_k + \int \omega(\tau)v(KT - \tau)\,d\tau + \int h_1(\tau)c(kT - \tau)\,d\tau$$

$$+ \int \omega(\tau)\left[\int h_2(\tau_1)C(kT - \tau_1 - \tau)\,d\tau_1\right]d\tau$$

The PSD of n_k^o is

$$S_{no}(z) = \frac{N_0}{2} + \left(\frac{N_1}{2}\right)|W^o(z)|^2 + \left(\frac{N_c}{2}\right)|H_1(z) - H_2(z)W^o(z)|^2$$

$$= \frac{N_0}{2} + \left|\frac{|S(z)|^2 z}{(1 + |S(z)|^2)}\right|^2 + \frac{N_c}{2}\left[\frac{1}{(1 + |S(z)|^2)^2}\right] \tag{4.19a}$$

when $\hat{d}_k = 0$, there is no reference input,

$$\tilde{n}_k^o = n_k + u_k$$

The PSD of this noise output is

$$S_{\tilde{n}_o}(z) = \frac{N_0}{2} + \frac{N_c}{2}|S(z)|^2 \qquad (4.19b)$$

The comparison of Eqs. (4.19a) and (4.19b) shows that the output noise power is reduced by the noise canceler. If we ignore thermal noise density N_0 and N_1, then the output noise power spectral density with noise canceler is reduced one-fourth assuming that signal is a point target with $|S(z)|^2 = 1$.

4.7 RADAR SIGNAL-TO-NOISE RATIO EQUATION

The maximum detection and tracking range of a radar system is primarily a function of three parameters: (1) transmitted power, (2) antenna gain, and (3) receiver sensitivity. Increasing the transmitted power will increase the radiated energy, which in turn will result in a stronger target return. The antenna gain is a measure of the radiated energy in the direction of the target as compared to uniform radiation of energy. Receiver sensitivity is a measure of the capability of the receiver to detect target returns. When the power radiated by the radar is denoted as P, and this power is assumed to radiate uniformly in all directions, the power per unit area (or the power density) will be

$$\text{Power density} = \frac{P}{4\pi R^2} \qquad (4.20)$$

where $4\pi R^2$ represents the area of a sphere with radius R. Since most radars have a directive antenna (that is, power is radiated in a specific direction), this is accounted for by incorporating the antenna gain in the above equation. The antenna gain at a particular angle θ is defined as the ratio of the radiated intensity at θ to the radiation intensity of a uniformly radiating antenna (see Fig. 4.20). At $\theta = 0$, the maximum gain of an antenna is related to its physical area A by the equation

$$G = \frac{4\pi A}{\lambda^2} \qquad (4.21)$$

where A is the aperture area, and λ is the wavelength of the transmitted wave.[b] In Eq. (4.21) an antenna efficiency of 100 percent is assumed. When the antenna gain G is incorporated into Eq. (4.20), the power density in a particular antenna-pointing direction is obtained; that is,

$$\text{Radiated power density of directive antenna} = \frac{PG}{4\pi R^2} \qquad (4.22)$$

---

[b] See Section 4.13.

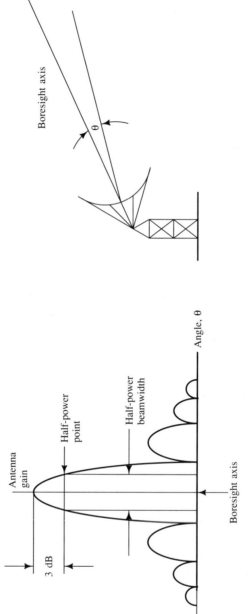

FIGURE 4.20 Typical antenna gain curves.

The target intercepts and reflects part of the radiated energy. The portion of the power reflected by the target in the direction of the receiving antenna will be a function of the ability of the target to "focus" the received energy in that direction. Qualitatively, this focusing ability is proportional to the target-radar cross section, σ, which has the units of area (square meters). Using this, the reflected power from the target becomes

$$\text{Reflected power from the target} = \frac{PG\sigma}{4\pi R^2} \tag{4.23}$$

When the reflected power given by the above equation is used as a uniform source of power located at the target, the power density of this energy at the radar (which is distance R away) can be calculated as follows:

$$\text{Power density of reflected target power at the radar} = \frac{PG\sigma}{(4\pi R^2)^2} \tag{4.24}$$

If we assume that the area capturing this energy is equal to the aperture area A of the transmitting antenna (it is assumed that both transmitting and receiving antennas are one and the same), then

$$\text{Power of reflected signal at the radar} = \frac{PG\sigma A}{(4\pi R^2)^2} \tag{4.25}$$

Solving Eq. (4.21) for A and substituting in Eq. (4.25), we get

$$P_r = \frac{PG^2\lambda^2\sigma}{(4\pi)^3 R^4} \tag{4.26}$$

where P_r designates target power received at the radar.

In the detection circuitry of the radar, this target return power will be processed in the presence of receiver noise. Receiver noise can originate within the receiver itself, or it may enter the receiver through the receiving chain, starting with the antenna and followed by the waveguides, amplifiers, and so on. One portion of the noise generated within the receiver is due to the thermal agitation of electrons in the receiver. This thermal noise, as given below, is proportional to the temperature T in kelvins, and the receiver noise band-width B_n in hertz; that is,

$$\text{Thermal noise} = kTB_n \tag{4.27}$$

where k is Boltzmann's constant and is equal to 1.38×10^{-23} J/K. Equation (4.27) represents only one portion of the receiver noise. To account for all of the receiver noise, Eq. (4.27) is multiplied by an experimentally determined constant, called the noise figure and designated by F. With this factor, the total receiver noise can be written

$$N = FkTB_n \tag{4.28}$$

where N represents the total receiver noise.

Denoting the target power received by the radar [Eq. (4.26)] by S and the noise power [Eq. (4.28)] by N, we obtain the signal-to-noise ratio (SNR), as follows:

$$\text{SNR} = \frac{PG^2\lambda^2\sigma}{(4\pi)^3 R^4 (FkTB_n)} \tag{4.29}$$

Equation (4.29) represents the radar equation in its simplest form. Note from this equation that target signal power is inversely proportional to the fourth power of range and directly proportional to the square of antenna gain.

Equation (4.29) was derived without incorporating the losses of energy that accompany transmission, reception, and the processing of electromagnetic radiation. Incorporating all of these losses in one term, we can write Eq. (4.29) as follows:

$$\frac{S}{N} = \frac{PG^2\lambda^2\sigma}{(4\pi)^3 R^4 (FkTB_n)L} \tag{4.30}$$

where L is the loss factor representing the sum total of transmission and reception losses.

4.7.1 Other Forms of the Radar Equation

If the transmit and receive antennas are not at the same location and do not have the same antenna gain, Eq. (4.30) can be written as

$$\frac{S}{N} = \frac{PG_t G_r \lambda^2\sigma}{(4\pi)^3 R_t^2 R_r^2 (FkTB_n)L} \tag{4.31}$$

where R_t and R_r are ranges between the target and the "transmit" antenna and the target and the "receive" antenna, respectively. The corresponding transmit and receive antenna gains are G_t and G_r.

In cases where a signal is transmitted from one radar and is received by another radar, the beacon (rather than the radar) equation applies. The beacon equation can be readily derived from the previously given information as follows:

$$\frac{S}{N} = \frac{PG_t G_r \lambda^2}{(4\pi)^2 R^2 (FkTB)_r L} \tag{4.32}$$

where

$$R = \text{range between transmitting and receiving radars}$$

$$G_t = \text{transmit antenna gain}$$

$$G_r = \text{receive antenna gain}$$

$$(FkTB)_r = \text{receiver noise level}$$

Note that in the case of Eq. (4.32), the signal-to-noise ratio is proportional to the second (rather than the fourth) power of range R.

4.7.2 Integration of Returned Pulses

Note that Eq. (4.30) was derived for a single returned pulse. In search radars where the antenna sweeps across the target, depending on the rate of sweep and the number of transmitted pulses, a great number of pulses can be integrated to improve the attained signal-to-noise ratio. Assuming the number of transmitted pulses per second to be designated by PRF and the on-target time by T_i, the number of target returned pulses can be written

$$N_p = T_i \cdot \text{PRF} \tag{4.33}$$

where N_p is the number of returned pulses, T_i is the on-target time, and PRF is the number of transmitted pulses per second. Using Eq. (33) in Eq. (4.30), we get

$$\frac{S}{N} = \frac{PG^2\lambda^2\sigma T_i(\text{PRF})}{(4\pi)^3 R^4 (FkTB_n)L} \tag{4.34}$$

For a transmitted pulsewidth of τ, the matched receiver filter bandwidth B_n is given by $1/\tau$ (see Fig. 4.21). Using this in Eq. (4.34), we get

$$\frac{S}{N} = \frac{PG^2\lambda^2\sigma T_i(\text{PRF})\tau}{(4\pi)^3 R^4 (FkT)L} \tag{4.35}$$

From Fig. 4.21 note that $P\tau$ represents transmitted per-pulse energy. Dividing this energy by the duration of the interpulse period $\bar{T}$, we can compute the average power

$$P_{\text{ave}} = \frac{P\tau}{\bar{T}}$$
$$= P\tau(\text{PRF}) \tag{4.36}$$

where P_{ave} is the average power, and $\bar{T} = 1/\text{PRF}$ is the interpulse period. Using Eq. (4.36) in Eq. (4.35), we get

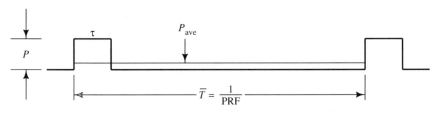

FIGURE 4.21   Transmitted pulse τ, interpulse period $\bar{T}$, and average power.

$$\frac{S}{N} = \frac{P_{\text{ave}} T_i G^2 \lambda^2 \sigma}{(4\pi)^3 R^4 (FkT)L} \tag{4.37}$$

Equation (4.37) represents the signal-to-noise ratio of a matched transmit-receive radar system. From Eq. (4.37), note that the SNR, as expected, is proportional to the product of the average power by the on-target time, which represents the amount of transmitted energy. As shown in Ref. 1, Eq. (4.37) applies to all types of radars—i.e., low-, medium-, and high-PRF systems.

4.8 SEARCH RADARS

Two parameters, scan coverage and detection range, dominate the design of search radars. In this section—starting with the signal-to-noise ratio given in Eq. (4.37)—parametric equations are derived describing the interrelation of radar parameters with these factors.

Using the antenna gain equation in the form

$$G = \frac{4\pi A}{\lambda^2} \tag{4.38}$$

where A is the aperture area and λ is the wavelength, in Eq. (4.37) we get

$$\frac{S}{N} = \frac{P_{\text{ave}} A^2 \sigma T_i}{(4\pi)\lambda^2 R^4 (kT)L} \tag{4.39}$$

Assuming a circular aperture with diameter D and substituting the value of $\pi D^2/4$ for one of the areas A in Eq. (4.39), we get

$$\frac{S}{N} = \frac{P_{\text{ave}} D^2 A \sigma T_i}{(16)\lambda^2 R^4 (kT)L} \tag{4.40}$$

For an angular scan coverage of Ω steradians and an antenna linear beamwidth of θ_{bw}, we can calculate the number of antenna beam positions n_p within the scan coverage

$$n_p = \frac{\Omega}{\theta_{\text{bw}}^2} \tag{4.41}$$

where θ_{bw}^2 represents the antenna beamwidth in steradians. Using the relation $\theta_{\text{bw}} = \lambda/D$, where the beamwidth is represented as a function of wavelength λ and diameter D, in Eq. (4.41) we get

$$n_p = \frac{D^2}{\lambda^2} \Omega \tag{4.42}$$

For a scan duration or radar frametime of t_s seconds, using Eq. (4.42) we can obtain the on-target time T_i as

$$T_i = \frac{t_s}{n_p}$$

$$= \frac{\lambda^2 t_s}{D^2 \Omega} \tag{4.43}$$

Substituting Eq. (4.43) into Eq. (4.40) for T_i, we get

$$\frac{S}{N} = \frac{P_{\text{ave}} A \sigma t_s}{(16) R^4 (kT) L \Omega} \tag{4.44}$$

Equation (4.44) gives the SNR for a search radar with scan coverage of Ω steradians and radar frame time (or scan time) of t_s seconds. Equation (4.44) is independent of radar transmit frequency. For a given set of radar parameters and scan coverage, the SNR depends on power-aperture product. Thus, for a given power-aperture product, Eq. (4.44) can be solved for range R in terms of radar parameters, as follows:

$$R = \left[\frac{P_{\text{ave}} A \sigma t_s}{(16)(kT) L (S/N) \Omega} \right]^{1/4} \tag{4.45}$$

From Eq. (4.45) note that the detection range of a search radar is inversely proportional to the area coverage Ω and directly proportional to radar scan time t_s. The value of S/N in Eq. (4.45) represents the required SNR for target detection. Observe further that the detection range of the search radar as given by Eq. (4.45) is independent of transmitting radar frequency.

4.9 TRACKING RADARS

Tracking radars "memorize" the past spatial positions of the target and predict future positions based on these past data; that is, they keep target tracks. Target tracking can be either continuous or discontinuous. Continuous tracking is achieved by keeping the radar beam on the target and extracting information so as to predict future target positions continuously. Discrete tracking is accomplished by using periodic and noncontinuous radar data (such as obtained in search radars) and tracking the targets in a computer. These radars, usually called track-while-scan (TWS) radars, scan a given search volume and illuminate several targets, periodically extracting space angles (azimuth and elevation), range, and range rate data from each target. This information is then processed in a digital computer to accurately calculate the spatial positions of the targets. For a complete discussion of multiple-target radar data processing see Ref. 5.

Several signal processing methods are used in continuous-tracking radars. Among them are sequential lobing, conical scan tracking, and amplitude-comparison monopulse.

4.9.1 Sequential Lobing and Conical Scan Tracking

In sequential loading, the antenna beam is switched between two positions in each vertical and horizontal plane. In each position, target strength is measured and converted to a voltage. The difference between these voltages is a measure of angular error between the actual position of the target and the position of the antenna lobing axis when the measurement was made. Figure 4.22a shows an example of this method in a single plane.

Note from the Fig. 4.22 that the target is illuminated at a higher point of antenna gain pattern in Position 1 than in Position 2. Moving the lobing axis to Position 1 and repeating the process will eventually reduce the error and put the lobing axis on the target direction. More sophisticated servo systems can be designed to utilize the magnitude and direction of the difference voltage between Positions 1 and 2 of the lobe and converge the lobing axis to actual target position more rapidly.

Conical scan tracking is the logical extension of sequential lobing where the antenna beam is rotated about the axis, as shown in Fig. 4.22b. In this method the target return is examined during one complete rotation of the antenna boresight axis around the axis of rotation. Using these data, azimuth

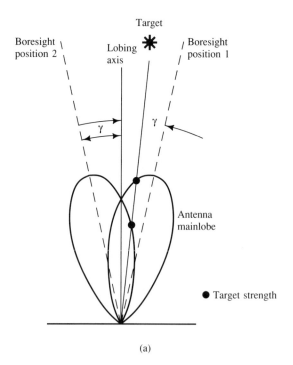

(a)

FIGURE 4.22   Single-target tracking antennas. (a) Sequential lobing. (b) Conical scan.

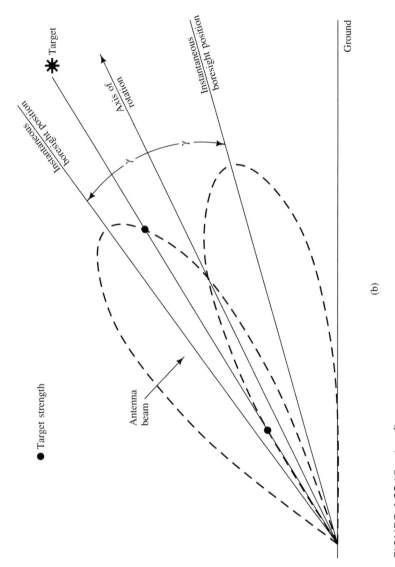

Target

Instantaneous
boresight position

Axis of
rotation

Instantaneous
boresight position

γ

γ

● Target strength

Antenna
beam

Ground

(b)

FIGURE 4.22 (*Continued*)

and elevation commands are generated by a servo system to make the axis of rotation of the antenna colinear with target direction.

4.9.2 Monopulse Tracking

In the continuous-tracking methods described above, more than one target return pulse is required to establish radar tracking error. In practice, the number of consecutive pulses is four, since two pulses are required in each of the vertical and horizontal directions. In the above methods, fluctuation of target size between pulses results in tracking errors. The target size fluctuation problem can be eliminated if a single pulse is used to determine tracking error. Tracking systems that use a single pulse to calculate tracking errors are called, appropriately, monopulse systems.[6] Angle errors in these systems are derived by using either amplitude or phase comparison of returned signals.

Figure 4.23 illustrates the basics of the amplitude comparison monopulse system. Two overlapping antenna beams are needed for each of the two coordinate directions (azimuth and elevation). These beams may be generated with a single reflector illuminated by four adjacent feeds. The sum pattern is used for transmission, whereas both the sum and the difference patterns are used for reception. The sum channel is used for range measurement and also as a reference to obtain the sign of the angular error. Figure 4.24 gives typical sum and difference gain curves as a function of the angle off boresight, θ. For small angles ($\theta < 2°$), the ratio of relative amplitudes of target returns from sum and difference channels will result in a voltage proportional to angular error, as illustrated in Fig. 4.24.

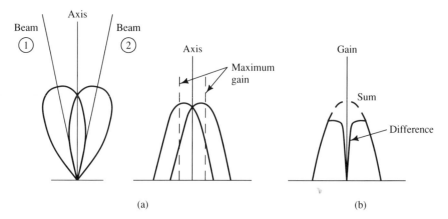

FIGURE 4.23   Amplitude comparison monopulse sum and difference beams and gain patterns. (a) Antenna beam and gain curves. (b) Sum and difference gains.

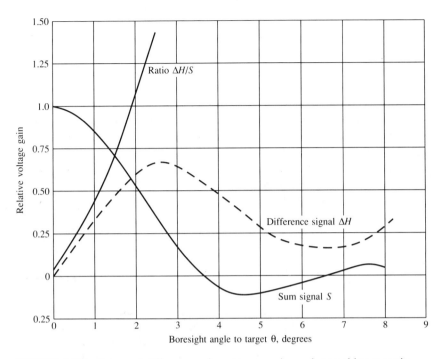

FIGURE 4.24 Sum and difference gain patterns and angular tracking error in amplitude comparison monopulse system.

Figure 4.25 shows a block diagram of a low-PRF amplitude comparison monopulse radar for a single (azimuth or elevation) channel. The duplexer switches the antenna between transmit and receive functions. Note that the transmission is through the sum channel. The mixers of the figure reduce the target received frequency from radio frequency (RF) to intermediate frequency (IF). The phase-sensitive detector of the figure determines the sign of the angular error—i.e., right and left for azimuth channel and up and down for

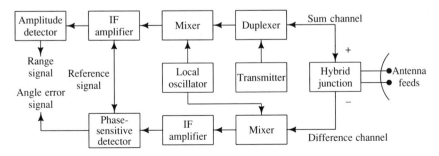

FIGURE 4.25 Amplitude comparison monopulse radar block diagram.

elevation channel. The final result is the detection of the range and angular error signal as shown.

4.9.3 Track-While-Scan Radars

Modern-day scanning radars are capable of obtaining target observations (range and angular position) for several targets within their scan illumination coverage. These radars can be used for multiple target tracking and airspace surveillance.[5]

Figure 4.26 shows a typical inertial coordinate system with the radar at the origin. The north and the east axes point to the north and east, respectively; the vertical axis is perpendicular to these axes. Denoting the respective direction cosines by N, E, and V, we define the position of each target with respect to the origin by range, R, and two out of three direction cosines. Target angular position can also be specified in terms of azimuth (η) and elevation (ε) angles rather than N, E, V as described above.

For a scanning radar, the target data rate will depend on the radar frame time (time between consecutive illuminations of a specific point in space). Thus, tracking of several targets becomes largely a matter of sorting target observations obtained in each radar frame and assigning them to proper target tracks. These tracks form the basis for estimating future target positions. The process of tracking targets based on discrete radar information obtained while the radar continues to scan the airspace is referred to as track-while-scan (TWS) and takes place in a digital computer.

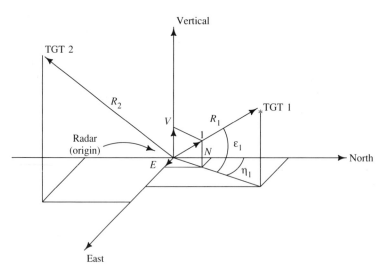

FIGURE 4.26   North, east, vertical (N, E, V) coordinate system (1 is the unit vector.)

Blocks

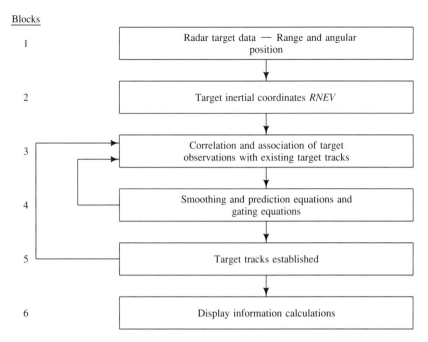

FIGURE 4.27   TWS data processing.

Figure 4.27 illustrates the data processing required for the computer's TWS operation. The angular position of each radar target observation is converted to the direction cosines N, E, V of the inertial coordinate system.

This inertial angular position, together with range R, specifies the inertial target position, $RNEV$, Block 2 in the figure. Note that although one of the direction cosines is redundant, all three are carried as target coordinates to ensure maximum tracking accuracy. These target observations are correlated and associated with existing tracks. The correlation and association process ensures that proper observations are assigned to each target track by comparing these observations with the predicted target positions. Subsequently (Block 4), these target observations are incorporated in the proper target tracks through smoothing and prediction equations which primarily use Kalman filtering methods (see Fig. 4.28). For each radar frame the acceptability region for correlating and associating target observations (the gates) is established around the predicted target positions for each of the four coordinates R and NEV. At Block 5, target tracks, together with predicted target positions for the next radar frame and the appropriate gates, are calculated and stored within the computer. This information is used by Blocks 3 and 4 for correlation, association, and gating. It may also be used as the basis for computing information to be displayed (Block 6).

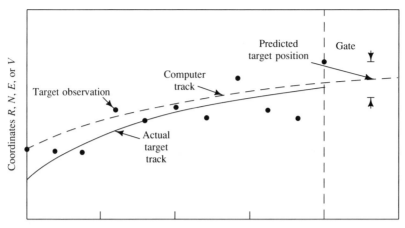

Time from initial target observation, seconds

FIGURE 4.28   Smoothing and prediction of target data.

4.10 MEASUREMENT AND TRACKING ACCURACY

An important consideration in search and tracking radars is the accuracy of the radar measurements and the accuracy with which targets are tracked. Measurement accuracy is primarily a function of radar characteristics such as range gate duration, Doppler filter bandwidth, antenna beamwidth, and radar signal processing methods. Tracking accuracy, on the other hand, is a function of the amount of data processing and smoothing performed on the measured data. Present-day radar systems use a variety of filtering and smoothing methods on measured data to obtain desired target tracking accuracy. These smoothing methods are typified by least-squares fit techniques and Kalman filtering methods. In this section, however, we shall concentrate only on radar measurement accuracies.

In low-PRF radars, range measurement accuracy is determined primarily by the size of the range gate. In most cases, a typical target return will fall into two adjacent range gates, and simple averaging of the corresponding range values of each gate will result in a range accuracy of

$$\Delta R = \frac{c\tau}{4} \tag{4.46}$$

where ΔR is the range measurement accuracy, c is the velocity of light, and τ is the pulse length (range gate duration).

Similarly, the accuracy with which a Doppler radar can measure radar-target range rate, $\dot{R}$, will depend on the frequency separation of two adjacent filters (Doppler filter bandwidth). When this separation is denoted by Δf_s, the range rate measurement accuracy becomes

$$\Delta \dot{R} = \frac{\lambda \, \Delta f_s}{4} \tag{4.47}$$

where $\Delta\dot{R}$ is the range rate measurement accuracy, λ is the wavelength, and Δf_s is the Doppler filter bandwidth.

The accuracy of radar angular measurements is a function of the antenna half-power beamwidth, θ_{bw}. For simply constructed antennas, the antenna beamwidth can be approximated as a function of wavelength and antenna size as follows:

$$\theta_{bw} = \frac{\lambda}{l} \text{ (radians)}$$

$$\approx \frac{60\lambda}{l} \text{ (degrees)} \tag{4.48}$$

where θ_{bw} is the half-power antenna beamwidth, λ is the wavelength, and l is the aperture dimension. In a circular antenna, l will equal the diameter.

Continuous-tracking radars, as shown in Fig. 4.22a and b, can always keep the target within one-half of their antenna beamwidths. Thus the angular measurement accuracy can be written as

$$\Delta\theta = \frac{\theta_{bw}}{2} \tag{4.49}$$

where $\Delta\theta$ is the angular measurement accuracy and θ_{bw} is the antenna beamwidth. In search radars, on the other hand, the angular measurement accuracy is usually twice the value given in Eq. (4.49).

Application of sophisticated signal processing methods and waveform analysis will show that (1) the behavior of measurement accuracies given by Eqs. (4.46) to (4.49) most accurately follow a Gaussian distribution, and (2) the limiting value of measurement accuracy is a function of the signal-to-noise ratio. Using these two effects in Eqs. (4.46) to (4.49), we can write typical measurement accuracies as

$$\sigma_R = \frac{c\tau}{4\sqrt{S/N}} \tag{4.50a}$$

$$\sigma_{\dot{R}} = \frac{\lambda \Delta f_s}{4\sqrt{S/N}} \tag{4.50b}$$

and

$$\sigma_\theta = \frac{\theta_{bw}}{2\sqrt{S/N}} \tag{4.50c}$$

where σ_R, $\sigma_{\dot{R}}$, and σ_θ are the standard deviations of range, range rate, and angular measurements, respectively. The square root of the S/N power ratio, representing the S/N voltage ratio, is used in Eq. (4.50). Note that the above accuracies are obtained from a consideration of signal processing and thermal noise characteristics and do not include errors from other measurement sources such as servo systems, quantization, and the like.

The above discussion of accuracies concerns measurement accuracies, where the values are obtained from a single illumination of the target. In cases where target tracks are employed, smoothing and filtering of the data can considerably improve the measured accuracies.

4.11 DESIGN EXAMPLE

4.11.1 Ground-Based Low-PRF Radar

Assume that there is a requirement to design a ground-based low-PRF radar with the capability of tracking a 5-m^2 target at a range of 40 nmi. The design may begin by a study of existing radars of similar capability[7] and development of parameters for the required radar system. Using this procedure, one may arrive (after several iterations) at the following parameters:

Transmit frequency, f $\qquad$ = 6000 MHz (C-band)

Pulse-repetition frequency, PRF = 1500 Hz

Antenna diameter, l $\qquad$ = 4 ft

Receiver noise figure, F $\qquad$ = 4

Noise temperature, T $\qquad$ = 300 K

Transmit power, P $\qquad$ = 100 kW

Pulse length, τ $\qquad$ = 0.5 μs

Receiver noise bandwidth, B $\qquad$ = 2 MHz

System losses, L $\qquad$ = 8 dB = 6.3095

Target tracking SNR $\qquad$ = 4 dB = 2.5119

In the above, a matched-receiver bandwidth is selected. This bandwidth is approximately equal to $1/\tau$, where τ is the transmitted pulse length. From previous equations, we can calculate the antenna gain and wavelength as follows:

$$G = \frac{4\pi A}{\lambda^2} = 5.868 \times 10^3 = 37.685 \text{ dB}$$

$$\lambda = \frac{c}{f} = 0.164 \text{ ft}$$

(4.51)

Using Eq. (4.30), we can calculate the corresponding SNR

$$\text{SNR} = \frac{(100 \times 10^3)(5.868 \times 10^3)^2(0.164)^2(5)(3.2808)^2}{(4\pi)^3(R^4)(4)(1.38 \times 10^{-23})(300)(2 \times 10^6)(6.3095)}$$

$$= \frac{1.202 \times 10^{22}}{R^4}$$

(4.52)

where 3.2808 is the conversion factor from meters to feet. Since a signal-to-noise ratio of 4 dB or 2.5119 is required for target tracking, the target tracking range will be

$$R = \left[\frac{1.202 \times 10^{22}}{2.5119}\right]^{1/4} = 2.6301 \times 10^5 \text{ ft} = 43.25 \text{ nmi} \tag{4.53}$$

where a conversion factor of 6080 ft/nmi is used. Note, therefore, that this radar-target tracking range is close to the requirement of 40 nmi.

The PRF of 1500 Hz will result in an unambiguous range of

$$R_{ua} = \frac{c}{2\,\text{PRF}}$$

$$= \frac{3 \times 10^8 \times 3.2808}{2 \times 1500 \times 6080}$$

$$= 53.96 \text{ nmi} \tag{4.54}$$

which is in excess of the 40 nmi tracking range and is acceptable.

Consider the use of the radar parameters above in the search configuration, and assume that it is desired to implement a $360° \times 10°$ azimuth and elevation search in a 10-s radar frame time. Using Eq. (4.45),

$$R = \left[\frac{P_{ave}A\sigma t_s}{(16)(FkT)(S/N)L\Omega}\right]^{1/4} \tag{4.55}$$

we can compute the detection range of this search radar. The terms of Eq. (4.55) can be computed as follows:

$$P_{ave} = P\tau(\text{PRF}) = 75 \text{ W}$$

$$A = \frac{\pi D^2}{4} = 1.168 \text{ m}^2$$

$$\Omega = \frac{360 \times 10}{(57.3)^2} = 1.096 \text{ steradians} \tag{4.56}$$

Additionally, assume that an SNR of 15 dB (a factor of 31.62) is required for target detection; from Eq. (4.55) we have

$$R = \left[\frac{(75)(1.168)(5)}{(16)(1.38 \text{ E} - 23)(4)(300)(31.62)(6.3095)} \cdot \frac{10}{1.096}\right]^{1/4}$$

$$= 93.24 \text{ km}$$

$$= 50.30 \text{ nmi} \tag{4.57}$$

Observe that the detection range of this radar is slightly less than its unambiguous range corresponding to a PRF of 1500 pps as given by Eq. (4.54), as it should be in a well-designed radar.

In this example an SNR of 15 dB was assumed for detection. This value can be computed from more sophisticated analyses, which include probability of detection, threshold setting, and target cross-section fluctuations as discussed at the end of this chapter.

Additional radar characteristics can also be obtained from the given date. Using the antenna diameter of 4 ft, we get the antenna beamwidth from Eq. (4.48); that is,

$$\theta_{bw} = \frac{(60)(0.164)}{4} = 2.46° \tag{4.58}$$

From Eq. (4.49), the angular measurement accuracy will be 1.23°, which is one-half the antenna beamwidth value given in Eq. (4.58). The range measurement accuracy of a low-PRF radar is given by Eq. (4.46) as

$$\Delta R = \frac{3 \times 10^8 \times 3.2808 \times 0.5 \times 10^{-6}}{4}$$

$$= 123 \text{ ft} \tag{4.59}$$

4.11.2 Pulse Doppler Radar

As a variation of the above design, consider a high-PRF pulse Doppler radar with applicable parameters of the radar in Section 4.11.1 that uses frequency modulation ranging and a Doppler filter bandwidth of 1000 Hz. The list of parameters of this radar is given below. (Changed items from the previous low-PRF radar design are indicated with an asterisk.)

Transmit frequency, f	= 6000 MHz (C-band)
*Pulse-repetition frequency, PRF	= 200,000 Hz
Antenna diameter, l	= 4 ft
Receiver noise figure, F	= 4
Noise temperature, T	= 300 K
*Transmit power, P	= 10 kW
*Pulse length, τ	= 2 μs
*Receiver noise bandwidth (assumed equal to Doppler filter bandwidth), B	= 1000 Hz
System losses, L	= 8 dB = 6.3095
Target tracking SNR	= 4 dB = 2.5119

The PRF is increased from 1500 to 200,000 pps, inasmuch as the interpulse period is no longer used to obtain radar-target range. Because of increased PRF, the transmitted power is decreased from 100 to 10 kW to avoid transmitter burnout or power dumps. The pulse length is increased to 2 μs from its low-PRF value of 0.5 μs since, in the case of high-PRF radar, the pulse length no longer determines radar-target range measurement accuracy. Note that a PRF of 200,000 corresponds to an interpulse period of 5 μs (that is, 2 μs for transmission and 3 μs for reception). In the case of Doppler radars, the effective transmitted peak power, called the centerline power, will be proportional to the square of the transmitter duty cycle—that is, $(2/5)\lambda^2$. The receiver noise power will be proportional to the receiver duty cycle, or 3/5.

The receiver noise bandwidth no longer needs to be matched to the transmitted pulse and can be decreased to 1000 Hz or below. The criterion is the availability of Doppler filters and the cost of implementation. The Doppler filter time constant, however, should be matched to the duration of the target energy return. Considering the effect of new parameters in the radar equation, we shall increase the target tracking range by the following factor:

$$\left[\frac{2 \times 10^6\,\text{Hz}}{1000\,\text{Hz}} \times \frac{10\,\text{kW}}{100\,\text{kW}} \times \frac{(2/5)\lambda^2}{(3/5)} \right]^{1/4} = 2.70 \tag{4.60}$$

where the last term accounts for transmit and receive duty cycles. Using this and Eq. (4.53), we note that the high-PRF Doppler radar will have a target tracking range of

$$43.25 \times 2.70 = 116.77\,\text{nmi} \tag{4.61}$$

As noted in the following discussion, this approximately threefold increase in tracking range is achieved at the expense of degraded range accuracy and sophisticated hardware design. Using frequency-modulation ranging, we can obtain the range quantization ΔR from Eq. (4.4) as follows:

$$\Delta R = \frac{cB}{2\dot{f}} \tag{4.62}$$

where B is the Doppler filter bandwidth, $\dot{f}$ is the modulation slope, and c is the velocity of light. For an assumed modulation slope of $\dot{f} = 10 \times 10^6$ Hz/s, Eq. (4.62) results in

$$\Delta R = \frac{3 \times 10^8 \times 3.2808 \times 1000}{2 \times 10 \times 10^6}$$

$$= 4.9212 \times 10^4\,\text{ft} \tag{4.63}$$

$$= 8.09\,\text{nmi}$$

Thus a high-PRF Doppler radar of equivalent design achieves higher target tracking ranges, but at the expense of more sophisticated hardware design (Doppler filters and transmitted frequency modulations) and degraded range accuracy. Note that this high-PRF radar will have a range rate measurement accuracy, from Eq. (4.47), of

$$\Delta \dot{R} = \frac{\lambda B}{4}$$

$$= \frac{0.164 \times 1000}{4}$$

$$= 41 \text{ ft/s} \tag{4.64}$$

In a number of cases, this range rate accuracy is utilized to track velocity of targets.

4.12 ANTENNA GAIN PATTERN AND ELECTRONIC SCAN

4.12.1 Antenna Gain Pattern

The antenna gain pattern can be derived from basic relations by considering the propagation of electric field from a set of radiating elements. Consider the radiating elements of Fig. 4.29, with each element radiating energy with the same amplitude and phase. At a point P in the plane of the figure, the total sum of the energy can be computed by obtaining the phase of emitted radiation from each element. Assuming the phase shift of the radiated energy from

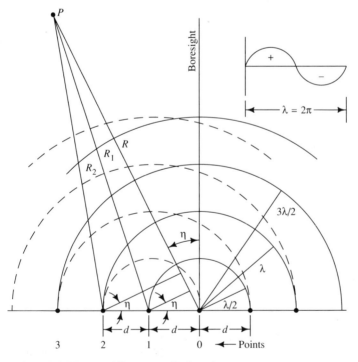

FIGURE 4.29 Equidistance radiating elements.

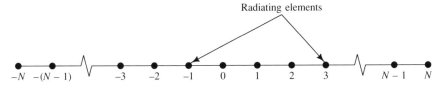

Radiating elements

FIGURE 4.30   Arrangement of $2N + 1$ radiating elements.

point zero (0) arriving at P to be zero, we have

Phase shift of point 0 at $P = 0$

Phase shift of point 1 at $P = R - R_1$

$$= d \sin \eta$$

$$= d \sin \eta 2\pi/\lambda \text{ (radians)}$$

Phase shift of point 2 at $P = R - R_2$

$$= 2d \sin \eta$$

$$= 2d \sin \eta 2\pi/\lambda \text{ (radians)}$$

Phase shift of point n at $P = R - R_n$

$$= nd \sin \eta$$

$$= nd \sin \eta 2\pi/\lambda \text{ (radians)}$$

In the above relations, one wavelength λ is set equal to 2π radians.

In general, field strength at P for N radiating elements, with each element radiating with an amplitude of unity, will be the sum of radiation from each element as follows:

$$E(\eta) = \sum_{n=0}^{N} \cos\left(\frac{2\pi}{\lambda} nd \sin \eta\right) \tag{4.65}$$

where N is the number of radiating elements and $E(\eta)$ is the field strength at P. In Eq. (4.65) a cosine wave propagation is assumed.

For symmetrically located radiating elements as shown in Fig. 4.30, the field strength $E(\eta)$ becomes

$$E(\eta) = \sum_{-N}^{N} \cos\left(\frac{2\pi}{\lambda} nd \sin \eta\right) \tag{4.66}$$

where the limits are from $-N$ to N. Normalizing the value $E(\eta) = 1$ for $\eta = 0$, we get

$$E(\eta) = \frac{1}{2N + 1} \sum_{-N}^{N} \cos\left(\frac{2\pi}{\lambda} nd \sin \eta\right) \tag{4.67}$$

where $1/(2N + 1)$ is the normalizing factor that will result in $E(0) = 1$.

The summation of Eq. (4.62) for an antenna consisting of a continuous radiator can be written in the integral form. Inherent in going from discrete summation to continuous integral is the assumption that the radiating elements are closely spaced. Thus, the subject integral becomes

$$E(\eta) = \frac{1}{2N + 1} \int_{-N}^{N} \cos nx \, dn \qquad (4.68)$$

where

$$x = \frac{2\pi}{\lambda} d \sin \eta \qquad (4.69)$$

and the variable of integration is n. Integration of Eq. (4.68) results in

$$E(\eta) = \frac{1}{2N + 1} \frac{2 \sin Nx}{x} \qquad (4.70)$$

Since $2N + 1 \approx 2N$ for large N, we get

$$E(\eta) = \frac{\sin Nx}{Nx} = \frac{\sin \alpha}{\alpha} \qquad (4.71)$$

where Nx is equal to α. Equation (4.71) represents the often quoted $(\sin x)/x$ behavior of the antenna pattern for a linear array of radiating elements. Note that Eq. (4.71) represents voltage distribution; converting this to power, we get

$$G = [E(\eta)]^2 = \left(\frac{\sin \alpha}{\alpha}\right)^2 \qquad (4.72)$$

where G is the antenna power gain. A plot of Eq. (4.72) is given in Fig. 4.31.

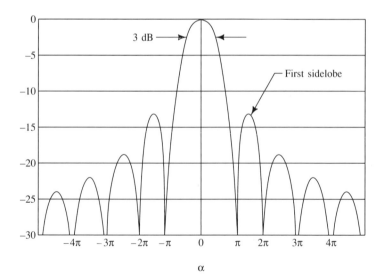

FIGURE 4.31   Antenna gain pattern $[(\sin \alpha)/\alpha]^2$.

Note that the value of α of Eqs. (4.71) and (4.72) is directly related to the physical dimensions of the antenna. From previously given relations, we have

$$\alpha = NX$$

$$= \frac{2\pi}{\lambda}(Nd)\sin\eta \tag{4.73}$$

For an antenna of length l, Nd will be $l/2$ and Eq. (4.73) becomes

$$\alpha = \frac{\pi l}{\lambda}\sin\eta \tag{4.74}$$

Thus the parameter α in the antenna gain, Eq. (4.72), is related to antenna length, angle off boresight η, and transmitted wavelength λ.

Figure 4.31 shows the half-power, or 3-dB, antenna beamwidth together with its sidelobe structure. Half-power beamwidth can be calculated as

$$\theta = \frac{51\lambda}{l}\text{(degrees)} \tag{4.75}$$

where θ is the half-power beamwidth and l is the antenna length. The first sidelobe in the case of Fig. 4.31 is 13.2 dB down from the mainlobe.

The above derivation of antenna gain pattern can be extended to two-dimensional cases where radiating elements are arranged in a planar array. A special and interesting case of planar-array antenna is the circular antenna. In this case the radiation field intensity involves first-order Bessel functions with the resulting half-power beamwidth

$$\theta = \frac{58.5\lambda}{D}\text{(degrees)} \tag{4.76}$$

where D is the antenna diameter. In this case the first sidelobe will be 17.6 dB (power) down from the antenna mainlobe instead of 13.2 dB of $(\sin\alpha)/\alpha$ radiation pattern.

4.12.2 Electronic Scan Antennas

The antenna gain pattern discussed above can be generated by a parabolic-dish antenna with a feed horn or, in the case of planar-array antennas, by a series of slots on the surface of the antenna that are backed by interconnected waveguides. These waveguides are constructed such as to produce uniform phasing of the radiated wavefront on the surface of the planar-array antenna. In both cases, a parabolic dish with a feed horn or the planar-array antenna, the antenna is mechanically moved through azimuth and elevation gimbals to spatially position its gain patterns. This movement, in effect, changes the position of a stationary (with respect to the antenna dish) gain pattern in order to obtain greater coverage. This coverage can also be obtained by keeping the antenna stationary and changing the position of the generated antenna gain pattern as is the case in electronic scanning antennas.

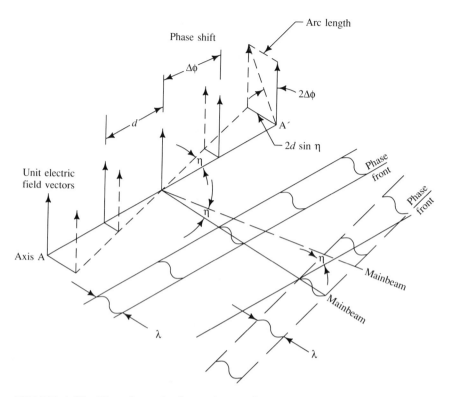

FIGURE 4.32   Phase fronts in electronic scanning antennas.

Electronic scanning antennas are mechanized by controlling the phase of generated waveform at each radiating element. By proper control of this phasing, the angular position of the antenna beam can be varied electronically as opposed to the mechanical motion of the antenna dish.[8]

Figure 4.32 shows an antenna with five radiating elements. For a given azimuth angle η, the phasing of emitted waveform from each element can easily be computed. Consider the case of equal phase increments $\Delta\phi$ between elements. For the second element from the center, we have

Distance from axis $AA' = 2d \sin \eta$

$$= 2d \sin \eta \, \frac{2\pi}{\lambda} \text{(radians)} \tag{4.77}$$

Equating this value to the phase increment at this location of $2 \, \Delta\phi$, we get

$$\Delta\phi = \frac{2\pi d}{\lambda} \sin \eta \tag{4.78}$$

For a desired location of antenna beam η, the phase increment $\Delta\phi$ of Eq. (4.78) can be computed.

Equation (4.78) was derived for a one-dimensional antenna. This equation can be extended to a two-dimensional array antenna, where radiating elements are located on a flat surface and each element is equipped with a phase shifter. For a given angular beam position in azimuth and elevation, the proper phase shift of each element is computed by a digital computer. This phase shift is then put in the appropriate phase shifter for the desired spatial beam position.

In addition to the elimination of mechanical motion of the antenna, in electronic scan radar systems the antenna beam can be moved quickly from one point to the next. This feature allows simultaneous tracking of a greater number of targets.

4.13 PROBABILITY OF DETECTION AND SIGNAL-TO-NOISE RATIO

In the previous discussion of the detection process, it was assumed that achieving a certain signal-to-noise ratio will be sufficient for detection. Actually, more sophisticated methods of calculating the probability of detection are available which relate SNR to threshold setting and result in the probability of detection of the target at a given radar scan. These calculations involve assumptions regarding the type of targets and the effective radar signal return from these targets. The threshold setting is directly related to the number of false target detections due to thermal noise of the radar receiver system, as will be seen in the following discussion.

4.13.1 Types of Radar-Target Returns

Radar targets can be divided into two general categories for the purposes of radar-target cross section (RCS) determination. Aircraft-type targets can be represented by a finite number of scattering points that intercept the transmitted electromagnetic energy and reflect portions of it toward the transmitting antenna. Assuming that each reflecting point on the target reflects energy (during radar target illumination time) with a Gaussian distribution, the total reflected energy from the target will be in the form of a Rayleigh distribution, as shown in Fig. 4.33 and discussed elsewhere.[9] Figure 4.33 shows experimentally measured samples of RCS values for a nose aspect aircraft-type target. The average value of these measurements is 6.93 m^2. It can be shown that the distribution in Fig. 4.33 can be approximated either by a Rayleigh or by an exponential probability distribution. This probability distribution of RCS is used in the probability of detection calculations. In the radar range equation, however, the average value of the RCS is used.

It can also be shown that targets with a dominating reflector will produce radar RCS values that have a Gaussian distribution with the average value

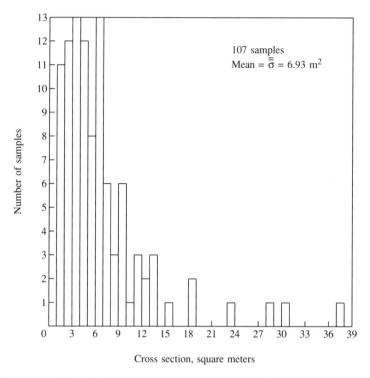

Cross section, square meters

FIGURE 4.33 Experimentally measured target radar cross section of an aircraft-type target.

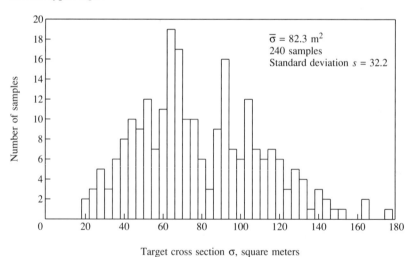

Target cross section σ, square meters

FIGURE 4.34 Target cross section data obtained from an augmented target.

410

closely approximating that of the dominating reflector. Figure 4.34 gives one such target, consisting of an augmenter that was used to enhance target RCS for detection and tracking purposes. Similar types of target returns are expected from satellites, which usually have a dominating spherical or cylindrical structure surrounded by peripheral equipment. The average RCS of Fig. 4.34 is 82.3 m^2, with a standard deviation of 32.2 m^2.

Note that the discussed target variations of Figs. 4.33 and 4.34 occur between radar scans of the target when the radar is in the search mode of operation. These variations are commonly referred to as *target fluctuations* or *target scintillation*.

4.13.2 Probability of Detection[9,10]

The above variations of target RCS values can be incorporated in the probability of detection of a sinusoidal signal in Gaussian noise. These calculations result in probability of detection versus signal-to-noise ratio curves of Figs. 4.35 and 4.36. These curves use the probability of false alarm P_n as a parameter.

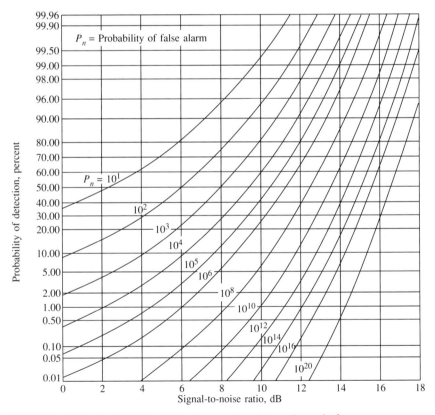

FIGURE 4.35   Probability of detection vs. signal-to-noise ratio for a nonscintillating target.

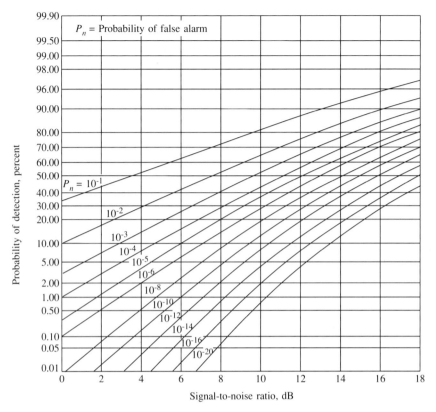

FIGURE 4.36   Probability of detection vs. signal-to-noise ratio for a scintillating target with exponential scintillation.

The value of P_n is utilized in computations relating to threshold setting and false-alarm rates. Figure 4.35 gives the probability of detection for a non-scintillating target. This curve can also be used for calculation where the target consists of a dominating reflector.

Figure 4.36 gives the probability of detection curves for a scintillating target with exponential scintillation. As mentioned before, exponential scintillation results from electromagnetic radiation return of targets with several "equal-strength" reflectors.

The threshold setting can be computed by using the 50 percent probability of detection signal-to-noise ratio of a nonscintillating target as given in Fig. 4.35. For example, a probability of false alarm P_n of 10^{-8} will require a threshold setting of 12.5 dB.

4.13.3 False-Alarm Rate

The threshold setting is usually computed from false-alarm requirements. The requirements are usually specified in terms of false alarms per unit time (i.e., per second or per minute). Consider a low-PRF radar system with a pulse duration of τ seconds and range gates of equal duration. The number of range gates examined per unit time will be

$$n = \frac{1}{\tau} \qquad (4.79)$$

where n is the number of range gates. Using P_n as the probability of false alarm in each range gate, the number of false alarms per unit time n_{fa} will be

$$n_{\text{fa}} = \frac{P_n}{\tau} \qquad (4.80)$$

Or, conversely, time between false alarms T_{fa} will be

$$T_{\text{fa}} = \frac{\tau}{P_n} \qquad (4.81)$$

The false-alarm rate of a high-PRF radar can be computed by considering the number of detection filters examined for the presence of target signal and the time constant of each filter; thus

$$T_{\text{fa}} = \frac{T_c}{NP_n} \qquad (4.82)$$

where T_c is the time constant of each filter and N is the number of Doppler filters.

4.13.4 Numerical Example

Consider a low-PRF radar system with a pulse duration of 0.1 μs. It is desired to have a time between false alarms of 10 minutes. Compute the probability of false alarm and the resulting threshold setting. From Eq. (4.81) we have

$$P = \frac{\tau}{T_{\text{fa}}}$$

$$= \frac{0.1\,\text{E} - 6}{(10)(60)}$$

$$= 1.66\,\text{E} - 10$$

$$\approx 2.0\,\text{E} - 10 \qquad (4.83)$$

The resulting threshold setting from Figure 4.36 will be 13.5 dB, which corresponds to a 50 percent probability of detection value with a probability of false alarm given by Eq. (4.83).

Furthermore, assume that a target SNR of 12 dB has been computed for a target-radar range of 100 km. Compute the probability of detection for this SNR for nonscintillating and exponentially scintillating targets.

From Fig. 4.35 for $P_n = 2 E - 10$ we will obtain a probability of detection of 15 percent for a nonscintillating target and from Fig. 4.43 we will obtain a probability of detection of 25 percent for an exponentially scintillating target.

In actual application of the probability of detection curves, the values of S/N are calculated as a function of radar-target range. The single look (at each radar scan) probability of detection values corresponding to each signal-to-noise (SNR) ratio is obtained from Figs. 4.35 or 4.36 depending on the type of target under consideration. These values will specify the probability of target detection as a function of range. Furthermore, the single-look probability values can be used to compute cumulative probability of detection values as the target-radar range decreases.

EXERCISES

1. Consider a pulsed radar with a PRF of 600 pulses per second. Show that the radar-target range for (a) $t = 320$ and (b) the maximum unambiguous range R_{max} are 48 km and 250 km, respectively.

2. A low-power radar uses a low-noise RF amplifier with a noise figure $F = 9$ dB. It is operating at 14 GHz. The antenna diameter is 1 m and the IF bandwidth is 500 KHz. The radar set is capable of detecting targets of 5-m^2 cross-sectional area at a maximum distance of 20 km. Find the peak transmitted pulse power.

3. *Beacon Radar*: The radar range can be extended by cooperative targets with appropriate receivers. Show that the signal-to-noise ratio at the correlator receiver (matched filter) at the cooperative target is given by

$$\frac{S}{N} = \frac{2E_T G_T G_R \lambda^2}{(4\pi)^2 R^2 K T_0 L}$$

where

$K =$ Boltzmann's constant $= 1.38 \times 10^{-23}$ J/K

$T_0 =$ temperature $= 290$ K

$L =$ system loss factor

$E_T =$ transmitted signal energy

$G_T =$ transmitter antenna gain

$G_R =$ receiver antenna gain

$R =$ distance between transmitter and receiver

$\lambda =$ signal wavelength $= c/f$

Note that the path from the radar to cooperative, known as interrogation or forward link, is independent of the return or response link from the cooperative target.

4. *Jamming Noise*: A radar operating in a hostile environment is subjected to jamming signal. Let a jammer radiate average P_j watts within bandwidth B_j from an antenna with power gains G_j. Show that the maximum range is given by

$$(R^2)_{max} = \frac{P_{ave} G_t^2 E_i(n)}{L_S} \cdot t_0 \cdot \frac{\sigma}{(S/N)_{min}} \cdot \frac{B_S}{P_j G_j}$$

where

$E_i(n) = $ efficiency in integrating n pulses

$\sigma = $ radar cross section

$(S/N)_{min} = $ minimum signal-to-noise ratio necessary to detect a target with a specified probability of detection and false alarm for a single pulse

$t_0 = $ signal integration time in seconds

$P_{ave} = $ average power in watts

$L_S = $ system losses

5. Let

$$S(t) = R_e[u(t)e^{i2\pi f_c t}]$$

where

$$u(t) = \left(\frac{2}{\pi T^2}\right)^{1/4} e^{-t^2/T^2}$$

Show that

$$X(\tau, f_d) = \exp\left[-\frac{1}{2}\left(\frac{t^2}{T^2} + \frac{f_d^2 T^2}{4}\right)\right]$$

6. Let the returned signal be

$$r(t) = Re[u(t-\tau)e^{i2\pi f_c t(t-\tau)}] + n(t)$$

where $n(t)$ is a white noise with variance $N_0/2$. Show that

$$var(\tau) = \frac{1}{\beta^2(2E/N_0)}$$

where β is the RMS bandwidth and ε is the energy of the signal.

7. Let the returned signal be

$$r(t) = Re[u(t)e^{i2\pi(f_c + f_d)t}] + n(t)$$

where $n(t)$ is a white noise with variance $N_0/2$. Show that

$$var(f_d) = \frac{1}{\alpha^2(2\varepsilon/N_0)}$$

where α is the RMS time duration and ε is the energy of the transmitted signal.

8. Prove (a) Eq. (4.13) and (b) Eq. (4.16b).

9. (a) Prove that the optimum filter for adaptive clutter canceler can be expressed as

$$W^o(z) = \frac{1}{H(z)[P(z) + 1]}$$

where

$$P(z) = \frac{S_{uu}(z)}{S_{nn}(z)|H(z)|^2}$$

(b) Prove Eqs. (4.17e) and (4.17f).

10. The observed signal is given by

$$y_k = x_k + n_k \qquad 1 \le k \le N \qquad t_0 < t_1 < \cdots < t_N$$

where n_k is a white Gaussian noise with mean zero and variance σ_k^2. A moving target is modeled as

$$x_k = x_0 + (k - a)v \qquad 1 \le k \le N$$

where a is the initial time and x_0 is the initial position; x_0 and v, the velocity, are unknown. Show that

(a) $\hat{x}_0 = \dfrac{\displaystyle\sum_{k=1}^{N} (y_k/\sigma_k^2)}{\displaystyle\sum_{k=1}^{N} (1/\sigma_k^2)}$

(b) $\hat{v} = \dfrac{\displaystyle\sum_{k=1}^{N} y_k(k - b)/\sigma_k^2}{\displaystyle\sum_{k=1}^{N} (k - b)^2/\sigma_k^2}$

where

$$b = \frac{\displaystyle\sum_{k=1}^{N} (k/\sigma_k^2)}{\displaystyle\sum_{k=1}^{N} (1/\sigma_k^2)}$$

(c) $\sigma_{\hat{x}_0}^2 = \left[\displaystyle\sum_{k=1}^{N} \left(\frac{1}{\sigma_k^2} \right) \right]^{-1}$

(d) $\sigma_{\hat{v}}^2 = \left[\dfrac{\displaystyle\sum_{k=1}^{N} (k - b)^2}{\sigma_k^2} \right]^{-1}$

REFERENCES

1. S. A. Hovanessian, *Radar System Design and Analysis*, Boston, Mass.: Artech House, 1984.
2. S. A. Hovanessian, "An Algorithm for the Calculation of Range in Multiple PRF Radars," *IEEE Trans. Aerosp. Electron Syst.*, vol. AES-12, 1976.
3. W. H. Long and K. A. Harriger, "Medium PRF for the AN/APG-66 Radar," Proc. *IEEE*, vol. 73, Feb. 1985.
4. B. L. Lewis, F. F. Kretschmer, and W. Shelton, *Aspects of Radar Signal Processing*, Boston, Mass.: Artech House, 1986.
5. S. S. Blackman, *Multiple-Target Tracking with Radar Applications*, Boston, Mass.: Artech House, 1986.

6. A. I. Leonov and K. I. Fomichev, *Monopulse Radar*, (translated by William F. Barton from the original 1984 publication in Russian), Boston, Mass.: Artech House, 1986.

7. T. J. Nessmith, "Range Instrumentation Radars," *IEEE Trans. Aerosp. Electron. Syst.*, vol. AES-12, November 1976.

8. E. Brookner, "Phased Array Radars," *Sci. Am.*, pp. 94–99, February 1985.

9. N. C. Mohanty, *Random Signals, Estimation and Identification: Analysis and Applications*, New York: Van Nostrand Reinhold, 1986.

10. N. C. Mohanty, *Signal Processing: Signals, Filtering and Detection*, New York: Van Nostrand Reinhold, 1987.

11. B. Widrow and S. D. Stearns, *Adaptive Signal Processing*, Englewood Cliffs, N.J.: Prentice-Hall, 1975.

LAURENCE B. MILSTEIN

An Introduction to Anti-Jam Performance of Spread Spectrum Communications

Spread spectrum communications consists of transmitting the desired signal in a bandwidth much larger than that of the information. This bandwidth expansion is sometimes accomplished by modulating the information sequence with a large-bandwidth-coded waveform, and such a system is referred to as a *direct-sequence (DS) spread spectrum system.* Alternatively, the spreading can be accomplished by a technique known as *frequency hopping (FH).* In an FH system, the carrier is periodically switched to a different frequency in a seemingly random manner. At any instant, then, the signal occupies only a small bandwidth (i.e., the information bandwidth). However, the total bandwidth of the system is determined by the number and spacing of the frequency slots, and this number can be made as large as necessary.

When DS is used, the modulation format is typically binary phase-shift keying (BPSK) or quadrature phase-shift keying (QPSK), and a coherent receiver is employed for detection. In FH, noncoherent reception is typically used, and the modulation format is often frequency-shift keying (FSK), either binary or *M*-ary. There are a variety of sequences that one can use to spread the spectrum, but probably the most common ones are the so-called *pseudo-noise (PN) sequences.* These sequences are generated from linear shift registers whose feedback connections are such as to produce a code with maximum period. As such, they are sometimes referred to as *maximal-length shift register sequences.*

A typical shift register might appear as shown in Fig. 5.1. The x_i are either 0 or 1. If the initial conditions on the above shift register are

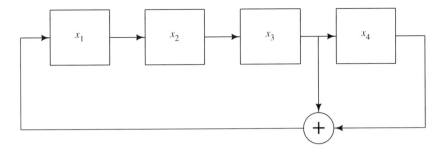

FIGURE 5.1   Maximum length shift register.

$$x_i = 1 \qquad i = 1, 2, 3, 4$$

then the sequence generated will be

1 1 1 1 0 0 0 1 0 0 1 1 0 1 0

Clearly, the output of such a device is periodic, and the maximum period is $2^{L_1} - 1$ for an L_1 stage shift register. This is because the all-zero state is a degenerate state (i.e., if the shift register entered that state, it would never leave it). Notice that every state except the all-zero state appears once and only once. This means, among other things, that in any period of a PN sequence, the number of zeros is exactly one less than the number of ones (e.g., in the above example, there are eight 1's and seven 0's).

The primary advantage of a spread spectrum communication system is the ability to reject interfering signals, such as intentional jamming, unavoidable interference from other users in a multiple-access situation, multipath interference, etc. In what follows we will concentrate on the former scenario, that of intentional jamming, and illustrate how both DS and FH perform in typical jamming environments.

5.1 PROCESSING GAIN

The idea of *processing gain* is, in a sense, the essence of spread spectrum communications, since it is the ability of spread spectrum waveforms to reduce the effect of interfering signals that motivates most of the interest in them. This interference attenuation ability is directly related to the amount of "spreading" in the system, which in turn is related to the processing gain.

It is not clear that there is a universally accepted definition of processing gain, but the one probably most often used, and the one to be used here, can be stated as follows: By definition, we will call the *processing gain* of a system the ratio of the radio-frequency (RF) bandwidth occupied by the spread waveform to the bandwidth occupied by the information-bearing signal itself.

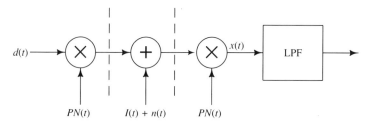

FIGURE 5.2 Direct sequence spread spectrum system.

An alternative, but very similar, definition that is sometimes used is the ratio of RF spread bandwidth to *baseband* information bandwidth. This definition differs from ours by a factor of two if double-sideband (DSB) signaling is used.

Because of the central role that processing gain plays in spread spectrum communications, it is used as a key system descriptor of virtually every spread spectrum system, in particular with respect to how much interference rejection capability the system processes. It is in this latter context that misconceptions and misunderstandings about the correct interpretation of processing gain most frequently arise. As we shall see below, processing gain as defined above almost never provides a precise quantitative description of by how much a given interfering signal is reduced, and at times, can even produce qualitatively misleading results.

Before delving into the details of specific types of spread spectrum systems to analyze the effect of processing gain, consider the simple system shown in Fig. 5.2. A binary information-bearing waveform $d(t)$ is multiplied by a so-called spreading sequence, denoted by $PN(t)$. This spreading sequence is a binary waveform composed of rectangular pulses, each of height ± 1 with a duration of T_c seconds (to conform with accepted terminology, each symbol of the spreading waveform will be referred to as a "chip"). The chip duration is taken to be much shorter than the symbol duration T of the information-bearing signal. This, of course, means that the bandwidth of the product $PN(t)d(t)$ is much larger than that of $d(t)$, since it is at least as large as the bandwidth of $PN(t)$ (recall that multiplication in the time domain corresponds to convolution in the frequency domain). In other words, the bandwidth of the transmitted waveform [that is, $PN(t)d(t)$] has been spread.

This waveform is now transmitted over a channel that adds an interfering signal $I(t)$ and additive white Gaussian noise (AWGN) of two-sided spectral density $\eta_0/2$. At the receiver, the sum of the three signals is multiplied by $PN(t)$ and then a low-pass filter (LPF) is used. Consider the input to the LPF. It is given by

$$x(t) = [d(t)PN(t) + I(t) + n(t)]PN(t)$$
$$= d(t) + PN(t)I(t) + PN(t)n(t) \tag{5.1}$$

since

$$[PN(t)]^2 = 1 \tag{5.2}$$

If $I(t)$ is not correlated with $PN(t)$, the bandwidth of $PN(t)I(t)$ will be much larger than that of $d(t)$. Therefore, if the bandwidth of the final LPF is set equal to the bandwidth of $d(t)$, most of the energy in $PN(t)I(t)$ will be filtered out.

It is this final filtering operation that provides the interference attenuation capability referred to above, and it is the attempt to quantify the amount of this attenuation for different modulation spreading formats that is the subject of this section. Notice that the removal of the spreading sequence from the desired signal, referred to as "despreading," in turn spreads the bandwidth of the interference.

To take this model one step further, consider the following scenario: Assume that the interference is arising from an intentional jamming signal. This jamming signal is assumed to be a band-limited white Gaussian noise process with zero mean and average power J. The strategy that the jammer employs is that of spreading the power of its signal uniformly across the bandwidth occupied by the transmitted spread spectrum waveform, and this latter bandwidth is usually taken to be twice the chip rate (that is, $2/T_c$). Hence the power spectral density of the jamming signal is given by

$$\eta_{0_J} = \frac{J}{2/T_c} = \frac{JT_c}{2} \tag{5.3}$$

and the power out of the LPF of Fig. 5.2, assuming the bandwidth of that filter is $B = 1/T$, is given by

$$P_J = \eta_{0_J} B = \frac{T_c}{2T} J = \frac{J}{2(\mathrm{PG})} \tag{5.4}$$

where PG stands for *processing gain* and equals

$$\frac{\text{RF spread BW}}{\text{RF information BW}} = \frac{2/T_c}{2/T} = \frac{T}{T_c} \tag{5.5}$$

Notice that the amount of power in the jamming signal at the output of the receiver has been reduced by a factor of 2PG compared to what it is at the input to the receiver. Also note that we can express the processing gain in terms of the "time-bandwidth product" of the system, since

$$\mathrm{PG} = \frac{T}{T_c} = \frac{TB_{\mathrm{RF}}}{2} \tag{5.6}$$

Although the above discussion was for a DS system, similar results can be shown to apply to other systems such as those employing FH. Assume, for example, that an FH system transmits one of PG frequencies (assumed to be an integer) every T seconds, and that each of these frequencies is separated by $2/T$ hertz. Then the jammer again has to spread its energy over an RF bandwidth of

$$\text{PG}\left(\frac{2}{T}\right) = \frac{2}{T_c} \tag{5.7}$$

if it wants to be certain of jamming every slot,[a] so its power spectral density is again given by Eq. (5.3) and its power as seen at the output of the final LPF is again given by Eq. (5.4).

The discussion up to now has been very elementary and has been intended to show in an intuitive manner what is the effect of processing gain on a spread spectrum system. In the next two sections the systems will be treated much more carefully, and thus the problems alluded to at the beginning of this section will become apparent. Furthermore, although the interference will be limited to intentional interference, or jamming, the same general concepts apply to the attenuation of other forms of interference (e.g., multipath). When spread spectrum systems are designed to mitigate the effects of an intentional jamming signal, they are often referred to as *anti-jam systems*. This terminology will be used in what follows.

5.2 ANTI-JAM PERFORMANCE OF DIRECT-SEQUENCE SYSTEMS

There are many modulation formats that can be used with DS spreading, although the two most common ones are BPSK and QPSK. With respect to jamming signals, again there are many types of jammers, noise and tone jammers probably being the most commonly occurring ones. Finally, since DS spreading is to be employed, it is possible to put an entire period of the spreading sequence in each symbol, or, as is essentially always done when secure communications is desired, to make the period of the spreading sequence much larger than the duration of one data symbol. In what follows in this section we will first analyze the performance of several DS systems using *short codes* and then examine the system performance when *long codes* are used.

5.2.1 Short Codes

Let us consider the average probability of bit error for a direct-sequence spread spectrum employing BPSK modulation. The receiver is a coherent detector employing matched filter detection, and perfect carrier, bit, and chip synchronization is assumed. The analysis to follow is taken from Ref. 1.

A block diagram of the system is shown in Fig. 5.3. Assuming that a PN code is used to spread the spectrum, the transmitted signal takes on the following form:

---

[a] It should be noted that this is not necessarily the best strategy for the jammer to take. This will be discussed below.

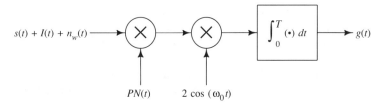

FIGURE 5.3   Correlation receiver for a DS-PSK signal embedded in white noise $n_w(t)$, in the presence of interferers $I(t)$. (From L. B. Milstein, S. Davidovici, and D. L. Schelling, "The Effect of Multiple-Tone Interfering Signals on a Direct Sequence Spread Spectrum Communication System, *IEEE Trans. Commun. Syst.*, vol. COM-30, March 1982. © 1982 IEEE.)

$$x(t) = a_i(t)PN(t)\cos\omega_0 t \tag{5.8}$$

where $a_i(t)$ is the information bearing waveform and $PN(t)$ is the PN code. Let

$$a_i(t) = a_i s(t) \tag{5.9}$$

where a_i takes on the values $\pm A$ for some constant A, and where $s(t)$ determines the pulse shape of each symbol. Each symbol is received in the presence of an intentional jamming signal plus additive white Gaussian noise of two-sided spectral density $\eta_0/2$.

Noise Jamming

Assuming, for simplicity, that the pulse shape $s(t)$ is rectangular; that is,

$$s(t) = \begin{cases} 1 & 0 \le t \le T \\ 0 & \text{elsewhere} \end{cases}$$

denoting by $g(T)$ the output of the detector of Fig. 5.3, and, assuming the jammer at the input is denoted by $I(t)$, then

$$g(T) = a_i T + \int_0^T 2PN(t)I(t)\cos\omega_0\, dt + N(T)$$

$$= a_i T + g_I(T) + N(T) \tag{5.10}$$

where

$$g_I(T) \triangleq 2\int_0^T PN(t)I(t)\cos\omega_0 t\, dt \tag{5.11}$$

and where $N(T)$ is a zero-mean Gaussian random variable with variance

$$\sigma^2 = \eta_0 T \tag{5.12}$$

The jamming signal $I(t)$ is assumed to be a zero-mean stationary Gaussian process independent of both the signal and the noise. It is assumed to have a constant power spectral density over the total bandwidth of the system. For ease of analysis, one common simplifying assumption is made. Since the product of $I(t)$ and $PN(t)$ spreads the bandwidth of $I(t)$ over a range significantly greater than the spread bandwidth of the system and since even just the system spread bandwidth is usually much larger than the information bandwidth, for the purpose of computing the variance of the term $g_I(T)$, the interference $I(t)$ will be modeled as truly white random process.

With this approximation, the contribution to $g(T)$ due to the jammer is a zero-mean Gaussian random variable with variance $\eta_{0_J} T$, where η_{0_J} is the one-sided power spectral density of the jammer and is taken to equal the jammer's power in a bandwidth equal to twice the PN code chip rate divided by that bandwidth. That is, if the jammer is assumed to have a power of J watts in a bandwidth of $2f_c$, where f_c is taken to be the chip rate, then

$$\eta_{0_J} = \frac{J}{2f_c} \tag{5.13}$$

Finally, since the thermal noise and the jamming noise are independent, the total variance of $g(T)$ is given by

$$\sigma_g^2 = (\eta_0 + \eta_{0_J})T \tag{5.14}$$

Under the above assumptions, if $E = A^2 T/2$ is the energy per bit of the transmitted signal (recall that A is the signal amplitude and T is the duration of a bit), then the average probability of error is given by the well-known expression

$$Pe_{\text{PSK}} = \phi\left(-\sqrt{\frac{2E}{\eta_{0_T}}}\right) \tag{5.15}$$

where

$$\eta_{0_T} \triangleq \eta_0 + \eta_{0_J} \tag{5.16}$$

and where

$$\phi(x) \triangleq \frac{1}{\sqrt{2\pi}} \int_{-\infty}^{x} e^{-\frac{y^2}{2}} dy \tag{5.17}$$

Single-Tone Jamming

Assume that initially the system is jammed with a single tone located at the carrier frequency of the transmitted signal. Denoting the jammer by $I(t)$, then

$$I(t) = \alpha \cos(\omega_0 t + \theta) \tag{5.18}$$

where θ is uniformly distributed between zero and 2π. For this situation, the output of the receiver shown in Fig. 5.3 is given by

$$g(T) = a_i T + \cos\theta \int_0^T PN(t)\,dt + N(T) \tag{5.19}$$

where $N(T)$ is again a zero-mean Gaussian random variable with variance $\sigma^2 = \eta_0 T$.

For ease of notation, define

$$\beta \triangleq \int_0^T PN(t)\,dt = -T_c$$

where the last equality followed from the fact that we are using a short code. With this result, the average probability of error for binary PSK is given by

$$P_e = \frac{1}{2\pi} \int_0^{2\pi} \phi\left(-\sqrt{\frac{2E}{\eta_0}} - \frac{\alpha\beta\cos\theta}{\sqrt{\eta_0 T}}\right) d\theta$$

$$\qquad + \frac{1}{2}\operatorname{erfc}(\sqrt{E/\eta_{0T}})$$

$$\operatorname{erfc} x = \frac{2}{\sqrt{\pi}} \int_x^\infty e^{-t^2}\,dt \tag{5.20}$$

If the interfering tone is offset from the carrier frequency of the desired signal by some amount $\Delta\omega \triangleq \omega_i - \omega_0$, where ω_i is the frequency of the interference, then the interference term at the output of system of Fig. 5.3 is given by

$$g_I(T) = \alpha\cos\theta \int_0^T PN(t)\cos\Delta\omega t\,dt - \alpha\sin\theta \int_0^T PN(t)\sin\Delta\omega t\,dt$$

$$\qquad = \alpha\beta_c \cos\theta - \alpha\beta_s \sin\theta \tag{5.21}$$

where

$$\beta_c \triangleq \int_0^T PN(t)\cos\Delta\omega t\,dt \tag{5.22}$$

and

$$\beta_s \triangleq \int_0^T PN(t)\sin\Delta\omega t\,dt \tag{5.23}$$

Therefore, the average probability of error is given by

$$P_e = \frac{1}{2\pi} \int_0^{2\pi} \phi\left(-\sqrt{\frac{2E}{\eta_0}} - \frac{\alpha[\beta_c\cos\theta - \beta_s\sin\theta]}{\sqrt{\eta_0 T}}\right) d\theta$$

$$\qquad = \frac{1}{2\pi} \int_0^{2\pi} \phi\left(-\sqrt{\frac{2E}{\eta_0}} - \frac{\alpha\gamma\cos\hat{\theta}}{\sqrt{\eta_0 T}}\right) d\hat{\theta} \tag{5.24}$$

where

$$\gamma \triangleq \sqrt{\beta_c^2 + \beta_s^2} \tag{5.25}$$

and

$$\hat{\theta} \triangleq \theta + \tan^{-1}\frac{\beta_s}{\beta_c} \tag{5.26}$$

Notice that because θ is uniformly distributed in $[0, 2\pi]$, and any angle is only defined modulo 2π, $\hat{\theta}$ is also uniformly distributed in $[0, 2\pi]$.

If we define

$$J \triangleq \frac{\alpha^2}{2}$$

to be the power in the interfering tone and let $S = E/T$ be the power in the transmitted signal, then Eq. (5.24) can be expressed in the more common form

$$P_e = \frac{1}{2\pi} \int_0^{2\pi} \phi\left(-\sqrt{\frac{2E}{\eta_0}}\left(1 - \sqrt{\frac{J}{S}}\frac{\gamma}{T}\cos\hat{\theta}\right)\right) d\hat{\theta} \tag{5.27}$$

where J/S represents the jammer-to-signal power ratio. Hence we see that the system performance is determined by three key quantities,

$$\frac{E}{\eta_0}, \quad \frac{J}{S}, \quad \text{and} \quad \frac{\gamma}{T} \tag{5.28}$$

Let us now examine the effect of offsetting the frequency of the interfering tone in an intuitive manner. This procedure can be done quite easily by considering the power that falls within the mainlobe of the final integrate-and-dump filter due to the jammer. The power spectrum of a PN sequence L chips in length that has a period of T_1 seconds is known to be given by

$$S(\omega) = \frac{2\pi}{L^2}\delta(\omega) + 2\pi\frac{L+1}{L^2}\left(\frac{\sin\frac{\omega T_1}{2L}}{\frac{\omega T_1}{2L}}\right)^2 \sum_{\substack{i=-\infty \\ i\neq 0}}^{\infty} \delta\left(\omega - \frac{i2\pi}{T_1}\right) \tag{5.29}$$

If $T_1 = T$, where T is the duration of a data symbol, then when the frequency of the jammer equals the carrier frequency, only the component of $S(\omega)$ centered at $\omega = 0$ [that is, $2\pi/L^2\delta(\omega)$] falls in the mainlobe of the integrate-and-dump, and hence the power is $1/L^2$. If, however, the jammer's frequency is the carrier frequency plus $2\pi/T$, then the terms of $S(\omega)$ at $\pm 2\pi/T$ will fall within the mainlobe, and the power contributed by the jammer will be approximately

$$\frac{L+1}{L^2} \approx \frac{1}{L} \tag{5.30}$$

which is a factor of L times as great as what it was when the jammer was

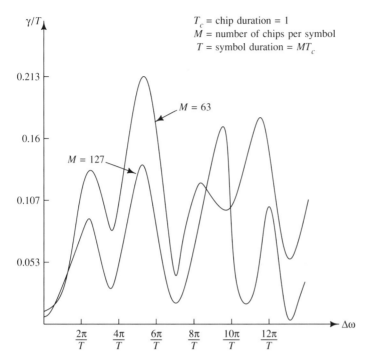

FIGURE 5.4   Variation of γ/T with frequency offset $\Delta\omega$. (From L. B. Milstein, S. Davidovici, and D. L. Schilling, "The Effect of Multiple-Tone Interfering Signals on a Direct Sequence Spread Spectrum Communication System, *IEEE Trans. Commun. Syst.*, vol. COM-30, March 1982. © 1982 IEEE.)

at the carrier frequency. Hence, we would certainly anticipate significant degradation in system performance when the jammer's offset is $2\pi/T$ rather than zero.

This degradation can be further seen quantitatively by examining the two curves of Fig. 5.4. These curves correspond to PN sequences of length 63 and 127. The ordinate of the graph is γ/T, and the abscissa is $\Delta\omega$, so the curves show, for the two specific PN sequences chosen, the actual attenuation experienced by a tone interfering signal. Notice that T_c, the chip duration, has been held constant, and therefore T, the symbol duration, is twice as large for the 127-chip sequence as it is for the 63-chip sequence.

This figure provides us with two important perspectives on the performance of DS systems using short spreading sequences. The first is the fact that the most favorable frequency at which the tone can be located (from the system's point of view) is $\Delta\omega = 0$ (unless $\Delta\omega \gg \pi/T$, which is not a realistic case). The second perspective is best illustrated with the following scenario. Suppose a DS system has two modes of operation, one at bit rate f_b with a spreading factor of 63, and the other at a bit rate of $f_b/2$, with a spreading factor of 127. The former bit rate would be used until the system became too heavily stressed by the tone jammer, at which point it would halve its data rate and double its processing gain, thereby keeping the total spread bandwidth constant.

Intuitively, one would think the system is better off in this latter mode, since now it has twice the processing gain with which to combat the jamming. Indeed, in most instances this situation is true. However, from Fig. 5.4 we see that this is not always the case because at times, for a fixed jammer location, the curve corresponding to $M = 63$ is actually lower than that corresponding to $M = 127$ (i.e., the system with the smaller processing gain is actually outperforming the one with the higher processing gain). This example illustrates the statement made at the beginning of this chapter that processing gain is not always a meaningful concept, at least not as it relates to interference rejection.

Multiple-Tone Interference

The performance of a DS system in the presence of multiple-tone interference can be treated as described in Ref. 2. It is no longer feasible to evaluate precisely the probability of error because, in general, each interfering tone has an independent phase associated with it; hence, for K tones, each of the K phases has to be averaged over.

The foregoing method[2] gets around this problem by using the Chernoff bound[3] to provide an upper bound on the average probability of error. The signal is again taken to be $a_i(t)PN(t)\cos\omega_0 t$, as in Eq. (5.8). Also, as before, it is assumed that an entire period of the PN code is contained in each T-second data symbol. The noise is AWGN with two-sided spectral density $\eta_0/2$, but the interference $I(t)$ is now given by

$$I(t) = \sum_{i=1}^{K} \alpha_i \cos(\omega_i t + \theta_i) \tag{5.31}$$

where α_i, ω_i, and θ_i are the amplitude, frequency, and random phase, respectively, of the ith tone. All the θ_i are uniformly distributed in $[0, 2\pi]$ and all are independent of each other and independent of both the signal and the noise.

Denoting by $g_I(T)$ the output of the system due to the interference, we have

$$g_I(T) \approx \sum_{i=1}^{K} \alpha_i \int_0^T \cos[(\omega_i - \omega_0)t + \theta_i]PN(t)\, dt \tag{5.32}$$

where the sum frequency terms have been ignored. Expanding Eq. (5.32) yields

$$
\begin{aligned}
g_I(T) &= \sum_{i=1}^{K} \alpha_i \cos\theta_i \int_0^T [\cos(\omega_i - \omega_0)t]PN(t)\, dt \\
&\quad - \sum_{i=1}^{K} \alpha_i \sin\theta_i \int_0^T [\sin(\omega_i - \omega_0)t]PN(t)\, dt \\
&= \sum_{i=1}^{K} \alpha_i \beta_{c_i} \cos\theta_i - \sum_{i=1}^{K} \alpha_i \beta_{s_i} \sin\theta_i
\end{aligned}
\tag{5.33}
$$

where, as in (Eqs. 5.22) and (5.23),

$$\beta_{c_i} \triangleq \int_0^T PN(t)\cos(\omega_i - \omega_0)t \, dt \tag{5.34}$$

and

$$\beta_{s_i} \triangleq \int_0^T PN(t)\sin(\omega_i - \omega_0)t \, dt \tag{5.35}$$

Assuming that a minus one is the transmitted data symbol, the test statistic $g(T)$ at the output of the integrator of Fig. 5.3 is given by

$$g(T) = -AT + g_I(T) + N(T) \tag{5.36}$$

where $N(T)$ is a zero-mean Gaussian random variable with random $\eta_0 T$. The probability of error is given by

$$P_e = P\{-AT + g_I(T) + N(T) > 0\} \tag{5.37}$$

By using the Chernoff bound, Eq. (5.37) can be upper bounded as follows [for simplicity, $N(T)$ will be expressed just as N]:

$$P\{N + g_I(T) > AT\} \leq e^{-\lambda_0 AT} E\{e^{\lambda_0[N + g_I(T)]}\}$$

where $E\{\cdot\}$ denotes expectation and where λ_0 satisfies the following equation:

$$E\{[N + g_I(T)]\exp[\lambda_0(N + g_I(T))]\} = ATE\{\exp[\lambda_0(N + g_I(T))]\} \tag{5.38}$$

From Eq. (5.38), the equation for λ_0 can be rewritten as

$$E\left\{Ne^{\lambda_0 N} \prod_{i=1}^K \exp[\lambda_0(\alpha_i \beta_{c_i} \cos\theta_i - \alpha_i \beta_{s_i} \sin\theta_i)]\right\}$$

$$+ E\left\{\sum_{i=1}^K [\alpha_i \beta_{c_i} \cos\theta_i - \alpha_i \beta_{s_i} \sin\theta_i]e^{\lambda_0 N}\right.$$

$$\left. \times \prod_{j=1}^K \exp[\lambda_0[\alpha_j \beta_{c_j} \cos\theta_j - \alpha_j \beta_{s_j} \sin\theta_j]]\right\}$$

$$= ATE\{\exp[\lambda_0(N + g_I(T))]\} \tag{5.39}$$

Finally, it can be shown that Eq. (5.39) reduces to

$$\sum_{i=1}^K \alpha_i \sqrt{\beta_{c_i}^2 + \beta_{s_i}^2} \frac{I_1(\lambda_0 \alpha_i \sqrt{\beta_{c_i}^2 + \beta_{s_i}^2})}{I_0(\lambda_0 \alpha_i \sqrt{\beta_{c_i}^2 + \beta_{s_i}^2})} = -\lambda_0 \eta_0 T + AT \tag{5.40}$$

where $I_0(x)$ and $I_1(x)$ are modified Bessel functions.

To solve Eq. (5.40) the following approximation to the Bessel functions, valid for large values of their arguments, will be used:[4]

$$I_v(z) \simeq \frac{e^z}{\sqrt{2\pi z}} \tag{5.41}$$

If Eq. (5.41) is used in Eq. (5.40), and if it assumed that each of the tone jammers has the same power ($\alpha_i = \alpha$, all i), then λ_0 can easily be shown to be given by

$$\lambda_0 = \frac{T\sqrt{\dfrac{S}{J}\displaystyle\sum_{i=1}^{K}\gamma_i}}{\sqrt{\dfrac{J}{2}T^2\left(\dfrac{\eta_0}{E}\right)\left(\dfrac{S}{J}\right)}} \tag{5.42}$$

where

$$\gamma_i \triangleq \sqrt{\beta_{c_i}^2 + \beta_{s_i}^2} \tag{5.43}$$

Finally, the bound on the average probability of error reduces to

$$P_e \leq \exp\left\{-\frac{E}{\eta_0}\left[1 - \frac{J}{S}\left(\frac{\displaystyle\sum_{i=1}^{K}\gamma_i}{T^2}\right)\right]\right\} \cdot \prod_{i=1}^{K} I_0\left[\frac{2E}{\eta_0}\sqrt{\frac{J}{S}\frac{\gamma_i}{T}}\left(1 - \sqrt{\frac{J}{S}}\frac{\displaystyle\sum_{i=1}^{K}\gamma_i}{T}\right)\right] \tag{5.44}$$

5.2.2 Long Codes

When a spreading sequence whose period is longer than the symbol duration is used, each symbol will typically be spread by a different subsequence of the code, and so the results of the previous section have to be modified somewhat.

The one exception is the case of Gaussian broadband noise jamming. Subject to our approximation that the despread noise jammer can be modeled as a white Gaussian process, the specific pattern of ± 1's comprising the spreading sequence of any given data symbol will have no affect on the probability of error of that symbol as long as the system is properly synchronized.

For tone jamming, one technique that can be used is to employ the Chernoff bound as described above on each individual symbol and then average the results. This procedure is followed in Ref. 2, and the results are presented below.

The essence of the technique is to treat the γ_i of Eq. (5.43) as random variables. Then Eq. (5.44) can be considered a conditional bound on the average probability of error, given a value (or set of values if $K > 1$) of the γ_i. The unconditional probability of error is then upper-bounded as follows: Denoting the left-hand side of Eq. (5.44) by $P_e(\gamma^{-(i)})$, then

$$P_e \leq \frac{1}{n_1}\sum_{i=1}^{n_1} P_e(\gamma^{-(i)}) \tag{5.45}$$

where $\gamma^{-(i)}$ is a K vector whose elements are the K values of the γ_j appropriate for the ith data symbol, and where n_1 is the number of distinct sets of $\gamma^{-(i)}$.

Another technique described in Ref. 2 for the special case of $K = 1$ and $\Delta\omega = 0$ is the use of the central limit theorem to approximate the effect of the interference on the final test statistic. For the conditions stated above, the output of the integrate-and-dump detection filter is given by

$$g(T) = AT + \alpha \cos\theta \int_0^T PN(t + lT_c)\,dt + N(T)$$

$$\triangleq AT + g_I(T) + N(T) \tag{5.46}$$

assuming that a "plus one" is transmitted and that lT_c is the starting phase of the PN sequence for the current symbol.

Consider the integral in the second term of the right-hand side of Eq. (5.46). Assuming there are M chips per symbol, and denoting by d_k the kth chip of the PN sequence, then

$$\int_0^T PN(t + lT_c)\,dt = T_c \sum_{i=1}^{M-1} d_{i+l} \tag{5.47}$$

If $1 \ll M \ll L$, where L is the period of the PN sequence, the terms in the sum of the right-hand side of Eq. (5.47) can be approximated as independent, zero-mean, binary random variables, and hence the sum can be approximated as having Gaussian statistics by the central limit theorem. The average value of the sum is zero, and the second moment of the sum is given by $T_c^2 M$.

From these results, $g_I(T)$ is a conditional Gaussian random variable with zero mean value and variance, conditioned upon θ, given by

$$\text{var}(g_I(T)|\theta) = \alpha^2 T_c^2 (\cos^2\theta) M \tag{5.48}$$

With the above approximation for the density of $g_I(T)$, the average probability of error, conditioned upon θ, is given by

$$P(e|\theta) = P\{N(T) + g_I(T) < -AT\}$$

$$= \phi\left\{\frac{-AT}{[\eta_0 T + \alpha^2 T_c^2 M \cos^2\theta]^{1/2}}\right\}$$

$$= \phi\left\{\frac{-1}{\left[\dfrac{\eta_0}{2E} + \dfrac{J}{S}\left(\dfrac{1}{M}\right)\cos^2\theta\right]^{1/2}}\right\} \tag{5.49}$$

The unconditional probability of error is given by

$$P_e = \frac{1}{2\pi} \int_0^{2\pi} P(e|\theta)\,d\theta \tag{5.50}$$

Using the results derived above, we can now see how the system performs. Figure 5.5 shows the results of evaluating Eq. (5.50) for the ratio M/L equal to 0.1. However, as another perspective on these results, consider the results predicted by the Chernoff bound of Eq. (5.45). Figure 5.6 shows the results, where $M = 255$, $L = 2047$, and $J/S = 10$ dB, and when all possible starting

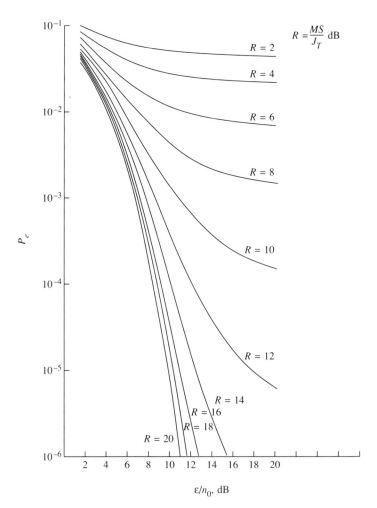

FIGURE 5.5   Probability of error for $M/L = 0.1$. (From L. B. Milstein, S. Davidovici, and D. L. Schilling, "The Effect of Multiple-Tone Interfering Signals on a Direct Sequence Spread Spectrum Communication System, *IEEE Trans. Commun. Syst.*, vol. COM-30, March 1982. © 1982 IEEE.)

positions of the code are averaged over. Also shown in Fig. 5.6 is the curve predicted from Eq. (5.50) corresponding to $M/L = 0.1$. It can be seen that at error rates greater than about 5×10^{-5}, the agreement between the two curves is fairly good. However, below 5×10^{-5}, the curve predicted by the Gaussian approximation starts to "bottom out," whereas the curve predicted by the Chernoff bound shows no such behavior.

This bottoming-out effect is inherent in the use of the Gaussian approximation and can be seen very easily from Eq. (5.49). Note that for a given J/S, as $E/\eta_0 \to \infty$ in Eq. (5.49), the argument of the $\phi(\cdot)$ function remains finite, and hence the average probability of error does not go to zero. On the other hand, the true probability of error is given precisely by

$$P_e = \frac{1}{2\pi} \int_0^{2\pi} \phi \left\{ -\frac{\sqrt{2E}}{\eta_0} \left[1 - \frac{J}{S} \left(\frac{\beta_c}{T} \right) \cos \theta \right] \right\} d\theta \qquad (5.51)$$

where

$$\beta_c \triangleq \int_0^T PN(t)\, dt \qquad (5.52)$$

Notice that if

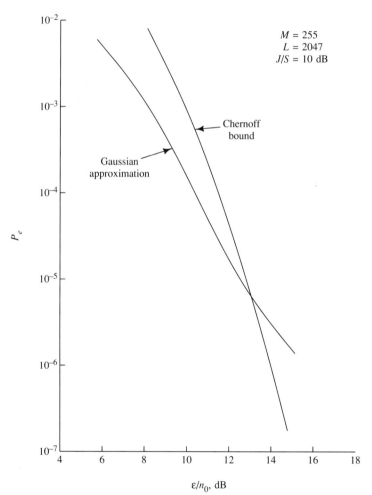

FIGURE 5.6   Comparison of Chernoff bound and Gaussian approximation for $M/L = 0.1$. (From L. B. Milstein, S. Davidovici, and D. L. Schilling, "The Effect of Multiple-Tone Interfering Signals on a Direct Sequence Spread Spectrum Communication System, *IEEE Trans. Commun. Syst.*, vol. COM-30, March 1982. © 1982 IEEE.)

$$\frac{J}{S}\left|\frac{\beta_c}{T}\cos\theta\right| < 1 \qquad (5.53)$$

the curve predicted by Eq. (5.51) can never bottom out, and hence it would be in disagreement with the approximate result of Eq. (5.51). Note also that this is true regardless of whether $M = L$ or $M < L$. However, it is much less likely to hold as $M \ll L$, since highly unbalanced sequences of length M are likely to result under this latter condition. In particular, if

$$L = 2^{L_1} - 1 \qquad (5.54)$$

(that is, if L_1 is the length of the shift register that generates the PN sequence) and if $M \le L_1$, then Eq. (5.53) will always be violated whenever $J/S > 1$ (i.e., $J/S > 0$ dB), since there will always be at least one sequence of M consecutive code chips that are all "minus one." In other words, there will be at least one sequence of code chips for which the code provides no processing gain whatsoever.

Therefore, it can be anticipated that Eq. (5.50) will yield more accurate results for lower values of M/L rather than higher values. Finally, since Eq. (5.50) resulted from a central limit type of argument, it must be interpreted subject to the usual reservations about the density of a sum of a finite number of independent random variables—namely, that the approximation loses its accuracy the further out one goes on the tail of the distribution.

5.3 ANTI-JAM PERFORMANCE OF FREQUENCY-HOPPED SYSTEMS

When FH spreading is employed, the most common modulation techniques are frequency-shift keying (FSK) and differential phase-shift keying (DPSK). In this section, the analysis is limited to binary FSK, and the results presented constitute a special case of those presented in Ref. 5. The system to be considered is shown in Fig. 5.7. If we assume that are $N_1 \triangleq 2^{L_1}$ frequencies over which to hop, we can specify the one we use with an L_1-bit word. The PN generator is used to determine the $L_1 - 1$-most significant digits of the L_1-bit word, and the data are then used to specify the least significant digit. This guarantees that the two frequencies in a given MARK/SPACE pair are always contiguous in frequency. The dehopper in the receiver is assumed to be in synchronism with the hopping pattern of the received signal, but no phase coherence is assumed on any individual hop. After the signal is dehopped, it is detected in a standard noncoherent FSK demodulator, shown in Fig. 5.8. If error-correction coding is used, the decoder follows the FSK demodulator.

The FSK-FH system will be evaluated in the presence of both partial-band noise jamming and partial-band tone jamming. In particular, the interferer will be allowed to jam K out of the N slots over which the signal can hop, with K an integer between 1 and N_1. As is well known, when a non-

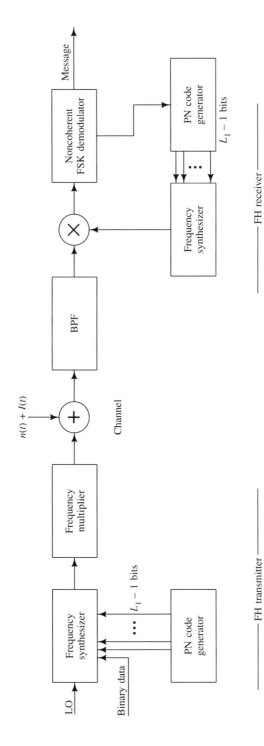

FIGURE 5.7 Frequency-hopped communication system. (From L. B. Milstein, R. L. Pickholtz, and D. L. Schilling, "Optimization of the Processing Gain of an FSK-FH System," *IEEE Trans. Commun. Syst.*, vol. COM-28, July 1980. © 1980 IEEE.)

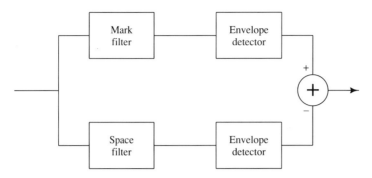

FIGURE 5.8   Noncoherent detection of FSK. (From L. B. Milstein, R. L. Pickholtz, and D. L. Schilling, "Optimization of the Processing Gain of an FSK-FH System," *IEEE Trans. Commun. Syst.*, vol. COM-28, July 1980. © 1980 IEEE.)

coherent binary FSK system is used over an AWGN channel, the probability of error is given by the simple exponential

$$P_e = \frac{1}{2}\exp\left[-\frac{E}{2\eta_0}\right] \tag{5.55}$$

where E is the energy per bit and η_0 is the one-sided noise spectral density. However, for the cases of interest here, (i.e., when jamming is present), the general results provided in Ref. 6 are needed, and those are as follows:

$$P_e = \frac{1}{2}[1 - Q(\sqrt{b}, \sqrt{a}) + Q(\sqrt{a}, \sqrt{b})] - \frac{\hat{A}}{2}e^{-(a+b)/2}I_0(\sqrt{ab}) \tag{5.56}$$

where

$$Q(a,b) \triangleq \int_b^\infty xe^{-(a^2+x^2)/2}I_0(ax)\,dx \tag{5.57}$$

$$I_0(x) \triangleq \frac{1}{2\pi}\int_0^{2\pi} e^{x\cos\theta}\,d\theta \tag{5.58}$$

$$\hat{A} \triangleq \frac{\sigma_2^2 - \sigma_1^2}{\sqrt{(\sigma_1^2 + \sigma_2^2)^2 - 4\sigma_1^2\sigma_2^2|\rho|^2}} \tag{5.59}$$

and

$$\begin{Bmatrix} a \\ b \end{Bmatrix} \triangleq \frac{1}{2}\left[\frac{(\sigma_1^2 + \sigma_2^2)(|\mu_1|^2 + |\mu_2|^2) - 4\sigma_1\sigma_2\,\text{Re}\{\mu_1\mu_2^*\rho\}}{(\sigma_1^2 + \sigma_2^2)^2 - 4\sigma_1^2\sigma_2^2|\rho|^2}\right.$$
$$\left.\mp \frac{|\mu_1|^2 - |\mu_2|^2}{[(\sigma_1^2 + \sigma_2^2)^2 - 4\sigma_1^2\sigma_2^2|\rho|^2]^{1/2}}\right] \tag{5.60}$$

In Eqs. (5.59) and (5.60), the five quantities μ_1, μ_2, σ_1^2, σ_2^2, and ρ are functions of the received signal, interference, and noise. In particular, if $z_2(t)$ and $z_1(t)$

are the complex envelopes of the outputs of the MARK and SPACE filters, respectively, then at the appropriate sampling instant T,

$$\tfrac{1}{2}E\{|z_i(T_s) - \mu_i|^2\} \triangleq \sigma_i^2 \tag{5.61}$$

$$E\{z_i(T)\} \triangleq \mu_i \qquad i = 1, 2, \ldots \tag{5.62}$$

and

$$\tfrac{1}{2}E\{(z_1(T) - \mu_1)^*(z_2(T) - \mu_2)\} \triangleq \rho\sigma_1\sigma_2 \tag{5.63}$$

where ρ is, in general, complex. However, for our purposes, ρ will always be zero. Finally, in any of the above definitions, $\text{Re}\{\cdot\}$ refers to the "real part of" and * denotes complex conjugate.

The jammer at any instant of time will jam K of the N_1 slots. Since K does not in general equal N_1, for a given pair of MARK and SPACE frequencies the jammer might be jamming both frequencies, one of the two frequencies, or neither frequency, and each case has to be examined separately.

Consider first the model for a noise jammer. The jammer in the ith slot will be taken to be

$$j(t) = j_{ci}(t)\cos\omega_i t - j_{si}(t)\sin\omega_i t \tag{5.64}$$

where ω_i is the center frequency of the ith slot and $j_{ci}(t)$ and $j_{si}(t)$ are independent zero-mean Gaussian random processes with power equal to J and power spectral density as shown in Fig. 5.9. The jammers in any two slots i and j will be taken to be independent of each other for all $i \neq j$. The jammer is assumed to have a total power of J_T, and that power is distributed equally among all K slots being jammed, so J, the jammer power per slot, equals J_T/K. Both the MARK and SPACE filters [denoted $H_2(\omega)$ and $H_1(\omega)$, respectively] are assumed to be ideal rectangular filters centered at the respective MARK and

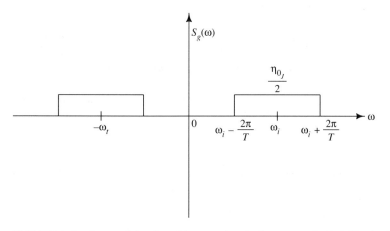

FIGURE 5.9 Spectral density of jammer in ith slot. (From L. B. Milstein, R. L. Pickholtz, and D. L. Schilling, "Optimization of the Processing Gain of an FSK-FH System, *IEEE Trans. Commun. Syst.*, vol. COM-28, July 1980. © 1980 IEEE.)

SPACE frequency and having bandwidth of $2f_c = 2/T$, where T is the symbol duration. Denoting the one-sided power spectral density of the jammer by η_{0_J}, we have

$$J = 2\eta_{0_J}f_c \tag{5.65}$$

Also, assuming the jammer is at the MARK frequency, then the total noise power at the output of $H_2(\omega)$ is given by

$$\sigma_2^2 = 2\eta_0 f_c + J = 2(\eta_0 + \eta_{0_J})f_c \tag{5.66}$$

Likewise, the noise power out of the SPACE filter is given by $\sigma_1^2 = 2\eta_0 f_c$. When the jammer is present in both the MARK and SPACE slots, the noise power of σ_2^2 equals σ_1^2, and both equal

$$\sigma_1^2 = \sigma_2^2 = 2[\eta_0 + \eta_{0_J}]f_c \tag{5.67}$$

For tone jamming, the various noise terms are always the same, independent of how many slots are being jammed. Also, the parameter $\hat{A}$ of Eq. (5.59) is zero for this case. Therefore, one only needs to be concerned with the quantities a and b. Finally, the jammers at the output of filters $H_1(\omega)$ and $H_2(\omega)$ are taken to be

$$j_1(t) = A_{J_1}\cos(\omega_1 t + \theta_1) \tag{5.68}$$

and

$$j_2(t) = A_{J_2}\cos(\omega_2 t + \theta_2) \tag{5.69}$$

respectively, where the A_{J_i}, $i = 1, 2$, are constants and the θ_i are independent random variables uniformly distributed in the interval $[0, 2\pi]$. Note that A_{J_i} will be zero if the jammer is not present in the respective filter. Equations (5.68) and (5.69), along with the signal components of $H_1(\omega)$ and $H_2(\omega)$, determine the parameters μ_1 and μ_2.

With the above models, the probability of error of the system can now be derived. The actual expressions for the parameters $\hat{A}$, a, and b for each of the two cases considered above are presented in Appendix A. The functional form of the equations representing the probability of error are given below, where all the conditional probabilities are given by Eq. (5.56) for appropriate values of the parameters $\hat{A}$, a, and b. These probability of error expressions are also explicitly given in Appendix A.

For noise jamming,

$$P_e = \frac{(N_1 - K)(N_1 - K - 1)}{N_1(N_1 - 1)}P \text{ (error|neither slot jammed)}$$

$$+ \frac{K(N_1 - K)}{N_1(N_1 - 1)}[P(\text{error|only MARK jammed})$$

$$+ P(\text{error|only SPACE jammed})]$$

$$+ \frac{K(K - 1)}{N_1(N_1 - 1)}P(\text{error|both slots jammed}) \tag{5.70}$$

For tone jamming,

$$P_e = \frac{(N_1 - K)(N_1 - K - 1)}{N_1(N_1 - 1)} P(\text{error}|\text{neither slot jammed})$$

$$+ \frac{K(N_1 - K)}{N_1(N_1 - 1)} \frac{1}{2\pi} \left[\int_0^{2\pi} P(\text{error}|\text{only MARK jammed}) \, d\theta_2 \right.$$

$$\left. + \int_0^{2\pi} P(\text{error}|\text{only SPACE jammed}) \, d\theta_1 \right]$$

$$+ \frac{K(K - 1)}{N_1(N_1 - 1)} \frac{1}{4\pi^2} \int_0^{2\pi} \int_0^{2\pi} P(\text{error}|\text{both slots jammed}) \, d\theta_1 \, d\theta_2$$

$$(5.71)$$

These expressions yield the performance of the system when no error correction coding is used. However, in order to obtain reasonable performance with an FH-FSK system in the presence of an intentional jammer, it is well known that some type of error-correction coding must be used. This can be seen very simply by assuming that the jammer is much stronger than the desired signal and that it chooses to put all its power in a single shot (i.e., the jammer jams one out of N_1 slots). Then, with no error-correction coding and assuming the transmitted signal is a MARK, the system will make an error (with high probability) every time the particular frequency being jammed is the SPACE corresponding to that MARK (or vice versa). This will happen, on the average, one out of every N_1 hops, so the probability of error of the system will be approximately $1/N_1$, independently of signal-to-noise ratio.

In light of this, results are presented for systems encoded with a [7,4] Hamming code,[b] and also for systems encoded with a [23,12] Golay code.[b] As is well known, the [7,4] Hamming code corrects all single errors, whereas the [23,12] Golay code corrects all combinations of three or fewer errors.

Since the addition of parity check bits required in error-correcting codes results in an increased bandwidth (assuming the information rate is constant), to make a meaningful comparison between the systems, the amount of spread spectrum processing gain is decreased by the same proportion that the error-correcting code expands the bandwidth. For example, if the uncoded FH-FSK system uses N_1 slots, the [7,4] Hamming encoded system will use $4N_1/7$ slots.

The final probability of bit error in the decoded sequence can be shown to be given approximately by[7]

$$P_b \simeq \frac{1}{n} \sum_{i=e+1}^{n} \binom{n}{i} P_e^i (1 - P_e)^{n-i} \tag{5.72}$$

where P_e is given by either Eq. (5.70) or Eq. (5.71) and e is the number of correctable errors of the code. Equation (5.72) assumes that the frequency to

---

[b] See Chapter 2.

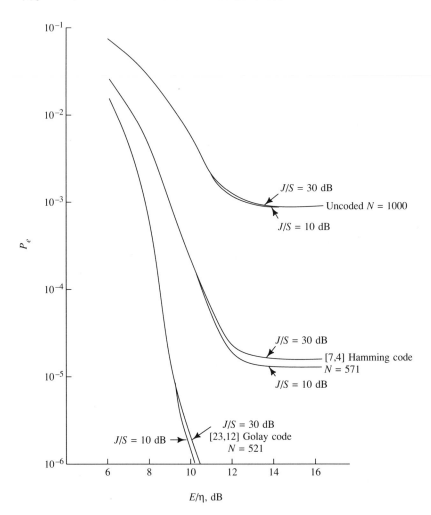

P_e

E/η, dB

FIGURE 5.10   Noise jammer, $K = 1$. (From L. B. Milstein, R. L. Pickholtz, and D. L. Schilling, "Optimization of the Processing Gain of an FSK-FH System," *IEEE Trans. Commun. Syst.*, vol. COM-28, July 1980. © 1980 IEEE.)

which the signal hops on the ith hop is independent of the frequency it was at on the jth hop, for all $i \neq j$.

The results evaluating Eqs. (5.70) to (5.72) are shown in Figs. 5.10 to 5.13. Figures 5.10 and 5.11 show the effect of partial band noise jamming when K, the total number of slots being jammed, is equal to 1 and 100, respectively. The total number of slots that were available for use by the uncoded system was $N_1 = 1000$. This resulted in 571 slots for the system employing the [7,4] Hamming code and 521 slots for the system using the [23, 12] Golay code. From Fig. 5.10 it can be seen that the difference in system performance when $J_T/S = 10\,\text{dB}$ from when $J_T/S = 30\,\text{dB}$ is negligible for each of the three coding

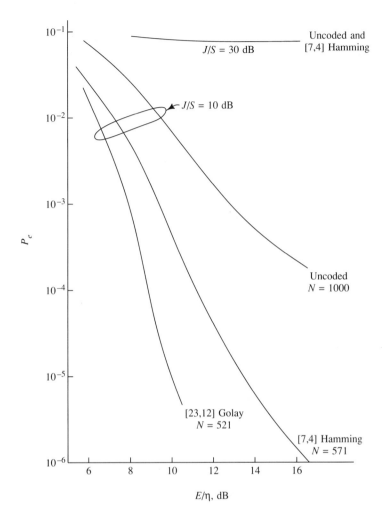

FIGURE 5.11   Noise jammer, $K = 100$. (From L. B. Milstein, R. L. Pickholtz, and D. L. Schilling, "Optimization of the Processing Gain of an FSK-FH System," *IEEE Trans. Commun. Syst.*, vol. COM-28, July 1980. © 1980 IEEE.)

situations. On the other hand, whereas the curves for the uncoded system flatten out at $P_e = 1/N_1 = 10^{-3}$, the coded systems perform markedly better. In particular, the advantage of using the three-error-correcting Golay code is evident.

From Fig. 5.11 one can see a clear qualitative difference from the results of Fig. 5.10. There is now a tremendous difference in the effect of the jammer when J_T/S is increased from 10 dB to 30 dB. In fact, for the $J_T/S = 30$ dB case, the Golay-encoded system was actually the poorest and resulted in an error rate of slightly greater than 10^{-1}. For this reason, that particular curve is missing from Fig. 5.11. Also, it is clear from Figs. 5.10 and 5.11 that for the

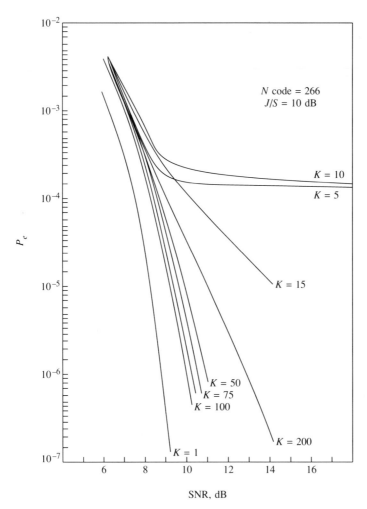

FIGURE 5.12 P_e variation with K, where $P/S = 10$ dB. (From L. B. Milstein, S. Davidovici, and D. L. Schilling, "The Effect of Multiple-Tone Interfering Signals on a Direct Sequence Spread Spectrum Communication System," *IEEE Trans. Commun. Syst.*, vol. COM-30, March 1982. © 1982 IEEE.)

same jammer power it is much more advantageous to the jammer to jam 100 of the slots rather than just a single slot.

If the noise jammers are replaced with tone jammers, analogous results are obtained. Also, note that the label on the abscissas of these figures is denoted SNR, where SNR is the average signal-to-noise ratio at the output of the MARK or SPACE filter (depending upon which signal is transmitted). Therefore, to compare meaningfully, say, the [23, 12] encoded curves and the uncoded curves, the coded curves have to be shifted by $10 \log(23/12) = 2.83$ dB to the right, since for equal average power and equal information rates in

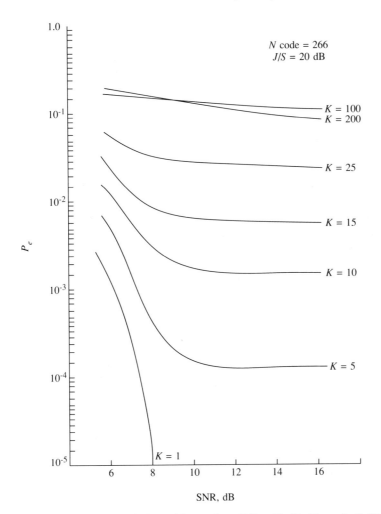

FIGURE 5.13 P_e variation with K, where $P/S = 20$ dB. (From L. B. Milstein, S. Davidovici, and D. L. Schilling, "The Effect of Multiple-Tone Interfering Signals on a Direct Sequence Spread Spectrum Communication System," *IEEE Trans. Commun. Syst.*, vol. COM-30, March 1982. © 1982 IEEE.)

the two systems, the energy per symbol of the coded system is 12/23 of the energy-per-bit symbol of the uncoded system.

Finally, the effect of varying K, the number of interfering signals, for the case of tone interference is explicitly illustrated in Fig. 5.12, in which $J_T/S = 10$ dB. It can be seen that $K = 10$ degrades the system performance the most. If J_T/S is increased to 20 dB, Fig. 5.13 shows the corresponding results. Now K in the vicinity of 100 corresponds to worst-case performance. This, is reasonable since, for large E/η_0, the interference power per slot does not have to be much greater than 0 dB (relative to the signal power) to cause an error with high probability. In other words, from the point of view of worst-case performance, the number of interfering tones should be roughly equal to the number that results in a J/S per slot of 0 dB.

The overview presented was meant to introduce the reader to the manner in which both DS and FH systems can be used to combat intentional interference. For those readers interested in pursuing the subject in greater depth, there are many sources of information available, including Refs. 8–12.

APPENDIX A
Probability of Error Expressions for FH-FSK

It is desired to determine the three parameters a, b, and $\hat{A}$ for each of the four possible jamming situations for both noise and the tone jammers, as well as expressions for the resulting probabilities of error.

Noise Jamming

For a jammer absent from both slots,

$$\hat{A}_0 = 0 \tag{A.1}$$

$$\begin{Bmatrix} a_0 \\ b_0 \end{Bmatrix} = \frac{1}{2}\left[\frac{2(2\eta_0 f_c)[A_m^2 + A_{sc}^2 + A_{ss}^2]}{D_0^2} \mp \frac{A_m^2 - (A_{sc}^2 + A_{ss}^2)}{D_0} \right] \tag{A.2}$$

where

$$D_0 \triangleq 4\eta_0 f_c \tag{A.3}$$

$A_m \triangleq$ amplitude of signal component out of MARK filter

$A_{sc}, A_{ss} \triangleq$ in-phase and quadrature amplitude of signal component out of SPACE filter.

The specific expressions for A_m, A_{sc}, and A_{ss} are given in Appendix B.

For jammer present in only the MARK slot,

$$\hat{A}_{11} = -\frac{2\eta_{0J} f_c}{D_1} \tag{A.4}$$

$$\begin{Bmatrix} a_{11} \\ b_{11} \end{Bmatrix} = \frac{1}{2}\left[\frac{[2(2\eta_0 f_c) + 2\eta_{0J} f_c][A_m^2 + A_{sc}^2 + A_{ss}^2]}{D_1^2} \right.$$
$$\left. \mp \frac{A_m^2 - (A_{sc}^2 + A_{ss}^2)}{D_1} \right] \tag{A.5}$$

where

$$D_1 \triangleq 4\eta_0 f_c + 2\eta_{0J} f_c. \tag{A.6}$$

For a jammer present in only the SPACE slot,

$$\hat{A}_{12} = -\hat{A}_{11} \tag{A.7}$$

$$\begin{Bmatrix} a_{12} \\ b_{12} \end{Bmatrix} = \begin{Bmatrix} a_{11} \\ b_{11} \end{Bmatrix} \tag{A.8}$$

For a jammer present in both slots,

$$\hat{A}_2 = 0 \tag{A.9}$$

$$\begin{Bmatrix} a_2 \\ b_2 \end{Bmatrix} = \frac{1}{2} \left[\frac{2[2\eta_0 f_c + 2\eta_{0,J} f_c][A_m^2 + A_{sc}^2 + A_{ss}^2]}{D_2^2} \right.$$

$$\left. \mp \frac{A_m^2 - (A_{sc}^2 + A_{ss}^2)}{D_2} \right] \tag{A.10}$$

where

$$D_2 \triangleq 2(2\eta_0 f_c + 2\eta_{0,J} f_c). \tag{A.11}$$

To determine the overall probability of error of the system, consider the following. The K slots that the jammer hits are assumed to be randomly chosen (with a uniform distribution) from the N_1 slots that the FSK signal can hop over. Hence we have the following probabilities:

P [neither MARK nor SPACE jammed]

$$= \left(1 - \frac{K}{N_1} \right) \left(1 - \frac{K}{N_1 - 1} \right)$$

$$= \frac{(N_1 - K)(N_1 - K - 1)}{N_1(N_1 - 1)} \tag{A.12}$$

P [MARK jammed or SPACE jammed (but not both)]

$$= \frac{2K}{N_1} \frac{(N_1 - 1) - (K - 1)}{N_1 - 1} = \frac{2K(N_1 - K)}{N_1(N_1 - 1)} \tag{A.13}$$

$$P \text{ [both MARK and SPACE jammed]} = \frac{K(K - 1)}{N_1(N_1 - 1)} \tag{A.14}$$

Therefore, denoting the probability of error given by Eq. (5.56) by $P_e(a, b, \hat{A})$ and reserving P_e for the overall system probability of error, and assuming that a MARK has been transmitted, we have

$$P_e = \frac{(N_1 - K)(N_1 - K - 1)}{N_1(N_1 - 1)} P_e(a_0, b_0, 0)$$

$$+ \frac{1}{2} \frac{2K(N_1 - K)}{N_1(N_1 - 1)} [P_e(a_{11}, b_{11}, \hat{A}_{11}) + P_e(a_{12}, b_{12}, \hat{A}_{12})]$$

$$+ \frac{K(K - 1)}{N_1(N_1 - 1)} P_e(a_2, b_2, 0) \tag{A.15}$$

Tone Jamming

Since the parameter $\hat{A}$ of Eq. (5.60) is zero for this case, it just remains to determine the quantities a and b. The jammers at the output of filters $H_1(\omega)$ and $H_2(\omega)$ are given by

$$j_1(t) = A_{J_1} \cos(\omega_1 t + \theta_1) \qquad (A.16)$$

and

$$j_2(t) = A_{J_2} \cos(\omega_2 t + \theta_2) \qquad (A.17)$$

respectively, where the A_{J_i}, $i = 1, 2$, are constants and the θ_i are independent random variables uniformly distributed in the interval $[0, 2\pi]$. With this model, it can be shown that[5]

$$\begin{Bmatrix} a \\ b \end{Bmatrix} = \frac{1}{2} \left[\frac{2\sigma_n^2 [[A_m^2 + 2A_{J_1} A_m \cos\theta_1 + A_{J_1}^2] + [A_{sc}^2 + A_{ss}^2 + 2A_{J_2}(A_{sc}\cos\theta_2 + A_{ss}\sin\theta_2) + A_{J_2}^2]]}{D^2} \right.$$

$$\left. \mp \frac{A_m^2 + 2A_m A_{J_1}\cos\theta_1 + A_{J_1}^2 - A_{J_2}^2 - A_{sc}^2 - A_{ss}^2 - 2A_{J_2}(A_{sc}\cos\theta_2 + A_{ss}\sin\theta_2)}{D} \right] \qquad (A.18)$$

where

$$D = 4\eta_0 f_c \qquad (A.19)$$

Note that in the above equations, $\frac{1}{2}A_{J_i}^2$ equals the jammer power whenever the ith slot is jammed. If the jammer is not present in the ith slot, then $A_{J_i} = 0$.

The following quantities are now defined.

$$a_0 = a|_{A_{J_1} = A_{J_2} = 0} \qquad (A.20)$$

$$b_0 = b|_{A_{J_1} = A_{J_2} = 0} \qquad (A.21)$$

$$a_{11} = a|_{A_{J_2} = 0} \qquad (A.22)$$

$$b_{11} = b|_{A_{J_2} = 0} \qquad (A.23)$$

$$a_{12} = a|_{A_{J_1} = 0} \qquad (A.24)$$

$$b_{12} = a|_{A_{J_1} = 0} \qquad (A.25)$$

Using Eqs. (A.20) to (A.25) the conditional probability of error of the system, conditioned on the values of the random phases θ_1 and θ_2, is given by

$$P(e|\theta_1, \theta_2) = \frac{(N_1 - K)(N_1 - K - 1)}{N_1(N_1 - 1)} P_e(a_0, b_0, 0)$$

$$+ \frac{1}{2} \frac{2K(N_1 - K)}{N_1(N_1 - 1)} [P_e(a_{11}, b_{11}, 0) + P_e(a_{12}, b_{12}, 0)]$$

$$+ \frac{K(K - 1)}{N_1(N_1 - 1)} P_e(a_2, b_2, 0), \qquad (A.26)$$

where a_2 and b_2 correspond to neither A_{J_1} nor A_{J_2} equaling zero. The unconditional probability of error of the system is given by the average of $P(e|\theta_1, \theta_2)$ over all θ_1 and θ_2. That is,

$$P_e = \frac{(N_1 - K)(N_1 - K - 1)}{N_1(N_1 - 1)} P_e(a_0, b_0, 0)$$

$$+ \frac{1}{2\pi} \frac{K(N_1 - K)}{N_1(N_1 - 1)} \left[\int_0^{2\pi} P_e(a_{11}, b_{11}, 0) \, d\theta_1 + P_e(a_{12}, b_{12}, 0) \, d\theta_2 \right]$$

$$+ \frac{1}{4\pi^2} \frac{K(K - 1)}{N_1(N_1 - 1)} \int_0^{2\pi} \int_0^{2\pi} P_e(a_2, b_2, 0) \, d\theta_1 \, d\theta_2 \qquad \text{(A.27)}$$

These latter averages have to be evaluated numerically to obtain the final results.

APPENDIX B
Intermediate-Frequency Filter Output Signals for FH-FSK

It is desired to determine the outputs of MARK and SPACE filters when a MARK is transmitted. Using the well-known technique of low-pass equivalent filtering,[6] it is easily shown that the output of an ideal bandpass filter centered at ω_0 with two-sided bandwidth $2\omega_c$ when the input is the signal

$$S_i(t) = \begin{cases} A \cos(\omega_1 t + \theta) & 0 \le t \le T \\ 0 & \text{elsewhere} \end{cases} \qquad \text{(B.1)}$$

is given by

$$f_{\text{out}}(t) = \text{Re}\{S_o(t) e^{-j(\omega_1 t + \theta)}\} \qquad \text{(B.2)}$$

where $S_o(t)$ is the output of the low-pass equivalent of the band-pass filter. For an ideal brick-wall band-pass filter,

$$S_o(t) = \frac{1}{2\pi} \int_{\omega_0 - \omega_1 - \omega_c}^{\omega_0 - \omega_1 + \omega_c} A \sin \frac{\omega T/2}{\omega/2} e^{-j\omega T/2} e^{j\omega t} \, d\omega \qquad \text{(B.3)}$$

Choosing the sampling time $T_s = T/2$ yields the following for the outputs of the MARK and SPACE filters:

$$S_{o_M}\left(\frac{T}{2}\right) = \frac{2A}{\pi} Si\left(\omega_c \frac{T}{2}\right) \qquad \text{(B.4)}$$

and

$$S_{o_s}\left(\frac{T}{2}\right) = \frac{A}{\pi}\left[Si(\omega_1 - \omega_2 + \omega_c)\frac{T}{2} - Si(\omega_1 - \omega_2 - \omega_c)\frac{T}{2} \right] \qquad \text{(B.5)}$$

where

$$Si(x) \triangleq \int_0^x \frac{\sin y}{y} dy \tag{B.6}$$

and where ω_1 and ω_2 are the SPACE and MARK carrier frequencies, respectively. Letting

$$E = \frac{A^2 T}{2}$$

be the energy per bit of the signal yields the following for the quantities A_m, A_{sc}, and A_{ss} defined as follows:

$$A_m = \frac{2}{\pi} \sqrt{\frac{2E}{T}} Si\left(\omega_c \frac{T}{2}\right) \tag{B.7}$$

$$A_{sc} = \frac{1}{\pi} \sqrt{\frac{2E}{T}} \left[Si(\omega_1 - \omega_2 + \omega_c)\frac{T}{2} - Si(\omega_1 - \omega_2 - \omega_c)\frac{T}{2} \right]$$

$$\times \cos(\omega_2 - \omega_1)\frac{T}{2} \tag{B.8}$$

$$A_{ss} = \frac{1}{\pi} \sqrt{\frac{2E}{T}} \left[Si(\omega_1 - \omega_2 + \omega_c)\frac{T}{2} - Si(\omega_1 - \omega_2 - \omega_c)\frac{T}{2} \right]$$

$$\times \sin(\omega_2 - \omega_1)\frac{T}{2} \tag{B.9}$$

REFERENCES

1. D. L. Schilling, L. B. Milstein, R. L. Pickholtz, and R. W. Brown, "Optimization of the Processing Gain of an M-ary Direct Sequence Spread Spectrum Communication System," *IEEE Trans. Commun. Syst.*, vol. COM-28, pp. 1389–1398, August 1980.
2. L. B. Milstein, S. Davidovici, and D. L. Schilling, "The Effect of Multiple-Tone Interfering Signals on a Direct Sequence Spread Spectrum Communication System," *IEEE Trans. Commun. Syst.*, vol. COM-30, pp. 436–446, March 1982.
3. J. M. Wozencraft and I. M. Jacobs, *Principles of Communication Engineering*, New York: Wiley, 1965.
4. M. Abramowitz and I. A. Stegun, *Handbook of Mathematical Functions*, National Bureau of Standards, 1964.
5. L. B. Milstein, R. L. Pickholtz, and D. L. Schilling, "Optimization of the Processing Gain of an FSK-FH System," *IEEE Trans. Commun. Syst.*, vol. COM-28, pp. 1062–1079, July 1980.
6. M. Schwartz, W. R. Bennett, and S. Stein, *Communications Systems and Techniques*, New York: McGraw-Hill, 1966.
7. *Error Control Handbook*. Final report prepared by Linkabit Corporation for U.S. Air Force, Contract No. FYY 620-76-C-0056, July 15, 1976.
8. J. K. Holmes, *Coherent Spread Spectrum Systems*, New York: Wiley, 1982.
9. C. E. Cook, F. W. Ellersick, L. B. Milstein, and D. L. Schilling, *Spread Spectrum Communications*, New York: IEEE Press, 1983.

10. D. J. Torrieri, *Principles of Secure Communication Systems*, Boston, Mass.: Artech House, 1985.
11. M. K. Simon, J. K. Omura, R. A. Scholtz, and B. K. Levitt, *Spread Spectrum Communications*, vols. I–III, Rockville, Md.: Computer Science Press, 1985.
12. R. E. Ziemer and R. L. Peterson, *Digital Communications and Spread Spectrum Systems*, New York: Macmillan, 1985.

CHARLES L. RINO

6

New Developments in Propagation Theory and Their Impact on Communication and Radar Signal Survivability

The propagation of radio waves in random media has been studied intensively for many decades now. Under the conditions of narrow-angle scattering that apply in the propagation environments of concern to satellite system designers, a complete and tractable mathematical theory exists; moreover, satellite scintillation data have been accumulated to verify structure models and the implementations of the theory as it is used for the performance evaluation of transionospheric radar and communication systems.

Considerable progress has also been made in understanding the physics of ionospheric irregularities that cause radio-wave scintillation. When a localized plasma enhancement is convected across magnetic field lines, ion-neutral collisions cause polarization fields that act to enhance perturbations. This $E \times B$ or gradient-drift instability mechanism had been known in laboratory plasmas for some time. Linson and Workman[1] first suggested that it might be responsible for the structuring of the plasma clouds created by ionospheric barium releases. Woodman and La Hoz[2] proposed a variant of the gradient-drift mechanism to explain equatorial spread F as observed in radar back-scatter but known to be associated with intense radio-wave scintillation. In situ rocket probes of bottomside equatorial spread F showed a one-dimensional spectrum with a k^{-2} power-law form that seemed to be consistent with the steepened gradients associated with the $E \times B$ instability.[3]

Nonetheless, the simultaneous occurrence of equatorial scintillation from

450

very high frequencies (VHF) to gigahertz frequencies could not be explained by a simple power-law spectral model. Indeed, the discovery of gigahertz scintillation by Craft[4] and Westerlund was completely unexpected. Later, in situ rocket probes[5] and satellite data[6] revealed that the irregularities primarily responsible for the scintillation have a two-component power-law structure. By taking this two-component structure into account, a consistent explanation of equatorial gigahertz scintillation can be obtained.[7,8]

Numerical simulations of the instability process have played an important role in our understanding of irregularity development and their spectral characteristics;[9,10] however, only recently has a consistent picture emerged.[11,12] The two-component spectrum is observed only when an in situ probe cuts *across* the principal axis of the irregularities, which lies along the direction of convection. The single-component k^{-2} power-law spectrum is observed only when the scan direction lies strictly *along* the direction of cross-field elongation. The steepened gradients occur mainly at the tips of the elongated structure. Because most of the fluctuation power resides in the modes supporting the larger-scale structure, the two-component spectrum dominates the propagation effects. Thus, in collisional plasmas the scintillation-producing irregularities have directionally dependent spectral characteristics in addition to being highly elongated along the magnetic field direction.

Auroral-zone and polar-cap scintillation are caused by the same basic mechanism, but the presence of a highly conductive auroral E layer, and the complicated electrodynamics of these regions causes a much more complicated pattern. The highly inclined magnetic field geometry further complicates the scintillation characteristics.[13]

The consequences for propagation effects of these most recent findings have just begun to be explored. We have already noted that the two-component model is necessary to explain the observed wavelength dependence of equatorial scintillation, but models that fully accommodate what is now known about ionospheric irregularities have yet to be implemented. The propagation theory itself has not changed, but considerable progress has been made in understanding the behavior of intensity scintillation in power-law environments. In the remainder of this chapter, we shall briefly review these findings and discuss their ramifications for systems analysis.

6.1 PROPAGATION THEORY

A more detailed discussion of the material presented in this section and Section 6.2 can be found in Refs. 4, 8, and 15 and the references cited therein. A general characterization of the propagation environment for systems analysis is straightforward. Upon traversing the quiescent ionosphere, a signal at frequency f_c well above the ionospheric critical frequency admits the representation

$$S(t) = \mathrm{Re}[v(t - \tau_d) \exp\{2\pi i(f_c - f_d)\}] \qquad (6.1)$$

where τ_d is the propagation delay

$$\tau_d = \frac{R}{c} + \frac{r_e c N_P}{2\pi f_c^2} \tag{6.2}$$

and $f_d = \partial\phi/\partial t$ is the Doppler shift, where

$$\phi = \frac{2\pi f_c R}{c} - r_e c \frac{N_P}{f_c} \tag{6.3}$$

The free-space path length is R and N_P is the path integral of the electron density. The remaining parameters r_e and c represent the classical electron radius and the velocity of light. For simplicity, we have made an assumption that the frequency dependence of τ_d and ϕ are negligible over the bandwidth of the signal.

When irregularities are present, each Fourier component of the signal acquires a time-dependent complex modulation. Thus, $v(t)$ in Eq. (6.1) is replaced by

$$v_r(t) = \int_{-F/2}^{F/2} \hat{v}(f) h(t, f + f_c) \exp\{2\pi i f t\}\, df \tag{6.4}$$

where $\hat{v}(f)$ is the Fourier transform of $v(t)$ and $h(t, f)$ is a complex random transfer function that characterizes the propagation medium. For *frequency flat* fading, $h(t, f)$ is constant over the frequency band of the signal, whereby $v_r(t) \approx v(t) h(t, f_c)$. This is the most commonly encountered situation in which the amplitude and phase of the signal acquire random perturbations. The time variation is primarily due to the relative motion of the ionospheric irregularities across the propagation path, in which case $h(t, f) = u_f(\mathbf{v}_{\text{eff}} t, z)$, where $u_f(\rho i, z)$ is the *response* of the ionospheric channel to a sinusoidal excitation at frequency f.

The spatially varying transfer function $u(\rho; z)$ is obtained by solving the parabolic-wave equation

$$\frac{\partial u}{\partial z} = \overbrace{-i\frac{k}{2}\nabla_T^2 u}^{\text{Diffraction}} - \overbrace{i r_e \lambda \delta N_e u}^{\text{Phase}} \tag{6.5}$$

Over small propagation paths, the signal acquires a simple phase perturbation; however, diffraction will ultimately convert phase structure to amplitude perturbations. Indeed, from Ref. 5 it follows that in the absence of the phase term each Fourier component in the spatial wavenumber spectrum propagates by acquiring the phase change $\exp\{-iK^2(\lambda z/4\pi)\}$. The spatial wave number K is the magnitude of the transverse component. Because the magnitude of the Fourier component is k, we have the relation $\mathbf{k} = (\mathbf{K}, k_z)$, where $k_z = \sqrt{k^2 - K^2}$. The propagation factor is simply the quadratic approximation to k_z. Thus,

$$u(\rho, z) = \int\!\!\int \hat{u}(\mathbf{K}, z_0) \exp\{i(KZ)^2\} \exp\{i\mathbf{K}\cdot\boldsymbol{\rho}\} \frac{d\mathbf{K}}{2\pi} \tag{6.6}$$

The quantity $Z = \sqrt{\lambda z/4\pi}$ is called the Fresnel radius. At a distance z, an irregularity comparable to the Fresnel radius will begin to cause amplitude variations.

The essential elements of this process are preserved in a simple model in which the effects of the medium are replaced by an equivalent phase screen that imparts the phase perturbation

$$\phi(\rho) = -r_e\lambda \int \delta N_e(\rho, z)\,dz \tag{6.7}$$

In Eq. (6.6), $\hat{u}(\mathbf{K}, z_0)$ is replaced by the two-dimensional Fourier transform of

$$u(\rho, z_0) = \exp\{i\phi(\rho)\} \tag{6.8}$$

For a Gaussian random process,

$$R(\Delta\rho) = \langle u(\rho, z_0)u^*(\rho', z_0)\rangle$$
$$= \exp\{-D(\Delta\rho)\} \tag{6.9}^a$$

where

$$D(\Delta\rho) = \langle [\phi(\rho) - \phi(\rho')]^2\rangle \tag{6.10}$$

is the phase structure function.

If we let $u_{1,2}$ represent the signals at frequencies f_1 and f_2, it is readily shown by direct computation from Eq. (10.6) that

$$\langle u_1(\rho, z)u_2^*(\rho', z)\rangle$$
$$= \int\!\!\int \Phi_{12}(\mathbf{K})\exp\{-i(KZ)^2(\delta f/f_c)\}\exp\{-i\mathbf{K}\cdot\Delta\rho\}\frac{d\mathbf{K}}{(2\pi)^2} \tag{6.11}$$

where $\Phi_{12}(\mathbf{K})$ is the Fourier transform of $R(\Delta\rho)$, $\delta f = f_2 - f_1$ is the frequency separation, and f_c is the mean frequency. If the two frequencies are equal, $\delta f = 0$, and thus Eq. (6.11), which is now the mutual coherence function for the wave field, is equal to $R(\Delta\rho)$ independently of the propagation distance z. This very old result is strictly true only for homogeneous wavefields, whereas slow background variations are always present to some degree.

The two-point two-frequency coherence function defined by Eq. (6.11) provides a sufficient characterization of the ionospheric channel to evaluate both the signal coherence time and its average loss of frequency coherence over the signal bandwidth. With appropriate refinements, this characterization leads to the generalized transfer function for the channel, which is the

---

[a] $\langle x\rangle \triangleq E(x)$, the expectation operator.

basis for both simulations and performance evaluation. The characterization is strictly complete, however, only if the channel is Gaussian. Under strong scatter conditions, one expects the fading statistics to be not only Gaussian, but Rayleigh. This condition is generally referred to as saturation, where the scintillation index defined as

$$S_4 = \frac{\sqrt{\langle I^2 \rangle - \langle I \rangle^2}}{\langle I \rangle} \tag{6.12}$$

where

$$I = |u(\rho, z)|^2 \tag{6.13}$$

is unity. Thus, understanding intensity scintillation under conditions of strong scatter is crucially important.

Before describing these results, we note that ionospheric irregularities can be characterized by their spatial wave-number spectrum. Even in the simplest situation, however, we must consider multiple power-law segments of the form $\Phi_N(\mathbf{q}) = C_s q^{-(2v-1)}$, where C_s is the *turbulent strength* for the power-law segment. The relation between the spectrum of phase fluctuations as defined by Eq. (6.7), and the in situ spectrum is particularly simple, namely,

$$\Phi_\phi(\mathbf{K}) = r_e^2 \lambda^2 l_p \Phi_N(\mathbf{q})|_{q_z=0} \tag{6.14}$$

At the very least, the effects of multiple power-law propagation environments must be considered. At the next level of complexity, the ramifications of directionally dependent spectral characteristics should be evaluated. In effect, the power-law index v depends on both the wave number and its direction.

6.2 INTENSITY STRUCTURE

To summarize the results that have been developed recently, we shall use a simplified one-dimensional model in which

$$\Phi_i(K_x) = \int \langle II' \rangle \exp\{iK_x \Delta x\} \, d\, \Delta x \tag{6.15}$$

represents the intensity spectral density. It follows that

$$\int \Phi(K_x) \frac{dK_x}{2\pi} = S_4^2 + 1 \tag{6.16}$$

From the phase screen theory, it can be shown that

$$\Phi_i(K_x) = \int \exp\left\{-g\left(\eta, K_x \frac{z}{k}\right)\right\} \cos(K_x \eta) \, d\eta \tag{6.17}$$

where

$$g(\eta, \xi) = \int \Phi_\phi^1(K_x) \sin^2\left(\frac{\eta K_x}{2}\right) \sin^2\left(\frac{\xi K_x}{2}\right) \frac{dK_x}{2\pi} \tag{6.18}$$

The one-dimensional form of the phase spectrum is obtained by integrating the two-dimensional spectrum as defined by Eq. (6.14) over one of the spatial wave number components. The result of this integration for a single power law with an outer-scale cutoff q_0 takes the form

$$\Phi_\phi^1(K_x) = \frac{C_p}{(q_0^2 + K_x^2)^{p/2}} \tag{6.19}$$

where $p = 2v$ is the one-dimensional phase spectral index as it is usually measured. For example, a one-dimensional in situ index of 2 corresponds to $v = 1.5$ or $p = 3$.

Substituting this spectral form into Eq. (6.18) and changing variables gives the result

$$g(\eta, K_x l_f^2) = U'8 \int (R^2 + \omega^2)^{-p/2} \sin^2\left(K_x l_f \frac{\omega}{2}\right) \sin^2\left(\eta \frac{\omega}{2l_f}\right) \frac{d\omega}{2\pi} \tag{6.20}$$

where the parameters are defined as follows:

$$l_0 = C_p^{-1/(p-1)}$$
$$l_f = \sqrt{z/k}$$
$$R = q_0 l_f$$
$$U' = (l_f/l_0)^{p-1}$$

The critical parameters are l_0, which is the coherence scale for the complex wave field itself, the Fresnel scale $l_f = \sqrt{2Z}$, the ratio of the Fresnel scale to the outer scale R, and the strength parameter U. The two-dimensional problem can be formulated in exactly the same way. It is easily seen that Eq. (6.20) is well defined for $R = 0$. Thus, there is a scale-free or fractal limit in which the scintillation structure is determined entirely by l_0, l_f, and p. It happens that, in this regime, the scintillation behavior is critically dependent on the power-law index p.

In the weak scatter limit $U' \ll 1$, the well-known result

$$\Phi_i(K_x) \approx 4\Phi_\phi^1(K_x) \sin^2(l_f^2 K_x^2/2) \tag{6.21}$$

is readily demonstrated; moreover $S_4^2 \propto U'$. The general behavior of S_4^2 is summarized in Fig. 6.1. The solid curves in the lower portion of the figure show the fractal limits for $p = 2$ and $p = 4$. The dashed curves show how the transition to scale dependent behavior occurs. The important point to note is that in propagation environments with steeply sloped power-law regimes, sustained departures from saturated scintillation can take place. Indeed, whenever there is a characteristic scale in the medium that is comparable to the Fresnel radius, at least a localized departure from Rayleigh fading—strong focusing—will occur. In multiple-power law environments the observed scintillation characteristics depend very critically on the Fresnel scale relative to the scale at which transition occurs.

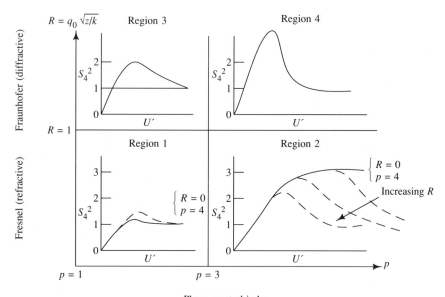

FIGURE 6.1   Behavior of S_4 in power-law environments.

In the ionosphere, naturally occurring irregularities with scale sizes larger than ≈ 800 m follow a power law with $p < 3$; moreover, at the maximum perturbation levels that occur, saturated scintillation only occurs at frequencies and observing distances that place the Fresnel radius within this power law range. This is Region 1 in Fig. 6.1, and it explains why strong focusing is rarely observed in naturally occurring ionospheric scintillation. The more steeply sloped power-law segment below 800 m generally does not carry sufficient strength to cause strong saturation when the Fresnel radius is smaller than 800 m. Of course, this could change dramatically with artificially generated irregularities.

In conclusion, progress in the physics of ionospheric irregularities has shown that the dominant instability mechanism leads to structures that must be characterized by multiple power-law environments. We know, moreover, that simple saturation at the Rayleigh limit only occurs in shallowly sloped power-law environments. This behavior evidently dominates naturally occurring ionospheric scintillation, but it is different in artificially generated structures. Thus, a significant effort in predictive structure modeling has gone into determining where the break scales occur. Once this is established, the means of predicting the scintillation behavior in detail is well established. A good example can be found in the two-dimensional simulations performed by Martin and Flatté.[16] Beyond that, the ramifications of directionally dependent spectra should be investigated.

REFERENCES

1. L. M. Linson and J. B. Workman, "Formation of Striations in Ionospheric Plasma Clouds," *J. Geophys. Res.*, vol. 75, pp. 249–259, 1970.
2. R. F. Woodman and C. La Hoz, "Radar Observations of F Region Equatorial Irregularities," *J. Geophys. Res.*, vol. 81, pp. 5447–5466, 1976.
3. M. C. Kelley, G. Haerendel, H. Kappler, A. Valenzuela, B. B. Balsley, D. A. Carter, W. L. Ecklund, C. W. Carlson, B. Haausler, and R. Torbert, "Evidence for a Rayleigh-Taylor Type Instability and Upwelling of Depleted Density Regions During Equatorial Spread F," *Geophys. Res. Lett.*, vol. 3, pp. 448–460, 1976.
4. H. D. Craft, "Scintillation at 4 and 6 GHz Caused by the Ionosphere," *AIAA Paper 72-179*, 1972.
6. C. L. Rino, R. T. Tsunoda, J. Petriceks, R. C. Livingston, M. C. Kelley, and K. D. Baker, "Simultaneous Rocket-Beacon and In Situ Measurements of Equatorial Spread F—Intermediate Wavelength Results," *J. Geophys. Res.*, vol. 86, pp. 2411–2420, 1981.
6. S. Basu, S. Basu, J. P. McClure, W. B. Hanson, and H. E. Whitney, "High Resolution Topside In Situ Data of Electron Density and VHF/GHz Scintillations in the Equatorial Region," *J. Geophys. Res.*, vol. 88, pp. 403–415, 1983.
7. S. J. Franke, C. H. Liu, and D. J. Fang, "Multifrequency Study of Ionospheric Scintillation at Ascension Island," *Radio Sci.*, vol. 19, pp. 695–706, 1984.
8. C. L. Rino and J. Owen, "Numerical Simulations of Intensity Scintillation Using the Power Law Phase Screen Model," *Radio Sci.*, vol. 19, pp. 891–908, 1984.
9. S. L. Ossakow, "Ionospheric Irregularities," *Rev. Geophys. Space Phys.*, vol. 17, pp. 521–533, 1979.
10. B. B. Fejer and M. C. Kelley, "Ionospheric Irregularities," *Rev. Geophys. Space Phys.*, vol. 18, pp. 401–454, 1980.
11. S. Zargham and C. E. Seyler, "Collisional Interchange Instability. 1. Numerical Simulations of Intermediate-Scale Irregularities," *J. Geophys. Res.*, vol. 92, pp. 10073–10088, 1987.
12. M. C. Kelley, C. E. Seyler, and S. Zargham, "Collisional Interchange Instability. 2. A Comparison of the Numerical Simulations with the In Situ Experimental Data," *J. Geophys. Res.*, vol. 2, pp. 10089–10094, 1987.
13. R. T. Tsunoda, "High-Latitude F Region Irregularities: A Review and Synthesis," *Rev. Geophys. Space Phys.* In press.
14. C. L. Rino, "On the Application of Phase Screen Models to the Interpretation of Ionospheric Scintillation Data," *Radio Sci.*, vol. 4, pp. 855–867, 1969.
15. C. L. Rino, "The Applications of Strong-Scatter Theory to Transionospheric Radio Propagation," in *Multiple Scattering of Waves in Random Media and Random Rough Surfaces, Proc. Int. Symp., Penn. State Univ.*, pp. 311–326, 1985.
16. J. M. Martin and S. M. Flatté, "Intensity Images and Statistics from Numerical Simulation of Wave Propagation in 3-D Random Media," *Appl. Opt.*, vol. 27, pp. 2111–2126, 1988.

7

DENNIS L. KNEPP
and
L. WILLIAM BRADFORD

· ·

Scintillation Effects on Space Radar

The performance of a space-based radar (SBR) is highly dependent upon the characteristics of the propagation channel through the ionosphere, since even small fluctuations in received power can result in degradation. Spatial irregularities in the ionospheric electron density can produce rapid random fluctuations in the amplitude, phase, and angle of arrival of propagating electromagnetic waves. These fluctuations, called scintillation, have been observed in VHF and UHF satellite communication links (Pope and Fritz, 1971; Skinner et al., 1971; Taur, 1976; Fremouw et al., 1978) as well as in VHF and UHF radar observations (Towle, 1980) through the ambient ionosphere. Strong scintillation has occasionally been observed at frequencies as high as the L band. To increase survivability the effects of signal fluctuations due to propagation through disturbed ionospheric channels should be considered in the initial stages of SBR design.

Although lesser degrees of scintillation are possible in a highly disturbed propagation environment, the scintillation in the amplitude of a signal may be described by Rayleigh amplitude statistics. Such conditions are likely to apply after high altitude nuclear explosions (Arendt and Soicher, 1964; King and Fleming, 1980), or after chemical releases (Davis et al., 1974; Wolcott et al., 1978). Increased electron densities and irregularities in ionization structure after nuclear detonations can lead to intense Rayleigh signal scintillation at frequencies as high as the 7- to 8-GHz SHF band (Knepp, 1977). An SBR using VHF to X band, with the potential to operate in highly disturbed environments, must be designed with the effects of scintillation in mind.

Earlier work (Knepp and Dana, 1982; Dana and Knepp, 1986) investigated the performance of an SBR operating in an environment characterized by a severely disturbed propagation channel. It was shown that strong scintillation causes severe degradation in detection performance. In particular, systems relying upon coherent integration of many pulses are susceptible to fast fading, which can dramatically reduce SBR detection performance.

In this chapter an inverse measure of scintillation severity is used—namely, the signal decorrelation time, τ_0. Large values of τ_0 correspond to slow fading conditions, whereas small values of τ_0 indicate fast fading. Under fast fading conditions, the signal decorrelation time is small with respect to the coherent processing time so that the coherent integration process experiences a loss relative to its performance under slow fading conditions. This loss is caused by the destructive addition of successive radar pulses, which may become decorrelated in amplitude and phase. The average received power at the output of the integrator is then reduced.

As the signal decorrelation time decreases, the relative power output of the integrator decreases. If a look, consisting of relatively large number of pulses (of the order of 100–500), is divided into bursts, over which the coherent integration takes place, the coherent integration loss per look will be reduced. By noncoherently adding the power from each burst, the signal output per look is expected to be improved over the output of a look that has not been subdivided into bursts.

The purpose of the work reported here is to determine the trade-offs involved in a detection technique that adaptively varies the number of coherent integration processes (bursts) within a look. It will be shown that some improvement in performance is available if such a technique is adopted; however, some increased complexity in the design of an SBR is required.

7.1 RECEIVED-SIGNAL STATISTICS

In this chapter it is assumed that an SBR must operate in an environment where the one-way signal propagation channel is so disturbed that the channel output signal is described with Rayleigh amplitude statistics. That is, a transmitted signal with constant amplitude will be received, after one-way propagation through the channel, with amplitude fluctuations whose first-order amplitude statistics possess a Rayleigh probability density function. Rayleigh statistics are worst case, in the sense that no matter how much more severe the environment becomes, the received signal does not deviate from the worst case Rayleigh probability distribution. This is true for the propagation of electromagnetic signals over a wide frequency range in many different kinds of random media including laser propagation in turbulent air (Fante, 1975), VHF propagation through the ionosphere (Fremouw et al., 1978), and through striations composed of barium ions (Marshall, 1982). For propagation channels disturbed by high-altitude nuclear bursts, Rayleigh statistics describe the signal fading characteristics (Wittwer, 1980).

7.1.1 Received-Signal First-Order Statistics

Any actual instantaneous received radar signal power, S_r, may be expressed as

$$S_r = S_0[C/\langle C \rangle]S \tag{7.1}^a$$

where S_0 is the mean power returned by a point target and S is the fractional change in the power due to the propagation channel. The effect of a fluctuating target cross section C is also included, normalized to the mean target cross section, $\langle C \rangle$. The mean signal level is contained within the factor S_0, so $\langle S \rangle$ may be set to unity. In this formulation, received signal fluctuations are due both to the fluctuations in target cross section and to fluctuations caused by the disturbed propagation channel.

7.1.2 Target Statistics

In this chapter the statistics that describe fluctuations in target cross section are assumed to be those applying to a Swerling type 2 target. That is, the target cross section varies independently from burst to burst but is constant for all the pulses comprising the coherent burst. The radar cross section C therefore obeys a negative exponential probability distribution

$$p(C) = \frac{1}{\langle C \rangle} \exp(-C/\langle C \rangle) \tag{7.2}$$

where $\langle C \rangle$ is the mean cross section. Equation (7.2) fully describes the first-order statistics of the target cross section. To complete the description of the first-order statistics of the received power, the specification of the fluctuations due to the disturbed propagation channel are required.

7.1.3 Propagation Channel Statistics

For monostatic SBR operation, the transmitted and received signals traverse the same path, and therefore the irregularities in the intervening medium are identical for both. In this case, the received voltage is proportional to the square of the voltage for one-way propagation. The received power is similarly proportional to the square of the received power after one-way propagation. The two-way or monostatic probability distribution for received power (Dana and Knepp, 1986) is then given by

$$p_m(S) = \frac{1}{\sqrt{2S\langle S \rangle}} \exp\{-\sqrt{2S/\langle S \rangle}\} \tag{7.3}$$

where $\langle S \rangle$ is the mean received power.

The combined effect of fluctuations in the propagation channel and target cross section is obtained by multiplying the two probability density functions.

---

[a] $\langle x \rangle \triangleq E(x)$.

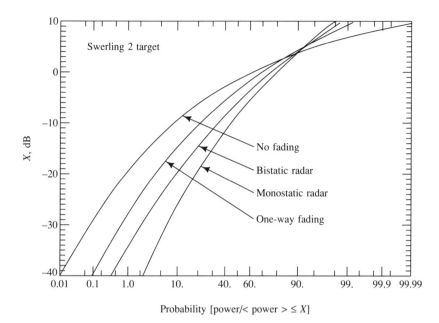

Probability [power/< power > ≤ X]

FIGURE 7.1   Cumulative probability distribution function of the received power for a Swerling 2 target and different propagation geometries.

The resulting function can then be used to obtain the cumulative probability function, which provides a means of assessing the probability of occurrence of deep fades, which are detrimental to radar performance. The cumulative distribution for the received power S_r in the monostatic case is given by

$$p_m(S_r) = 1 - \int_0^\infty \exp\left[-u - \frac{2S_r}{u^2 S_0}\right] du \tag{7.4}$$

where S_0 is the mean received power.

Figure 7.1 shows the cumulative probability functions for the combined Swerling 2 target and monostatic radar propagation channel. For comparison, the figure also shows the cumulative probability distributions for three other cases. The curve marked "no fading" applies to the Swerling 2 target with a constant (nonfluctuating) propagation channel. The curve marked "one-way fading" applies to the bistatic radar case where one of the one-way propagation paths is undisturbed. The remaining curve, marked "bistatic radar" applies to the case of a bistatic radar where both one-way propagation channels exhibit Rayleigh fluctuations. The distinction between the monostatic and the bistatic curves is that, whereas in the monostatic case the electromagnetic wave propagates through the same irregularities on each of the two one-way propagation paths, in the bistatic radar case the two one-way paths are independent, giving the statistics shown in the figure (Dana and Knepp, 1986).

The Swerling 2 target, with no fading, shows a probability of 10^{-4} for cross section levels 40 dB or more below the mean value. For a Swerling 2 target and severe fading in the monostatic radar geometry, the probability of fades exceeding 40 dB below the mean level is 0.02. It has been shown that such fading causes degradation of SBR target detection performance.

The fading conditions described so far have assumed the worst-case situation, where Rayleigh statistics characterize the one-way propagation channel. Other first-order or amplitude statistics may be used to characterize the received signal in less severely disturbed propagation environments. However, Rayleigh first-order amplitude statistics are generally associated with large spatial regions in a nuclear environment and thus provide a useful basis for SBR design.

7.1.4 Received-Signal Second-Order Statistics

The second-order fading statistics are specified by the correlation function of the received complex voltage. For the case of one-way propagation of an initially constant signal through a severely disturbed ionospheric channel, the autocorrelation function of the received voltage is given as the two-position, two-frequency mutual coherence function (Knepp, 1983). For cases where the scintillation is not so severe as to cause time-delay jitter in the received waveform (i.e., pulse distortion), the two-position, single-frequency mutual coherence function is the correlation function of the signal. The effective velocity of the line of sight of the radar signal through the ionospheric irregularities can be utilized to convert the spatial coordinates of the mutual coherence function into temporal coordinates, thereby obtaining the correlation function of signal fluctuations due to ionospheric irregularities. This procedure is described by Knepp (1983).

7.1.5 Signal Decorrelation Time

For worst-case Rayleigh fading, the correlation function of the received complex voltage $E(t)$ always has the Gaussian form

$$\langle E(t)E^*(t + \tau)\rangle = S_0 \exp(-\tau^2/\tau_0^2) \tag{7.5}{}^b$$

where τ_0 is the decorrelation time (the fading rate is $1/\tau_0$) for fluctuations over the two-way propagation path from the transmitter to target and back. The actual value of τ_0 is a function of radar geometry and of the irregularity structure and intensity of the disturbed ionospheric channel. Large values of τ_0 correspond to slow fading conditions and small values to fast fading.

As a concrete example, consider the case of a radar and target separated by a layer of ionization as might occur in the case of a space-based radar

---

[b] $\langle\ \rangle$ stands for expectation operator, that is, $\langle x\rangle \triangleq E(x)$.

observing a target on the ground. For a K^{-4} in situ power spectrum of three-dimensional ionization irregularities between outer scale L_0 and inner scale l_i, the decorrelation time, τ_1, associated with the one-way propagation path is

$$\tau_1 = \frac{\sqrt{2}L_0}{\ln(L_0/l_i)\sigma_\phi v_L} \tag{7.6}$$

where

$\qquad v_L$ = the velocity of the line-of-sight through the center of the ionized layer

$\qquad \sigma_\phi^2 = 2(r_c\lambda)^2 L_0 L \overline{\Delta N_c^2} \text{ (rad}^2)$

$\qquad \lambda$ = RF wavelength

$\qquad r_c$ = classical electron radius $(2.82 \times 10^{-15} \text{ m})$

$\qquad L$ = thickness of the ionized layer

$\qquad \overline{\Delta N_c^2}$ = variance of electron density irregularities

For the monostatic SBR propagation geometry, the decorrelation time of the received signal is related to the decorrelation time for each of the one-way propagation channels according to the relationship

$$\tau_0 = \tau_1/\sqrt{2} \tag{7.7}$$

where τ_1 is the decorrelation time of the one-way propagation path.

In the work reported here it is assumed that τ_0 is large with respect to the duration of the transmitted pulse. The received signal is then coherent during the pulse duration, which is typically of the order of several tens of microseconds.

7.2 RADAR SYSTEM CHARACTERISTICS

Since an SBR must operate with relatively low transmitter power, and with targets at very long ranges, the received signal-power-to-noise ratio (SNR) per pulse will be lower than that of a comparable ground-based system. If, in addition, the SBR must operate in a disturbed ionospheric environment, where the signal propagation channel is subject to strong scintillation, greater demands are made on the radar system design.

A low signal-to-noise ratio per pulse implies the need for long coherent integration times, but target motion may cause constructive and destructive interference from many scattering centers, resulting in rapid changes in the target cross section. To compensate, the total energy in a look is partitioned into a number of bursts consisting of a number of coherent pulses, which are coherently integrated upon reception. If the radar frequency is also varied sufficiently from burst to burst, target cross section samples will be indepen-

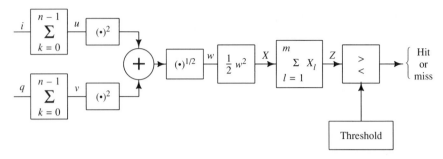

FIGURE 7.2   Block diagram of an SBR receiver.

dent; in addition, a form of protection from jamming is provided. The detected amplitudes of all the bursts that comprise a look are then noncoherently combined in a postdetection integration process. The resulting signal is then compared to a threshold level (which varies with the number of bursts per look), to determine whether a target has been detected during the look.

Figure 7.2 shows a simplified block diagram of a generic SBR receiver. The complex signal input contains amplitude, phase, and Doppler information from target, clutter, and thermal noise sources. An appropriately weighted discrete Fourier transform (DFT) algorithm can serve both as a clutter rejection filter, and as a means for coherently integrating the pulses in a burst. The input to the filter can be represented by a complex-valued voltage. In Fig. 7.2 the in-phase and quadrature components are denoted by i and q, respectively. The output from the filter is also a complex voltage, with components u and v. These components are coherently added, to form the amplitude w. After squaring and normalization, the output is placed in an accumulator, where the noncoherent integration takes place. Once all the bursts in a look are received, the resulting amplitude is compared to a threshold and a "hit" or "miss" is declared, based upon the results of that comparison. To obtain results for detection performance, these results are collected as a function of the signal-to-noise ratio, the number of bursts per look, and the scintillation environment description (that is, τ_0).

7.2.1 Coherent Doppler Processing

Coherent Doppler processing is utilized in modern radar systems to achieve coherent integration of the received target voltage and to suppress unwanted clutter returns. The Doppler filters are implemented using a weighted discrete Fourier transform (DFT). In general, the complex weights can be functions of both the Doppler filter number and the pulse number. The number of Doppler filters implemented is less than or equal to the number of pulses coherently processed per burst. In this chapter where the probability of detection in the absence of clutter is calculated, unity weights are assumed.

The input voltage to the Doppler filters from a single transmitted pulse is represented as

$$V(t) = E_r(t) + E_N(t) \tag{7.8}$$

where $E_r(t)$ is the voltage received from the target after conversion to baseband. The voltage $E_N(t)$ is the additive white Gaussian noise that exists in a radar receiver regardless of the presence of a target voltage. The noise voltage is given as the sum of two quadrature components

$$E_N(t) = n_i + in_q \tag{7.9}$$

where n_i and n_q are independent from each other and from pulse to pulse. The probability density function of n_i is

$$p(n_i) = \frac{1}{\sqrt{2\pi\sigma_N^2}} \exp(-n_i^2/2\sigma_N^2) \tag{7.10}$$

with an identical expression holding for $p(n_q)$. The total noise power per pulse is then

$$\langle E_N(t)E_N^*(t)\rangle = \langle n_i^2\rangle + \langle n_q^2\rangle = 2\sigma_N^2 \tag{7.11}$$

Therefore the mean signal-to-noise ratio per pulse at the coherent integration input is

$$\text{SNR}_{in} = \frac{S_0}{2\sigma_N^2} \tag{7.12}$$

where S_0 is the mean power received from a point target.

The assumption that there is no clutter and that the target is in the zero-Doppler filter allows the representation of the coherent integrator output as

$$y(t) = \sum_{l=0}^{n-1} V(lT) \tag{7.13}$$

without loss of generality, where n is the number of pulses integrated coherently and T is the interpulse period.

In the absence of propagation effects or in the slow fading limit the amplitude, phase, and Doppler frequency of the received voltage $E_r(t)$ are constant during the burst. In this case let $E_r(t) = ae^{i\phi}$, so the output voltage from the coherent integrator is

$$y(t) = nae^{i\phi} + \sum_{l=0}^{n-1} E_N(lT) \tag{7.14}$$

Since the I and Q components of the noise voltage are independent from pulse to pulse, the correlation function can be written

$$\langle E_N(lT)E_N^*(kT)\rangle = 2\sigma_N^2\delta_{l,k} \tag{7.15}$$

where δ is the Kronecker delta function. Thus the mean power at the output of the coherent integrator becomes

$$\langle y(t)y^*(t) \rangle = n^2 a^2 + \sum_{l=0}^{n-1} \sum_{k=0}^{n-1} \langle E_N(lT)E_N^*(kT) \rangle$$

$$= n^2 a^2 + 2n\sigma_N^2 \qquad (7.16)$$

On the average, the mean received target power is S_0, and thus $a^2 = S_0$ and the signal-to-noise ratio at the output of the coherent integrator is

$$\text{SNR}_{\text{out}} = \frac{a^2 n^2}{2n\sigma_N^2} = \frac{nS_0}{2\sigma_N^2}$$

$$= n \, \text{SNR}_{\text{in}} \qquad (7.17)$$

which is n times the input signal-to-noise ratio.

7.2.2 Fast Fading

Under fast fading conditions, the signal decorrelation time is small with respect to the coherent processing time, so the coherent integration process experiences a loss relative to its performance under slow fading conditions. This loss is caused by the destructive addition of radar pulses which are uncorrelated in amplitude and phase with previous and following pulses during the integration process or burst. Under these conditions the average target power at the output is then a function of the decorrelation time and is given by

$$P(\tau_0) = \sum_{k=0}^{n-1} \sum_{l=0}^{n-1} \langle E_r(kT)E_r^*(lT) \rangle$$

$$= S_0 \sum_{k=0}^{n-1} \sum_{l=0}^{n-1} \exp \left[\frac{-(k-l)^2 T^2}{\tau_0^2} \right] \qquad (7.18)$$

where the correlation function of the complex electric field is obtained from Eq. (7.5). In the slow fading limit, when the signal remains constant during the coherent integration time, τ_0 is large, so $P(\tau_0)$ becomes $n^2 S_0$, as before. The coherent integration loss may then be defined as

$$L_{\text{CI}} = P(\tau_0 \to \infty)/P(\tau_0) \qquad (7.19)$$

which is a measure of the loss caused by decorrelation during the coherent processing of n pulses. L_{CI} is easily calculated from Eqs. (7.18) and (7.19) as

$$L_{\text{CI}} = \frac{n}{1 + \dfrac{2}{n} \sum_{l=1}^{n-1} (n-l) \exp \left[\dfrac{-(l/n)^2 T_{\text{CI}}^2}{\tau_0^2} \right]} \qquad (7.20)$$

where

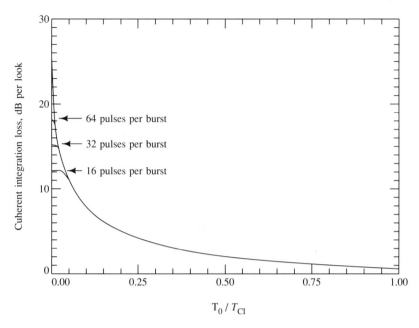

FIGURE 7.3   Coherent integration loss as a function of the number of pulses per burst.

$$T_{CI} = nT \tag{7.21}$$

is the coherent integration time. The loss L_{CI} is then a function of n and of the ratio τ_0/T_{CI}.

For very fast fading conditions, where $\tau_0 \ll T_{CI}$, L_{CI} has its maximum value of n. For very slow fading conditions, L_{CI} approaches unity for $\tau_0 \gg T_{CI}$. In the former case the target signal is completely decorrelated from pulse to pulse, and the target signal integrates in the same manner as noise, so the SNR at the integration output is equal to the SNR at the input. It is assumed that τ_0 is still much greater than the pulse width, so the power in a single pulse is not affected by loss of coherency.

Figure 7.3 shows a plot of coherent integration loss in dB as a function of τ_0/T_{CI}, where n, the number of pulses coherently integrated, takes on the values 16, 32, and 64. The coherent integration loss exceeds 3 dB for $\tau_0/T_{CI} \leq 0.4$ and exceeds 1 dB for $\tau_0/T_{CI} \leq 0.9$. For values of τ_0/T_{CI} greater than unity the coherent integration loss due to signal fading is less than 0.8 dB.

7.3 PROBABILITY OF DETECTION

The major goal in this chapter is to determine the amount of improvement in target detection performance that would result if the SBR had knowledge of the fading rate and was able to adjust the coherent integration time. Given knowledge of the fading rate, it is assumed for simplification that the radar processing strategy would consist of simply changing the number of bursts per look while maintaining a constant total power on a target. The time duration of the coherent burst (coherent integration time) would be chosen to be of the order of, or greater than, the signal decorrelation time so as to avoid the coherent integration loss as shown in Fig. 7.3. Although a reduction in the coherent integration time will certainly affect other aspects of radar performance (e.g., Doppler resolution, unless the pulse repetition frequency is changed), these issues are the topic of another effort.

In this discussion, SBR target detection performance is obtained by utilizing a computer simulation of a space-based radar. This simulation can be operated in a Monte Carlo fashion to collect received pulses repetitively from a Swerling 2 radar target and to combine them into a preselected number of bursts and looks as discussed in the following.

In order to assess the improvement in detection performance, the simulation results for radar detection performance during scintillation are compared to the detection performance in an undisturbed propagation environment. For the case of an undisturbed propagation environment, the probability of detection may be obtained analytically as follows.

7.3.1 Undisturbed Propagation Environment

Under undisturbed propagation conditions, the output amplitude z of the noncoherent integration shown in the block diagram of Fig. 7.2 has as its probability density function the expression (Dana and Knepp, 1986)

$$p(z) = \frac{\{z/[n\sigma_N^2(1 + \langle SNR \rangle)]\}^{m-1} \exp\{-z/[n\sigma_N^2(1 + \langle SNR \rangle)]\}}{n\sigma_N^2(1 + \langle SNR \rangle)\Gamma(m)} \quad (7.22)$$

where

σ_N^2 = noise power per pulse in each complex component channel

n = the number of coherently integrated pulses

m = the number of bursts per look

$\Gamma(m)$ = the gamma function $(m - 1)!$

$\langle SNR \rangle$ = mean signal-to-noise ratio per burst

The probability of detection is dependent upon the value set for the detection threshold, against which the received, combined amplitude is compared. The threshold value is in turn dependent upon the desired probability of false alarm. For this chapter, the probability of false alarm, p_{fa}, is set at 10^{-6}, a typical value. The probability of false alarm is expressed as

$$p_{fa} = \int_T^\infty p(z)\,dz = \frac{[\Gamma(m, T/(n\sigma_N^2))]}{\Gamma(m)} \quad (7.23)$$

where $\Gamma(m, x)$ is the incomplete gamma function, and n is the number of coherently integrated pulses. Equation (7.23) may be numerically inverted to find T.

Once the threshold T has been obtained, the probability of detection is computed from

$$p_d = \int_T^\infty p(z)\, dz \tag{7.24}$$

which is the probability that a received amplitude z of signal plus noise exceeds the detection threshold. The probability of detection may be evaluated and expressed as

$$p_d = \frac{\Gamma\{m,\, T/[n\sigma_N^2(1 + \langle SNR \rangle)]\}}{\Gamma(m)} \tag{7.25}$$

Equation 7.25 expresses the probability of detecting a Swerling 2 target as a function of the signal-to-noise ratio per burst and of the number of bursts per look. The probability of detection is not a function of the number of pulses per burst as long as the received signal remains constant from pulse to pulse.

The probability of detection is shown in Fig. 7.4 for a Swerling 2 target in an undisturbed propagation environment as a function of the mean signal-to-noise ratio per look, $m\langle SNR \rangle$. The number of bursts noncoherently integrated is 1, 2, 4, 8, or 16. The probability of false alarm is fixed at 10^{-6} in these results.

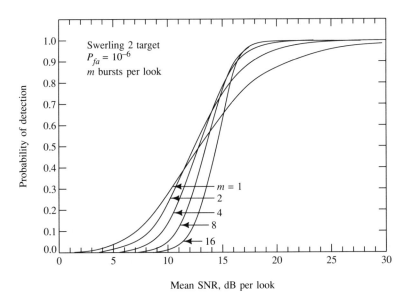

FIGURE 7.4   Probability of detecting a target using noncoherent combining of m bursts per look.

From the figure, for signal-to-noise ratios less than about 15 dB per look, the number of bursts per look that minimizes the SNR required to achieve a given probability of detection depends on the probability of detection. If an SBR could always operate at SNR above 15 dB per look, then the optimum number of bursts per look is the largest number possible. In general, however, it is desirable that the detection performance of the radar degrade as gracefully as possible as the SNR falls until the point where the probability of detection falls below the minimum (say 50 percent) required to maintain a radar track. At a 50 percent probability of detection, the 2-bursts-per-look case requires a signal-to-noise ratio of only 12.5 dB per look whereas the 16-bursts-per-look case requires 14.5 dB per look. The optimum number of bursts per look will therefore be arbitrarily defined by finding the detection curve which, for probabilities of detection above 0.5, minimizes the difference between the SNR that it requires to achieve a given probability of detection and the minimum SNR required to achieve that same probability of detection. For the detection curves plotted in Fig. 7.4 the optimum number of bursts per look is 4.

The results presented above show that the optimal number of bursts per look is about 4 when the transmission frequency is changed from burst to burst. Otherwise, a single burst per look is optimal. However, these waveform design considerations are limited by many practical constraints. The maximum duration of a coherent burst is limited, for example, by the time duration of radar cross section fluctuations caused by target motion. On the other hand, the minimum number of pulses per burst or equivalently the minimum duration of a burst is limited, for example, by Doppler resolution and clutter attenuation requirements. Therefore, it is not possible to choose the number of bursts per look merely on the basis of target detectability considerations. In addition, an SBR may be required to operate at lower probabilities of detection than ground-based radars because of power constraints. Decreasing probability of detection with decreasing signal-to-noise ratio might then drive the waveform design to 2 or 3 bursts per look.

7.3.2 Slow and Fast Fading

It is useful to make a distinction between slow and fast fading that is an invaluable aid in understanding the effects of scintillation on SBR coherent integration performance.

Slow fading conditions may be defined as occurring when the duration of signal fluctuations is very long compared to the coherent integration (or burst) time. Equivalently, we say that the signal decorrelation time is much greater than the burst time. In this case, the signal amplitude and phase are relatively constant over the burst time, so only burst-to-burst or look-to-look signal amplitude fluctuations affect SBR target detection performance. Examination of Eq. (7.5) shows that, with τ representing the burst time and τ_0 the signal decorrelation time, the correlation function is near unity, with no decorrelation from pulse to pulse. This result indicates that first-order amplitude statistics are sufficient to determine the effect of slow fading on target detection performance.

Fast fading occurs when the duration of signal fluctuations is very short compared to the coherent integration (or burst) time. We then say that the signal decorrelation time is less than the burst time duration. In fast fading, many of the pulses that comprise the coherent burst may be independent.

Under slow fading conditions, it is somtimes possible to obtain analytic expressions for the probability of detection (Dana and Knepp, 1986). However, under severe fast fading conditions it is usually more convenient to measure radar detection performance with the aid of Monte Carlo simulations.

7.4 RESULTS

Previous work (Knepp and Dana, 1982; Dana and Knepp, 1986) has indicated that there is a possibility for increased radar detection performance if some of the coherent integration loss sustained under fast fading conditions can be avoided by reduction of the coherent integration time.

In this section, this hypothesis is investigated over a range of fading rates and values of the number of bursts per look. It is desirable to obtain results that will apply to a generic SBR design; because of this constraint and of the large variation in the degree of disturbance in a nuclear environment, it is useful to consider a wide range of fading rates in this work. The results presented here add to our understanding of the effects of scintillation on radar detection performance and may thereby aid future SBR design and evaluation studies.

7.4.1 Radar Detection Simulation

To obtain results for the probability of detection, a Monte Carlo simulation of a space-based radar is used. A Swerling 2 target is utilized, and the propagation, reception, and processing of pulse trains that have propagated through a strongly disturbed ionized environment is accomplished. To simplify the calculation, perfect Doppler tracking is assumed, and there is no clutter. A block diagram of the receiver design used in this simulation is shown in Fig. 7.2. Here a number of coherent bursts are transmitted, each at a different radar frequency to ensure that Swerling 2 target statistics apply and that the propagation environment is decorrelated between bursts. Upon reception, the pulses that comprise each burst are integrated using a discrete Fourier transform that additionally provides for target Doppler discrimination. In this particular calculation, the target Doppler is known, and no sidelobe suppression is required because the calculation is performed in the absence of clutter. The amplitude in the target range cell is then summed (noncoherently) over all the bursts that comprise a look. This amplitude is then compared to a threshold (based on the noise power) and a hit or a miss is declared. The probability of detection is obtained simply by dividing the number of hits by the total number of looks.

To represent the fluctuating propagation channel caused by a nuclear detonation, a numerical technique known as statistical signal generation (Knepp and Wittwer, 1984) is used to generate realizations of the impulse response of the disturbed propagation channel. These signal realizations are then used by the Monte Carlo code to provide the samples of the received signal for each pulse that include the effects of scintillation. Statistical signal generation is applicable only to signal propagation through strongly ionized turbulence, where the one-way signal amplitude statistics are Rayleigh.

To obtain the generic SBR detection probability results presented here, five values of the ratio τ_0/T_{LOOK} were used. These values were 0.03, 0.1, 0.2, 0.5, and 1.0. For each of these values of τ_0/T_{LOOK}, the probability of detection was obtained for five values of the number of bursts per look—namely, 1, 2, 4, 8, and 16. For a constant value of τ_0/T_{LOOK}, then, the effect of an increase in the number of bursts per look is to increase the value of the ratio of the signal decorrelation time to the coherent integration time (τ_0/T_{CI}) at each value of τ_0/T_{LOOK}. Therefore, from Fig. 7.3 the coherent integration loss per burst would decrease from what it might be otherwise. This range of values represents a wide range in the underlying values of the signal decorrelation time and should be useful for many applications involving pulse Doppler SBRs.

In general, the result of the coherent integration of a finite number of pulses is dependent upon the number of pulses. Therefore, the coherent integration loss (as shown in Fig. 7.3), is also dependent upon the number of pulses integrated. However, for a large number of pulses per burst, the coherent integration *loss* is independent of the number of pulses comprising the burst, except for extremely small values of τ_0/T_{CI}. In the results to be presented, special precautions were taken to remove the dependence upon the number of pulses per burst. This precaution is performed in the work here by increasing the number of pulses per burst in accordance with the value of the actual decorrelation time. In many of the simulations, 128 pulses per look are used, but for fast fading conditions and large numbers of bursts per look, it is required that 512 pulses per look be used to ensure that the probability of detection remains independent of the number of pulses per burst.

The need to use 512 pulses per look to determine the coherent loss under fast fading conditions is also indicated by examination of Fig. 7.3. In the figure, coherent integration loss is shown as a function of the ratio of signal decorrelation time to coherent integration time. The curve is plotted for several values of the number of pulses that comprise the coherent burst. The curve plotted for 16 pulses in a burst flattens out at 12 dB for short decorrelation times. Similarly, the curve for 32 pulses in a burst flattens out at 15 dB, but does so at even smaller decorrelation times. This flattening of the curve is the direct result of the use of a discrete number of pulses in a burst and corresponds to the case of independent pulses where the signal power gain is equal to the noise power gain, with no improvement in resulting output signal-to-noise ratio. These flat areas are avoided in these calculations in order to render the results independent of the number of pulses per burst and, therefore, more useful.

For example, from the figure, if the ratio in signal decorrelation time to coherent integration time is 0.04 or less, the use of only 16 pulses in a burst would lead to a coherent integration loss of 12 dB, whereas if many pulses were used, the actual value would be larger. For a value of τ_0/T_{CI} of 0.03, 32 pulses per burst would give results that would not change even if there were more than 32 pulses. Whenever there are 16 bursts per look, 512 pulses are required in a look period, giving 32 pulses per burst for this case.

It is noted that the actual number of pulses per burst used in a modern pulse doppler radar may be several hundred or more. However, to obtain the results here that are independent of the number of pulses per burst it is sufficient to use a reduced number as described above.

Figures 7.5 to 7.9 show the major results of this study in the form of plots of the probability of detection for a Swerling 2 target as a function of the mean signal-to-noise ratio per look for each of the five different values of the number of bursts per look. In all cases, the probability of a false alarm is 10^{-6}, as obtained through the use of Eq. (7.23).

Each figure contains two sets of five curves. The first set applies to the case of a Swerling 2 target with no fading, and always lies on the left-hand side of the figures at the lower values of the SNR. Five curves are shown for the five values of the number of bursts per look: 1, 2, 4, 8, and 16. This first

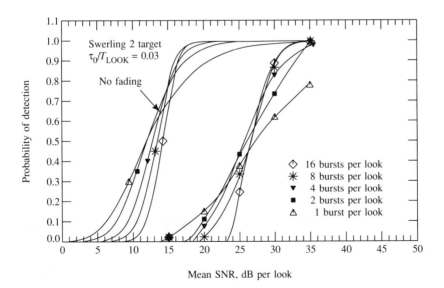

FIGURE 7.5   Probability of detection versus mean signal-to-noise ratio as a function of τ_0/T_{LOOK} for 1 burst per look.

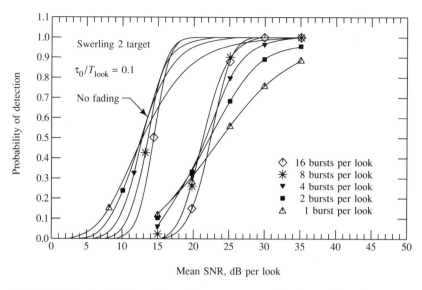

FIGURE 7.6 Probability of detection versus mean signal-to-noise ratio as a function of τ_0/T_{LOOK} for 2 bursts per look.

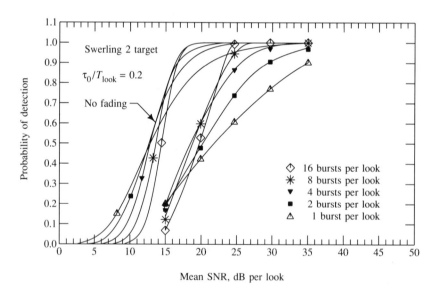

FIGURE 7.7 Probability of detection versus mean signal-to-noise ratio as a function of τ_0/T_{LOOK} for 4 bursts per look.

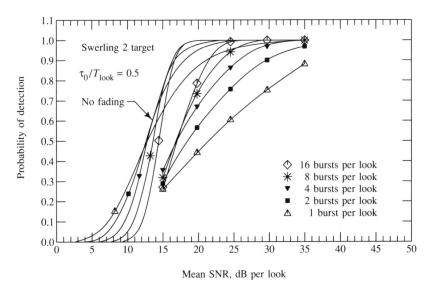

FIGURE 7.8 Probability of detection versus mean signal-to-noise ratio as a function of τ_0/T_{LOOK} for 8 bursts per look.

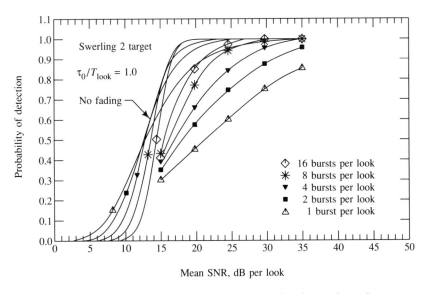

FIGURE 7.9 Probability of detection versus mean signal-to-noise ratio as a function of τ_0/T_{LOOK} for 16 bursts per look.

set of curves is identical in each of the five figures, and is provided to serve as a reference for the results pertaining to severe scintillation.

The other set of five curves in each of the figures was obtained for the case of severe fading for the same five values of the number of bursts per look—that is, 1, 2, 4, 8, and 16. In each of the five figures the ratio τ_0/T_{LOOK} takes on a different value. The ratio τ_0/T_{LOOK} has the values 0.03, 0.1, 0.2, 0.5, and 1.0 in Figs. 7.5 to 7.9, respectively. These figures represent a wide range of fading rates for a space based radar. In Fig. 7.5, where the results for the fastest fading is presented for a value of τ_0/T_{LOOK} of 0.03, it can be seen that the two sets of five curves fall far apart, indicating a large loss in detection performance during fast fading, even after an increase in the number of bursts per look. However, it is apparent from Fig. 7.5 that an increase in the number of bursts per look does indeed give improved performance. In Fig. 7.9, for the slowest fading considered here, where τ_0/T_{LOOK} is equal to 1.0, it is seen that an increase in the number of bursts per look can greatly improve radar detection performance. The measured probabilities of detection in Figs. 7.5 to 7.14 have an average error of roughly 0.05, associated with the finite number of Monte Carlo measurements.

Figures 7.10 to 7.14 present the same results, but in a different format where each figure depicts the probability of detection for a fixed value of the number of bursts per look. One value of this format is that it gives the radar designer measurements of the performance of a system with a fixed number of bursts per look as a function of the fading rate or τ_0. The appropriate curve for an undisturbed propagation environment is also shown in each figure.

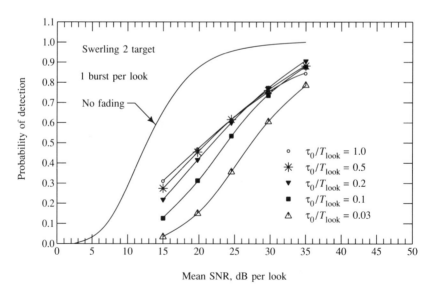

FIGURE 7.10   Probability of detection versus mean signal-to-noise ratio as a function of the number of bursts per look, $\tau_0/T_{\text{LOOK}} = 0.03$.

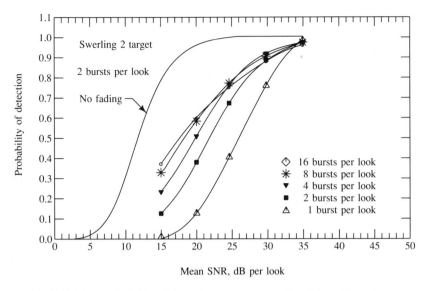

FIGURE 7.11   Probability of detection versus mean signal-to-noise ratio as a function of the number of bursts per look, $\tau_0/T_{\text{LOOK}} = 0.1$.

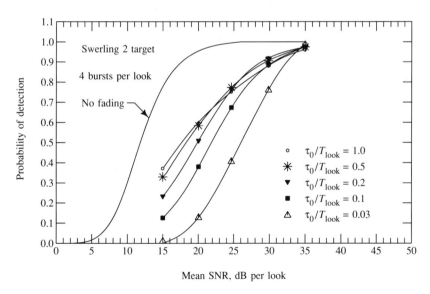

FIGURE 7.12   Probability of detection versus mean signal-to-noise ratio as a function of the number of bursts per look, $\tau_0/T_{\text{LOOK}} = 0.2$.

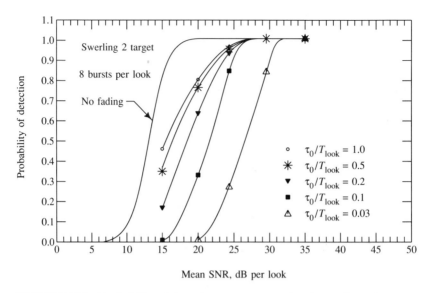

FIGURE 7.13 Probability of detection versus mean signal-to-noise ratio as a function of the number of bursts per look, $\tau_0/T_{LOOK} = 0.5$.

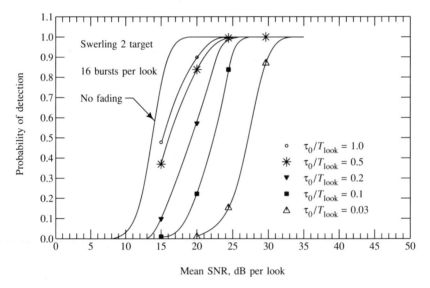

FIGURE 7.14 Probability of detection versus mean signal-to-noise ratio as a function of the number of bursts per look, $\tau_0/T_{LOOK} = 1.0$.

It is immediately obvious that, if the fading rate increases (τ_0 decreases), the SNR required to obtain a given probability of detection increases. By increasing the number of bursts per look, the SBR designer can offset this somewhat. In other words, by an appropriate choice of the number of bursts per look, the SNR required to achieve a given probability of detection can be minimized.

7.4.2 Optimal Number of Bursts per Look

One rule of thumb to determine the optimum number of bursts per look is described above. The optimum number of bursts per look may be determined by finding that detection curve which, for probabilities of detection above 0.5, minimizes the difference between the SNR required to achieve a given probability of detection and the minimum SNR required to achieve that same probability of detection. For an undisturbed propagation environment, the optimum number of bursts per look is four. In part, this choice is predicated on another criterion, which requires that the detection performance degrade gracefully until the SNR falls below the minimum value required to maintain a radar track. This minimum value is assumed to be 0.5 in this chapter. Essentially, this requirement specifies the sensitivity of the probability of detection to changes in the SNR. This same criterion for choosing the optimum number of bursts per look may be used for design purposes in a fading environment.

It is clear from an examination of the figures that the SNR minimization process described above does not necessarily result in the choice of a detection curve that gives the smallest decrease in probability of detection with decreasing signal-to-noise ratio. Although we do not take this effect into account in choosing the optimal number of bursts per look, an SBR designer with specific design requirements may have a need to do so.

In order to choose the optimal detection curve for each of the five values of τ_0/T_{LOOK}, the following graphical procedure was adopted. For each value of τ_0/T_{LOOK} or each of the five figures numbered from 5 to 9, values of probability of detection of 0.5, 0.6, 0.7, 0.8 and 0.9 were examined.

Then for each of the five different values of number of bursts per look, the difference (in decibels) in SNR between the individual curves and the best curve was determined. These differences are presented in Table 7.1 for each of the five values of τ_0/T_{LOOK}. Only values of 4, 8 and 16 bursts per look are presented here since these curves were always superior to those corresponding to 1 and 2 bursts per look. The table then consists of values of the deviation of the SNR from the optimum value measured at five different values of the probability of detection. The optimum number of bursts per look is chosen for which the mean deviation (averaged in decibels over the five values of probability of detection shown in the table) from the best value is least. Table 7.1 lists the deviations for the five values of τ_0/T_{LOOK}, as well as their mean. An

TABLE 7.1 Difference Between SNR (dB) and the Minimum SNR Measured at Several Probabilities of Detection

τ_0/T_{LOOK}	Bursts per Look	Probability of Detection					Mean
		0.5	0.6	0.7	0.8	0.9	
0.03	4	0.0	0.0	0.0	0.4	1.6	0.4*
0.03	8	1.0	0.7	0.2	0.1	0.0	0.4*
0.03	16	1.0	0.5	0.0	0.0	0.0	0.3*
0.1	4	0.5	1.0	1.1	1.4	1.7	1.1
0.1	8	0.0	0.0	0.0	0.0	0.0	0.0*
0.1	16	1.1	1.0	0.8	0.7	0.2	0.8
0.2	4	0.0	0.0	0.5	1.1	2.3	0.8
0.2	8	0.2	0.0	0.0	0.0	0.4	0.1*
0.2	16	1.1	0.7	0.4	0.2	0.0	0.5
0.5	4	0.2	0.5	1.4	2.7	4.1	1.8
0.5	8	0.0	0.0	0.3	0.9	1.6	0.6
0.5	16	0.2	0.0	0.0	0.0	0.0	0.04*
1.0	4	1.3	2.3	3.2	5.5	5.8	3.6
1.0	8	0.4	1.2	1.3	1.5	1.7	1.2
1.0	16	0.0	0.0	0.0	0.0	0.0	0.0*

* Optimum values.

TABLE 7.2 The Number of Bursts per Look Resulting in the Minimum SNR

τ_0/T_{LOOK}	Probability of Detection					Best Choice
	0.5	0.6	0.7	0.8	0.9	
0.03	4	4	4/16	16	8/16	4/8/16
0.1	8	8	8	8	8	8
0.2	4	4/8	8	8	8	8
0.5	8	8/16	16	16	16	16
1.0	16	16	16	16	16	16

asterisk at the end of a line corresponding to a fixed number of bursts per look in Table 7.1 indicates the value of the number of bursts per look that is chosen as optimum. Note that for τ_0/T_{LOOK} equal to 0.03, the difference in the means was not great enough to warrant choosing between 4, 8, or 16 bursts per look.

Table 7.2 summarizes the results for the optimum number of bursts per look at each of the values of τ_0/T_{LOOK} as a function of the probability of detection. The overall optimum choice for the number of bursts per look based on the average SNR deviation from optimum at each probability of detection is also included as the final column in Table 7.2. It is evident that minimization of SNR requirements during fading requires dynamic variation of the number of bursts per look.

Note that the rule for the choice of the best number of bursts per look was made for probabilities of detection between 0.5 and 0.9. Another choice of probabilities of detection might yield somewhat different results. Also, weighting the choices of probability of detection might result in a different set of choices.

Recall that the optimum number of bursts per look in an ambient, unperturbed propagation environment was found to be 4. For this reason, it is useful to measure any performance increases (decreases in SNR) relative to the SNR that would be required if the bursts per look were held constant at 4. Table 7.3 summarizes this information. The results in this table can be obtained from Table 7.1 by comparing the values for 8 and 16 bursts per look with those for 4 bursts per look. In most table entries, the SNR required at 4 bursts per look is greater than that required using the optimum choice, which is usually 8 or 16 bursts per look. If the SNR required at 4 bursts per look is smaller than that required for the optimum choice, then a zero is entered in the table.

In Fig. 7.15 the "improvement" of Table 7.3, measured in dB, is plotted as a function of the ratio of signal decorrelation time to look time, for three different values of probability of detection. The data for these curves were obtained by visual examination of Figs. 7.5 to 7.9, and is not exact. The presence of the slight bump at 0.1 on the abscissa is probably due to a combination of measurement error and to the fact that the curves in Figs. 7.5 to 7.9 tend to intersect at different points in each figure.

It is apparent from Fig. 7.15 that it is possible to obtain an improvement in the detection performance of the system by varying the number of bursts per look. This increase in performance can be as much as 6 dB, for probabilities of detection of about 0.9. At a more modest probability of detection of 0.7, an increase in performance of nearly 3 dB is attainable.

TABLE 7.3 Improvement in SNR (dB) Between the Minimum SNR at a Given Probability of Detection and the SNR for 4 Bursts per Look

	Probability of Detection				
τ_0/T_{LOOK}	0.5	0.6	0.7	0.8	0.9
0.03	0.0	0.0	0.0	0.4	1.6
0.1	0.5	1.0	1.1	1.4	1.7
0.2	0.0	0.0	0.5	1.1	2.3
0.5	0.2	0.5	1.4	2.7	4.1
1.0	1.3	2.3	3.3	5.5	5.8

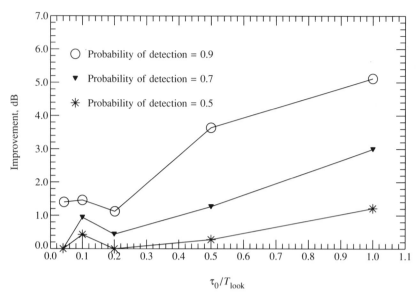

FIGURE 7.15   Improvement factor versus τ_0/T_{LOOK} for three values of probability of detection.

7.4.3 The Influence of Statistics

It is useful to examine the shape of the curves in Figs. 7.5 to 7.9, and consequently of Fig. 7.15, in order to understand the reasons for the improvements obtained. It is evident from this study that increasing the number of bursts per look is more effective in slow fading than in fast fading.

The differences in the shape of the detection curves for different numbers of bursts per look at slow fading seem to indicate that the probability distributions change with the number of bursts per look. To check this hypothesis, the Monte Carlo simulation was temporarily modified to yield the probability distributions of the received signal power. Figure 7.16 shows some of the results, in the form of cumulative probability distribution functions of the power at the output of the noncoherent integrator. The ordinate shows the value of the received signal power divided by the mean power. To obtain these results for the cumulative power distributions of the signal alone, the simulation was utilized in a zero-noise condition. Two values of τ_0/T_{LOOK} are shown in the figure; 0.03 and 1.0, respectively. These represent the slowest and fastest fading rates considered in this study. For each of these values of τ_0/T_{LOOK}, the number of bursts per look are taken as 1, 4, and 16. The two curves labeled 2 and 3 in the figure for a value of τ_0/T_{LOOK} of 0.03 are seen to fall close together, indicating that there is very little change in the statistics at the output of the noncoherent integrator for this value of τ_0/T_{LOOK} for values of 4 to 16 bursts per look. This result is simply a verification of results presented previously that show little change in probability of detection as the number of bursts is increased. The other three curves—labeled 4, 5, and 6, for slow fading condi-

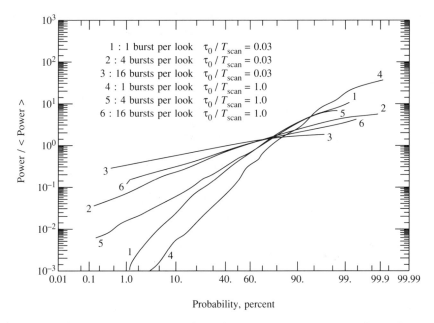

FIGURE 7.16 Cumulative probability distribution of output power for various τ_0/T_{LOOK} and numbers of bursts per look.

tions where τ_0/T_{LOOK} is 1.0—show a change toward a probability distribution with less variation as the number of bursts per look is increased. A horizontal line on this figure at an abscissa value of unity would indicate a constant signal, and variation in the signal is seen as cumulative distribution curves with larger slopes.

An interesting feature in Fig. 7.16 is that the curve representing 16 bursts per look at slow fading ($\tau_0/T_{\text{LOOK}} = 1.0$) is nearly the same as the curve for 4 bursts per look at fast fading ($\tau_0/T_{\text{LOOK}} = 0.03$). The implication is that under fast fading conditions, the noncoherent summation of the power from many bursts does not affect the output signal statistics significantly. This is evident because of the similarity of curves 2 and 3 in Fig. 7.16. Thus, in fast fading conditions, since the output power variation is small, a decrease in the signal threshold may be effective in yielding increased detection performance, albeit with a greater false-alarm probability. In slow fading, the noncoherent integration of the power from multiple bursts improves detection performance because it reduces the variability of the output power.

REFERENCES

Arendt, P. R., and H. Soicher, "Effects of Arctic Nuclear Explosions on Satellite Radio Communication," *Proc. IEEE*, vol. 52, pp. 672–676, June 1964.

Dana, R. A., and D. L. Knepp, "The Impact of Strong Scintillation on Space Based Radar Design. II—Noncoherent Detection," *IEEE Trans. Aerosp. Electron. Syst.*, vol. AES-22, January 1986.

Davis, T. N., G. J. Romick, E. M. Westcott, R. A. Jeffries, D. M. Kerr, and H. M. Peek, "Observations of the Development of Striations in Large Barium Clouds," *Planet. Space Sci.*, vol. 22, p. 67, 1974.

Fante, R. L., "Electromagnetic Beam Propagation in Turbulent Media," *Proc. IEEE*, vol. 63, pp. 1669–1692, December 1975.

Fremouw, E. J., R. L. Leadabrand, R. C. Livingston, M. D. Cousins, C. L. Rino, B. C. Fair and R. A. Long, "Early Results from the DNA Wideband Satellite Experiment—Complex Signal Scintillation," *Radio Sci.*, vol. 13, pp. 167–187, January–February 1978.

King, M. A., and P. B. Fleming, "An Overview of the Effects of Nuclear Weapons on Communications Capabilities," *Signal*, pp. 59–66, January 1980.

Knepp, D. L., *Multiple Phase-Screen Propagation Analysis for Defense Satellite Communications System*, DNA 4424T, MRC-R-332, Mission Research Corporation, September 1977.

Knepp, D. L., "Analytic Solution for the Two-Frequency Mutual Coherence Function for Spherical Wave Propagation," *Radio Sci.*, vol. 18, pp. 535–549, July–August 1983.

Knepp, D. L., and R. A. Dana, "Impact of the Propagation Environment on Space Based Radar Design," AIAA-82-0424, AIAA 20th Aerospace Sciences Meeting, Orlando, Florida, January 1982.

Knepp, D. L., and L. A. Wittwer, "Simulation of Wide Bandwidth Signals That Have Propagated Through Random Media," *Radio Sci.*, vol. 19, pp. 303–308, January–February 1984.

Marshall, J., "PLACES—A Structured Ionospheric Plasma Experiment for Satellite System Effects Simulation," AIAA 20th Aerospace Sciences Meeting, Orlando, Florida, January 1982.

Pope, J. H., and R. B. Fritz, "High Latitude Scintillation Effects on Very High Frequency (VHF) and S-Band Satellite Transmissions," *Indian J. Pure Appl. Phys.*, vol. 9, pp. 593–600, August 1971.

Skinner, N. J., R. F. Kelleher, J. B. Hacking, and C. W. Benson, "Scintillation Fading of Signals in the SHF Band," *Nature (Phys. Sci.)*, vol. 232, pp. 19–21, 5 July 1971.

Taur, R. R., "Simultaneous 1.5- and 4-GHz Ionospheric Scintillation Measurements," *Radio Sci.*, vol. 11, pp. 1029–1036, December 1976.

Towle, D. M. "VHF and UHF Radar Observations of Equatorial *F* Region Ionospheric Irregularities and Background Densities," *Radio Sci.*, vol. 15, pp. 71–86, January–February 1980.

Wittwer, L. A., *A Trans-Ionospheric Signal Specification for Satellite C^3 Applications*, DNA 5662D, Washington, DC: Defense Nuclear Agency, December 1980.

Wolcott, J. H., D. J. Simons, T. E. Eastman, and T. J. Fitzgerald, "Characteristics of Late-Time Striations Observed During Operation STRESS," in *Effect of the Ionosphere on Space and Terrestrial Systems*, J. M. Goodman, Ed., pp. 602–613, Washington, DC: U.S. Government Printing Office, 1978.

J. GREGORY ROLLINS
and
JOHN CHOMA, JR.

8

. . . .

Radiation Effects on Semiconductors

In certain applications, integrated circuits are required to operate in hostile environments. The space environment is one such area because of radiation from natural and man-made sources. If integrated circuits are to function properly in such an environment, provisions must be made either to allow for radiation-induced degradation or to attempt hardening measures to reduce the degradation produced by radiation.

In either case, production of a radiation tolerant design normally extracts a price in both the cost of the system and system performance. In order to reduce these penalties it is essential that radiation issues be considered early in the design process, either during the design of any custom integrated circuits, or during the selection of off the shelf components. Proceeding with a design and assuming that radiation hardening can be performed later (for example, by the addition of shielding) is a dangerous practice.

This chapter provides an introduction to radiation effects in integrated circuits. The treatment given is by no means exhaustive but provides descriptions of the main areas of concern. Excellent analyses of radiation effects can be found in Refs. 1–4.

Radiation energy is similar to that of heat and visible light. This energy can interact with the materials of an integrated circuit (IC) in a number of complex ways that are deleterious to the operation of the integrated circuit. The effects of radiation are normally classified as to the duration of the disruption. Transient radiation effects are caused by bursts of radiation (as produced by a nuclear explosion) or by the impact of a cosmic ray. Transient effects result in large impulse-like currents that normally do not last for more than a microsecond. These currents can scramble computer memories, produce false logic signals, and in extreme cases can overheat metal lines, resulting in permanent failure.

The other classification, total dose damage, involves a slow degradation of the semiconductor and insulator structures, accumulating over a period of months or years. This damage can degrade the performance (speed, logic levels, fanout, power dissipation) of an integrated circuit below acceptable limits.

Radiation arises because of natural or man-made processes. Some examples of natural radiation are the following:

1. *Cosmic Rays.* These are composed of the nuclei of single atoms accelerated by complex processes deep in space, to relativistic speeds. Cosmic rays have kinetic energies of from 1 MeV to 100 GeV and a distribution of isotopes similar to those found in earth. Cosmic rays cause a localized transient process in spaceborne integrated circuits known as single-event upset (SEU). Continued exposure to cosmic rays also contributes to total dose damage.

2. *Alpha Particles.* These are helium nuclei, typically with energies of from 1 MeV to 5 MeV, which can be ejected from certain isotopes found in ceramic IC packages. These nuclei are another source of SEUs in sensitive circuits.

3. *Energetic Electrons and Protons.* Both are produced by the sun. The high-energy protons can cause nuclear reactions, which can produce SEUs. Both the protons and electrons cause total dose damage to the silicon lattice or oxide layers. Both can become trapped in the earth's magnetic field.

Man-made radiation results from the explosion of an atomic bomb or operation of a nuclear reactor. Man-made radiation has the following effect:

1. The explosion produces a short (50-ns) pulse of gamma- and X-ray radiation. Note that gamma rays are photons produced by a nuclear process with energies from 0.1 MeV to 10 MeV, whereas X rays are photons produced by an electronic process with energies of from 1 keV to 0.1 MeV. On the ground, these energetic photons are quickly thermalized by interaction with air molecules to heat and visible light. In space, however, the gamma and X-ray pulse can travel for hundreds of miles. The gamma pulse causes severe electrical disturbances in an IC because of the energy it imparts and its ability to penetrate metal shields. This form of transient upset is referred to as dose-rate upset or gamma dot upset.

2. Neutrons emitted by the blast produce nuclear reactions and severe damage to the Si lattice, which tends to destroy bipolar devices. As with the gamma pulse, neutrons are absorbed in the atmosphere, but have further effects reaching into space.

3. A nuclear explosion in space at the proper altitude excites charged particles trapped in the Van Allen belts. These excited particles greatly enhance total dose effects on satellites in orbits that intersect the belt.

4. Gamma and X rays can knock electrons free from atoms, resulting in a large current away from the point of detonation. This current creates an electromagnetic pulse with a frequency spectrum ranging from the gigahertz region down to almost 0Hz. Near the point of detonation, the electric field strength can be as large as 10^5 V/m.

8.1 GAMMA AND X-RAY RADIATION

Gamma rays and X rays are photons with energy in the range of a few thousand to several million electron volts. The gamma photon typically interacts with an electron in a material via the Compton effect, photoelectric effect, or electron-positron production to produce one or more high-energy electrons and a new, lower-energy photon as shown in Fig. 8.1. The kinetic energy T of the ejected electron and new photon depends on the energy of the original photon and how tightly the electron is bound to the atom. In dense, high-Z (atomic number) material, the photons are absorbed more rapidly because of the larger number of electrons. In addition, with the photoelectric effect, each collision extracts more energy from the incident photon. This occurs because the inner electrons of a high-Z material are more tightly bound to the atom than in a low-Z material. The process by which the energy of one high-energy photon is spread over many electrons is known as thermalization.

As the beam of photons penetrates a material, the loss in intensity I is proportional to the distance traveled, giving rise to an exponential dependence on the depth of penetration x:

$$I(x) = I_0 \exp(-ux) \tag{8.1}$$

where u is the absorption coefficient (with units of cm^{-1}).

Since the gamma photon has very high energy to start with, it may take many collisions to absorb all the photon energy, and gamma photons are very penetrating. For example 12 cm of iron are needed to reduce the flux of 1-MeV gamma rays by $1/e$ ($u = 0.083$), compared to visible light, which is absorbed in a few micrometers in Si). Examples of absorption in aluminum and lead are given in Fig. 8.2. The units given in Fig. 8.2 (cm^2/gm) can be converted to cm^{-1} by multiplying by the density (2.70 g/cm^3 for Al and 11.36 g/cm^3 for Pb).

In Fig. 8.2 two sets of curves are shown, one set for absorption and one for attenuation. The difference between the two is that in the case of absorption, *all* of the photon energy is transferred to the lattice. Attenuation, on the other hand means that only the original photon is stopped. This difference is important mainly with Compton scattering. For example a 1-MeV photon may be scattered (attenuated), producing a high-energy electron and new photon that leaves the vicinity, carrying a portion of the energy with it.

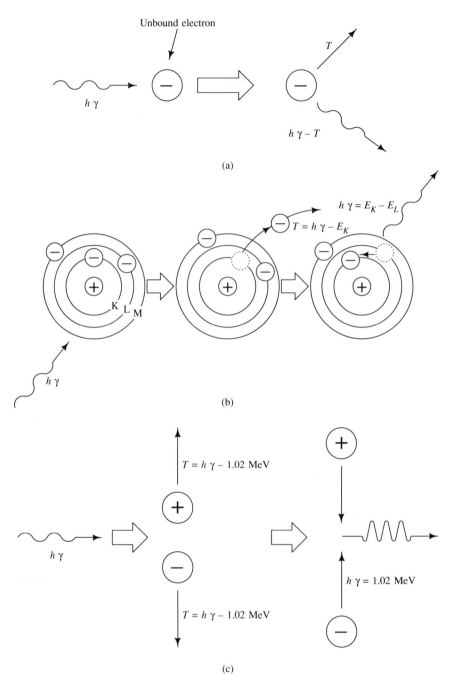

FIGURE 8.1 Transfer of energy from photon to electron. (a) Compton effect. (b) Photoelectric effect. (c) Pair production.

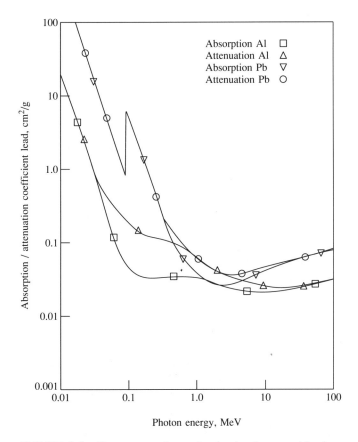

FIGURE 8.2   Gamma ray absorption in aluminum and lead.

The flow of energy (either as photons or energetic electrons) into a material determines the "dose" of radiation. The cascade of electrons and lower-energy photons created by the interactions complicates the process of calculating the dose a particular material may see. In particular, if a high-Z layer is close to a low-Z layer, electrons are scattered into the low-Z material from the high-Z layer. This can cause the low-Z material to receive a greater dose of radiation than if the high-Z material were not there. This effect is known as "dose enhancement." At gold-silicon interfaces, dose enhancement can result in a factor-of-20 dose increase in the silicon layer for 75-keV to 150-kEV X rays, and a factor-of-2.0 increase of 1-MeV gamma rays.

The main way in which radiation pulses produce transient effects in a microcircuit is by the generation of electron-hole pairs. The case of dose rate upset by a gamma pulse will be considered first. It is the absorbed energy in a slice of material that is of the greatest interest since the absorbed energy generates the electron-hole pairs. Since the dimensions of a transistor are on the order of a few micrometers, while the coefficient of absorption for the gamma flux is measured in cm (10^4 μm), the gamma flux, and EHP generation may be considered constant throughout the device.

A unit to describe the absorbed energy has been developed. This is the rad, which is the absorption of 100 ergs per gram of material. Note that since the absorption rate is different in different materials, the same gamma pulse will produce different generation rates in different materials. For example, a piece of GaAs will absorb more energy than a piece of Si the same size, owing to the greater density of the GaAs. In transient upset, the rate at which the energy is delivered is important since it determines the current generated. Transient upset sensitivity is therefore measured in rads per second, with typical upset rates being between 10^6 and 10^6 rad/s. Since the radiation pulse is short, the total dose will be small (less than 1000 rad) and not likely to degrade performance. Total dose failure levels are based on the total amount of radiation (or energy) a device receives. Total dose exposure is measured in rads with typical failure levels on the order of 10^3 to 10^6 rad.

In an atmospheric burst, a large portion of the gamma-ray energy is absorbed by the atmosphere. The following empirical expression gives the magnitude of the gamma pulse:[4]

$$D = 3.3 \times 10^6 \frac{(Y + 6Y^3)}{R^2} \exp(-pR/0.326) \qquad (8.2)$$

where D is the total dose in rads (Si), Y is the bomb yield in megatons of TNT, p is the density of the atmosphere in grams per liter (1.1 at sea level), and R is the distance from the explosion in kilometers. In the case of an exoatmospheric blast, there is no atmospheric attenuation of the gamma pulse, and only a $1/R^2$ dependence results. The following expression may be used:

$$D = (8.0 \times 10^5) \frac{Y}{R^2} \qquad (8.3)$$

Comparing the above expressions it is found that a 1-MT bomb exploded at sea level gives an exposure of about 100 rad at a distance of 3 km; in space, at the same distance, the dose exceeds 10,000 rad. An estimate of the dose rate can be obtained by dividing the total gamma dose D by the gamma pulsewidth. In most cases, the prompt gamma pulse is over in about 50 ns, so the dose rate can be obtained by multiplying the above expressions by 2×10^7.

8.1.1 Dose Rate Upset and Survivability

The term *upset* refers to a momentary disturbance in the operation of a system. For example, a radiation pulse may scramble the memory of a digital computer, necessitating a restart using information stored in a nonvolatile memory. *Survivability* implies that the hardware components of the computer and information stored in the nonvolatile memory remain intact, so continued operation is possible after the nuclear event. An overview of the dose rate upset mechanism in a solid-state circuit will now be given.

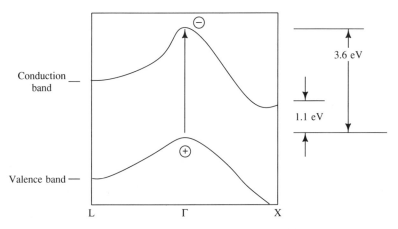

FIGURE 8.3   Electron-hole pair (EHP) generation in silicon.

In semiconductors or insulators, absorbed radiation kicks electrons from the valence band to the conduction band. Since photons have little momentum, the transition is nearly vertical. The energy required in silicon is 3.6 eV per electron-hole pair (EHP) as shown in Fig. 8.3. It can be shown that 1 rad of absorbed radiation produces 4.2×10^{13} EHP/cm^3. In other materials, the varying band gap energies and other effects produce different generation rates—for example in SiO2, 8.0×10^{12}; in GaAs, 7.0×10^{13}, and in Ge, 1.2×10^{14}.

If the EHPs are generated deep within a wafer, they can recombine and the energy will be lost as heat (phonons). If a PN junction is nearby, however, the built in electric field of the junction causes the carriers to move producing a current. If the generation rate is not too large, EHPs are swept from the depletion region as fast as they are formed, and, using the continuity equation for electrons, the current is

$$0 = \frac{\partial n}{\partial t} = \frac{\partial J_n}{\partial x} + G(t) - R(t) \tag{8.4}$$

where J_n is the prompt component of the collection current.

Assuming that carriers are collected faster than they can recombine, $R = 0$ and

$$J_n = W_d G(t) \quad \text{or} \quad I_p = A W_d G(t) \tag{8.5}$$

where W_d is the junction depletion width, A is the junction area, and $G(t)$ is the generation rate in EHPs/cm^3/s.

Current I_p is referred to as the "prompt" component of the collection current. Carriers can also diffuse up to the junction and be collected resulting in current which continues to flow after the gamma pulse has ended. This is the delayed component and can be calculated from the current density equation by assuming zero electric field outside the depletion region:

$$\frac{\partial P}{\partial t} = D_p \frac{\partial^2 P}{\partial x^2} - \frac{P - P_{no}}{\tau_p} + G(t) \tag{8.6}$$

$$J_p(x, t) = q D_p \frac{\partial P(x, t)}{\partial x} \tag{8.7}$$

If the radiation pulse is rectangular with duration T and $T \ll \tau_n$ or τ_p the solution is given by

$$I_d = q A G T \left[\frac{\sqrt{D_n} \exp(-t/\tau_n) + \sqrt{D_p} \exp(-t/\tau_p)}{\sqrt{\pi t}} \right] \tag{8.8}$$

Here τ_n and τ_p are the electron and hole minority carrier lifetimes, D_n and D_p are the electron and hole diffusion constants (note that $D_n = 0.026 U_n$, $D_p = 0.026 U_p$ through the Einstein relation). The total current is the sum of I_p and I_d. At very high generation levels, the assumption of zero electric field outside the depletion region starts to break down. The result is enhanced charge collection, above that produced by diffusion. A model that accounts for this effect can be found in Ref. 6. A diode model with a photocurrent source is given in Fig. 8.4.

The response of a bipolar transistor is more complex. In an NPN device, electrons generated within the base (W_b) and collector depletion region (W_{bc}) are collected at the collector junction giving rise to a current

$$I_{ce} = GA(W_b + W_{bc}) \tag{8.9}$$

Holes are blocked from leaving the base by the electric field at the collector junction. This same electric field, however, extracts electrons from the base. This causes new electrons to be injected into the base from the emitter to maintain charge neutrality. At the same time some holes leave the base at the emitter junction. Since the ratio of electrons entering the base to holes leaving at the emitter is The current gain, Hfe, the current due to the holes is

$$I_{ch} = GA(W_b + W_{bc}) \text{Hfe} \tag{8.10}$$

and the total current is the sum of I_{ch} and I_{ce}, as shown in Fig. 8.4. Note the

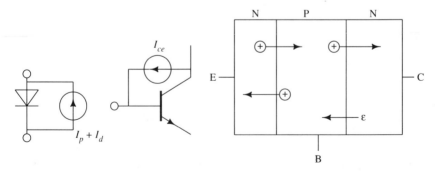

FIGURE 8.4   Diode and transistor with photocurrent source.

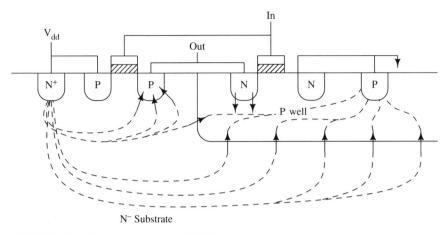

FIGURE 8.5   Photocurrents in a CMOS process.

multiplication of the current by Hfe, this multiplied component is referred to as a secondary photocurrent.

Of course, this model is highly simplified. As was seen in the diode, charge can also be collected from outside the base by diffusion and time-dependent effects become involved. A model of this type is described in Ref. 5. The main effect of these photocurrents is to cause conduction in reverse-biased junctions. Consider the complementary metal oxide semiconductor (CMOS) cross section of Fig. 8.5. Photocurrents will flow at the two drain substrate junctions, and may cause a change in the electrical state of the output. Currents flow at the P-well–substrate junction and must be supported by V_{dd}. The large power supply currents that result can cause voltage drops to develop along the supply rails, and the supply voltage will be reduced to certain portions of the chip (rail span collapse).[7,8] Under more intense radiation, the large current can cause metal lines to melt, producing complete failure of the chip.

8.1.2 Dose Rate Upset Hardening

Chips can be hardened at the process level by several means. One method is to fabricate the CMOS on the epitaxial layer (Epi) as shown in Fig. 8.6. This reduces the diffusion component since the minority carrier lieftime and diffusion length will be short in the heavily doped substrate. In addition, power can be distributed through the conductive substrate, easing the burden on the metalization. Minority carrier lifetime reduction in MOS devices also reduces the photocurrent by reducing the diffusion lengths. Neutron irradiation is sometimes used for this purpose.

Fabrication on insulators is another popular option. This approach greatly reduces the PN junction area and therefore the magnitude of the photo-current. Although the insulating substrate also conducts during irradiation due to formation of electrons and holes within the insulator, mobilities and

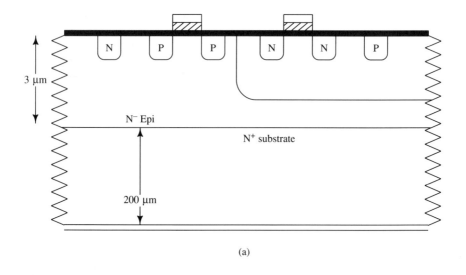

(a)

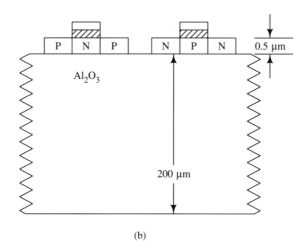

(b)

FIGURE 8.6 Use of (a) Epi and (b) silicon oxide semiconductor (SOS) to suppress photocurrents and latchup.

lifetimes in the insulator are short and the problem is much less severe than in silicon. The main problem with silicon-on-insulator (SOI) is that it is difficult to grow high-quality thin films on an insulating substrate since there is no seed crystal. SOI processes have additional advantages, since latchup becomes impossible and parasitic junction capacitance is reduced.

Circuit hardening is also possible. The type of measures used depend on whether it is necessary to "operate through" the nuclear event, or simply survive the event. If survivability is all that is required, the main concern is limiting device currents to prevent overheating of the transistors and metal lines. This can be accomplished by a number of means. The simplest is to include resistors at strategic locations in series with the power bus. Other measures include "crowbar" circuits, which short circuit the power bus to ground at the start of the gamma pulse. Resistors and crowbarring also have the advantage of suppressing any latchup that may be initiated by the gamma pulse.

If it is necessary to operate through the event, then process level hardening is the best area to start. Reducing the power supply impedance is also important due to rail span collapse. Multiple V_{dd} and ground pins are often used, and capacitors are often included as part of the IC package. Compensation transistors can be used to shunt photocurrents to ground. An example of such a design can be found in Fig. 8.7. The dose rate upset sensitivities of a number of technologies are illustrated in Fig. 8.8.[4]

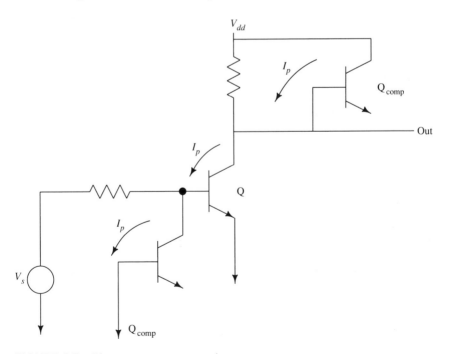

FIGURE 8.7   Photocurrent compensation

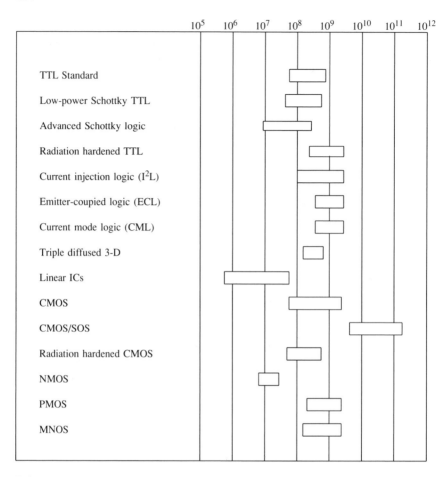

FIGURE 8.8   Comparision of technologies with regard to dose rate upset sensitivity.

8.2 SINGLE-EVENT UPSET

Single-event upset (SEU) is similar to dose rate upset in that electrical disturbances are produced by electron hole pairs generated by radiation. In SEU a thin line of charge is produced as a cosmic ray (CR) passes through a piece of Si. The cosmic ray typically takes less than a picosecond to pass through, and the charge track is less than a micrometer in diameter. The events are called "single events" since in space the probability of a CR striking a typical PN junction is less than 10^{-4} per day. The intensity of the charge track is conveniently measured in units of pC/μm. Since the generation of charge arises from the loss of the CR's energy as it passes through, the charge density can also be expressed as the L_{et} or "linear energy transfer" of the CR, and 25 MeV/μm corresponds to 1 pC/μm. Here, once again, the 3.6 eV/EHP comes into play.

It can easily be calculated that a charge density of 1 pC/μm inside a charge column 1 μm in diameter produces a carrier density in excess of 10^{-18}/cm^3. This density is greater than the doping density in most devices and is therefore capable of producing electrical disturbances. The resistance of the SEU track can be found from

$$R = \frac{1}{qL_{et}(U_n + U_p)} \tag{8.11}$$

where R is the resistance per unit length, and U_n and U_p are the electron and hole mobilities. Typical resistance values are less than 100 Ω/μm. During a strike on a typical reverse biased PN junction, the charge track completely shorts out the junction. This process produces current pulses that can be as large as 20 mA, but are of very short duration, typically less than 1 ns. The result is that the junction potential drops to zero. If the junction is part of a memory element, a bit flip or logic error can occur.

Consider the four-transistor memory cell of Fig. 8.9. Initially node 1 is high and node 2 low. The current pulse produced if the drain of Q_1 is struck can pull 1 down and cause the cell to change state, with 1 low and 2 high after the "upset" (see Fig. 8.9). If this cell were storing data, such as the program counter of a microprocessor, serious problems can result. Whether or not the cell will flip depends on whether the current pulse produced is large enough, and lasts long enough to pull node 1 all the way to ground and keep it there long enough for the two inverters to change state.

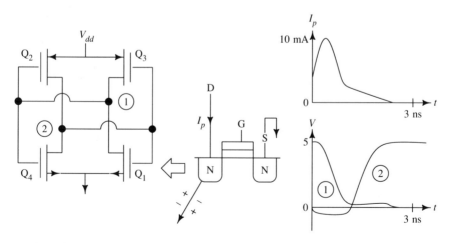

FIGURE 8.9   Static random access memory (SRAM) cell and upset waveforms.

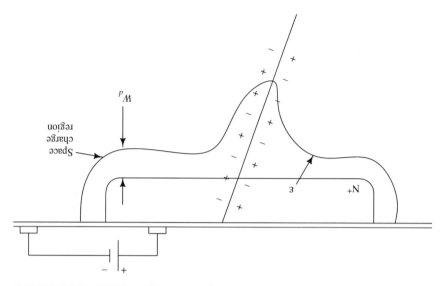

FIGURE 8.10   SEU funneling mechanism.

A charge collection model was derived by Hu[12] to determine the size of the charge collection pulses. The model operates as follow (see Fig. 8.10):

1. Assume an N + P junction.

2. Along the charge column, near the junction, the electron current is (ignoring diffusion) $J_n = qU_n nE$, and the hole current is $J_p = qU_p pE$.

3. Since the electric field is the same for electrons and holes and $n = p$ since equal numbers of electrons and holes were generated,

$$J = J_n + J_p = J_p(1 + U_n/U_p) \quad \text{or} \quad J_p = \frac{J}{1 + U_n/U_p} \tag{8.12}$$

4. The junction will remain shorted and current will flow as long as holes remain inside the depletion region and neutralize the fixed acceptors. Holes are removed from the depletion region by J_p, which is flowing down. They are not replaced since charge is transported by electrons in the heavily doped N^+ material. The number of holes that must be removed is $L_{et}W_d$, and charge collection will stop when

$$\int_0^t J_p(t)\, dt = L_{et}W_d \tag{8.13}$$

5. Observe that the total collected charge is just

$$Q = L_{et}W_d(1 + U_n/U_p) \tag{8.14}$$

Note that the collected charge is dependent upon the junction voltage and substrate doping through $W_d = \sqrt{2\varepsilon V/qN_a}$, and increased doping ($N_a$) reduces the collected charge. Several other charge collection models have also been developed.[13]

Most of the methods of preventing dose rate upset such as epitaxial processes and SOI technologies also help prevent SEU by cutting down the funnel length $W_d(1 + U_n/U_p)$. Circuit level hardening can also be applied. The most popular method is to use decoupling resistors as illustrated in Fig. 8.11.

The resistors harden the cell by slowing down the switching speed of the two invertors. This gives the restoring current I_r more time to bleed off the charge deposited by the ion. The disadvantage is that some operating speed is also lost. Note that the charge needed to upset the cell is given crudely by

$$Q_c = 2I_rR(C_n + C_p) \tag{8.15}$$

where C_n and C_p are the gate capacitance of the N and P field-effect transistors (FETs).

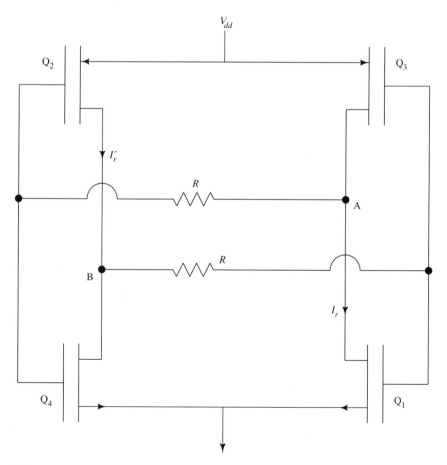

FIGURE 8.11   SRAM cell hardened with decoupling resistors.

$$C = \frac{\varepsilon_{ox}LW}{t_{ox}}$$ (8.16)

where

ε_{ox} = the dielectric constant of SiO_2

L = the channel length

W = the channel width

t_{ox} = the oxide thickness

It can be seen that as the device gets smaller, larger values of hardening resistance are needed since the capacitance is reduced. Increasing the value of I_r can be accomplished by increasing V_{dd}. Note that I_r is the drain current of Q_3. I_r may therefore be estimated as the saturation current of Q_3 with Q_3 fully *on*, or

$$I_r = \frac{U_p W_p E_{ox}}{t_{ox}} (V_{dd} - V_{th})^2$$ (8.17)

Figure 8.12 gives a simplified layout of a static random access memory

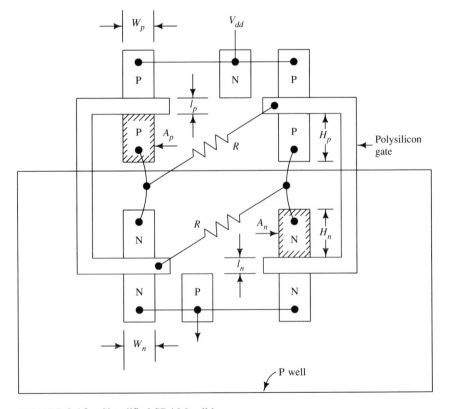

FIGURE 8.12   Simplified SRAM cell layout.

(SRAM) cell. Observe that there are two sensitive regions A_n and A_p, these being the drains of Q_1 and Q_2. When the drain of Q_1 is struck, node A is pulled down to ground potential. The drain current of Q_3 (which is still on) supplies the restoring current I_r, which fights against the SEU current pulse and tries to pull node A back up. If the drain of Q_2 is struck, then node B (which is initially 0 volts *Low*) is pulled up to V_{dd} potential by the SEU pulse. In this case, I_r', the restoring current that tries to pull node B back down is supplied by Q_4.

The critical L_{et} required in each case is different because:

1. I_r is different since, at node A, I_r comes from a P-FET but at node B, I_r' comes from an N-FET.

2. The charge collection mechanism is different. At Q_1, an N + P junction is struck, but at Q_2 a P + N junction is struck. For N + P the collected charge is $L_{et}W_d(1 + U_n/U_p)$; for P + N the charge is $L_{et}W_d'(1 + U_p/U_n)$.

Therefore two critical values for L_{et} exist:

$$L_{et} = \frac{2I_r R(C_n + C_p)}{W_d(1 + U_n/U_p)} \quad \text{at } Q_1 \tag{8.18}$$

$$L_{et}' = \frac{2I_r' R(C_n + C_p)}{W_d'(1 + U_p/U_n)} \quad \text{at } Q_2 \tag{8.19}$$

$$W_d = \sqrt{\frac{2\varepsilon V_{dd}}{qN_a}} \tag{8.20}$$

$$W_d = \sqrt{\frac{2\varepsilon V_{dd}}{qN_d}} \tag{8.21}$$

Here, N_a is the doping of the P well that surrounds Q_1 and N_d is the doping of the N well that surrounds Q_2.[18] Note that below the critical L_{et} no upsets occur. A graph of the target area versus L_{et} for this circuit is given in Fig. 8.13. Here A_n is the area of Q_1's drain and A_p is the area of Q_2's drain. Note that Q_3 and Q_4 are not sensitive since their drain-substrate junctions are at zero bias (assuming node A is high and B low). If the cell were in the opposite state (A low, B high), then Q_3 and Q_4 would be sensitive and Q_1 and Q_2 would be insensitive.

Once a target versus L_{et} curve is obtained, it can be used with an environmental model to predict the upset rate (errors per bit per day) in space. Several programs such as CREME (for "cosmic-ray effects on microelectronics")[16] have been written to do this. The L_{et} versus target curve can also be measured experimentally using a particle accelerator such as a cyclotron.[14]

The SEU sensitivity of a given part is dependent on a number of variables. From Eq. (8.15), it can be seen that the SEU resistance is proportional to capacitance and power supply voltage. Chip feature size, speed, and power consumption per bit are also dependent proportional to capacitance and

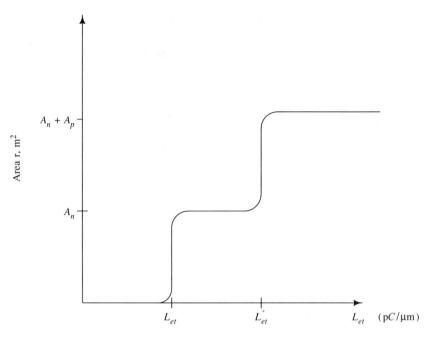

FIGURE 8.13   Cross section vs. L_{et} plot for SRAM cell.

power supply voltage. It can therefore be stated that in a given technology, faster, low-power, high-density parts are more susceptible to SEU than slower, high-power parts. The most SEU resistant devices are listed first:[15]

1. CD4000 series CMOS

2. Other CMOS logic (LS, HC, HCT, SC)

3. Standard bipolar (54XXX)

4. Other bipolar logic (low power, FAST, Schottky)

5. Low-power Schottky bipolar (54LSXXX)

6. Advanced LS bipolar (54ALSXXX)

8.2.1 SEU Environmental Model

In the space environment, cosmic rays corresponding to all 92 natural elements are found, in approximately the same distribution as on earth. The heavier elements produce the largest L_{et}, and the L_{et} of a given ion species depends upon the energy of the ion. The highest L_{et} is normally found at low energy just as the ion is grinding to a halt. Due to the high energy of the ions, shielding is difficult since the ions can penetrate several centimeters of aluminum. Models of the cosmic-ray environment have been developed to calculate the number and energy of each type of ion. The result of such a model is the integral L_{et} spectrum. The spectrum is a function of the space craft orbit, year (because of solar cycles), and shielding. An example of an L_{et} spectrum is given in Fig. 8.14.

A simple approximation often used in place of an environmental model is Petersen's equation:[17]

$$U = KWHT^2Q_c^2 \qquad\qquad (8.22)$$

where K is an empirical constant and H, W, and T are the dimensions of the

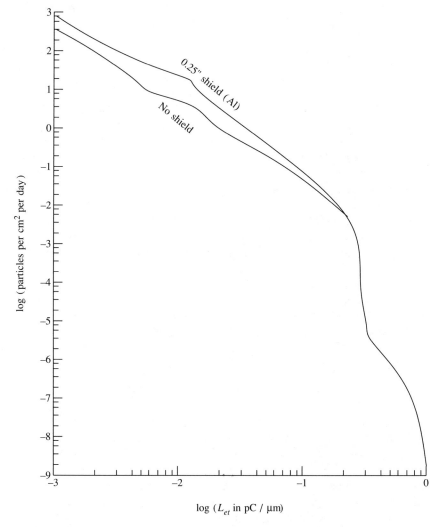

FIGURE 8.14   Cosmic-ray L_{et} spectra for geosynchronous orbit.

"target volume" through which the ion must pass to produce upset. The target volume is usually the depletion region of a reverse-biased PN junction or the base of a bipolar transistor. Note that $T \ll H$ and W. Constant K is highly dependent upon the spacecraft orbit and shielding. A typical value for a geosynchronous orbit with 0 degrees inclination is between 3×10^{-11} and 5×10^{-10}. H, W, and T are in micrometers, Q_c is in picocoulombs per micrometer, and U is in upsets per bit per day. In the example of Fig. 8.12, dimensions H and W have been indicated for transistor Q_1. Dimension T is the charge collection length, into the page of the figure, $W_d(1 + U_n/U_p)$ given by Hu's model. The total upset rate for the cell is the sum of the rates for Q_1 and Q_2.

If only cyclotron test data are available, Petersen's equation may be rearranged. Note that $HW = A_n$ and $L_{et}T = Q_c$; therefore

$$U = KA_n/L_{et}^2 \tag{8.23}$$

Single-event upset may also be caused by indirect means. Neutrons or light nuclei such as protons may have sufficient energy to initiate nuclear reactions with silicon atoms. In this case a portion of the reaction energy is transferred to the silicon atom, causing it to recoil. The recoiling silicon atom may then, by virtue of its larger mass, be able to produce SEU even though the original proton could not. This effect is important since protons are abundant in the cosmic-ray spectra.

In addition to circuit and process hardening, hardening at the systems level is possible. These measures usually take the form of redundant bits and error-correcting codes. Since random access memory (RAM) chips make up the largest portion (in terms of bits) of a digital computer, hardening is most often applied there. SEU has not been a major concern in space based systems of the past for two reasons. First, old systems simply did not use that many bits. Even a soft RAM chip only has probability of upset of 10^{-5} per bit per day, and thus old systems with only a few thousand bits have a small probability of upset. The second reason is that older chips were larger in size resulting in larger capacitances and critical charge. These chips were therefore difficult to upset.

8.3 TOTAL-DOSE EFFECTS IN BIPOLAR DEVICES

Total dose occurs when radiation slowly degrades the performance of a solid-state device to the point where it can no longer function. In bipolar devices the main mechanism is the formation of traps and defects in the silicon lattice and the formation of inversion regions under oxide layers. Traps and defects occur because Si atoms are knocked off their lattice sites, forming interstitial-vacancy pairs. The defects, in turn, form recombination centers that reduce the minority-carrier lifetime, and degrade the mobility. The formation of inversion layers will be discussed in the section on MOS devices.

8.3.1 Neutrons

In bipolar devices, neutrons can be the worst offenders. They are produced during the explosion of a fission or fusion bomb. As with gamma rays, neutrons are produced with a wide range of energies. The neutron energy is important to determining the way the neutron interacts with the semiconductor device. In the following discussion, the energy spectrum produced by a thermonuclear explosion will be assumed. The following empirical expression has been developed to describe the flux of neutrons produced during a ground-level burst:[9]

$$\phi = \frac{(2 \times 10^{15})Y}{R^2} \exp\left(\frac{-pR}{0.238}\right) \tag{8.24}$$

In the above equation, ϕ is the neutron flux, in neutrons per cm^2, Y is the bomb yield in megatons of TNT, p is the density of air in grams per liter (1.1 at sea level), and R is the distance from the blast in km. In the space environment there is no atmospheric absorption, so the exponential term is unity.

A neutron can interact with the nucleus of an Si atom via a number of reactions. Energy is released (or the neutron may have high energy to start with) and the Si atom "recoils." Recoiling Si atoms can possess energies of a few keV up to 1 MeV. Since it takes only 6 eV to knock a Si atom off it's lattice site the result is that each recoil produces a "track" containing several thousand defect sites. Fortunately, a large percentage of the interstitials and vacancies recombine and over time the damage may be repaired or annealed out. This repair process is greatly accelerated at higher temperatures ($>200°C$).

It has been determined experimentally that the minority-carrier lifetime can be approximated as

$$\frac{1}{t} = \frac{1}{t_0} + \frac{F}{K} \tag{8.25}$$

Here t is the new lifetime, t_0 is the lifetime before radiation, F is the neutron flux per square centimeter and K is a constant (s/cm^2). The constant K is dependent on resistivity and injection level and arises because both the minority carriers and therefore the position of the Fermi level and the Shockley-Read-Hall (SRH) recombination mechanism. A plot of K is given in Fig. 8.15.[4]

In bipolar junction transistors (BJTs) the reduced lifetime is important in the base, the emitter and base-emitter depletion region. The overall effect is a reduction in Hfe. Recall that Hfe is proportional to the square root of the emitter lifetime. The following empirical expression has also been developed to predict the Hfe reduction in low-speed, unhardened transistors.

$$1/\text{Hfe} = 1/\text{Hfe}_0 + \frac{\phi K}{2\pi F_t} \tag{8.26}$$

In this expression, Hfe_0 is the preradiation value of Hfe, ϕ is the neutron flux

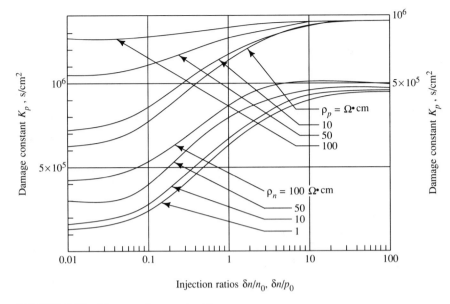

FIGURE 8.15   Neutron damage in silicon.

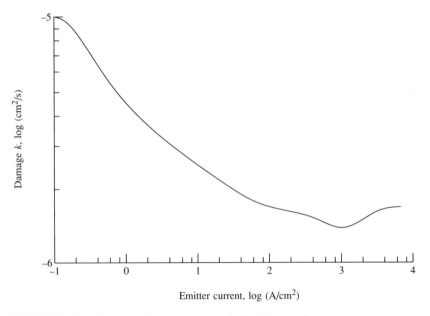

FIGURE 8.16   Neutron damage constant in bipolar transistors.

in neutrons/cm, F_t is the transistor bandwidth, and K is a constant, given in Fig. 8.16,[3] which depends on the emitter current density. The minimum value for K in GaAs in 1.7×10^{-6}.

To compensate for radiation damage in BJTs it is best to:

1. Use feedback circuits to minimize Hfe dependence.

2. Be sure sufficient base drive is available if transistors are used in the switching (saturated) mode.

3. Derate the frequency response and Hfe.

8.3.2 Total Dose in MOS Devices

Since MOS devices utilize majority carriers, lifetime reduction is not as important and MOS devices are resistant to neutrons. In fact, CMOS devices are often deliberately irradiated with neutrons to suppress latchup by reducing the Hfe of parasitic BJTs.

Total dose radiation is measured in rad, and an expression for the prompt dose following the explosion of a bomb was given in Eq. 8.2. In space, there is a significant amount of natural radiation present due to trapped charged particles in the Van Allen belts. These particles may be further excited by an exoatmospheric nuclear explosion, and the radiation greatly increased. Figure 8.17 shows the natural and theoretical maximum bomb-enhanced radiation

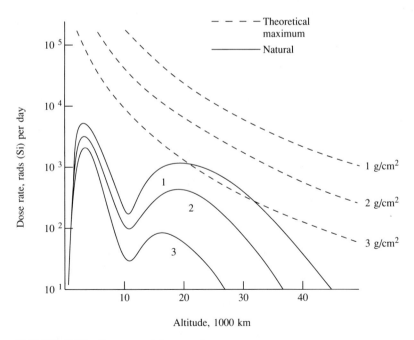

FIGURE 8.17   Space total dose environment.

as a function of altitude and shielding.[2] The enhanced radiation may persist for hundreds of days following a nuclear event. Since this radiation is due to charged particles of moderate energies, shielding is effective at reducing the exposure. Note that a shield of 1 g/cm^2 of aluminum is 0.146 in. thick.

In MOS devices, the main area where damage takes place is in the oxide layers. Here gamma and charged-particle radiation are the main culprits. The mechanism is as follows:

1. Gamma rays produce electron-hole pairs in the oxide layer, 8.8 eV/EHP are needed due to the large bandgap of SiO_2.

2. Many of these EHPs recombine; however, some of the holes become trapped at the oxide-Si interface.

3. The trapped holes alter the threshold voltage of the MOSFETs. This can have two effects: (a) It may become impossible to turn an N-FET off since the trapped holes attract electrons and keep the channel in an inverted state, thus often making it impossible for the output of an inverter to reach a high state. (b) Silicon under field oxide may become inverted, and undesirable leakage current can result.

The process is complicated by the effect of gate bias. It is the gate bias that drives the holes to the Si-SiO_2 interface as indicated in Fig. 8.18. As a result the threshold voltage shift is greatest in N-FETS when they are biased *on*. P-FETs are not as sensitive since the holes drift toward the gate electrode, as shown in Fig. 8.19. In addition to trapped holes, negatively charged surface

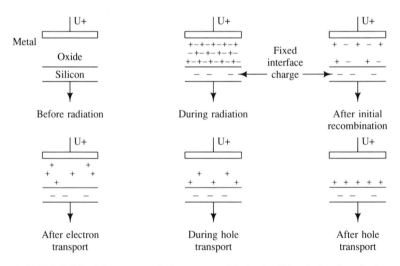

FIGURE 8.18   Movement of electrons and holes in SiO_2 during irradiation.

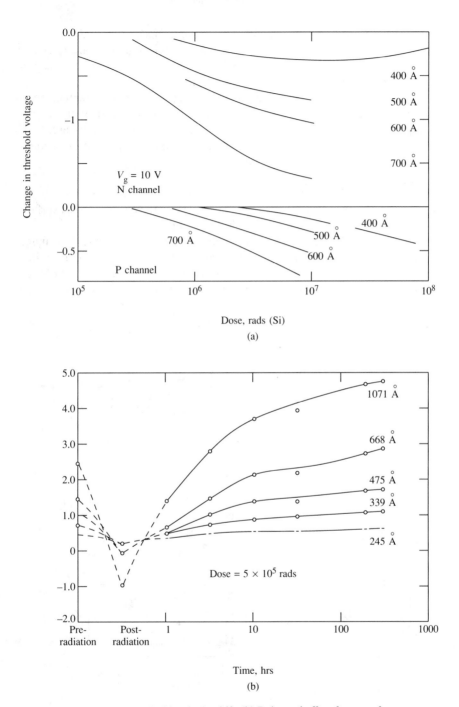

FIGURE 8.19 (a) Threshold voltage shift. (b) Rebound effect from surface states.

states are also produced, which compensate the trapped holes and reduce the total threshold shift. However, as the device anneals, the holes are removed (by electrons tunneling into the oxide) but the surface states remain, thereby producing a rebound effect,[10] as illustrated in Fig. 8.19.[11]

The way in which the oxide was fabricated (wet or dry), temperature, thickness, impurities, etc. can affect the radiation properties of the oxide. It appears that thinner oxides are better. The two main reasons are: first, a thinner oxide presents a smaller volume to hold trapped holes; second, $V = Q/C$ and, since $C = A\varepsilon/t_{ox}$, then

$$\Delta V_{th} = -\varepsilon_{ox}(Q_h - Q_{int})/T_{ox} \tag{8.27}$$

where Q_h is the charge due to trapped holes, Q_{int} is the charge due to surface states, and ΔV_{th} is the change in threshold voltage.

Inversion under field oxide is important since it can contribute to leakage currents. These currents can affect the logic levels in an MOS circuit and/or cause increased power consumption as shown in Fig. 8.20. In "isoplanar"-type bipolar processes inversion of deep oxide layers can contribute to leakage

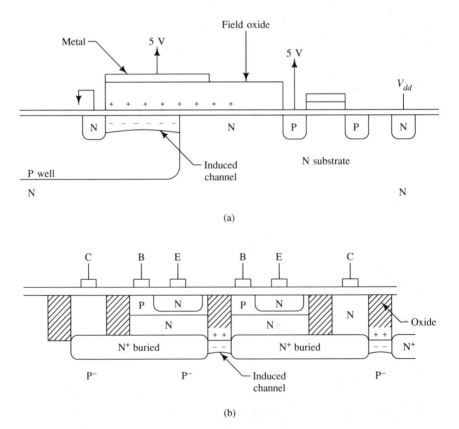

FIGURE 8.20   (a) Leakage in CMOS and (b) leakage in bipolar transistors.

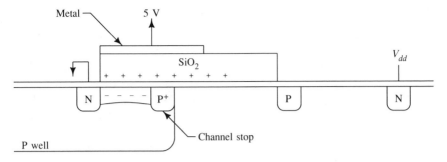

FIGURE 8.21   Use of channel stop to prevent leakage.

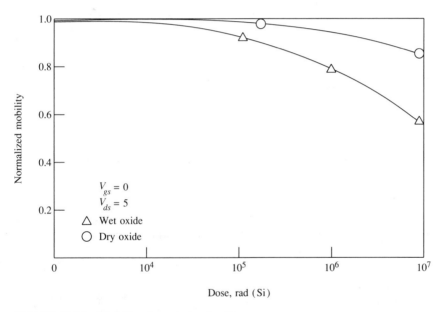

Dose, rad (Si)

FIGURE 8.22   Mobility degradation in silicon.

between buried layers and, in certain processes, collector-emitter leakage. The most common way to prevent field oxide leakage is to dope the underlying silicon heavily, as shown in Fig. 8.21.

Silicon-on-sapphire (SOS) devices, which have advantages during transient radiation, can develop problems with total dose, due to a "back gate" effect, whereby holes become trapped within the sapphire and induce a channel that causes drain-source leakage. The other effect of radiation on all MOS devices is damage to the silicon. Although the decrease in minority-carrier lifetime is not a problem, the reduction in mobility causes a decrease in the transconductance (g_m) since g_m is proportional to the mobility (see Fig. 8.22). This decrease causes a reduction in operating speed.

8.4 ELECTROMAGNETIC EFFECTS

During a nuclear explosion, the gamma/X-ray pulse interacts with electrons in a material via the Compton effect. Since momentum (of photons and the electron) must be conserved, the electron is always scattered forward along the path of the radiation. This motion of electrons constitutes a current, and the current may interact with an electronic system in a number of ways. If the "material" is the atmosphere, the current generates an electromagnetic pulse (EMP). This pulse has frequency components which cover the complete range from dc up to several gigahertz. Electric field strengths near the blast, can be as large as 100 kV/m. If the material the gamma/X-ray pulse interacts with is the body of a satellite, then currents are generated inside the satellite itself and the effect is known as system-generated EMP or SGEMP.

The large voltages generated by these EMP effects can be picked up by antennas and cables. The currents developed near a blast are often strong enough to melt a metal cable. Obviously, semiconductor devices are particularly sensitive to overstress and burnout. Modeling EMP effects is complex, and a further discussion will not be given here. The reader is encouraged to consult the following Refs. 3, 4, 19, and 20.

Methods of hardening against EMP effects are mainly concerned with bypassing the pulse to ground. This is often accomplished with a band-pass filter to reject the majority of the EMP frequencies and/or with clipping diodes to limit the excursion of voltage at a particular node. Electromagnetic shielding is also of value against EMPs.

8.5 LATCHUP

Latchup[21] is a particularly serious problem in which parasitic bipolar transistors are activated and form a low impedance from V_{dd} to ground. Current flowing in the low-impedance path can cause overheating and result in chip failure. Latchup is related to radiation effects because certain transient effects, such as the gamma pulse and SEU,[22] can initiate the latchup process.

Latchup can occur in CMOS or bipolar ICs but is most common in CMOS bulk devices. A cross section of a CMOS device with parasitic bipolar devices and resistance elements is shown in Fig. 8.23. During normal IC operation, the bipolar transistors are off and no latchup current flows. However if photocurrent source I1 (from P well to substrate) is activated, the latchup process can begin. Observe that the collector of Q_n supplies the base current for Q_p and vise-versa. Therefore if resistors R_n and R_p are momentarily ignored (open-circuited), then

$$I_{cn} = I_{bp} \quad I_{cp} = I_{bn} \tag{8.28}$$

$$I_{cn} = \mathrm{Hfe}_n I_{bn} \quad I_{cp} = \mathrm{Hfe}_p I_{bp} \tag{8.29}$$

$$I_{cn} = \mathrm{Hfe}_n I_{cp} = \mathrm{Hfe}_n \mathrm{Hfe}_p I_{bp} = \mathrm{Hfe}_n \mathrm{Hfe}_p I_{cn} \tag{8.30}$$

The above equations imply that if the product $\mathrm{Hfe}_n * \mathrm{Hfe}_p = 1$, then I_{cn} can take on any value, or, if $\mathrm{Hfe}_n * \mathrm{Hfe}_p > 1$, then I_{cn} will increase without bound.

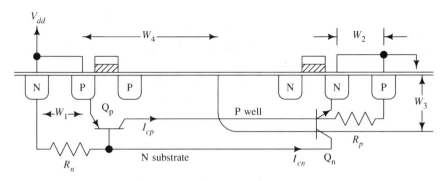

FIGURE 8.23   Latchup paths in a CMOS bulk inverter.

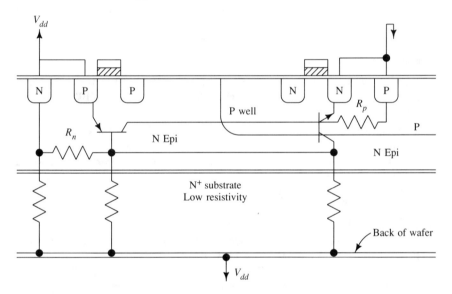

FIGURE 8.24   Use of Epi process to reduce probability of latchup.

In reality, however, I_{cn} increases only until high injection effects reduce the Hfe values of the transistors and the Hfe product falls to unity. (Note the two transistors essentially form a positive-feedback loop.) The resistors R_n and R_p have the effect of limiting the minimum current needed to start the latchup process. In order to conduct current, Q_n and Q_p need $V_{be} \geq 0.8$ V. When $V_{be} < 0.8$ V, most of the current flows through the resistor instead of into the base and the above equations do not apply. With the resistors included, two additional conditions for latchup become necessary:

$$I_1 > 0.8/R_n \quad \text{and} \quad I_1 > 0.8/R_p \tag{8.31}$$

In terms of the device layout, it can be shown that

$$R_n \propto W_1 \quad R_p \propto W_2 \quad \text{Hfe}_n \propto 1/W_3 \quad \text{Hfe}_p \propto 1/W_4 \tag{8.32}$$

Therefore the most sensitive device is the one with the smallest W_3 and W_4

and the largest W_1 and W_2. Fabrication on Epi reduces the probability of latchup since the Epi conductance shunts R_n (as shown in Fig. 8.24) and thereby increases the initiating current I_1. If R_n can be made small enough, latchup can be prevented entirely, since Hfe$_n$ and Hfe$_p$ roll off at high current due to high injection effects and Hfe$_n$ * Hfe$_p$ will become <1.

EXERCISES

1. A 0.2 megaton bomb is exploded at sea level.
   (a) Calculate the total gamma dose and neutron flux at distances of 2 km and 20 km.
   (b) If the average gamma energy is 1.1 MeV, find the gamma dose behind 1 cm of aluminum.
   (c) Find the dose rate assuming a 50-ns gamma pulse.

2. Repeat Exercise 1 for an exoatmospheric blast.

3. A bipolar transistor has an emitter area of 10^{-6} cm^2 Hfe $= 150$, and $F_t = 300$ MHz, before radiation. Calculate the new Hfe after exposure to 10^{12}, 10^{13}, and 10^{14} neutrons/cm^2.

4. A silicon N + P junction diode is reverse-biased at 5 V and has the following parameters: $A = 10^{-6}$ cm^2, $N_a = 10^{15}$/cm^3, $N_d = 10^{19}$/cm^3, $U_n = 850$ cm^2/V·s, $U_p = 430$ cm^2/V·s, $T_n = T_p = 10$ ns, $\varepsilon = 1$ pF/cm. Plot the prompt and delayed photocurrent pulses for
   (a) a 50-ns 10^8-rad/s gamma pulse
   (b) a 50-ns 10^{10}-rad/s gamma pulse.

5. The static RAM cell of Fig. 8.25 has the following parameters: $T_{ox} = 700$ Å, $R = 100$ kΩ, $U_n = 850$ cm^2/V·s, $U_p = 420$ cm^2/V·s, $N_a = 10^{15}$/cm^3, $N_d = 10^{16}$/cm^3, $V_{dd} = 5.0$ V, $\varepsilon_{ox} = 0.33$ pF/cm $V_{thn} = 1.0$ V, $V_{thp} = -1.0$ V.
   (a) Find W_n, W_p, L_n, L_p, I_r, C_n, C_p.
   (b) Find the two critical values of critical charge.
   (c) Calculate the two critical values of L_{et} and construct a plot similar to Fig. 8.13.
   (d) Use Petersen's equation and calculate the upset rate in geosynchronous orbit.

6. The transistors in the RAM cell of Exercise 5 are subjected to total dose radiation and the threshold voltages changes as indicated in Fig. 8.19a.
   (a) Based on Eq. (8.17), at what radiation level does the leakage current start to increase?
   (b) What is the static power consumption per cell after exposure to 10^7 rad (Si)?

7. Consider the latchup circuit of Fig. 8.26. The resistances may be approximated by the following equations:

$$R_n = \frac{1}{N_d q U_n h} \qquad R_p = \frac{W_2}{N_a q U_p h W_3} \tag{8.33}$$

In the above equations, N_d is the substrate doping (10^{15}) and N_a is the average P-well doping (5×10^{15}). Assuming $U_n = 850$ cm^2/V·s, $U_p = 420$ cm^2/V·s, and $V_{dd} = 5$V, find the minimum dose rate (RADs/s) necessary to initiate latchup (consider the prompt photocurrent only). Dimension h is the depth of the device into the page.

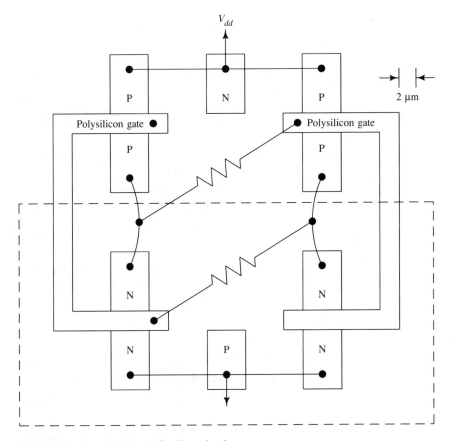

FIGURE 8.25 SRAM cell for Exercise 5.

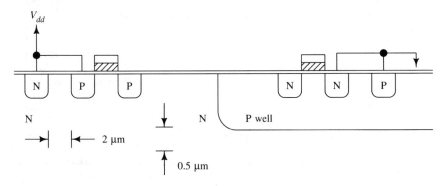

FIGURE 8.26 Latchup diagram for Exercise 7.

REFERENCES

1. S. Glasstone and P. J. Dolan, *The Effects of Nuclear Weapons*, Washington, DC: U.S. Departments of Defense and Energy, 1977.
2. *The Trapped Radiation Handbook*, DNA 2524H, Washington, DC: Defense Nuclear Agency, 1971.
3. J. E. Grover "Basic Radiation Effect in Electronics Technology," Sandia National Laboratories, February 1984.
4. G. C. Messenger and M. S. Ash, *The Effects of Radiation on Electronic Systems*, New York: Van Nostrand Reinhold, p. 296, 1986.
5. J. L. Wirht and C. S. Rodgers, "Transient Response of Transistors and Diodes to Ionizing Radiation," *IEEE Trans. Nucl. Sci.*, vol. NS-11, pp. 24–38, November 1964.
6. E. W. Enlow and D. R. Alexander, "Photocurrent Modeling of Modern Microcircuit PN Junctions," *IEEE Trans. Nucl. Sci.*, vol. NS-35, pp. 1467–1474, December 1988.
7. R. L. Woodruff, D. A. Nelson, and S. Scherr, "Predicting Transient Upset in Gate Arrays," *IEEE Trans. Nucl. Sci.*, vol. NS-34, pp. 1426–1430, December 1987.
8. M. R. Acherman, R. E. Mikawa, L. W. Massengill, S. E. Deihl, "Factors Contributing to CMOS Static RAM Upset," *IEEE Trans. Nucl. Sci.*, vol. NS-33, pp. 1524–1529, December 1986.
9. H. L. Brode, "Close-in Weapon Phenomena," *Ann. Rev. Nucl. Sci.*, vol. 8, pp. 153–202, 1969.
10. A. J. Lelis, H. E. Boesch, T. R. Oldham, and F. B. Mclean, "Reversibility of Trapped Hole Annealing," *IEEE Trans. Nucl. Sci.*, vol. NS-35, pp. 1186–1191, December 1988.
11. J. Schwank, W. R. Dawes, "Irradiated Silicon Gate MOS Device Bias Annealing," *IEEE Trans. Nucl. Sci*, vol. NS-30, pp. 4100–4104, 1983.
12. C. Hu, "Alpha-Particle-Induced Field and Enhanced Collection of Carriers," *IEEE Electron Devices Lett.*, vol. EDL-3, no. 2, pp. 31–34, 1982.
13. F. B. Mclean and T. R. Oldham, "Charge Funneling in N- and P-type Si Substrates," *IEEE Trans Nucl. Sci.*, vol. NS-29, pp. 2018–2024, December 1982.
14. R. Koga and W. A. Kolasinski, "Heavy Ion-Induced Single Event Upsets of Microcircuits: A Summary of The Aerospace Corporation Test Data," *IEEE Trans. Nucl. Sci.*, vol. NS-31, pp. 1190–1195, December 1984.
15. D. K. Nichols, W. E. Price, W. A. Kolasinski, R. Koga, J. C. Pickel, J. T. Blandford, and A. E. Waskiewicz, "Trends in Parts Susceptibility to SEU from Heavy Ions," *IEEE Trans. Nucl. Sci.*, vol. NS-32, pp. 4189–4194, December 1985.
16. J. H. Adams Jr., R. Silberberg, C. H. Tsao, J. R. Letlaw, and D. F. Smart, "Cosmic Ray Effects on Microelectronics Parts I–III", NRL Memorandum Reports 4506 (August 25, 1981), 5099 (May 26, 1983), and 5402 (August 9, 1984). Washington, DC: Naval Research Laboratory.
17. E. L. Petersen, P. Shapiro, J. H. Adams, Jr., and E. A. Burke, "Calculation of Cosmic-Ray Induced Soft Upsets and Scaling in VLSI Devices," *IEEE Trans. Nucl. Sci.*, vol. NS-29, pp. 2055–2063, December 1982.
18. J. G. Rollins, "Single Event Upset in Silicon Integrated Circuits," Ph.D. dissertation, University of Southern California, 1989.
19. J. W. Palchefsky, "A Two-Port Equivalent Model for a Transmission Line with Distributed Sources," *IEEE Trans. Nucl. Sci.*, vol. NS-34, pp. 1488–1491, December 1987.
20. M. J. Schmidt, "Elementary External SGEMP Model for System Engineering Design," *IEEE Trans. Nucl. Sci.*, vol. NS-32, pp. 4295–4299, December 1985.
21. D. B. Estreich, "The Physics and Modeling of Latchup in CMOS Integrated Circuits," G201-9, Standard Electronics Laboratory, Stanford University, November 1980.
22. J. G. Rollins, "Numerical Simulation of SEU Induced Latch-Up," *IEEE Trans. Nucl. Sci.*, vol. NS-33, pp. 1565–1570, December 1986.

ALFONSO MALAGA

9

Characterization of Nuclear Scintillation Effects and Mitigation Techniques for Satellite Communications

Extensive research indicates that satellite radio signals that propagate through an ionosphere that has been disturbed by a high-altitude nuclear explosion suffer from fading or scintillation effects.[1-4] The effects will persist for periods lasting from minutes to many hours, depending on the proximity of the nuclear blast and the frequency of the radio signals.

As a result of this work, mathematical models of received signal characteristics such as the scattering loss, decorrelation distance, decorrelation time, and coherence bandwidth have been developed. The signal characteristics have been derived by different means.[1,2] One approach has consisted of solving the parabolic-wave equation, given the second-order statistics of the random ionization irregularities. At the same time, research into the phenomenology of nuclear weapons effects has resulted in the development of elaborate computer programs that predict the height-area contours of constant ionization levels (electron densities) produced by a high-altitude nuclear detonation as a function of the *time after burst* (TAB). The predicted ionization levels and the mathematical models have been used to generate TAB versus area coverage contours for a number of the signal characteristics of interest to the communications system design.

In this chapter, the author will attempt to summarize the mathematical models of the nuclear scintillation satellite channel in such a way that they may prove useful to the communications system designer. In particular, we will show that all signal characteristics of interest can be expressed in terms

517

of the beamwidths of the incident energy and the receiving antenna. The beamwidth of the incident energy, in turn, is directly proportional to a single physical parameter that completely characterizes the severity of the nuclear scintillation phenomena. This parameter is the root-mean-square (RMS) phase fluctuations of the radio signal after propagation through the disturbed iono-sphere. This summary will be followed by a characterization of the earth-satellite communications channel as a time-varying linear system. We con-clude with a brief discussion of the most common techniques that have been used to combat fading and scintillation in other communications media such as troposcatter, microwave line-of-sight, high-frequency (HF) radio, and underwater acoustics communications.

9.1 PROPAGATION THROUGH A NUCLEAR-DISTURBED IONOSPHERE

A high-altitude detonation will generate a layer of cylindrically shaped irreg-ularities aligned with the earth's magnetic field. Satellite communications signals that propagate through a region of the ionosphere containing such irregularities are scattered and experience multipath propagation. This implies that the signal energy that arrives at the receiving site is centered about a cone of angles distributed about the line-of-sight direction as shown in Fig. 9.1. The mean signal and the angular distribution of the mean signal energy incident on the receiving antenna has been shown to be[2]

$$S = \iint p(\theta, \phi) \sin \theta \cos \theta \, d\theta \, d\phi \tag{9.1}$$

with

$$p(\theta, \phi) = \frac{S}{2\pi\sigma_x\sigma_y} \exp\left[-\frac{\sin^2\theta}{2}\left(\frac{\cos^2\phi}{\sigma_x^2} \rightarrow \frac{\sin^2\phi}{\sigma_y^2}\right)\right] \tag{9.2}$$

where

S = the mean signal power

θ and ϕ = the off axis and azimuth angles

σ_x = the *half-beamwidth* of the incident energy in the direction perpendicular to the earth's magnetic field (geomagnetic east-west)

σ_y = the half-beamwidth in the plane of the earth's magnetic field (geomagnetic north-south direction)

The incident-energy half-beamwidths at the earth station, σ_x and σ_y, are given by[2]

$$\sigma_x = \frac{\lambda}{\sqrt{2\pi l_0}} = \frac{\lambda z_t \sigma_\phi}{\sqrt{2\pi(z_t + z_r)L_0}} \tag{9.3a}$$

$$\sigma_y = \frac{\sigma_x}{\gamma} \tag{9.3b}$$

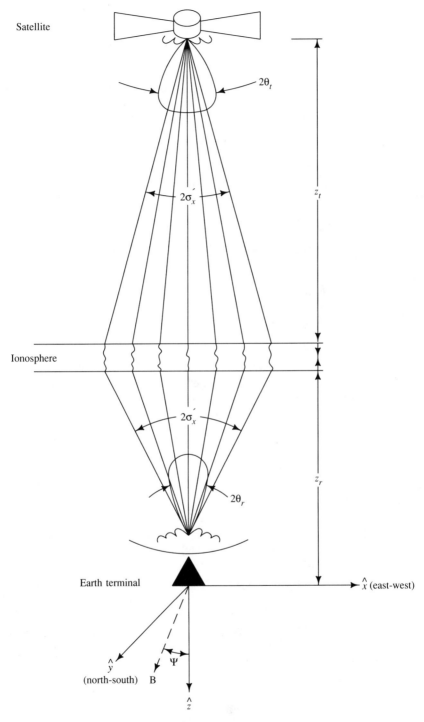

FIGURE 9.1 Geometry for satellite communications through a disturbed
ionosphere.

519

with

$$\gamma^2 = \cos^2 \psi + q^2 \sin^2 \psi \tag{9.4}$$

where

z_t = the distance from the satellite to the ionospheric layer of irregularities

z_r = the distance from the ionosphere to the earth station

λ = the wavelength

L_0 = the outer scale of the diameter of cylindrical shaped irregularities (about 10 km)

q = the ratio of the length to diameter (about 15 to 70)

ψ = the angle between the line of sight and the earth's magnetic field at the height of the irregularities (penetration angle)

σ_ϕ = the RMS phase fluctuation

The parameter l_0 is the minimum decorrelation distance of the incident energy.

Equations (9.3a) and (9.3b) indicate that greater multipath propagation is experienced when the phase fluctuations are large. The RMS phase fluctuations are proportional to the mean-squared fluctuations in the ionosphere's electron density, $\overline{\Delta N_e^2}$, along the line of sight. Assuming a fourth-power-law spectrum for the size distribution of the irregularities, the phase fluctuations are[1]

$$\sigma_\phi^2 = \frac{2(\lambda r_e)^2 q L_0 \overline{L \Delta N_e^2}}{\gamma} \tag{9.5}$$

where r_e is the classical electron radius (2.82×10^{-15} m) and L is the thickness of the layer of irregularities (about 500 to 1000 km). The RMS electron density fluctuations $\overline{\Delta N_e^2}$ can be of the same order of magnitude as the background electron density ($\approx 10^4$ to 10^8 e/cm^3). Note also that since the RMS phase fluctuations are directly proportional to the wavelength, the incident energy beamwidth is proportional to the square of the wavelength.

When propagation is perpendicular to the earth's magnetic field, the beamwidth in the geomagnetic north-south direction can be significantly smaller than in the east-west direction by a factor that depends on the elongation factor q of the irregularities and the penetration angle ψ. When $\psi = 0$ (north-south propagation along the magnetic-field lines), the two beamwidths are equal.

The incident energy half-beamwidths at the satellite, σ_x' and σ_y', differ from those at the earth station and can be found from Eq. (9.3) by interchanging z_t and z_r to obtain

$$\sigma'_x = \frac{z_r}{z_t}\sigma_x \quad \text{and} \quad \sigma'_y = \frac{z_r}{z_t}\sigma_y \tag{9.6}$$

For geostationary satellites $(z_t \gg z_r)$, the energy incident on the satellite antenna arrives within a cone of angles that is much smaller than the satellite antenna beamwidth. Hence the satellite antenna effects can be neglected.

9.1.1 Received-Signal Statistics

As a consequence of the multipath, the received signal experiences random amplitude and phase variations, also known as scintillation. The amplitude of the received signal has Rayleigh statistics; that is,

$$p(a|S) = \frac{2a}{S}e^{-a^2/S} \quad a \geq 0 \tag{9.7}$$

where

$a = $ the instantaneous signal amplitude

$S = $ the average received signal power

$p(a|S) = $ the probability density function of the signal amplitude

There is a minimum value σ_{ϕ_c} that the phase fluctuations σ_ϕ must exceed in order for Eq. (9.7) to be valid. The minimum RMS phase fluctuations for Rayleigh statistics depends on geometrical factors and the wavelength (or frequency), and is as follows:[3]

$$\sigma_{\phi_c} \simeq \sqrt{\frac{2\pi L_0^2}{\lambda}\left(\frac{z_t + z_r}{z_t z_r}\right)} \tag{9.8}$$

Multipath propagation with Rayleigh statistics persists as long as $\sigma_\phi > \sigma_{\phi_c}$. Hence, Eq. (9.8) shows that Rayleigh fading conditions prevail for shorter periods of time at higher frequencies because σ_{ϕ_c} is inversely proportional to the square root of the wavelength. When the phase fluctuations decrease to vaiues below the minimum-angle phase variance, the received signal amplitude has Rician or lognormal statistics.

9.1.2 Average Received Signal Power

The average received signal power is given by

$$S = \frac{P_t G_t G_r \lambda^2}{(4\pi z)^2} \exp[-0.115(A_g + A_R + A_a + A_{sc})] \tag{9.9}$$

where

$P_t = $ the transmitted power in watts

$S = $ the received power, also in watts

G_t and G_r = the satellite and earth station antenna gains over an isotropic radiator (numerical ratio)

λ = the wavelength in meters

$z = z_t + z_r$ = the distance between transmitter and receiver in meters

A_g = the gaseous attenuation in the atmosphere in dB

A_a = the absorption loss in the ionosphere in dB

A_R = the attenuation due to rain in dB

A_{sc} = the scattering loss in dB

Of the above loss factors, only the ionospheric absorption loss and the scattering loss are associated with a high-altitude nuclear detonation. The remaining loss factors are those encountered in typical satellite communication channels. Methods for estimating these losses can be found elsewhere.[5,6]

9.1.3 Ionospheric Absorption Loss and Sky Temperature

The ionospheric absorption loss is negligible at frequencies above a few GHz but can be significant at lower frequencies. At ultrahigh frequency (UHF) it causes a total blackout of the communication signal. The frequency dependence of the ionospheric absorption loss is given by

$$A_a = \frac{13.43 N L v \sec \theta_i}{f^2}$$

where

A_a = the absorption loss in dB

N = the electron density in e/m^3, which can be a few orders of magnitude greater than for a normal ionosphere

L = the thickness of the absorbing layer in meters

v = the electron-neutral collision frequency in Hz

θ_i = the angle of incidence between the line of sight and the lower part of the ionosphere where absorption takes place

f = the radio-wave frequency in Hz

The absorption of energy in the ionosphere contributes to the sky temperature according to the relationship

$$T_{sky} = T_{cs} e^{-0.115 A_a} + T_m (1 - e^{-0.115 A_a}) \tag{9.10}$$

where

T_{cs} = the clear-sky temperature in kelvins

T_m = the radiating temperature in the ionosphere (about 3000 to 9000 K)

A_a = the absorption loss in dB

The increase in the sky temperature during high-altitude nuclear blasts degrades the system noise temperature and must be taken into account. The degradation, L_{sky}, is given by

$$L_{sky} = 10a \log_{10}\left(\frac{T_{sky} + T_{sys}}{T_{sys}}\right) \qquad (9.11)$$

where T_{sys} in the system temperature.

9.1.4 Scattering Loss

The scattering loss is the loss of energy when the antenna beamwidths are smaller than the beamwidth of the energy incident on the antennas. The mathematical formulation for calculation of this loss has been developed in Ref. 2. Simple expressions can be obtained by assuming Gaussian beam patterns. The scattering loss for a receiving beam with pointing angles (θ_0, ϕ_0) relative to the line of sight is given by

$$A_{sc} = 5 \log_{10}(\alpha_x \alpha_y) + 4.343\left[\frac{\sin^2 \theta_0}{\theta_r^2} - \beta_x - \beta_y\right] \qquad (9.12)$$

with

$$\alpha_x = 1 + \frac{2\sigma_x^2}{\theta_r^2} + \left(\frac{z_r}{z_t}\right)^2 \frac{2\sigma_x^2}{\theta_t^2} \qquad (9.13a)$$

$$\alpha_y = 1 + \frac{2\sigma_y^2}{\theta_r^2} + \left(\frac{z_r}{z_t}\right)^2 \frac{2\sigma_y^2}{\theta_t^2} \qquad (9.13b)$$

$$\beta_x = \frac{2\sigma_x^2}{\theta_r^4} \frac{\sin^2 \theta_0 \cos^2 \phi_0}{\alpha_x} \qquad (9.13c)$$

$$\beta_y = \frac{2\sigma_y^2}{\theta_r^4} \frac{\sin^2 \theta_0 \sin^2 \phi_0}{\alpha_y} \qquad (9.13d)$$

where σ_x and σ_y are the half-beamwidths of the energy incident on the receiving earth station antenna; θ_t and θ_r are the beamwidths of the satellite antenna and the earth station antenna, respectively; and z_t and z_r are the distances from the satellite and the earth station to the layer of irregularities, respectively.

The first term in Eq. (9.12) represents the loss when the receive antenna beam is pointing at the satellite. The second term in brackets represents the additional loss when the receiving beam is pointing away from the line of sight. Note that this incremental loss is largest when the incident energy beamwidths

σ_x and σ_y are much smaller than the receiving antenna beamwidth θ_r. If the incident energy beamwidths σ_x and σ_y are much greater than the receiving antenna beamwidth θ_r, the incremental loss resulting from off-axis pointing is negligible.

9.1.5 Delay Spread

The energy received off-axis arrives after the energy received along the line of sight because of longer propagation delays to and from the scatterers. The power received per unit delay τ in the direction (θ, ϕ) can be expressed as[2]

$$p(\theta, \phi, \tau) = \frac{f_0 G(\theta, \phi)}{(2\pi)^{1/2}\sigma_x\sigma_y C_1} \exp$$

$$\times \left[-\frac{\sin^2\theta}{2}\left(\frac{\cos^2\phi}{\sigma_x^2} + \frac{\sin^2\phi}{\sigma_y^2}\right) - \frac{1}{2C_1^2}\left(2\pi f_0\tau - \frac{\sin^2\theta}{2\sigma_x^2}\right)^2 \right]$$

(9.14)

with

$$f_0 = \frac{z_t c}{2\pi(z_t + z_r)z_r\sigma_x^2}$$

(9.15a)

$$C_1 = \frac{\pi(z_t + z_r)L_0^2}{\lambda z_t z_r \sigma_\phi} = \frac{\sigma_{\phi c}^2}{2\sigma_\phi}$$

(9.15b)

where $G(\theta, \phi)$ is the receive antenna pattern and c is the speed of light (3×10^8 m/s). When the antenna pattern is approximated by a Gaussian beam, the mean-squared delay spread, defined as

$$\sigma_\tau^2 = \iiint (\tau - \tau_{av})^2 p(\theta, \phi, \tau) \sin\theta \cos\theta \, d\theta \, d\phi \, d\tau$$

(9.16)

is given by

$$\sigma_\tau^2 = \left(\frac{C_1}{2\pi f_0}\right)^2 + \left(\frac{1}{2\pi f_0}\right)^2 \frac{\alpha_x(1 + 4\beta_y) + \gamma^4\alpha_y(1 + 4\beta_x)}{2\gamma^4\alpha_x^2\alpha_y^2}$$

(9.17)

which includes the effects of antenna beamwidth through α_x and α_y, the beam-pointing angle through β_x and β_y, and the effects of propagation direction relative to the earth's magnetic field through γ. The first factor in Eq. (9.17) is negligible when there is multipath propagation, since σ_ϕ is large. This factor accounts for delay fluctuations due to the dispersive nature of the ionosphere.

When the antenna beamwidth θ_r is large compared to σ_x (that is, $\alpha_x \approx \alpha_y \approx 1$) and the antenna is pointing at the line of sight ($\beta_x = \beta_y = 0$), the delay spread σ_τ is given by

$$\sigma_\tau \approx \frac{1}{2\pi f_0} = \frac{(z_t + z_r)z_r\sigma_x^2}{z_t c} \qquad \text{if } \gamma \approx 1$$

(9.18a)

or

$$\sigma_\tau \approx \frac{1}{\sqrt{2}\,2\pi f_0} = \frac{(z_t + z_r)z_r\sigma_x^2}{\sqrt{2z_t c}} \qquad \text{if } \gamma \gg 1 \qquad (9.18b)$$

which shows that the delay spread is inversely proportional to the square of the beamwidth of the incident energy. When the antenna beamwidth θ_r is smaller than σ_x, the delay spread is significantly smaller due to the loss of energy which arrives at off-boresight angles beyond the antenna half-beamwidth. When the antenna beam is pointing off-axis, the delay spread is greater than for an antenna pointing directly at the satellite.

9.1.6 Coherence Bandwidth

The coherence bandwidth is defined as the frequency separation between two continuous radio waves that experience multipath fading with correlation equal to e^{-1}. The coherence bandwidth f_{coh} is inversely proportional to the delay spread and is defined as

$$f_{coh} = \frac{1}{2\pi\sigma_\tau} \simeq f_0 \frac{\sqrt{2\gamma^2\alpha_x\alpha_y}}{[\alpha_x(1 + 4\beta_y) + \gamma^4\alpha_y(1 + 4\beta_x)]^{1/2}} \qquad (9.19)$$

If the antenna beamwidth θ_r is large compared to σ_x, the antenna is pointing at the line of sight, and propagation is along the magnetic-field lines, then $f_{coh} = f_0$.

A signal whose bandwidth is smaller than the coherence bandwidth is said to experience flat Rayleigh fading. A signal whose bandwidth is greater than the coherence bandwidth experiences frequency-selective fading, since frequency components whose frequency separation is greater than the coherence bandwidth fade independently.

9.1.7 Space Diversity Decorrelation Distance

Space diversity reception is often used to combat fading effects. If we assume two receiving antennas with Gaussian beams pointing at the satellite, the correlation of the fading, ρ_s, is given by

$$\rho_s = \exp\left[-\left(\frac{\delta_x}{l_x}\right)^2 - \left(\frac{\delta_y}{l_y}\right)^2\right] \qquad (9.20)$$

where δ_x is the spacing between the antennas in the geomagnetic east-west direction, δ_y is the spacing in the north-south direction, and

$$l_x = l_0\sqrt{\alpha_x} = \frac{\lambda\sqrt{\alpha_x}}{\sqrt{2}\pi\sigma_x} \qquad (9.21)$$

$$l_y = l_0\gamma\sqrt{\alpha_y} = \frac{\lambda\sqrt{\alpha_y}}{\sqrt{2}\pi\sigma_y} \qquad (9.22)$$

The parameters l_x and l_y are decorrelation distances in the east-west and north-south directions, respectively. The decorrelation distances are inversely proportional to the beamwidths of the incident energy. Antennas whose beamwidths are smaller than the beamwidth of the incident energy ($\alpha_x > 1$ and $\alpha_y > 1$), require larger spacings than broader-beamwidth antennas to achieve the same degree of decorrelation.

To achieve maximum decorrelation during an entire event, the optimum antenna spacing must exceed the largest decorrelation distance expected during multipath propagation with Rayleigh statistics. Substituting Eq. (9.8) into Eqs. (9.3a) and (9.3b), and then substituting for σ_x and σ_y in Eq. (9.19), we get

$$(l_x)_{max} = \sqrt{\frac{\lambda z_r (z_t + z_r)\alpha_x}{2\pi z_t}} \tag{9.22a}$$

$$(l_y)_{max} = \gamma \sqrt{\frac{\lambda z_r (z_t + z_r)\alpha_y}{2\pi z_t}} \tag{9.22b}$$

Equations (9.22a) and (9.22b) show that smaller antenna spacings are needed as the frequency increases.

9.1.8 Angle Diversity Decorrelation

An alternative to space diversity consists of using a single antenna with multiple beams. Each beam is aimed to illuminate a different part of the scattering region. This technique is referred to as angle diversity. The correlation of the fading of the signals received by two different beams is a function of the overlapping of the two beam patterns. When the beams have equal beamwidths and their patterns can be approximated by Gaussian beams, the correlation is given by

$$\rho_A = \exp\left[-\frac{\sigma_x^2}{2\alpha_x \theta_r^4}(\sin\theta_1 \cos\phi_1 - \sin\theta_2 \cos\phi_2)^2 \right.$$
$$\left. -\frac{\sigma_y^2}{2\alpha_y \theta_r^4}(\sin\theta_1 \sin\phi_1 - \sin\theta_2 \sin\phi_2)^2 \right] \tag{9.23}$$

where (θ_1, ϕ_1) and (θ_2, ϕ_2) are the pointing angles of the two beams, θ_r is their beamwidth, and σ_x and σ_y are the half-beamwidths of the incident energy.

The decorrelation angle for two beams aimed in the east-west direction ($\phi_1 = \phi_2 = 0$) is given by

$$\theta_A(0) = \frac{\sqrt{2\alpha_x \theta_r^2}}{\sigma_x} \tag{9.24a}$$

The decorrelation angle for two beams aimed in the north-south direction ($\phi_1 = \phi_2 = 90°$) is greater by at least a factor γ since

$$\theta_A(90°) = \frac{\sqrt{2\alpha_y \theta_r^2}}{\sigma_y} = \frac{\sqrt{2\alpha_y \theta_r^2}\gamma}{\sigma_x} \tag{9.24b}$$

These expressions show that the decorrelation angle is inversely proportional to the half-beamwidth of the incident energy and directly proportional to the square of the receive antenna beamwidth.

9.1.9 Coherence Time

The signal coherence time is a function of the velocity of the transmitter and receiver as well as the motion of the irregularities. If one assumes that the irregularities move in unison across the line of sight, the signal coherence time τ_{coh} is given by

$$\tau_{coh} = \sqrt{\frac{l_x^2}{v_x^2 + v_y^2} \cdot \frac{v_x^2}{v_x^2 + v_y^2} + \frac{l_y^2}{v_x^2 + v_y^2} \cdot \frac{v_y^2}{v_x^2 + v_y^2}} \tag{9.25}$$

where l_x and l_y are the decorrelation distances in the geomagnetic east-west and north-south directions defined in Eq. (9.22) and v_x and v_y are effective velocities of the irregularities and the terminals along these two directions. Antenna effects are included in l_x and l_y.

9.1.10 Delay-Doppler Generalized Power Spectrum

The fading spectrum (integral of delay-Doppler power spectrum overall delays) is Gaussian. When the received signal bandwidth is small compared to the coherence bandwidth (flat fading conditions), the delay-Doppler power spectrum can be approximated as follows:[4]

$$\Gamma_n(v, \tau) = \frac{1.864\tau_{coh}\delta(\tau)}{(1 + 8.572\tau_{coh}^2 v^2)^2} \tag{9.26}$$

where

$$v = \text{the Doppler frequency}$$

$$\tau_{coh} = \text{the coherence time}$$

When the signal bandwidth is greater than the coherence bandwidth, the fading is frequency-selective and the fading spectrum differs for the various delayed components of the received signal. The delay-Doppler power spectrum is of the form

$$\Gamma_n(v, \tau) = \sqrt{2\pi C_1}\, f_0 \tau_0 \gamma \exp\left\{-(\pi v \tau_{coh})^2\right.$$
$$-\frac{1}{2C_1^2}[2\pi f_0 \tau - (\pi v \tau_0)^2]^2\right\} \int_{-\infty}^{\infty} \exp\left\{-\frac{x^4}{2} - \frac{1}{C_1}[(\pi v \tau_0)^2\right.$$
$$- 2\pi f_0 \tau]x^2 - \frac{C_1}{2}\left(\frac{\alpha_x v_y^2}{v_x^2 + v_y^2} + \frac{\alpha_y \gamma^2 v_x^2}{v_x^2 + v_y^2}\right)x^2$$
$$\left. + 2\sqrt{C_1}(\alpha_x - \alpha_y \gamma^2)\frac{v_x v_y}{v_x^2 + v_y^2}(\pi v \tau_0)x\right\} dx \tag{9.27}$$

where

$$\tau_0 = \frac{l_0}{\sqrt{v_x^2 + v_y^2}} \tag{9.28}$$

Equation (9.27) reduces to the delay-Doppler power spectrum specified in [4] when the antenna beamwidth effects are negligible ($\alpha_x = \alpha_y = 1$) and the irregularities are assumed to be isotropic, that is, $\gamma = 1$, so that $l_x = l_y = l_0$ and $\tau_{coh} = \tau_0$.

9.2 DIVERSITY CHANNEL CHARACTERIZATION

The nuclear-disturbed satellite to earth station communication channels with diversity reception can be modeled as a time-varying linear system whose time-varying impulse response is defined as

$$r_n(t) = \int_{-\infty}^{\infty} h_n(\bar{\mathbf{p}}_n, t, \tau) s(t - \tau)\, d\tau \tag{9.29}$$

where

$$s(t) = \text{transmitted signal}$$

$$r_n(t) = n\text{th diversity received signal}$$

$h_n(\bar{\mathbf{p}}_n, t, \tau) = $ randomly time-varying impulse response for nth diversity channel located at distance $\bar{\mathbf{p}}_n$ from center of reference location

In some cases, it is more convenient to characterize each diversity channel in terms of the time-varying transfer function of the delay-Doppler spread function, which is related to the time-varying impulse response as follows:

$$H_n(\bar{\mathbf{p}}_n, t, f) = \int_{-\infty}^{\infty} h_n(\bar{\mathbf{p}}_n, t, \tau) e^{j2\pi f \tau}\, d\tau \tag{9.30}$$

$$V_n(\bar{\mathbf{p}}_n, v, \tau) = \int_{-\infty}^{\infty} h_n(\bar{\mathbf{p}}_n, t, \tau) e^{j2\pi v t}\, dt$$

$$= \iint_{-\infty}^{\infty} H_n(\bar{\mathbf{p}}_n, t, f) e^{j2\pi(vt - f\tau)}\, dt\, df \tag{9.31}$$

where

$H_n(\bar{\mathbf{p}}_n, t, f) = $ randomly time-varying transfer function for the nth diversity channel

$V_n(\bar{\mathbf{p}}_n, v, \tau) = $ delay-Doppler spread function for the nth diversity channel

The statistics of each diversity channel are complex Gaussian; that is, the amplitude of the received signal is Rayleigh distributed and its phase is uniformly distributed. The statistics at different delays are uncorrelated; therefore it follows that

$$\bar{h}_n = \bar{H}_n = \bar{V}_n = 0 \tag{9.32a}$$

$$\overline{V_n V_m} = 0 \tag{9.32b}$$

$$\overline{V_n(\bar{\rho}_n, v, \tau) V_m^*(\bar{\rho}_m, v', \tau')} = \Gamma_{nm}(\bar{\rho}_n - \bar{\rho}_m, v, \tau)\delta(v - v')\delta(\tau - \tau') \tag{9.32c}$$

where

$\Gamma_{nn}(0, v, \tau) =$ scattering function for the nth diversity channel

$\Gamma_{nm}(\bar{\rho}, v, \tau) =$ cross-channel scattering function for nth and mth receivers separated by distance ρ

$\delta(x) =$ Dirac delta function

The signals received by each diversity receiver are completely described by the scattering function for each diversity channel $\Gamma_{nn}(0, v, \tau)$, where the v dependence describes the Doppler smearing (fading) introduced by the time-varying nature of the channel and the τ dependence describes the delay spreading caused by multipath propagation. The correlation of the fading between space diversities (or other type of diversities such as angle diversity) is described by the cross-channel scattering function $\Gamma_{nm}(\bar{\rho}_n - \bar{\rho}_m, v, \tau)$. The quasi-stationary nature of the statistics of the channel implies that the channel scattering functions are slowly varying functions of time. This time dependence has been dropped in order to simplify the notation.

9.2.1 Impulse Response Function

The time-varying impulse response of each diversity channel is given by

$$h_n(\bar{\rho}_n, t, \tau) = \sqrt{P_t G_t G_r} \left(\frac{\lambda}{4\pi z}\right) \exp\{-0.115 K_A + j2\pi v_A t\} \tilde{h}_n(\bar{\rho}_n, t, \tau - \tau_A) \tag{9.33}$$

where

$z =$ propagation path length

$K_A = A_s + A_R + A_a =$ clear-air attenuation plus rain attenuation plus absorption loss in dB

$v_A =$ Doppler shift introduced by the rate of change in the path length and the mean total electron content of the ionosphere

$\tau_A =$ mean delay due to propagation through the background ionosphere

$\tilde{h}_n(\bar{\rho}_n, t, \tau) =$ spatial and time-varying impulse response function that describes the delay (τ variable) spreading, temporal fluctuations (t variable) and two-dimensional spatial fluctuations ($\bar{\rho}$ variable) in the nth diversity channel

The Doppler shift and mean delay are given by

$$v_A = \frac{f_c}{c}\frac{dz}{dt} - \frac{cr_e}{2\pi f_c}\frac{dN}{dt} \tag{9.34a}$$

$$\tau_A = \frac{z}{c} + \frac{cr_e N}{2\pi f_c^2} \tag{9.34b}$$

where

f_c = carrier frequency (Hz)

c = speed of light (3×10^8 m/s)

N = mean total electron content (elec/m^2)

r_e = classical electron radius (2.82×10^{-15} m)

z = path length (m)

The Doppler shift and mean delay introduced by changes in the background ionosphere vary slowly in time. These variations, in themselves, may in some instances—in the low ultrahigh-frequency (UHF) band—be significant sources of Doppler and delay spreading.

9.2.2 Scattering Functions

The cross-channel scattering functions are defined as

$$\Gamma_{nm}(\bar{\mathbf{p}}_n - \bar{\mathbf{p}}_m, v, \tau)$$

$$= \frac{1}{(2\pi)^2}\iint \tilde{\Gamma}(\bar{\mathbf{K}}, v, \tau)g_n(\bar{\mathbf{K}} - \bar{\mathbf{K}}_n)g_m^*(\bar{\mathbf{K}} - \bar{\mathbf{K}}_m)e^{j\bar{\mathbf{K}}\cdot(\bar{\mathbf{p}}_n - \bar{\mathbf{p}}_m)}\,d^2\bar{\mathbf{K}} \tag{9.35}$$

where

$\tilde{\Gamma}(\bar{\mathbf{K}}, v, \tau)$ = angular delay-Doppler power spectrum of the incident energy (angular scattering function)

$\bar{\mathbf{K}} = (2\pi/\lambda)\cos\theta[(\cos\phi)\hat{\mathbf{x}} + (\sin\phi)\hat{\mathbf{y}}]$ = wave number (direction)[a] of wave incident on x-y plane aperture at off-boresight polar angles (θ, ϕ)

$\mathbf{K}_n$ = boresight direction (wave number) of the nth receive aperture or beam

$\bar{\mathbf{p}}_n$ = coordinates of center of the nth receive aperture relative to center of the receive array

$g_n(\mathbf{K})$ = voltage directional pattern for the nth receiver

---

[a] $\hat{\mathbf{x}}$ and $\hat{\mathbf{y}}$ are unit vectors.

The cross-channel scattering function for two apertures spaced a distance $\bar{\rho}_n - \bar{\rho}_m$, which have the same aperture size and are aimed in the same direction, is given by Eq. (9.35), with $g_n(\bar{K}) = g_m(\bar{K})$. Similarly, the cross-channel scattering functions for two beams generated by the same aperture and with boresights aimed in different directions is also given by Eq. (9.35) with $\bar{\rho}_n = \bar{\rho}_m$ and $g_n(\bar{K} - \bar{K}_n)$ and $g_m(\bar{K} - \bar{K}_m)$ representing the voltage patterns of the two different beams.

The scattering function for the nth receive aperture or nth receive beam can be obtained from Eq. (9.35) by setting $m = n$ to obtain

$$\Gamma_{nn}(0, v, \tau) = \frac{1}{(2\pi)^2} \int\int \tilde{S}(\bar{K}, v, \tau)|g_n(\bar{K} - \bar{K}_n)|^2 \, d^2\bar{K} \tag{9.36}$$

When the receive aperture has an omnidirectional antenna pattern—that is, when it is a point source with $g_n(\bar{K}) = 1$—Eq. (9.36) reduces to the scattering function of the incident energy $\Gamma(v, \tau)$. In effect,

$$\Gamma(v, \tau) = \frac{1}{(2\pi)^2} \int\int \tilde{\Gamma}(\bar{K}, v, \tau) \, d^2\bar{K} \tag{9.37}$$

The angular delay-Doppler power spectrum function, $\tilde{\Gamma}(\bar{K}, v, \tau)$, represents the average received power due to energy (waves) incident at angles in the interval $(\bar{K}, \bar{K} + d\bar{K})$, having delays in the range $(\tau, \tau + d\tau)$ and Doppler shifts in the range $(v, v + dv)$.

If one assumes that the field-aligned scatterers (irregularities) move in unison across the direction of propagation at a constant velocity, then the angular delay-Doppler power spectrum takes the form

$$\tilde{\Gamma}(\bar{K}, v, \tau) = \Gamma_0(\bar{K}, \tau)\delta(v + \bar{K} \cdot \bar{V}) \tag{9.38}$$

where

$$\bar{V} = v_x\hat{x} + v_y\hat{y} = \text{effective velocity of the transmitter, the receiver, and the irregularities}$$

$$\Gamma_0(\bar{K}, \tau) = \text{angular delay power spectrum}$$

The delta function in Eq. (9.38) implies that there is a unique relationship between the angle of arrival (wave number) of the received energy and its Doppler spectrum.

The angular delay power spectrum was introduced earlier (see Section 9.1.4) and is of the form

$$\Gamma_0(K, \tau) = \frac{f_0}{\sqrt{2\pi}\sigma_x\sigma_y c_1} \exp\left(-\frac{K_x^2}{2K_c^2\sigma_x^2} - \frac{K_y^2}{2K_c^2\sigma_y^2}\right)$$
$$\cdot \exp\left\{-\frac{1}{2c_1^2}\left[2\pi f_0\tau - \frac{K_x^2 + K_y^2}{2K_c^2\sigma_x^2}\right]^2\right\} \tag{9.39}$$

where all parameters have been defined except for K_x, K_y, and K_c, which are defined as

$$\bar{\mathbf{K}} = (K_x, K_y) = K_c \sin\theta[(\cos\phi)\hat{\mathbf{x}} + (\sin\phi)\hat{\mathbf{y}}]$$

$$K_c = \frac{2\pi}{\lambda} = \frac{2\pi f_c}{c}$$

where f_c is the carrier frequency.

In arriving at Eq. (9.39) it has been assumed that the field-aligned scatterers are in the y-z plane of a cartesian coordinate system where the line-of-sight is the z axis.

The angular delay power spectrum function, $\Gamma_0(\bar{\mathbf{K}}, \tau)$, physically represents the average received power arriving in a cone of angles corresponding to wave numbers in the interval $(\bar{K}, \bar{K} + d\bar{K})$ and arriving in the delay interval $(\tau, \tau + d\tau)$. Equation (9.39) also indicates that there are two sources of delay spreading of the received energy: one is delay spreading at a fixed angle of arrival (θ, ϕ) [Gaussian-shaped dependence on τ in Eq. (9.39)] caused by the large-scale irregularities; the other is delay spreading resulting from energy arriving at a range of different angles of arrival caused by the diffraction (scattering) effects of small-scale irregularities. The first mechanism is dominant when the parameter $C_1 > 1$ in Eq. (9.39), whereas the second mechanism is dominant when $C_1 \ll 1$.

When the antenna patterns can be approximated by Gaussian-shaped beams, the nth-channel scattering function defined by Eq. (9.36) reduces to the generalized delay-Doppler power spectrum introduced earlier in Section 9.1.10.

9.3 COMMUNICATION TECHNIQUES

Nuclear scintillation effects can be mitigated by employing a number of different techniques. The choice of an appropriate or effective modulation technique is a function of the product of the data and the coherence time, $R_b \tau_{coh}$. The choice of a mitigation technique is a function of the modulation technique and the ratio of data rate to coherence bandwidth R_b / f_{coh}.

9.3.1 Modulation Selection Criteria

When $R_b \tau_{coh} < 1$, the channel fades at a rate faster than the information transmission rate. In this case noncoherent modulation techniques such as M-ary frequency-shift keying (MSK) are effective, whereas coherent modulation techniques such as phase-shift keying (PSK) are ineffective. When $R_b \tau_{coh} > 1$ the channel fades at a rate comparable to the information transmission rate. In this case differentially coherent modulation techniques such as differential phase-shift keying (DPSK) and noncoherent MFSK are effective. Finally, if $R_b \tau_{coh} \gg 1$, coherent PSK, DPSK, and MFSK are all effective.

9.3.2 Mitigation Techniques

When $R_b/f_{coh} \ll 1$, the fading is flat or non–frequency-selective. Space diversity, angle diversity, frequency diversity, coding with interleaving, and spread spectrum modulation in the form of coded frequency hopping or direct sequence spreading with adaptive matched filter reception can all be used to combat the effects of fading. The selection of the appropriate technique depends on the availability of bandwidth versus space for multiple antennas and/or multiple receivers. Diversity combining techniques include selection diversity, equal-gain combining, or maximal-ratio combining. Selection and equal-gain combining are often used when the demodulation is noncoherent. Maximal-ratio combining is appropriate when the demodulation is coherent.

If $R_b/f_{coh} > 1$, the fading is frequency-selective. In this case it is necessary to mitigate the effects of intersymbol interference (ISI) in addition to those of fading. If the modulation of choice is noncoherent MFSK and there is sufficient bandwidth available, then frequency hopping with coding can be used to combat the effects of fading and ISI. If the channel fades slowly compared to the data rate so that coherent PSK modulation can be used effectively, then adaptive equalization can be used to combat ISI, and any combination of space, angle, or frequency diversity can be used to combat the effects of fading. Adaptive equalization with space, frequency, and angle diversity reception, with maximal-ratio combining has been used successfully in troposcatter channels that experience similar frequency-selective fading conditions as the nuclear scintillation satellite channel.[7]

9.4 FREQUENCY BAND IMPLICATIONS

The coherence time, coherence bandwidth, decorrelation distance, decorrelation angle, scattering loss, absorption loss, and sky temperature are frequency-dependent parameters. Table 9.1 summarizes the first-order frequency dependence of the signal characteristics.

In general, absorption loss and sky temperature effects are significant at frequencies below 2 GHz. Thus, at these frequencies, increasing the transmitter power is also necessary in order to maintain communications. At frequencies above 2 GHz, multipath fading and scintillation are the dominant effects. In general, when the channel coherence bandwidth is small, the channel fades faster (short coherence time). Thus, smaller bandwidths are needed to achieve a frequency diversity gain. However, coherent modulation techniques can be used only if the data rate is sufficiently high. As the ionosphere recovers, the coherence bandwidth and coherence time increase. The coherence bandwidth and coherence time also increase with frequency. Thus, larger bandwidths and interleaving amounts are needed at higher frequencies to achieve the necessary diversity to combat fading. Furthermore, rain attenuation and clear-sky absorption effects also become significant at higher frequencies.

TABLE 9.1 Frequency Dependence of Signal Characteristics

Signal Characteristic	Frequency Dependence
Absorption loss in dB, A_a	f^{-2}
Incident energy beamwidth, σ_x and σ_y	f^{-2}
Delay spread, σ_τ	f^{-4}
Coherence bandwidth, f_{coh}	f^4
Decorrelation distance, l_x and l_y	f
Maximum decorrelation angle, θ_A	f^0
Coherence time, τ_{coh}	f

REFERENCES

1. L. A. Wittwer, "Radio Wave Propagation in Structured Ionization for Satellite Applications," Defense Nuclear Agency Report 5304D, December 1979.
2. D. L. Knepp, "Aperture Antenna Effects After Propagation Through Strongly Disturbed Random Media," *IEEE Trans. Antennas Propagat.*, vol. AP-33, pp. 1074–1084, October 1985.
3. D. L. Knepp and L. A. Wittwer, "Simulation of Wide Bandwidth Signals That Have Propagated Through Random Media," *Radio Sci.*, vol. 19, pp. 303–318, January 1984.
4. L. A. Wittwer, "A Trans-ionospheric Signal Specification for Satellite C³ Applications," Defense Nuclear Agency Report DNA 5662D, December 1980.
5. CCIR Study Group 5 Doc. Report 721-1, vol. 5, "Attenuation by Hydrometeors, in Particular, Precipitations and Other Atmospheric Particles," 16th Plenary Assembly, Geneva, 1982.
6. R. K. Crane, "Prediction of Attenuation by Rain," *IEEE Trans. Commun. Technol.*, vol. COM-28, pp. 1717–1733, September 1980.
7. L. Ehrman and P. Monsen, "Troposcatter Test Results for a High-Speed Decision-Feedback Equalizer Modem," *IEEE Trans. Commun. Technol.*, vol. COM-25, pp. 1499–1504, December 1977.

MARVIN GANTSWEG

10

SATCOM Link Performance in the Presence of Nuclear Scintillation and Jamming

This chapter provides an introductory overview of satellite link scintillation testing methods, areas of applicability, and a discussion of test results for several SATCOM systems. It begins with a brief review of the satellite nuclear channel model. Methods are reviewed for creating an equivalent channel in terms of a hardware simulator. A few specific examples are briefly described, including both flat and frequency-selective fading simulators.

Selected results from past tests of military satellite links are then presented and reviewed. Included are comparisons of theoretical and hardware channel simulation testing results. The specific systems to be discussed and, especially, the testing results, are classified at the SECRET/NOFORN level.

The impacts of the testing results on subsequent development/upgrade programs are reviewed. The chapter then highlights the nuclear scintillation performance parameters to be expected for future SATCOM systems with emphasis on 40/20 GHz uplinks and downlinks through the ionosphere, and 60-GHz cross links. It concludes with a brief discussion of additional areas where expanded use of link scintillation testing may prove worthwhile.

10.1 NUCLEAR WEAPON COMMUNICATION EFFECTS

Nuclear weapon detonations in or near the ionosphere can degrade satellite communication links over large areas for significant time periods, as illustrated in Fig. 10.1. The resultant ionospheric irregularities produce random

535

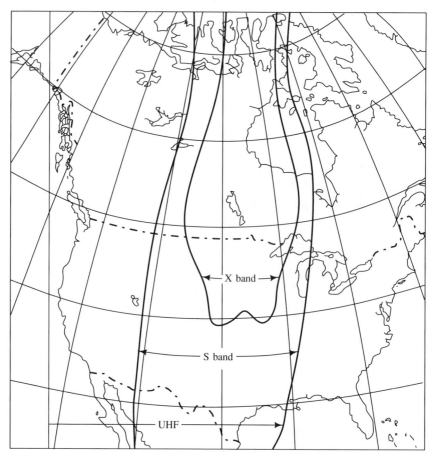

FIGURE 10.1   Spatial extent of strong scintillations 33 minutes after a high-altitude burst.

fluctuations, or scintillations, in amplitude, phase, frequency, and time delay of signals propagating through the disturbed medium. Figure 10.2 shows a typical depolyment scenario in which the propagation paths are perturbed by the disturbed ionosphere.

Communication link degradation is produced by both nonscintillation and scintillation effects. The nonscintillation effects, which are secondary effects because of their short time duration, include absorption and phase-only effects such as phase advance, time delay wander, Doppler shift and its derivatives, and Faraday rotation. Scintillation can be thought of as a multi-path phenomenon with a large number of random paths, as illustrated in Fig.

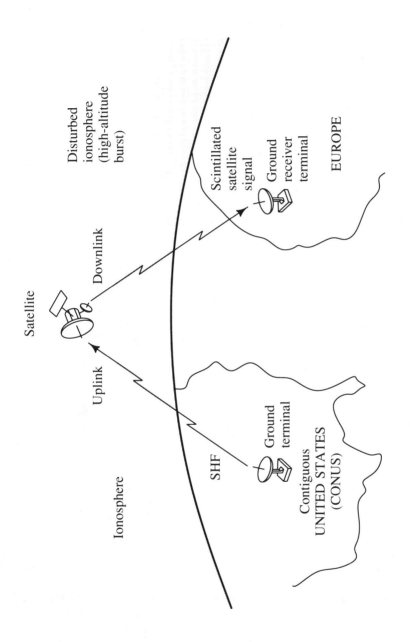

FIGURE 10.2   AN/USC-28/DSCS-III propagation scenario.

537

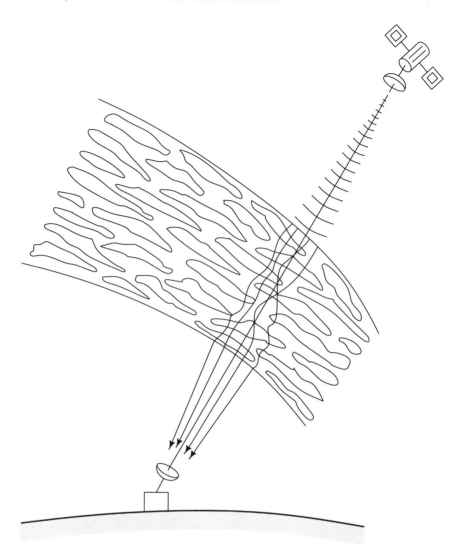

FIGURE 10.3   Multipath propagation through the striated plasma.

10.3. Both the phase and amplitude of the signals received over each path vary in a random manner, and when summed at the receiving-site antenna, result in a time-varying signal structure, as depicted in Fig. 10.4. The amplitude envelope of the received SATCOM signal carrier becomes, in the strong scintillation limit, Rayleigh distributed, and this condition persists for up to several hours after burst.

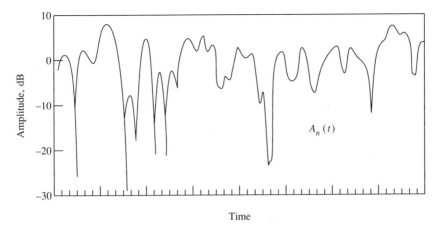

Amplitude fade depths bounded by Rayleigh distribution

(a)

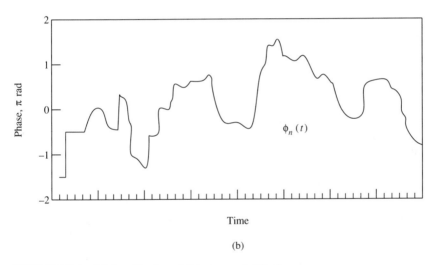

(b)

FIGURE 10.4   (a) Amplitude and (b) phase scintillation.

The Rayleigh amplitude probability distribution function is shown in Fig. 10.5. The signal phase is uniformly distributed over the carrier period. The signal perturbations are also correlated in time over a period defined by τ_0, the signal decorrelation time, which is an approximate measure of the average fade duration. The power spectral density (PSD) of the signal perturbations (the Fourier transform of its autocorrelation function) is approximated by

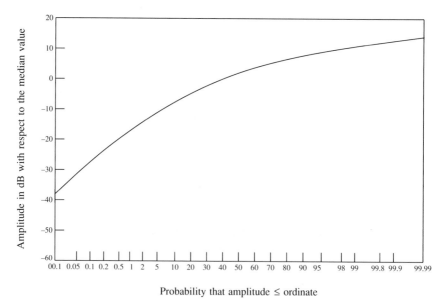

FIGURE 10.5   Probability distribution of a Rayleigh fading carrier.

$$S(f) = \frac{1.864\tau_0}{[1 + 8.572(\tau_0 f)^2]^2} \qquad (10.1)$$

This spectrum falls off at approximately 12 dB per octave, which makes it easy to simulate with simple low-pass filter designs. The results of this flat fading condition are depicted in Fig. 10.6a, which shows the scintillated envelope of the time-domain waveform received from a transmitted triangular pulse initially having a total width of two chips. Notice that the signal amplitude exhibits Rayleigh-like fading at a somewhat periodic rate, and the received energy is contained within the two-chip delay width.

In summary, for a flat fading channel, *the important perturbed signal features are a Rayleigh amplitude distribution, a uniform phase distribution, and a signal decorrelation time τ_0.* Flat fading occurs whenever the signal band-width is less than the frequency-selective bandwidth of the channel.

When the signal bandwidth is comparable to, or greater than, the frequency-selective bandwidth, the received signal exhibits *frequency-selective fading.*[a] Different frequency components of the signal exhibit different phase and amplitude scintillations, and significant pulse spreading and intersymbol interference can occur. This effect is depicted in Fig. 10.6b. Here the received energy is spread over many time-delay chips.

---

[a] See Chapter 1.

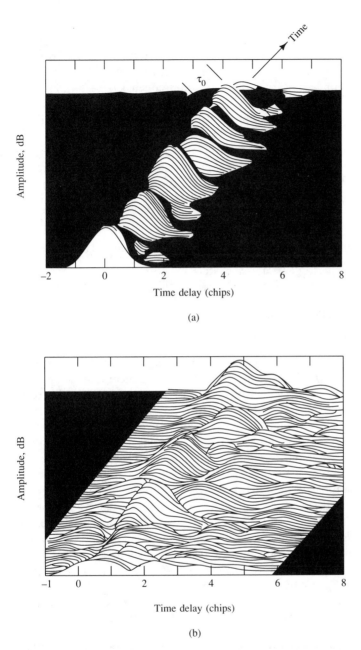

FIGURE 10.6 Received signal modulation. (a) Flat fading. (b) Frequency-selective fading.

10.2 TESTING OF LINK PERFORMANCE

Hardware testing involves the use of link simulators to determine link performance with actual transmitter and receiver hardware. This can be done at baseband, intermediate-frequency (IF), or radio frequency (RF), although an IF approach is typically chosen.

A very simple approach to implementing a hardware channel simulator applies when the link modulation bandwidth is much less than the channel frequency-selective bandwidth ($B_M \ll f_0$). Under such *flat fading* conditions, the delay spread is negligible, the link does not suffer intersymbol interference or related forms of modulation waveform distortion, and the channel model can therefore be written as

$$E_R(t) = H(t, \tau) E_T(t) \qquad \tau = 0 \tag{10.2}$$

In terms of the real signals—transmitted (E_T) and received (E_R)—this is

$$\text{Re}[E_R(t)] = h_I(t)\,\text{Re}[E_T(t)] + h_Q(t)\,\text{Re}(E_T(t)]_{90°} \tag{10.3}$$

where

$$h_I(t) = \text{Re}[h(t, \tau)] \qquad \tau = 0$$

$$h_Q(t) = \text{Im}[h(t, \tau)] \qquad \tau = 0$$

and where $\text{Re}[E_T(t)]_{90°}$ is the real transmitted signal with a 90° phase shift.

The quantities $h_I(t)$ and $h_Q(t)$ are independent, zero-mean, equal-variance Gaussian random processes with a PSD suitably approximated by Eq. (10.1). The f^{-4} power spectrum specified by Eq. (10.1) can be easily implemented by passing the output of an additive white Gaussian noise (AWGN) source through a two-pole low-pass filter. Figure 10.7 provides a block diagram for a flat fading hardware channel simulator of this design. The elements enclosed within the dashed box, a 90° phase shifter and two biphase scalar modulators, comprise a vector (phasor) modulator. The delay provides independent noise sources for the I and Q signals. Each biphase scalar modulator can also be implemented in terms of a 180° phase shifter with two voltage-controlled attenuators. Such channel simulators have been built and used for a number of different satellite link development and testing applications.

One approach for providing a *frequency-selective hardware channel simulator* employs a tapped-delay line implementation of the *channel impulse response* (CIR) model, with separately controlled vector modulators for each tap. The scintillation portion of the channel model is approximated as

$$E_R(t) = \Delta\tau \sum_{n=0}^{N-1} \eta(t, n\,\Delta\tau) E_T(t - n\,\Delta\tau) \tag{10.4}$$

where $N\,\Delta\tau \gg f_0^{-1}$ and, to prevent aliasing errors, $\Delta\tau \ll B_M^{-1}$. Taking the real part of $E_R(t)$, this becomes

$$\text{Re}[E_R(t)] = \sum_{n=0}^{N-1} h_{IN}(t)\,\text{Re}[E_T(t - n\,\Delta\tau)] + h_{Qn}(t)\,\text{Re}[E_T(t - n\,\Delta\tau)]_{90°} \tag{10.5}$$

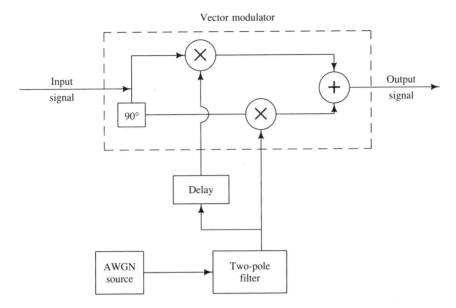

FIGURE 10.7   Block diagram of a flat fading channel simulator.

where $h_{In}(t)$ and $h_{Qn}(t)$ are the real and imaginary components of $h(t, n \Delta \tau)$. This discrete approximation of the CIR model has been implemented under Defense Nuclear Agency (DNA) sponsorship in a 48-tap delay line frequency-selective channel simulator that has been used successfully for development and testing of several wideband satellite link modems. The I and Q vector modulator voltages for each tap are derived from the DNA channel impulse response function (CIRF) code. A simplified block diagram of this frequency-selective fading simulator is shown in Fig. 10.8.

Other alternatives for hardware testing include: (1) the use of SAW chirp Fourier transform pairs[1] in a frequency-domain implementation of the frequency-selective fading channel, and (2) the use of baseband digital signal processing implementations of the channel model to perturb the sampled input modulation signal, followed by digital-to-analog conversion and modulation of the transmitter signal by the perturbed modulation signal. These approaches are currently under development.

10.3 EXPECTED PERFORMANCE OF 20/40-GHz SATCOM

Figures 10.9 and 10.10 provide some insight into the scintillation of 20/40-GHz propagation through a nuclear-disturbed ionosphere. First, absorption times are to be measured in fractions of a second, not in tens of minutes or seconds as occurs in the ultrahigh frequency (UHF) and superhigh frequency (SHF) ranges, respectively. Second, the range of decorrelation values is more constrained.

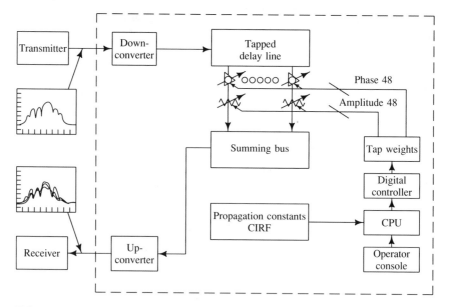

FIGURE 10.8   DNA/GE nuclear effects simulator functional diagram.

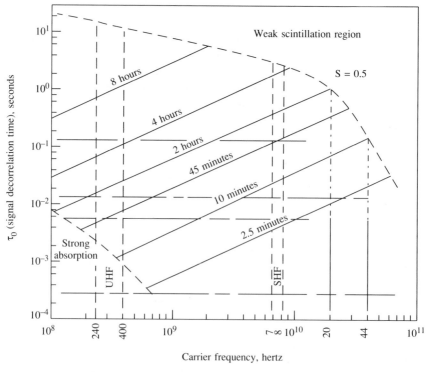

FIGURE 10.9   Decorrelation time for various frequencies and times.

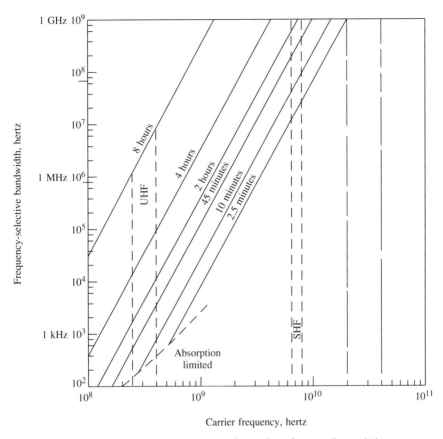

FIGURE 10.10   Decorrelation bandwidth for various frequencies and times.

The coherent bandwidth f_0 at 20/40 GHz is several orders of magnitude larger than at UHF, for which f_0 is in the kHz region for tens of minutes to a few hours after burst, and much larger than at SHF, for which f_0 is in the hundreds of kHz region at TAB of tens of seconds, increasing to a few MHz up tens of MHz at minutes to tens of minutes after burst. For 20/40 GHz, f_0 varies from tens of MHz at early times (up to a few seconds) to GHz values several minutes after burst.

10.4 EXPECTED PERFORMANCE OF CROSS LINKS

Unless the 60-GHz cross links traverse disturbed ionospheric paths as depicted in Fig. 10.11, no significant scintillation effects are to be expected. Although there are increases of electron density at higher altitudes, even beyond synchronous, these field-aligned striated regions are very sparsely populated.

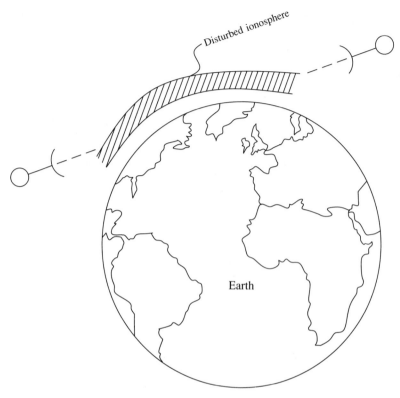

FIGURE 10.11   Disturbed ionosphere.

REFERENCE

1. N. C. Mohanty, *Signal Processing, Signals, Filtering and Detection*, New York: Van Nostrand. Reinhold, 1987, page 68.

11

S. BRUCE FRANKLIN
and
K. T. WOO

. . . .

Strategic Defense System Communications

The Strategic Defense System (SDS) is a major system being designed to protect the United States from a ballistic missile attack. Conceptually, the SDS contains three major functions: (1) sensors, (2) battle management, command, control and communications (BMC^3) and (3) weapons. Clearly, the sensor function is to sense the launch and progress of ballistic missiles directed at the United States. The SDS weapon function is to destroy or negate the attacking missiles. The BMC^3 element function is to provide supporting communications between all major SDS platforms, to provide the command and control function, and to manage the battle for the military commanders.

The SDS will contain a significant communications system to provide message connectivity between the many space and ground based platforms supporting sensors, command and control facilities and weapons. (Here, the term "platform" is a generic term for a significant entity participating in the SDS. It may be an orbital satellite, a ground-based or mobile facility, or a significant node in a ground-based communications network.) The SDS communications system must be able to operate effectively before, through, and after a nuclear war. Accordingly, it must be capable of operating in the most severe environment that will occur or be created by a multitude of natural and nuclear events and in the presence of potentially significant communication jamming.

Attacking missiles are assumed to progress through a number of phases: (1) boost, (2) postboost, (3) midcourse, and (4) terminal. The characteristics of the weapons and the defense mission are such that a family of sensors and weapons are used for cooperatively detecting and defending against the attack.

547

The Phase One SDS will contain the following major elements:

- Boost Surveillance and Tracking System   (BSTS)
- Space Surveillance and Tracking System   (SSTS)
- Space-Based Intercept System   (SABIR)
    Interceptor Carrier Vehicles   (SABIR-CV)
    Interceptors   (SABIR-IV)
- Ground Surveillance Tracking System   (GSTS)
- Ground-Based Radar   (GBR)
- Exoatmospheric Reentry Interceptor System (ERIS)

The space-based platforms will be in Keplerian orbit, and will be arranged in constellations selected for most effective support of their assigned function. The communications system hosted by the many platforms in the six elements, and connecting the ground-based command and control centers is functionally illustrated in Fig. 11.1.

BSTS sensors designed for sensing the boost phase are capable of detecting the exhaust plume of a weapon against an earth background. Accordingly, they may be located in synchronous orbit. When a missile is no longer thrusting, it is much harder to detect, and sensors designed for the later phases are most effective when viewing a region having a sky background. Accordingly, the SSTS sensors are orbited at lower altitudes and in constellations that provide adequate sensor coverage.

Space-based weapons are intended to intercept attacking weapons from as early as possible until there are no further SBI-CVs (or simply CVs) close enough that their interceptors (IVs) are capable of being used effectively. Note that an attacking weapon becomes a target to an interceptor. Later we will see that one of the messages to an IV is a target object map (TOM) for its sensor. Accordingly, they are located in relatively low-altitude orbits, with several CVs in each of several inclined planes.

Ground-based sensors (GSTS, GBR) are intended to detect incoming missiles in the late midcourse and terminal phases, and to support the launch and guidance of ERIS interceptors. There is a need to provide communications support to the GSTS and the ERIS from launch to the completion of their flight mission. Both may be supported by a ground-based communications system, but if and when they go beyond the communications line of sight from their launch site, alternate paths must be provided. One option is for this support to be provided by the space-based constellations.

Clearly, the SDS communications system is a global communications system in the truest sense of the word. The BSTS, SSTS, and SABIR platforms have communications networks within each element's constellation. Platforms in the same orbital planes are netted together to form a "carousel" that extends around the complete orbital ring. Additionally, each element (BSTS, SSTS, or SABIR) contains links that connect the individual orbital planes together to make an extensive intraelement network. The element networks are also tied together with interelement cross links. Space platforms within

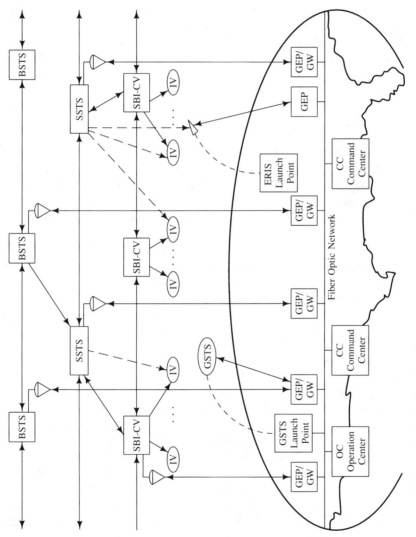

FIGURE 11.1   SDS communications.

view of the contiguous United States (CONUS) support communications between the space constellations and ground platforms.

The ground-based element platforms are distributed throughout the CONUS and must be connected with a communications system capable of operating before, through, and after a nuclear war. One option is to use an extensive Fiber-Optic Network (FON) to interconnect all of the ground-based platforms. Thus, the FON provides an extensive, relatively error-free, ground-based connectivity and makes possible the wide distribution of Ground Entry Points (GEPs) with gateways to interconnect the Space-Based Network with the Ground-Based Network.

This chapter is primarily concerned with the space-to-space links and the space-to-ground links, as affected by nuclear enhanced scintillation and jamming.

11.1 SYSTEM CONCEPTS

The SDS space-based network is hosted by three major constellations: the BSTS, SSTS, and SABIR-CV platforms. The SDS communications network's primary function is to support interplatform message traffic. To a first order, in wartime the sensors generate 90 percent of the BSTS or SSTS message traffic. Each sensor's observation of a missile, decoy, or similar object leads to a data message that becomes paired with similar messages from other sensors to generate a position in space. A sequence of such messages leads to a track. Tracks in turn are used by the weapon assignment process and lead to messages that either assign weapons to targets or control weapons on the way to intercepting targets. Messages are also routed to ground commanders performing the command and control function—to select battle states, to monitor operations, and to intervene as necessary. They are also sent to the ground gateways for routing to the various ground sites (GBR, ERIS) for early warning and support to their element functions.

Clearly, the SDS message traffic cannot be preordained; the communications system must be responsive to a high degree of uncertainty and variability in source-destination routing requirements. Constellations are constantly changing as satellites move through their orbits and sensors or weapons most likely to generate or need data move into or out of potential battle arenas.

The SSTS sensors, which sense targets in postboost and midcourse phases, generate the greatest amount of data, with interplatform data rates expected to be of the order of a few tens of megabits per second (Mbps). The BSTS sensors generate much less data during the boost phase, with data rates expected to be on the order of 1 Mbps or so. SABIR-CV platforms generate very little data, and the interplatform communications system must only support a few tens of kilobits per second (kbps). Additionally, the links that control the SABIR-IV and the ERIS are time-division multiple-access (TDMA) links with low burst-data rates, of the order of tens of kbps.

Platforms in all three constellations must communicate with the ground when in view of the CONUS. Space-to-ground data rates are expected to be commensurate with space-to-space data rates from each platform—that is, of the order of Mbps from BSTS, tens of Mbps from SSTS, and tens of kbps from SABIR.

It is evident from the preceding discussion that the SDS space based communications system has the following characteristics:

- Extensive but constantly changing network connectivity
- Extensive and highly variable message traffic
- Constantly changing source-destination pairs.
- Very low and moderately high data rates.

All of these considerations lead to the fact that the SDS communications network must be a packet-based network and must have many of the characteristics of modern packet networks.

11.1.1 System Design Assumptions

Different teams of designers have arrived at different packet network concepts as a result of differing starting assumptions about the difficulty in achieving link designs with very low bit-error rates (BERs) and packet-error rates (PERs). If one assumes that it is relatively easy to obtain low BER/PERs on the many space links, there is less concern about allowing long messages to be passed through the network. Conversely, a designer who assumes that it will be very difficult to obtain low BER/PERs on the many space links will be very concerned about the length of individual packets to be delivered error-free. These concerns reflect directly on the composition of individual messages, how long messages are handled, and the associated protocols that may be adopted. The SDS communications network must be designed to perform well in the presence of nuclear-enhanced scintillation. Network and link designs must consider the effects of fast and slow Rayleigh fading, whichever is worse, path absorption, very high sky temperatures, and considerable reduction in antenna gains that can occur.

As a practical matter, the SDS network is a very complex interconnection of tens to hundreds of space-based platforms. The deleterious effect of nuclear scintillation will be felt by only a moderate fraction of individual links. However, the total network must be designed with the performance of these few links uppermost in mind. The net effect is that the space-based network will be quite different from terrestrial peacetime local or even wide area networks in that the former are designed for relatively low bit- and packet-error rate conditions, whereas the SDS network must be designed for relatively poor BER and PER conditions and for long packet delays, and must be tolerant to degradation or outright loss of individual nodes.

11.1.2 Packet Network Model

It is convenient to adapt the International Standards Organization's Open System Interconnect (ISO/OSI) reference model to describe the major functions of the SDS Communications System.

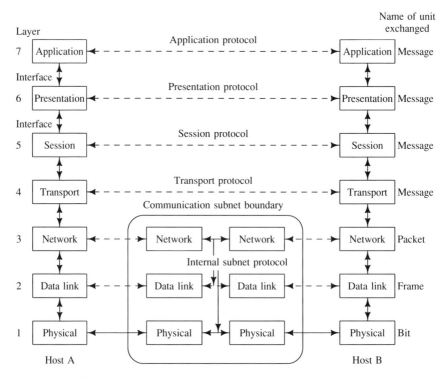

FIGURE 11.2   ISO-OSI network model.

The ISO-OSI model is described in detail in a number of references (Day and Zimmerman, 1983; Tannenbaum, 1981; Cerf and Cain, 1983; Rauch-Hindrin, 1985; Ananda and Srinivasan, 1984). Nevertheless, a brief review is in order here. The ISO-OSI model is illustrated in Fig. 11.2 for a situation in which application software in one processor must establish a connection through a network with a counterpart application software module in a different processor. Data messages work their way downward from level 7 to 6, down to level 3 and on to level 1, through the physical medium to a relay node and onto the destination node, where the message works its way upward to level 7 and interfaces with the receiving software application. (Messages may be fragmented into packets or may fit entirely within a single packet. Here, we use the term "message" to refer to an entire message, and "packet" to a data unit that passes through the transmission medium.) Conceptually, the system operates on a peer-to-peer basis, with software at each level behaving as if it were talking to a counterpart at the same level at the other end of the communication path. Software at each level actually interfaces with the adjacent upper or lower level but has no need to understand or be involved in the operations of other higher or lower levels.

Traditionally, levels 7 through 4 are contained within the source processor. Level 7 is the interface with the processor's application software. Level 6 simply performs format conversions to make the message suitable for transmission. Level 5 is often not used, but when it is, it is used to establish the communications session in a manner similar to what happens when a secretary places a call for someone. Level 4 is the interface with the network and provides a number of functions, including:

- Negotiating service classes with the session level.
- Multiplexing and demultiplexing between several higher-level (e.g., session level) entities.
- Data field encryption and decryption.
- Message assembly and ordering (if required).
- End-to-end acknowledgment services (if required).

The interface between levels 4 and 3 is comparable to the output of a personal computer modem and the telephone network. The network level arranges for all routing functions and ensures that the packet traffic is forwarded to its ultimate destination. Level 2 is a single one- or two-way data connection between two nodes in a communications network, and provides the protocols necessary to establish and maintain this single link. Level 1, the physical level, provides the signaling medium, be it a telephone wire or a pair of one-way RF channels between two terminals.

In this set of definitions, level 1 includes all processes necessary to prepare a packet for transmission and processes the received data; that is, it includes channel encrypting, encoding, interleaving, chipping, modulation, demodulation, and the inverse receiving functions necessary to outputting a decrypted bit stream. Level 2 includes packet synchronization, packet error checks, and provides for single-hop acknowledgment/no acknowledgment (ack/nak) services.

In a full-up OSI model, messages may (or may not) be fragmented into data units and packets and provided with a certain amount of header information with source and destination parameters, and other necessary information. At each downward step (level n to $n - 1$) at the transmitter, additional header (and tail) information may be appended until the physical level (level 1) is reached. As the message works its way through the network, it flows upward or downward between the physical (level 1) and network (level 3) levels until it reaches its destination. At that point, the message continues to flow upward until level 7 is reached, and the source data is input to the receiving application software.

Figure 11.3 illustrates the application of the generic OSI model of Fig. 11.2 to the SSTS network. The left-to-right flow across the top of the figure illustrates the relay functions. Data enter the flow from level 7, pass through levels $6 \rightarrow 5 \rightarrow 4$ to a routing switch at level 3. From there, it passes from platform to platform via level $3 \rightarrow 2 \rightarrow 1 \rightarrow 1 \rightarrow 2 \rightarrow 3$ processes. Ultimately, it is routed from the routing switch of level 3 through level $4 \rightarrow 5 \rightarrow 6$ to a software application process at level 7 shown toward the bottom of the figure for each platform.

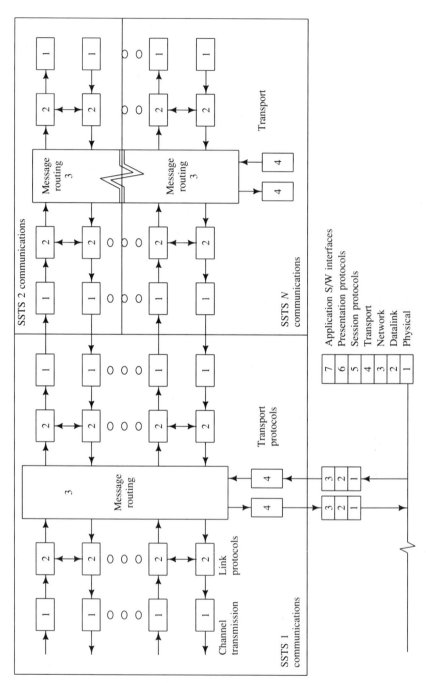

FIGURE 11.3   Example of OSI model imposed on SSTS.

11.1.3 Message Traffic Categories

Eight different categories of traffic have been identified for the SDS. Each category may have as many different message formats (message types) as necessary, but they will nominally be in the range of 32 to 256 types per category. The seven message categories are:

1. Engagement command and control

2. Sensor/track messages

3. Assessment

4. Engagement

5. Operational readiness

6. Network control

7. Telemetry, tracking, and command

8. Algorithm data bases

Clearly, categories 1, 3, 4 and 8 are extremely important messages for which "guaranteed delivery" is required, necessitating packet acknowledgment and retransmission techniques. In wartime, category 2 is the dominant message type, and will very likely comprise 90 percent of the total traffic supported by the network.

11.1.4 Quality of Service

Various classes of service may be desired of a communications system. There are other sets of classes, with the CCITT and the ISO agreeing on five classes. Of these, the NBS has adopted two. The following list includes the range of service classes that are commonly considered. At this time a specific set of classes has not been adopted for the SDS.

CLASS I TRANSMISSION SERVICE

- Transmission guaranteed complete (ack/nak)
- Error-free transmission (ack/nak)
- Data not duplicated in transmission
    Data delivered in sequence
    Delivered data may be out of sequence

CLASS II TRANSMISSION SERVICE

- Transmission may be incomplete (outages)
- Error-free transmission (discard packets in error)
- Data not duplicated in transmission
    Data delivered in sequence
    Delivered data may be out of sequence

CLASS III TRANSMISSION SERVICE

- Transmission may be incomplete (outages)
- Transmission errors possible
- Data not duplicated in transmission
    Data delivered in sequence
    Delivered data may be out of sequence

CLASS IV TRANSMISSION SERVICE

- Transmission may be out of sequence
- Transmission errors possible
- Data may have been duplicated in transmission
- Delivered data may be out of sequence

A primary concern is whether packets and/or messages are delivered in error, and whether messages that have been segmented into separate and distinct packets are reassembled before delivery and, if not, whether they are delivered in order. Some of these issues relate to protocols and processing and are not discussed further in this chapter. Other issues relate to probabilities of error-free packet or message delivery. These latter issues are addressed in more detail in subsequent sections of this chapter.

11.1.5 Network Routing Options

Two major and different approaches to routing communications packets have been proposed: preferential and flood routing. Preferential routing consists of selecting the most direct path for a given packet to take when being moved from the source to the destination node (i.e., an OSI level 3 function), Lookup tables are maintained in each node and are periodically updated to reflect the health and status of the other nodes in the network. Destination information is taken from each packet's header, and used as to enter the lookup table to retrieve the identification of the output path to which to route the packet. Flood routing consists of arbitrarily forwarding each packet over every output path from a given relay platform except back to the nearest neighbor from which the packet was received.

Each packet header contains enough information to uniquely identify that packet from all other packets. In order to keep the packet from circulating through the network indefinitely, each network switch keeps a log of the identity of each packet that it has seen. If a packet is received a second time via an alternate route, the switch recognizes that it has already been through the node and does not send it on again. Hybrids are also possible, with some packets given preferential routing, others given flood routing. The message category placed in a packet header may be used to select the correct option.

Figure 11.4 serves to illustrate the alternatives. As an example, assume that a packet is to be routed from node C to node M. With preferential routing, the packets would pass through nodes D, I, and H, or through nodes B, G,

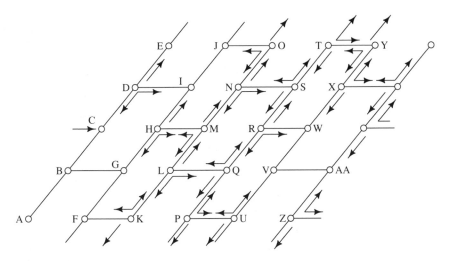

FIGURE 11.4   Network connectivity.

and H. With flood routing, the packet would pass from node C to *all* neighboring nodes until it arrives at node M. Because there are multiple routing opportunities, the packet need not traverse every path successfully.

11.1.6 Channel Capacity-Time Delay Trade-offs

Packet communications networks must be designed to operate with reserve capacity to accommodate the bursty nature of the packet traffic. With RF links, it is impractical to design for a variable data rate since demodulators, bit synchronizers, and related components are relatively inflexible, with most such components having uncommanded tracking ranges of 1 or 2 percent at most. Accordingly, the standard practice is to design for a constant data rate and to insert random data or dummy packets when there is not useful "information" to be transmitted. If possible, it is desirable that all packets have the same length to facilitate packet synchronization and using selected packets for other purposes such as ranging. In this event, dummy packets would be used to fill in the voids between "useful information" packets.

Clearly, if a channel is designed to carry only the average data rate, each node must have data buffers in which packets are stored so that the data flow into each channel can occur at a constant rate regardless of the packet arrival and local data generation statistics.

It is generally understood that when the channel capacity equals the average data capacity, some packets will have an infinite time delay while waiting their turn for a place on the link, and the local buffers will be exceedingly large (i.e., 100 percent utilization efficiency implies infinite time delays for some messages).

Typical rule-of-thumb practices are to design individual channels to have a peak capacity that is twice the average data rate (a 2 : 1 peak-to-average ratio). In some cases, a 10 : 1 peak-to-average ratio might be more appropriate, but extensive analysis and simulations are usually performed before adopting this extreme.

The Strategic Defense System communications network will consist of nodes that are connected to many other nodes. For example, typically, the SSTS will interface with three other SSTS, a SABIR-CV and possibly a ground entry point. On occasion, one of the links to a different SSTS will be redirected towards a BSTS. The BSTS and SABIR-CVs have similar connectivities.

Consider a cross link from an SSTS to an adjacent SSTS. In addition to being a destination node with its own processors, each receiving SSTS may also relay data to either one or two other SSTSs, and to other platforms as well. All SSTS-to-SSTS channels will have the same capacity. Clearly, then, if the links were designed so that each carried only useful traffic, with no empty or dummy packets, it would be impossible for each outgoing channel to accept all traffic received from multiple incoming channels. Either that, or there would have to be an extremely unique combination of packets that terminate in a node, packets that pass through a node, and packets that originate in the node so as to avoid saturating the outgoing links. The likelihood of the latter happening with a network carrying probabilistic traffic is nil. It follows that it is impractical to design a communications network so that all channels have a capacity equal to the average traffic load that must be carried.

For most space-based links, there is a regime in which it is relatively easy to provide adequate channel capacity. Once the edges of this regime are exceeded, however, it becomes incrementally more difficult to provide adequate capacity.

It is clear that selecting the capacity of each communications link is an extremely important task. Too little channel capacity results in a system that requires large data buffers and introduces long packet delays with some packets potentially being lost because of saturation. Too large a channel capacity imposes a need for a more capable channel, with larger antennas or power amplifiers than absolutely necessary.

Packet Routing Considerations

The techniques used to route packets through the space-based network are significant divers on the design of individual links. It has been shown that flood routing, in which every received packet is routed to all adjacent nodes except the source node, provides an extremely robust network. End-to-end packet delivery probability is extremely high, even when individual links have relatively poor error rates; for example, end-to-end packet-error rates of 2 percent can be achieved in the presence of single-hop packet-error rates as poor as 10 percent.

On the other hand, preferential routing techniques have been postulated in which packets are only distributed over the most direct paths. Unfortunately, the resulting network has much poorer packet delivery performance for the same individual link (single-hop) packet-error rate. For example, if two hops are necessary, a single-hop packet-error rate of 10 percent yields a 19 percent packet delivery rate; for three hops it is 27 percent. It is clear that for the same end-to-end delivery rates, the individual links must have a significantly higher fidelity.

Preferential routing is claimed to minimize the total data that must pass through each link and thereby permit individual links to have smaller capacities. This claim is generally valid for networks for which source data are relatively uniformly generated by all communications nodes and need to traverse only a few hops to the desired destination. The SDS is somewhat different, however, in that the majority of traffic is generated by the sensor platforms within view of the battle space and because sensor data often need to be passed to platforms many hops away. When the possibility of lost platforms is considered, it follows that it is always possible to find some individual links through which all traffic must pass. As a result, preferential routing techniques do not significantly reduce the necessary capacity of individual links; moreover, they lead to more complex data routing hardware and algorithms and to a less robust network.

Network Simulation

The problems described in the preceding paragraphs describe a difficult statistical analysis problem. Except for a very few simplistic models, it is virtually impossible to develop analytic models that will forecast such critical network behavior and interrelationships as

- Channel loading rates
- Network node buffer sizes
- Network delivery time delays

In general, for the type of problem faced by the SDS, it is necessary to use computer-based simulations. (A discussion of network simulations is beyond the scope of this chapter. As a first order, channel capacities can be sized for peak-to-average ratios of 2:1. More detailed simulations are required to develop accurate estimates of buffer sizes and time delays.)

Typical network simulations contain an extensive library of "primitives," which can be assembled to represent node macros, which in turn can be integrated to form complete communications networks. Simulation primitive libraries should include a large number of models, including buffers, time delays, routing and switching algorithms, probabilistic source distributions, probabilistic source data rates, flow control models, and so forth.

11.2 THE STRESSED ENVIRONMENT

The stressed environment in which the SDS communication links are required to operate include both natural and man-made threats. The ground-to-space uplinks and downlinks are EHF links, with 44.5-GHz uplinks and 40- (or 20-) GHz downlinks. The exact choice of downlink frequency depends on trade-offs between performance in the nuclear environment, operational scenarios, and the costs involved in developing 40-GHz terminals, and has not been determined at the time of writing of this chapter. Nevertheless, the weather losses at 44, 40, or 20 GHz are significant, and can exceed 30 dB in regions with heavy rain. Spatial diversity with widely separated CONUS sites, to be selected by the satellite according to weather patterns, is necessary to provide the necessary diversity gains on the downlinks from the SDS satellites to the ground entry points for protection against excessive rain losses.

Man-made threats include both jamming and nuclear effects. Ground-based, airborne, and/or spaceborne jamming can be present prior to and during the offensive attacks. The nuclear-enhanced environment is a multiple-burst turbulent environment. It includes a large number of high-altitude (above 500 km), high-yield (kilotons to megatons) detonations over the North American continent, and is created by a number of attacks, including defense suppression attacks against communications, sensors, and radars, as well as salvage fusings of enemy reentry vehicles when they detect that they have been hit by SDS interceptors.

High electron densities, up to 10^9 and 10^{10} electrons cm^3, which are orders of magnitude higher than normal (10^4 to 10^5 electrons/cm^3), will persist throughout the duration of the attack, creating severe impacts on radio-frequency communications. These nuclear scintillation effects can last a relatively long period of time (hours) after a nuclear blast, and can affect a large area (1000 to 2000 km in width and height).

Even in a single-burst scenario the electron density can reach 10^9 to 10^{10} electrons/cm^3 in the initial phase of the blast. But this level of electron density only lasts about 2 minutes, decreasing to 10^8/cm^3 after 2 minutes in a typical high-yield, high-altitude, single-blast scenario. The dominating effect in the single-burst scenario is nuclear scintillation, caused by variations in the electron density in a striated region as the signal propagates through it.

In the multiple-burst scenario, however, the environment is very turbulent. High electron densities persist in the affected regions as multiple detonations are made at different instants of time. The result is increased total electron content in the path and bulk effects, such as high absorption and increased sky temperatures, throughout the entire scenario and whenever the propagation path passes through the affected region. The signal decorrelation time will, however, be short during the times when the propagation path passes through the turbulent region with high electron densities. Slow fading will still be experienced at the end of the multiple-burst scenario. However, at that time the bulk effects caused by total electron content in the path will be sufficiently reduced.

11.3 COMMUNICATIONS DELIVERY SUCCESS PROBABILITY

Communication performance varies for coded and uncoded links that pass through a nuclear scintillating region. Because of the non-Gaussian characteristics of the scintillating environment, bits in error are not randomly distributed for uncoded data, and errors in decoded data do not follow the usual statistics of postdecoding errors.

Because the SDS uses packet communications, there is a need to identify whether certain critical packet fields are received in error. Clearly, this identification is essential for all data in packets containing category 1, 3, 4, and 8 messages, for which a single bit in error could have catastrophic results. Tracking filters operating on sensor data can tolerate missing data far easier than data in error. Other categories may be less affected by errors in the data fields. As a practical matter, for the SDS it has become an accepted practice to check each packet or message and to discard any found to contain an error.

Identifying packets with errors may be done on an end-to-end basis, or on a hop-by-hop basis. When done on an end-to-end basis, it becomes necessary for the original source to keep copies of all messages requiring guaranteed delivery until an acknowledgment is received from the furthest destination. Identifying packets with errors on a hop-by-hop basis is equally valid for most situations and simplifies the ack/nak procedure.

A side benefit of checking each packet for errors on a hop by hop basis is that it becomes feasible to develop analytical relationships or computer simulation results that relate E_b/N_0 (energy per bit over noise power spectral density ratio) and single-hop packet-error rates (PERs). This can be done in a manner which includes the effect of packet error distribution statistics. Once a packet has a single error, it could afford to have many more without affecting the single-hop PER. Thus, for the same bit-error rates (BERs), it would be convenient if all link errors were grouped together instead of randomly distributed. In that way there would be more error-free packets, and the PER would actually be better.

For these many reasons, it is assumed that the important parameter of interest is the probability that a complete packet will be received free of errors; that is, *PER is more important than BER* since the latter does not consider error distribution statistics within packets and cannot easily be converted to a meaningful communications delivery performance measure.

Single-Hop Packet-Error Rates

This section presents comparative performance as a means of providing insights helpful to the next section on routing alternatives. Additional information on modulation options and E_b/N_0 requirements is presented later when waveform considerations are discussed.

Two cases are illustrated here. In the first, packets are Reed-Solomon (RS) coded and interleaving is used; in the second, packets are either RS coded

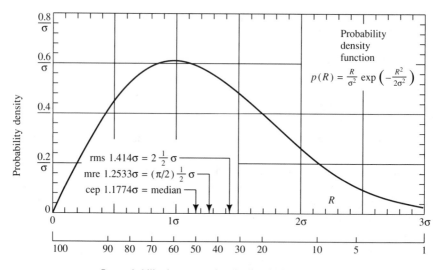

C = probability in percent that R exceeds the value on the R axis

$$P(R) = \frac{R}{\sigma^2} \exp\left(-\frac{R^2}{2\sigma^2}\right)$$

Let $\mu = \dfrac{R^2}{2\sigma^2}$

$P(\mu) = \exp(-\mu)$

Message Fading:

$$
\begin{aligned}
\mathrm{MER}(N) &= 1 - \int_\mu P(\mu) P(c\,|\,\mu)\, d\mu \\
&= 1 - \int_\mu P(\mu)\left[1 - P(e\,|\,\mu)\right]^N d\mu \\
&= 1 - \int_0^\infty \exp(-\mu)\left[1 - 0.5\exp\left(-\mu\frac{E_b}{N_0}\right)\right]^N d\mu
\end{aligned}
$$

Symbol Fading:

$$
\begin{aligned}
\mathrm{MER}(N) &= 1 - \left[1 - \int P(\mu) P(e\,|\,\mu)\, d\mu\right]^N \\
&= 1 - \left[1 - \frac{1}{2}\int_0^\infty \exp(-\mu)\exp\left(-\mu\frac{E_b}{N_0}\right) d\mu\right]^N
\end{aligned}
$$

Definitions: R = received signal envelope in fading

σ^2 = mean-squared value of R

N = number of DPSK symbols in a message

MER = Message Error Rate

FIGURE 11.5 Message error rates with Rayleigh fading.

without interleaving or not coded at all. In both cases, the data symbols are 5-bit differential phase-shift keying (DPSK) symbols. In the first case, packets are RS (31/13) block coded. Thus, each block contains 65 bits, with multiple blocks forming a complete packet.

Both cases assume that only amplitude fading affects performance; both assume that bit synchronizers are fully locked and are jitter-free, and that the demodulation process is fully effective and dependent only upon received signal strength. Thus the results presented below are optimistic. They do, however, illustrate relative performance very well. The important distinction is that in the first case all symbols within a packet are considered to be independent after deinterleaving, insofar as fading effects are concerned. In the second case the symbols are considered to be sufficiently close together that they are not independent; complete packets fade together.

The key equations for message-error rate (MER) governing these two cases are given in Fig. 11.5. Comparative performance results are presented in Fig. 11.6. Clearly, for all but the very poorest packet-error rates, performance with RS encoding in the first case, with adequate interleaving (symbol fading), requires much less E_b/N_0 than in the second case, with either inadequate interleaving (block fading) or with no coding at all. The importance of adequate interleaving is such that even with coding, the PER performance without interleaving (block fading) is essentially identical to the uncoded case.

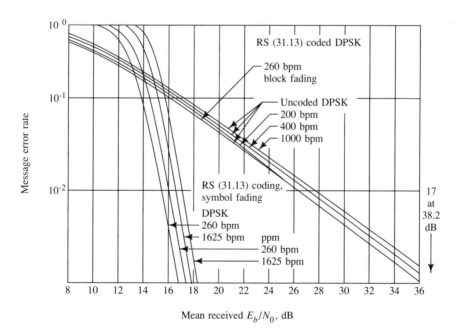

FIGURE 11.6   Message error rate vs. mean received E_b/N_0 with Rayleigh fading.

TABLE 11.1 Required E_b/N_0 vs. Packet-Error Rate

| | Required E_b/N_0 (dB) | |
PER	Rayleigh Fading	RS Coded with Interleaving*
10%	17.1	13.7
1%	27.1	15.3
0.1%	37.1	16.8

* This is not the most optimum coding, or modulation, but serves to illustrate the nature of the issue.

To compare, consider Table 11.1 for 260-bit messages with DPSK modulation. Clearly, adequate interleaving is very helpful. Aside from the additional degradations inherent in the demodulation and bit synchronization process, it is unlikely that either of the above two cases illustrated in Fig. 11.6 will actually exist for all time. The second requires adequate interleaving; the first requires none at all, or inadequate interleaving.

As a practical matter, the hardware will be designed for a specific signal structure, whereas the environment will change. Although a certain amount of interleaving will be provided to assist against jamming, the amount required for antijam purposes will be relatively small. However, the amount of interleaving required to offset scintillation degradations depends on the duration of the fades. Clearly, very intense scintillation is also very brief and rapid. Thus, the interleaving required for use against jammers would be adequate for short, intense scintillation fades. However, as the scintillation intensity diminishes, the fades become longer and further apart. As this happens, deinterleaving loses effectiveness in randomizing data, and coding gains diminish. The effect is for performance to move from the coded fully interleaved case of Fig. 11.6 toward the uncoded case—or to force the designer to provide a great deal of interleaving until interleaving/deinterleaving buffers and time delays become significantly large.

11.3.1 Robust Networks

The SDS communications system provides an extended and complex connectivity that permits consideration of a variety of routing options. Collectively, the SDS communications network can be extremely robust! However, it must be made that way to begin with.

The two routing options, flood and preferential, have a significant impact on network robustness. It is useful to consider the characteristics and benefits of both of the major approaches. First, however, some aspects of the SDS extended network merit further discussion.

Message Categories and Routing Options

The capacity of the internal networks for the three major SDS elements differ significantly. The SSTS network must be capable of supporting tens of Mbps, the BSTS network of supporting bit rates of the order of 1 Mbps, and the SABIR-CV network of supporting tens of kbps. This results in the SSTS internal network effectively providing the backbone of the SDS space network, with the other element networks principally providing data distribution within the element itself.

As a consequence of the architectural concept, SSTS sensor data do not need to flow through the BSTS or SABIR-CV networks, nor can the networks support them. Similar observations may be made about other message types. In order to route data selectively, the network switch in the nodes of each of the three elements must be capable of selectively admitting certain packet categories to the interelement links while not admitting other types. This goal is feasible, and primarily involves a logic function in the network routing switch.

Given the selective routing function, a comparison between routing approaches is equally valid for all packet types within a given category. The principal differences, then, lie with the extensiveness of the network through which a given packet category flows. It is assumed, for example, that all packets sent from the SSTS network to the SABIR-CV network will be routed within the CV net as well as within the SSTS network.

Performance Comparison of Preferential and Flood Routing

Referring to Fig. 11.4, in order for the packet to arrive at node M with preferential routing, it must traverse each and every intervening hop without an error. Thus, for n hops, the probability of successfully arriving is given by:

$$P_c = (1 - P_e)^n \qquad (11.1)$$

where P_e is the single-hop PER, and each hop has the same PER.

The connectivity of Fig. 11.4 is such that it is almost impossible to develop an analytical expression for the probability of successful packet delivery for flood routing. Although the following analysis does not address the specific connectivity of Fig. 11.4, it is useful to consider a simpler model to illustrate comparative performance. Assume that there are k parallel paths from node C to M, and that each path has n hops. In order for a packet not to reach node M, it must not survive passing through any of the parallel paths. Thus, the probability of not reaching node M is given by:

$$P_e' = (1 - P_c)^k$$
$$P_c' = 1 - P_e'$$
$$= 1 - (1 - P_c)^k$$

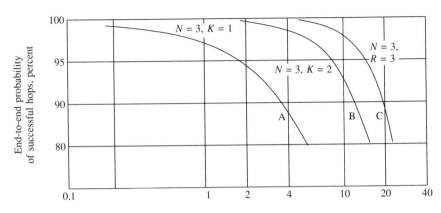

FIGURE 11.7   Preferential and food routing. N = number of series hops; K = number of parallel paths.

But P_c is given by Eq. (11.1). Thus

$$P_c' = 1 - [1 - (1 - P_e)^n]^k \qquad (11.2)$$

Equations (11.1) and (11.2) must now be inverted to determine the single-hop packet-error probability required to yield the same end-to-end probability of delivery success—i.e., to make $P_c = P_c'$. For preferential routing, this yields

$$P_e = 1 - P_c^{1/n} \qquad (11.3a)$$

For flood routing,

$$P_e' = 1 - [1 - (1 - P_c')^{1/k}]^{1/n} \qquad (11.3b)$$

These latter two equations are plotted as curves A, B, and C of Fig. 11.7. True flood routing performance, with additional paths between midpoints of each series string of the above model will be even better than shown by curves B and C of Fig. 11.7. Clearly, for the same end-to-end probability of successful delivery, flood routing permits the single-hop PERs to be much higher than for preferential routing. Thus the necessary E_b/N_0 will be lower for flood routing, and the performance burden on each link will be reduced.

Robustness Intangibles

The SDS networks must be capable of operating before, through, and after a nuclear battle. They must be robust and capable of performing well when one or many of the host satellites is destroyed or made inoperable, or when some of the links become incapable of supporting the design data rate because of severe scintillation or jamming degradations.

Individual links must be designed to meet performance requirements in the presence of nuclear scintillator, or in the presence of threat jammers. Additionally, the networks must be designed to operate when one or more complete satellites is missing.

Preferential Routing

Networks with preferential routing can be designed to adapt to the loss of one or more relay nodes. This is primarily accomplished by causing every relay platform to periodically (and frequently) transmit health and status messages to declare its continued existence and the quality of its performance in the network. All other platforms monitor these transmissions and update their on-board routing tables according to a common algorithm. When a platform is lost, its health and status messages will cease, and other platforms will stop receiving them. Other platforms will then deduce the loss of the platform for which it is no longer receiving messages.

Unfortunately, the process of recognizing that the network has changed is relatively slow and is dependent upon (1) not receiving a finite number of health and status messages and (2) how often health and status messages are generated. If the network suffers an outage, several seconds can elapse before the network recognizes the change and can change its routing structure. During this time, the packet traffic that would otherwise transit a lost platform suffers one of two fates. In one case, several seconds' worth of data are simply lost. In the second case, where the network nodes are designed with several seconds' worth of buffer memory, the stored packets are retransmitted over alternate routes when a lost platform is recognized. The size of the buffer memory is dependent upon the channel capacities, how much time is expected to elapse before a lost node is recognized, and how closely together (in time) multiple nodes are assumed to be lost.

Flood Routing

Networks with flood routing are inherently adapted to the loss of one or more nodes—without requiring special accommodations. The reason is that the nodes in a flood-routing network automatically transmit packets in all possible directions (except back to a neighbor from which a packet was received). Thus, the recovery process that a preferentially routed network must go through (i.e., reroute, or transmit over a different route) is continuously happening. It follows that the time delay before the loss of a relay node is realized has no adverse effect on network performance.

With flood routing, the only way a packet can be kept from a destination node is for the packet to be stopped in every possible path to the destination. This result can only occur with a combination of lost packets because of scintillation and jamming, or if multiple neighbors are made inoperable. Putting probabilities aside for the moment, if all links are perfect, the only way a packet can be kept from a destination is for a complete string of connecting platforms to be taken out, with one platform inside of the string, the other outside. We conclude that a network based on flood-routing concepts is much more robust than a network based on preferential routing.

11.3.2 Special Traffic Considerations

Guaranteed Delivery

Guaranteed delivery service is achieved by providing acknowledge/no acknowledge (ack/nak) service, and by repeatedly transmitting a packet until it is positively known to have been received. Ack/nak service can be used with both preferential and flood-routing options. Hop-by-hop ack/nak service can be implemented as an OSI level 2 function, and/or it can be implemented on an end-to-end basis as a transport (level 4) function.

Simply put, the ack/nak process provides for the receiver to acknowledge the receipt of packets requiring ack/nak service. The sender keeps each of these packets in a buffer and periodically retransmits each packet until delivery acknowledgment is received. As a practical consideration, each link will consist of a number of information and dummy packets. Thus, when there are no more information packets to be sent, the sender sends dummy packets to maintain the link. If packets requiring ack/nak service are available, they may be sent in lieu of dummy packets until there are no more such packets whose receipt has not been acknowledged. Since packets requiring acknowledgment are expected to be a small percent of the total traffic, repeat transmission does not materially increase link packet traffic.

Long Messages

It is desirable that all packets have the same length and that the length be short—256 to 768 bits per packet is a reasonable goal. Messages as large as 1536 bits would exceed the most likely size of the largest percentage of messages—sensor messages.

These larger messages may be handled in a variety of ways. One is to segment the message, and to put a part of it in each of several successive packets. A second is to allow a mixture of packet sizes, with a goal that they all be a multiple of a common integer. A third is to allow different size packets and not require they have a common divisor. Other approaches include using frames, with each frame containing several simplified packets and the frame header containing common data that would otherwise have to be repeated in each of the frames.

During peacetime, long packets should present no particular bit-error or packet-error problems other than data handling and memory management problems. This observation is especially true if each link is designed to operate through a scintillating environment, and in the presence of jammers. The major concern is packet handling under stressed conditions.

Traffic Mix Considerations and Design Options

The nature of the traffic to be carried by the SDS communications permits considering a variety of design options. The BSTS and SSTS carousels and the space-to-ground links will carry a very high percentage (perhaps from 90 to 95 percent) of sensor traffic. This category of traffic is somewhat forgiving in that it is generated on a regular basis and is smoothed by various tracking algorithms. As a consequence, very good packet delivery probabilities are not required for this category.

Although all traffic will receive the same error-correction coding as it passes through the RF links, it is possible to design the links for the majority traffic category and to deal with the other categories in different ways. One of these ways includes using repeat transmissions with ack/nak services until delivery is actually received and the source receives delivery acknowledgment. Another way is simply to repeat transmit packets at intervals spaced far enough to ensure independence. Repeat transmission will reduce the probability of nondelivery to the power of n. Thus, if a single packet has a 10 percent probability of arriving with an error, three repeat (total of four) transmissions will reduce this probability to 0.01 percent.

Normally, repeat transmission is a poor channel coding technique because it increases required channel capacity by a factor of 2 or more if applied to all channel traffic. In the case of the SDS, the amount of traffic requiring very good packet delivery probability is small; hence it becomes feasible to use repeat transmission techniques with a very small increase in required channel capacities.

11.4 WAVEFORM CONSIDERATIONS

The communication link designs between various SDS nodes must be robust to the highly stressed environment in which the SDS is expected to operate. A robust waveform design, sufficient link margin, and optimal signal processing techniques to mitigate the effects of nuclear scintillation and jamming are required to provide the link performance necessary to maintain network connectivity in times of hostility.

Besides robustness, which is the most important consideration in SDS waveform design, timely transmission of sensor, in-flight guidance updates, or command and control messages is also critical for the success of the SDS mission. Considering the fact that many packets have to pass through several hops between satellites, processing delays in packet transmission, relaying, and reception must be limited to a minimum without unduly sacrificing performance. An area of special concern is the interleaving span used to provide time diversity for scintillation mitigation. Long interleaving spans give better performances in slow fading. However, they must not be excessive because of the transmission delays introduced by interleaving and deinterleaving, and because of the size and weight of the required memory.

The complexity of any required signal processing is another important consideration in selecting waveforms for the SDS links. In the SDS application there are many space terminals. The number of ground terminals or ground entry points (GEPs), however, is small. Most of the links are one-to-one links. This application scenario is quite different from that of a communication relay satellite system that serves many user terminals with a small number of geosynchronous satellite repeaters. In the case of communication relay satellites it is cost-effective to place processing complexity on the spacecraft in order to simplify processing complexity in the large number of user terminals, so that the total system cost can be minimized. This is not the case in the SDS application. Because the number of SDS space terminals is much larger than the number of interfacing ground terminals, the complexity of each space terminal must be minimized in order to minimize total system cost. Thus the communication waveform design should be application-specific, providing data modes and functions only if they are necessary for the specific link application.

The SDS unique characteristics that should be considered by the waveform designer are listed below:

"YES" FEATURES	"NO" FEATURES
Node-to-node communication	No many-to-one FDM/TDM or multiple node to satellite
Packet switching	Not circuit switching
Each link has one fixed data rate	No need for multiple rates on one link
Basic function is a sensor satellite	Not a communication relaying satellite
Low altitude, with minimum contact time and small antenna ground footprint	Not multigeosynchronous operation with large antenna footprints

Most links are node-to-node links. There are no "many-to-one" arrangements requiring frequency-division multiplex (FDM) or time-division multiplex (TDM) channelizations. Packet switching, rather than circuit switching, is used throughout SDS. There is no need for the allocation of access and network control data fields in each data frame. Most satellites in SDS (e.g., the SSTS and SABIR-CV satellites) are low altitude satellites with minimum contact times and small antenna ground footprints. These small antenna footprints provide an inherent jamprotection on the links from the CONUS GEPs to the low-altitude SDS satellites. However, because of short contact times, these ground-to-space links must be acquired rapidly in order to maximize the useful connect time for data transmission.

Simplification of waveform functional and data modes does not imply suboptimal performance. In fact, for all required communication modes the waveform and demodulation processing design should maximize robustness to the stressed environment, so that the required effective isotopic radiated power (EIRP) and gain-to-noise temperature ratio (G/T) on SDS satellites can be minimized.

11.4.1 Jamming Considerations

The choice of radio frequency, link geometry, keep-out zones, antenna directivity, and spread spectrum combine to protect the SDS links from possible jamming effects. All RF cross links will be placed behind the 60-GHz oxygen absorption lines so that their vulnerability to ground-based or airborne jammers is minimized.

At 60 GHz the crosslink antennas can be made highly directional with small diameters. Sidelobe rejections of at least 40 dB can be expected for off-boresight jammers beyond the first 4 or 5 sidelobes from the main beam. Because of the line-of-sight geometries involved and the narrow beamwidths, in-beam jamming of crosslink receivers by coorbital spaceborne jammers is feasible only on a one-on-one basis; that is, it is infeasible for one jammer to jam more than one SDS receiving terminal. Since flood routing causes the same data to arrive at a given SDS platform from more than one direction, a separate jamming satellite must be deployed for each direction. Thus the deployment of several jamming satellites is required for each SDS platform to be jammed and therefore the threat of persistent in-beam crosslink spaceborne jammers is minimized.

In-beam jamming of a ground-to-space link is possible from an orbiting space jammer at a lower altitude. However, because it must be in a lower orbit, it cannot be at the same position for a prolonged period. In addition, keep-out zones can be established by the SDS satellites, maintaining a minimum distance (e.g., 500 km) between the receiving terminal and any space jammer. Frequency hopping is used on selected links to give additional protection against space jammers. With the above combination of link design provisions the SDS crosslinks are sufficiently protected against both airborne and spaceborne jamming.

The ground-to-space uplinks and downlinks are at 44-GHz and 40-GHz (or 20-GHz) links. They are not protected by the oxygen absorption lines. However, downlinks are protected by the fact that the GEPs are on CONUS, and that the ground receiving antenna is highly directional. Airborne jamming of space-to-ground downlinks will not be practical. With sufficient keep-out zones around the SDS's space transmitting terminals, the SDS GEPs can be jammed only through the ground antenna sidelobes by space jammers coorbiting with SDS satellites—if they knew which GEP to jam. Frequency hopping is also applied on selected downlinks to give sufficient jamming protection.

Frequency hopping will be used to protect the GEP uplinks. In addition to frequency hopping, the uplinks from GEPs to low-altitude SDS satellites such as the SSTS and SABIR-CV are also protected by antenna sidelobe rejection. The ground footprints of these low altitude satellites is small (approximately 20 km^2), because of their altitude and antenna directivity. Since these antennas will be CONUS-directed, it will not be possible for airborne or ground-based jammers to come into the mainlobe of these low-altitude satellites.

The BSTS satellites are in geosynchronous orbits. Even though their downlink antennas will be pointed toward a CONUS GEP, their ground footprints will be larger than those of the low-altitude satellites. Thus the GEP-to-BSTS uplink does not have the same amount of antenna discrimination protection against ground jamming as the GEP-to-SSTS uplink, for example. Sufficient hopping bandwidth will be required to protect this uplink from ground or airborne jammers.

11.4.2 Nuclear Effects

In addition to jamming, the stressed environment of the SDS links is significantly influenced by nuclear-enhanced effects. Nuclear scintillation mitigation is one of the most important considerations for the waveform design of the SDS communication links. Under severe scintillation conditions, the signal incident at the receiver can vary randomly in amplitude, phase, time of arrival, and angle of arrival. The random, time-varying amplitude and phase modulations on the received signal is characterized by Rayleigh fading in severe scenarios. The effect of Rayleigh fading can be from severe to very severe, depending on the rate of fading relative to the signal symbol rate. Slow fading (when the decorrelation time τ_0 is much greater than the signal symbol time T (e.g., when $\tau_0 = 100T$) will significantly degrade bit-error rate performances from an exponential dependence on E_b/N_0 to that of a linear relationship. Fast fading, when it is of the order of symbol time T, will further degrade bit-error rates. Irreducible bit rates occur in fast fading even at very high E_b/N_0's. Communication can be completely disrupted when the irreducible error rate approaches 0.5.

Nuclear-enhanced scintillation is discussed throughout this book, and is not extensively reviewed here. Nevertheless, it is useful to briefly review some of the major effects with a layman's eyes in order to appreciate, in a simple way, the effects upon communication signals. (See also Knepp, 1985, and Middlestead et al., 1987, for further information.)

The plasma created by a high-altitude nuclear burst introduces a turbulent medium with a characteristic that essentially behaves as a time-variant inhomogeneous dielectric having both bulk and random components. Consider the random components first. Plasma turbulence combined with $V \times B$ effects causes the plasma to become striated, with the major axis oriented along the earth's magnetic field lines. (V and B are the electromagnetic parameters that cause a charged particle in motion to spiral around the magnetic-field lines.) The major effect of the dielectric striations is to cause the propagating electromagnetic plane waves to split into a multitude of refracted waves and to take diverse paths, each of which experiences different path delays, and some of which ultimately converge on a receiving antenna. The combined effect of this convergence is to create a multipath environment in which received rays combine in a random phase relationship and behave as additive or subtractive interference.

The effect upon communications depends on the relative, or "scale," sizes of the dielectric striations and on the wavelength of the propagating signals. There is relatively little effect when the wavelength is much smaller than the striation scale factor. Under these circumstances, the interference ranges from nonexistent to relatively mild, and the signal has a random and a specular component. With moderate to severe scintillation, the striations are much smaller and the effect becomes more pronounced as the striation scale factor approaches the RF wavelength. Under these conditions, this effect leads to an amplitude-time distribution that is Rayleigh in nature.

The effect depends on the relative size of the striations, L, compared to the Fresnel zone size, λZ, where λ is the RF wavelength and Z is the distance between the scattering region and the receiver. When $\lambda Z < L$, then the wave is refracted and weakly scattered, and the amplitude statistics are non-Rayleigh. (The actual amplitude distribution is complicated and cannot be accurately described by a straightforward function such as a simple Rician distribution.) When $\lambda Z > L$, the wave is strongly scattered with Rayleigh amplitude fluctuations as a result. In the limit that $\lambda > L$, large-angle scattering will occur, and a diffusion process describes the propagation. In this case, signal energy will be scattered in the backward direction, returning to the transmitter. The signal propagating in either the forward or backward direction will have statistics similar to thermal noise.

The intensity of the amplitude variations vary with distance, time, and frequency. The parameters l_0, τ_0, and f_0 are commonly used to characterize the correlation distance, time, and bandwidth. To a first order each of these is analogous to:

l_0 = correlation distance; that is, if a receiver moves in position, the received signal strength will change (i.e., increase or decrease) by a factor of about e^{-1} in a single correlation distance.

τ_0 = correlation time; that is, if the received signal is in a very deep fade, it could be considered to change (i.e., increase or decrease) by a factor of about e^{-1} in a single correlation time interval.

f_0 = correlation bandwidth (i.e., the separation of two carrier frequencies for which a correlation of received signal strengths correspond to e^{-1}).

The physical phenomena causing these variations is primarily related to the multipath nature of the received signals.

Correlation distance is roughly the distance over which the characteristics of a received wavefront change by a factor of e^{-1}, which is analogous to walking behind a dielectric picket fence. In some locations, the vector sum of the received signals is relatively large, corresponding to constructive addition of the received rays. In others, destructive interference is taking place and causes deep fades to occur. When the received amplitudes are strong, the vector summation of the individual rays is near a maximum and is slow to change position on a vector phase plane. Under these conditions the relative phase of the vector sum is slowly varying. During fades, the vector summation of the many rays causes vector cancellation, with a very small resultant near the origin of the vector phase plane. Small changes in the phase angle of individual vectors can cause large increases in the magnitude of the resultant vector, and its phase can swing wildly on the phase plane. Thus, the phase of the received signal changes rapidly during fades and comparatively slowly when the signal is strong. The significance of this phenomenon is that the modulations that are most effective must be capable of dealing with the rapid phase variations during deep fades. Typically, then, short symbol intervals relative to fade durations are essential so that phase variations are small during a symbol interval—or noncoherent modulations such as M-ary frequency-shift keying (MFSK) be used.

In addition to decorrelation in time, the received signal can also suffer decorrelation in frequency, or frequency-selective, scintillation when the signal bandwidth exceeds the frequency-selective, or coherence, bandwidth of the channel. As a result, different frequency components in the signal bandwidth do not necessarily vary identically with time, and the result is statistical decorrelation at different frequencies within the signal bandwidth. This effect is characterized by severe intersymbol interferences in the received waveform. This degradation is of concern to all high-data-rate SDS links. Of particular concern is the 20-GHz downlink, since the channel coherence bandwidth will be narrower at lower frequencies.

Correlation time interval is related to correlation distance and depends upon the relative velocity of the two terminals and the nature of the scintillation striations. For two terminals in Keplerian orbit, each terminal is moving relatively rapidly with respect to the time variations of the striation scale factor. The effect upon the received signal is very much like the effect one perceives when walking behind a picket fence and receiving light transmitted through the pickets; the rate of variation becomes much more rapid, and the correlation time is effectively shortened, as follows:

$$\tau_0 = \tau_0' \left(\frac{v_p}{v_p + v_r} \right) \tag{11.4}$$

where

v_p = plasma velocity

v_r = communications platform relative velocity

11.4.3 Aperture Antenna Effects

Angle-of-Arrival Loss

When the correlation distance in the plane of an antenna aperture is short, the gain of the antenna is affected by deviations from a uniform phase front. Isotropic antennas sum all multipath signals arriving over a full range of angles. Aperture antennas normally receive a plane wave from a single source from a single direction. With severe multipath, an aperture antenna forms a weighted summation of the plane waves received over an angular spread about a mean. Two effects occur: first, the received rays no longer form a planar phase front; second, the normal antenna directionality tends to discriminate against rays arriving at larger angles from boresight. These two effects are such that the net gain of the antenna is reduced and the correlation bandwidth is increased.

With no scintillation, the correlation distance is large. As the path's total electron content increases and becomes more structured, correlation distance becomes smaller and the deviation from planar wavefronts becomes proportionally more significant. Dana (1986) has shown that antenna decorrelation loss for uniformly illuminated circular and square apertures is as follows:

$$\text{Loss} = 1 + 0.265(D/l_0)^2 \quad \text{circular aperture}$$
$$= 1 + 0.358\,(D/l_0)^2 \quad \text{square aperture} \tag{11.5}$$

When $D = l_0$, antenna scattering loss is about 1 dB for an antenna with a circular aperture. For larger D/l_0 ratios, increasing D increases loss as rapidly as it increases antenna gain until the two effects cancel.

When both the transmitting and receiving antennas are directional, Siegel (1987) has shown that a worst-case estimate of scattering loss may be determined from Eq. (11.5) when the D^2 term is replaced by D'.

$$D' = \left[\left(\frac{Z_t}{Z_r + Z_t}\right)D_r^2 + \left(\frac{Z_r}{Z_r + Z_t}\right)D_t^2\right] \tag{11.6}$$

where

Z_t = distance from the source transmitter toward the receiver to the beginning of nuclear-enhanced plasma

Z_r = distance from the receiving antenna to the far side of the nuclear enhanced plasma, or to the transmitting antenna, whichever is closer

In a link synthesis process, a designer will find that there is an upper bound to the useful size of the link antennas, insofar as increasing received signal strength is concerned. Increasing antenna diameters beyond this limit leads to increased angle-of-arrival loss and cancels out the increased gain expected by virtue of its larger diameter.

Correlation Bandwidth

Aperture antennas can also increase the correlation bandwidth of the received signals. Qualitatively, multipath can be viewed as a time-delay process with some signals taking the direct or shortest path. Other signals start out at an angle away from the direct path, but are scattered back toward the receiver, again arriving from an off-direct-path direction. If scattering effects are sufficiently strong, the signal traveling the off-nominal paths can be considerably delayed and can introduce envelope distortion and intersymbol interference. This interference is nominally the cause of reduced correlation bandwidths.

The antenna pattern of directional antennas create a spatial filtering effect that discriminates against the multipath rays arriving at large angles off-nominal. The net result is that the correlation bandwidth is not reduced as much as it would be with omnidirectional antennas. Alternatively, with directional antennas, the correlation bandwidth is wider than it would be with omnidirectional antennas.

Dana has shown that as long as the antenna apertures produce far-field patterns that are closely approximated by a Gaussian pattern, the relative bandwidth f_A/f_0 also obeys Eq. (11.5). (See Dana, 1986, sect. 3.2.3 and Figs. 3.5 and 3.6.) Thus, although it may not be helpful to employ larger antennas to increase received signal strength, they may be very useful for increasing correlation bandwidth. To illustrate, if $D = 5l_0$, antenna scattering loss of 7.8 dB almost cancels out the increased gain over a $D = l_0$ option, but the correlation bandwidth is 6 times larger.

Bulk Effects

In addition to the dynamic effects described above, there are bulk, or static, effects that are due mainly to the increase in the total electron content (TEC) in the line of sight propagation path. These include absorption loss, increase

TABLE 11.2 Frequency-Scaling Rules for Scintillation-Related Parameters

Rules	Parameters
Decorrelation time	$\tau_\perp \sim f$ $\tau_\parallel \sim f^3$
Frequency-selective bandwidth	$f_0 \sim f^4$
Perpendicular decorrelation distance	$l_0 \sim f$
Angle of arrival standard deviation	$\sigma_0 \sim f^{-2}$
Total electron content	$\dfrac{d_n}{dT^n}(\text{TEC}) = \text{constant}, n = 0, 1, 2, 3$
Signal phase shift and time derivatives	$\theta \sim f^{-1}$ for all
Faraday rotation angle	$\Omega \sim f^{-2}$
Absorption	$L_a \sim f^{-2} \, (\pm 10\%)$
Antenna noise temperature	Diminishes with increasing frequency; cannot be modeled by a power law

Source: Mission Research Corporation Report no. MRC-88-3087, September 30, 1988, p. 14.

in antenna temperature, time-delay and phase shifts due to TEC, and the associated Doppler shift and time-delay dynamics due to the rate of change of TEC in the propagation path. The magnitude of these effects depends upon the radio frequency and upon the intensity of the scintillation itself.

It has been estimated that the sky temperature decreases with time after a nuclear burst approximately according to the following equation. Presumably, absorption will change according to a similar equation

$$T = 10^4[1 - \exp(k/\tau_0^2)] \tag{11.7}$$

As an illustration, if the sky temperature is 4500 K for a τ_0 of 100 μs, it drops to a few degrees when the scintillation intensity diminishes so that τ_0 is 1 ms.

Nuclear scintillation effects are worse for lower RF values and become more tolerable at higher frequencies. Variations with frequency tend to follow the relations of Table 11.2. A related phenomenon is that the extent of the region over which nuclear scintillation is significant also diminishes for the higher RF values. Thus, for example, the region of very difficult communications for a single-burst case is of the order of 300 to 400 km in diameter when the RF carrier is 60 GHz, larger for the lower frequencies.

Nuclear Effects on Communications

A hypothetical nuclear-burst scenario has been postulated. Although it may be more severe than would be encountered in practice, it is useful for illustrating potential conditions. Figure 11.8 illustrates the electron density contours above CONUS for a hypothetical 600-burst scenario (Wittwer, 1987, 1988). In this scenario 600 bursts are detonated at 1-second intervals, occurring from left to right in 3 corridors and at altitudes corresponding to normal, lofted, and depressed trajectories. The peak density moves with the attack, and reaches a peak value of 10^{11} electrons/cm^3 within 7 minutes after the attack begins. High electron densities persist throughout the scenario. Densities as high as 10^{10}/cm^3 are present in large areas.

The effect of this electron concentration on a communication link will depend also on the link geometry, path length, and the relative velocities between the communicating terminals and the plasma. Tables 11.3 and 11.4 illustrate the propagation parameters associated with the above electron concentrations on some typical space links. They include absorption loss, increase in antenna temperature, signal decorrelation times ($\tau_{0_{min}}, \tau_{0_{ray}}$), frequency-selective bandwidth (f_0), signal decorrelation distance (l_0), and angular jitter (σ_θ). They are shown for both the single- and multiple-burst scenarios, and for several carrier frequencies. The recommended design levels correspond to the 10^{10} electrons/cm^3 level, which is the second contour in Fig. 11.8.

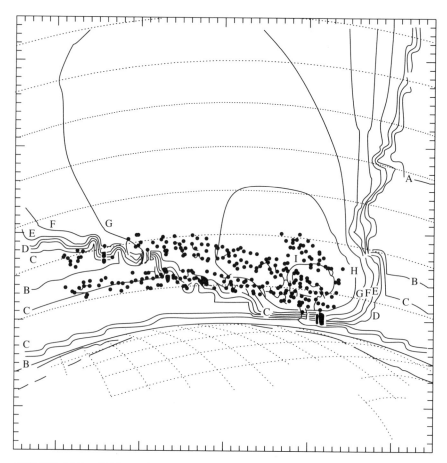

FIGURE 11.8   600-burst Scenario. (From L. A. Wittwer, "High Altitude Nuclear Effects," Defense Nuclear Agency, October 1988)

11.4.4 Waveform Design for Scintillation Effects

Since the main interest in this chapter is to describe the design considerations of the SDS communication system for scintillation effects, the emphasis in this and following subsections is mainly on waveform and demodulator processing design considerations for scintillation mitigation. The jamming protection aspects of the waveform design is not further discussed other than to mention that the SDS links will be antijam-protected by a combination of 60-GHz line hiding, frequency hopping, time permuting and chip rotation, antenna directivity, error-correction and error-detection coding, and interleaving. Coding and interleaving, which represent a form of time diversity, also serve to mitigate the effects of slow fading.

As discussed in earlier sections, there are three space communication networks in SDS: the BSTS, SSTS, and SABIR networks. In addition, the ground control centers are also connected by a CONUS fiber-optics network.

TABLE 11.3 Comparison of Single-Burst and Exotic Multiple-Burst Scenarios on Static Effects

	Single Large-Burst Scenario (reasonable worst case)			Exotic Multiple-Burst Scenario		
Frequency	1 GHz	8 GHz	44 GHz	8 GHz	30 GHz	60 GHz
				Near worst case:		
Absorption	30 dB	0.4 dB	0.01 dB	3900 dB	239 dB	55 dB
				Recommended design levels:		
				159 dB*	9.6 dB	2.2 dB
				Near worst case:		
Antenna temperature	10,000 K	857 K	24 K	10,000 K	10,000 K	10,000 K
				Recommended design levels:		
				10,000 K	9800 K	6800 K

Source: L. A. Wittwer, "Nuclear Environment Issues for Exotic Multi-Bursts," Defense Nuclear Agency, May 1987.
* Underlined values indicate design concerns.

TABLE 11.4 Comparison of Single-Burst and Exotic Multiple-Burst Scenarios on Dynamic Effects

	Single Large-Burst Scenario (reasonable worst case)			Exotic Multiple-Burst Scenario (near worst case and recommended design levels)		
Frequency	1 GHz	8 GHz	44 GHz	8 GHz	30 GHz	60 GHz
$\tau_{0_{min}}$	0.27 μs	2.1 ms	12 ms	1.9 μs*	7.1 μs	14 μs
				(9.5 μs)†	(36 μs)	(71 μs)
$\tau_{0_{ray}}$	2.7 s	0.43 s	48 ms	27 ms	8.7 ms	4.7 ms
				(27 ms)	(8.7 ms)	(4.7 ms)
f_0	500 Hz	2 MHz	1.9 GHz	0.67 kHz	133 kHz	2.1 MHz
				(17 kHz)	(3.3 MHz)	(53 MHz)
σ_θ	17°	0.27°	0.01°	4.2°	0.3°	0.07°
				(0.8°)	(0.06°)	(0.02°)
l_0	0.22 m	1.8 m	9.8 m	0.12m	0.43 m	0.87 m
				(0.58 m)	(2.2 m)	(4.3 m)

Source: L. A. Wittwer, "Nuclear Environment Issues for Exotic Multi-Bursts," Defense Nuclear Agency, May 1987.
* Near-worst-case values are given first.
† Recommended design levels are given next, in parentheses.
‡ Underlined values indicate design concerns.

The major traffic in the BSTS and SSTS networks is sensor data. The data rates are approximately 1 Mbps on BSTS, and 20 Mbps on SSTS. These data rates are also required on the links from these networks to the selected CONUS GEP. The cross links from BSTS to SSTS will be about 1 Mbps. The traffic on the SABIR network consists mainly of target state vectors, weapon status, in-flight guidance updates and in-flight homing views to interceptors, and command and control messages. The data rate on the SABIR network is much lower than that on the sensor networks. Current estimates place this data rate in the range of a few tens of kbps. The data rates on the SDS networks fall into two relatively widely separated ranges: on the order of tens of kbps and on the order of a few Mbps. In the following subsections the waveforms for these two data rate ranges will be separately discussed. A substantial amount of the conceptual developments in this section follow those described by Olsen (1989).

In these discussions the frequency-selective bandwidths will be assumed to be wide relative to the channel symbol rate, and therefore the Rayleigh fading channel is flat, a good assumption for the 60-GHz cross links (see Table 11.4). For the 20-GHz downlinks the assumption may not hold, depending on the exact link geometry and the nuclear scenario. Ground antenna directivity will broaden the correlation bandwidth, and that may be adequate—or can be made adequate by choosing oversize antennas. If this approach is not used, ground-based adaptive equalization may be suitable for the higher-data-rate space-to-ground-links applications.

11.4.5 Medium- to High-Data-Rate Links

Modulation Trade-Offs

The medium- to high-rate links support data rates in the range of 1 to 20 Mbps. With rate-(1/3) coding, for example, the symbol times are 0.33 μs on the BSTS-to-SSTS crosslink, and 16.7 ns on the SSTS-to-SSTS crosslinks. These symbol times are short compared to the minimum signal decorrelation times in the recommended design values for $\tau_{0_{min}}$ shown in Table 11.3. For the more severe multiple-burst scenario discussed earlier, the recommended design values for $\tau_{0_{min}}$ are 23 μs at 20 GHz and 71 μs at 60 GHz. Thus the BSTS-to-SSTS crosslinks experience a minimum decorrelation-to-symbol-time ratio $(\tau_{0_{min}}/T)$ exceeding 200. This ratio exceeds 4000 for the SSTS-to-SSTS crosslinks. The 20-GHz downlinks from SSTS and BSTS have smaller $\tau_{0_{min}}/T$ ratios. However, they still exceed 70 and 1000. If the downlink frequency is 40 GHz instead of 20 GHz, the corresponding ratios exceed 140 and 2000, respectively.

Figure 11.9 illustrates the bit error probability for uncoded DPSK for fast fading with various τ_0/T ratios. For small τ_0/T ratios (for example, < 10), the bit-error rate performance is significantly degraded relative to the slow fading performance, with irreducible bit-error probabilities that cannot be improved

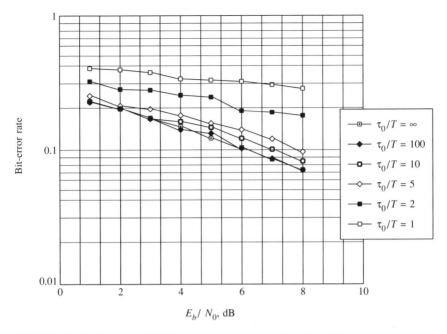

FIGURE 11.9   Uncoded DPSK performance in flat Rayleigh fading.

by increasing signal power. When τ_0/T approaches 100, on the other hand, the bit-error performance is almost identical to the slow fading limit. Considering the high τ_0/T ratios of the BSTS-to-SSTS and the SSTS-to-SSTS data links discussed above, only slow Rayleigh fading needed to be considered for these SDS links.

Figure 11.10 illustrates the transmitter and receiver functions of the medium- to high-data-rate SDS sensor data links. On the transmitter side, the data field in each packet is first provided with a cyclic redundancy check (CRC) tail and then encrypted. These packets are then input to the channel error-correction encoder. The error-correction encoder output is interleaved and assigned to hops (and also to subhops in a hop) in a data frame. Both interleaving and hop assignments are derived from a time-of-day (TOD) clock, which also controls the frequency-hopping Transec code generation and other functions such as hop time permutation and chip rotation. Phase reference bits are inserted into each subhop, on which chip rotations are performed. Sync hops are also added to each frame. The hops in a frame will be time-permuted. These baseband data will then be phase-modulated onto the hopping carrier.

The receiver's functions are the inverse of the transmitter's. First the received signal is dehopped. The dehopped signal is then converted to quadrature baseband signals and demodulated, with the aid of the phase reference bits, on each subhop. The soft-decision outputs are then chip derotated.

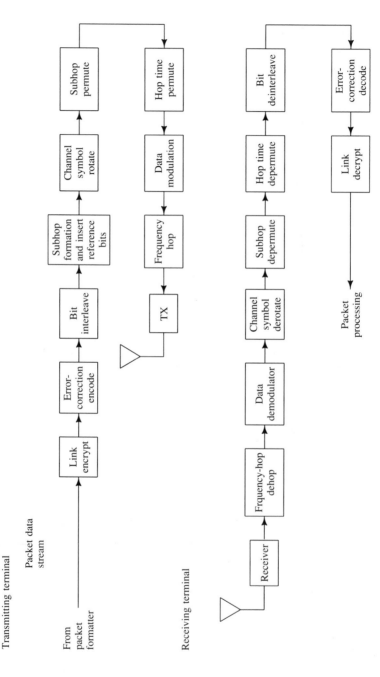

FIGURE 11.10 Terminal processing for medium- to high-data-rate links (DPSK or MSK).

The hops are time-depermuted. The rearranged soft-decision outputs in the data hops of each frame are then input to the deinterleaver. The deinterleaver output is soft-decoded by the decoder and link-decrypted. Packets are recovered from the decoded data, and error-detected, with erroneous packets discarded. Proper synchronization is required in each step of the data recovery process, and is maintained by a process that simultaneously addresses hop sync, bit sync, and TOD sync through detection of sync hop patterns transmitted during each frame. A typical data frame arrangement is illustrated in Fig. 11.11, which shows a possible format being considered for the SSTS-to-SSTS crosslinks.

Table 11.5 lists a number of binary, 4-ary, and M-ary coherent and noncoherent modulation techniques that are commonly used on satellite communication links. They are all constant-envelope modulations and are thus suitable for nonlinear traveling-wave-tube amplifiers (TWTAs) in the satellite applications.

The slow Rayleigh fading channel is characterized by fading variations that are very slow relative to the channel symbol rate, so that during each symbol time the amplitude and phase variations due to fading are basically constant. Since there are no specular components in the received signal, the envelope of the received signal has the Rayleigh probability density

$$p(A) = \frac{A}{\sigma_A^2} \exp\left(-\frac{A^2}{2\sigma_A^2}\right) \qquad A > 0 \tag{11.8}$$

where A is the envelope of the received signal, with first and second moments $\sigma_A\sqrt{\pi/2}$ and $2\sigma_A^2$, respectively.

The E_b/N_0 of the received signal is proportional to the square of the envelope and will thus have the exponential probability density

$$p(r) = \frac{1}{r_0} \exp\left(-\frac{r}{r_0}\right) \qquad r > 0 \tag{11.9}$$

where r stands for E_b/N_0 and r_0 is the mean E_b/N_0 received in the Rayleigh fading channel.

The bit-error probability of DPSK, for example, is given by $\frac{1}{2}\exp(-r)$ in the additive white Gaussian noise (AWGN) channel. With slow Rayleigh fading, the value of r is a random variable with a probability density given by Eq. (11.8). Its value is assumed constant for each bit duration. Thus the bit-error probability of DPSK in the slow Rayleigh fading channel is given by averaging the conditional probability of bit error r for a given E_b/N_0 over the density of r:

$$P_b = \int_0^\infty \frac{1}{2}\exp(-r)\frac{1}{r_0}\exp\left(-\frac{r}{r_0}\right) dr$$

$$= \frac{1}{2 + 2r_0}$$

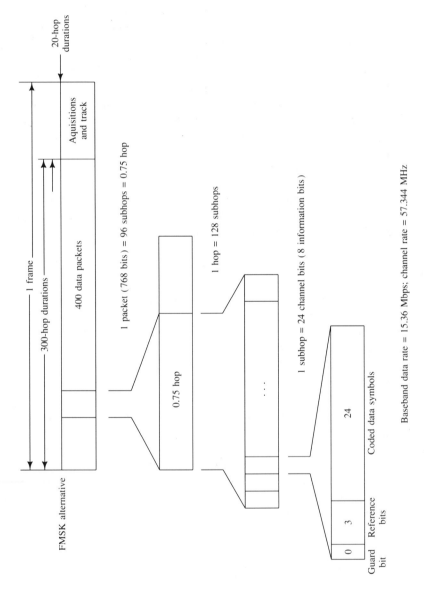

FIGURE 11.11 SSTS-to-SSTS link and SSTS-to-GEP link candidate numbering scheme.

Baseband data rate = 15.36 Mbps; channel rate = 57.344 MHz

TABLE 11.5 Slow Rayleigh Fading Performance of Various Modulation Techniques*

Modulation	Probability of Bit Error in AWGN	Probability of Bit Error in Slow Rayleigh Fading
DPSK	$\frac{1}{2}e^{-r}$	$\dfrac{1}{2(1+r_0)}$
BPSK, MSK	$\frac{1}{2}\operatorname{erfc}\sqrt{r}$	$\dfrac{1}{2}\left[1-\sqrt{\dfrac{r_0}{1+r_0}}\right]$
Coherent FSK	$\frac{1}{2}\operatorname{erfc}\sqrt{r/2}$	$\dfrac{1}{2}\left[1-\sqrt{\dfrac{r_0/2}{1+r_0/2}}\right]$
Noncoherent FSK	$\frac{1}{2}e^{-r/2}$	$\dfrac{1}{2(1+r_0/2)}$
QPSK	$\approx\frac{1}{2}\operatorname{erfc}\sqrt{r/2}$	$\approx\dfrac{1}{2}\left[1-\sqrt{\dfrac{r_0/2}{1+r_0/2}}\right]$
DQPSK	$\leq\frac{1}{2}\operatorname{erfc}\left(\sqrt{2r}\sin\dfrac{\pi}{8}\right)$	$\leq\dfrac{1}{2}\left[1-\sqrt{\dfrac{2r_0\sin^2\pi/8}{1+2r_0\sin^2\pi/8}}\right]$
MPSK	$\approx\dfrac{1}{\log_2 M}\operatorname{erfc}\left(\sqrt{r}\sin\dfrac{\pi}{M}\right)$	$\approx\dfrac{1}{\log_2 M}\left[1-\sqrt{\dfrac{\sin^2(\pi/M)r_0}{1+\sin^2(\pi/M)r_0}}\right]$
DMPSK	$\approx\dfrac{1}{\log_2 M}\operatorname{erfc}\left(\sqrt{2r}\sin\dfrac{\pi}{2M}\right)$	$\approx\dfrac{1}{\log_2 M}\left[1-\sqrt{\dfrac{2\sin^2(\pi/2M)r_0}{1+2\sin^2(\pi/2M)r_0}}\right]$
Noncoherent MFSK	$\left(\dfrac{M/2}{M-1}\right)\sum_{k=1}^{M-1}(-1)^{k+1}\binom{M-1}{k}\dfrac{1}{k+1}\exp[-kr/(k+1)]$	$\left(\dfrac{M/2}{M-1}\right)\sum_{k=1}^{M-1}(-1)^{k+1}\binom{M-1}{k}\dfrac{1}{kr_0+k+1}$

* $r = E_s/N_0$, $r_0 = $ mean E_s/N_0, $E_b/N_0 = r/\log_2 M$.

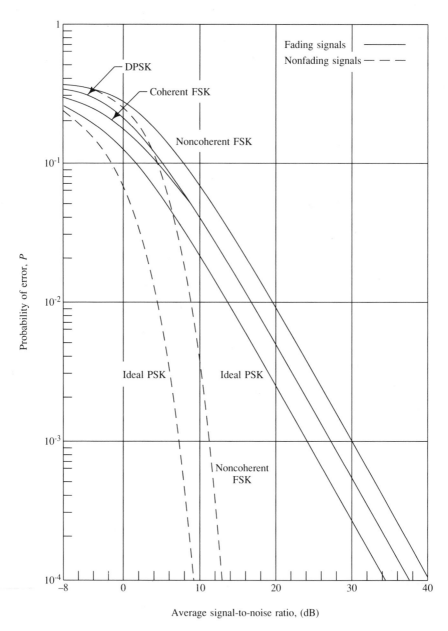

FIGURE 11.12   Probability of error for several binary systems in Rayleigh fading. (From S. Stein and J. Jones, Modern Communications Principles © 1967. Reprinted by permission of McGraw-Hill Publishing Co., New York.)

Table 11.5 gives the probabilities of error in AWGN and in slow Rayleigh fading for DPSK, BPSK, MSK, coherent BFSK, non-coherent BFSK, non-coherent MFSK, QPSK, MPSK, DQPSK, and DMPSK.

Figure 11.12 plots some of these bit-error probabilities for different E_b/N_0 values in AWGN, and for different mean received E_b/N_0 values in Rayleigh fading. The performances of all modulations in fading are significantly degraded from the equivalent AWGN case. From Fig. 11.12 it is seen that for a slow Rayleigh fading channel, among the binary schemes considered, BPSK (or MSK) is about 3 dB better than DPSK, and DPSK is 3 dB better than noncoherent BFSK. Thus PSK is at least 3 dB better than FSK, and it is possible to improve PSK bit-error performance by another 3 dB with coherent signal detection rather than differential detection. In the AWGN channel, however, BPSK is only about 0.7 dB better than DPSK at a bit-error rate of 10^{-5}.

The probability of bit errors in slow fading basically decreases linearly with E_b/N_0, rather than exponentially with E_b/N_0 as in the AWGN case. Since errors occur in pairs in differential detection, the bit-error rate for DPSK will basically double the BPSK rate. This explains the 3-dB gain in performance of BPSK (or MSK) over DPSK in slow Rayleigh fading.

The above performances assume ideal phase coherence in the detections of BPSK or MSK. In reality there will be phase error in the process of phase recovery either with an open-loop estimation technique (with the "two-pass demodulator," to be discussed in later paragraphs) or with a carrier recovery loop such as the Costas loop. Actual performance improvement will depend on the phase errors involved, which are influenced by noise, Doppler, and amplitude and phase scintillation effects on the received signal.

Figures 11.13 and 11.14 compare the power spectral densities of BPSK (or DPSK), QPSK (or DQPSK), and MSK, and their relative out-of-band signal powers. MSK has a smaller out-of-band spectral power density than BPSK, DPSK, or QPSK (see Fig. 11.14). This smaller SPD has advantages in a frequency-division multiplex application where crosstalk between different FDM channels must be minimized, particularly when the dynamic ranges of the received FDM signals are large, as a result of either terminal EIRP variations or range differences. There is no need for FDM in the SDS links, however. Thus, strictly speaking, either DPSK, BPSK, or MSK can be considered as candidate modulation techniques for the SDS sensor data links.

Another consideration is the frequency-selective bandwidth. Because of its narrower spectral occupancy, MSK will perform slightly better than BPSK in channels with a frequency-selective bandwidth, which is narrow compared to the channel data rate.

Figures 11.15 and 11.16 illustrate, respectively, the generation of serial MSK and its demodulation with a "two-pass" demodulator. In the two-pass demodulator the received carrier phase is estimated based on the observed quadrature outputs, with the aid of a number of reference bits (with known phase modulation) to resolve the polarity ambiguity, in the first pass over a

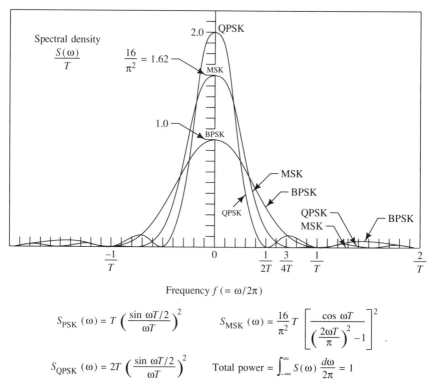

$$S_{PSK}(\omega) = T \left(\frac{\sin \omega T/2}{\omega T} \right)^2 \qquad S_{MSK}(\omega) = \frac{16}{\pi^2} T \left[\frac{\cos \omega T}{\left(\frac{2\omega T}{\pi} \right)^2 - 1} \right]^2$$

$$S_{QPSK}(\omega) = 2T \left(\frac{\sin \omega T/2}{\omega T} \right)^2 \qquad \text{Total power} = \int_{-\infty}^{\infty} S(\omega) \frac{d\omega}{2\pi} = 1$$

FIGURE 11.13   Spectral density functions for PSK, DPSK, and MSK.

subhop duration. The subhop duration must be short relative to the signal's decorrelation time and the reciprocal of the residual Doppler error in the channel. The recovered phase is then used to rotate the quadrature signals, in the second pass, to obtain phase coherence. The recovered data are detected in the in-phase channel.

Coding and Interleaving

Currently MSK with two-pass demodulation is a favored modulation candidate for the SDS sensor data links. However, regardless of MSK or DPSK, the uncoded performance of these modulations in scintillation is far from their respective performances in an AWGN channel. As illustrated in Fig. 11.12, the PSK (or MSK) performance in slow Rayleigh fading is 26 dB worse than in AWGN, at a bit error probability of 10^{-4}. The corresponding difference in DPSK is approximately 28 dB. Error-correction coding with sufficient bit interleaving must be applied to improve this performance.

Most error-correction codes (e.g., convolutional codes) are effective only if the errors occur randomly in time. However, burst errors result in the fading channel. Thus the symbol stream must be sufficiently interleaved over the fade

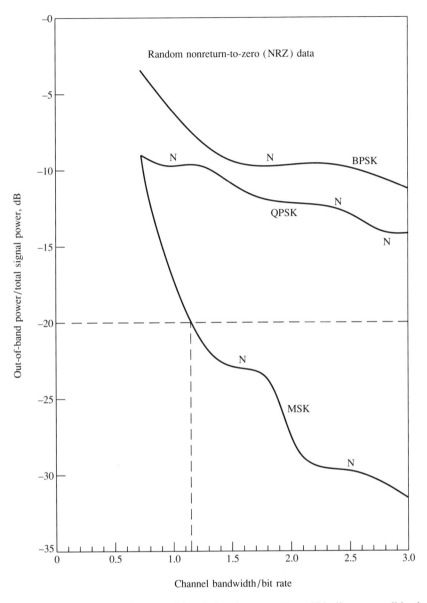

FIGURE 11.14 Relative out-of-band signal power. *Note*: N indicates a null in the power spectrum at that frequency.

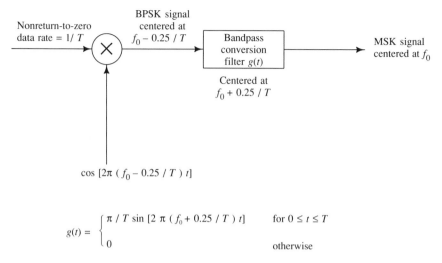

$$g(t) = \begin{cases} \pi / T \sin [2 \pi (f_0 + 0.25 / T) t] & \text{for } 0 \le t \le T \\ 0 & \text{otherwise} \end{cases}$$

FIGURE 11.15   Generation of serial MSK signal.

duration (which is proportional to the decorrelation time) to make the errors appear as random errors so that the decoder can correct the burst errors.

On the other hand, a very large interleaver memory can be required for the high-data-rate links. For example, at the SSTS-SSTS symbol rate of 60 Mbps, assuming rate-(1/3) coding, interleaving over a span of $10\tau_0$, with a τ_0 of 100 ms (i.e., with an interleaver span of 1 second), a convolutional interleaver/deinterleaver requires 60 Mbits of memory if hard-decision decoding is used. With 3-bit soft-decision decoding, the receiver's deinterleaver requires three times as much memory as the transmitter's. The total memory requirement for the transmitter and receiver for one terminal becomes 120 Mbits. Using 256K random access memories in hybrid packages the memory weight is approximately 1 lb per 10 Mbits. (This includes sparing and support circuits, printed circuit boards, and housing). Thus the weight for this interleaver-deinterleaver arrangement can be as much as 12 lb for a single SSTS-to-SSTS link, or 36 lb for three SSTS-to-SSTS links.

Two effects must be considered when selecting the interleaver spans. First, total network delays for multiple hops can become extensive. If a 1-s interleaver span is actually implemented for each hop, total network time delays can be several seconds long. Delays of this length are generally considered to be unacceptable, and less than optimum designs must be considered. Second, for long interleaver spans the rapid growth of interleaver weight becomes a significant factor. The amount of interleaving relative to signal decorrelation time must be carefully optimized with respect to performance. It is possible to trade E_b/N_0 for interleaver span, for example, by providing more EIRP or G/T, in order to reduce interleaver weight.

Either convolutional, block, or concatenated coding can be applied with interleaving to improve link performance in the slow fading channel. Con-

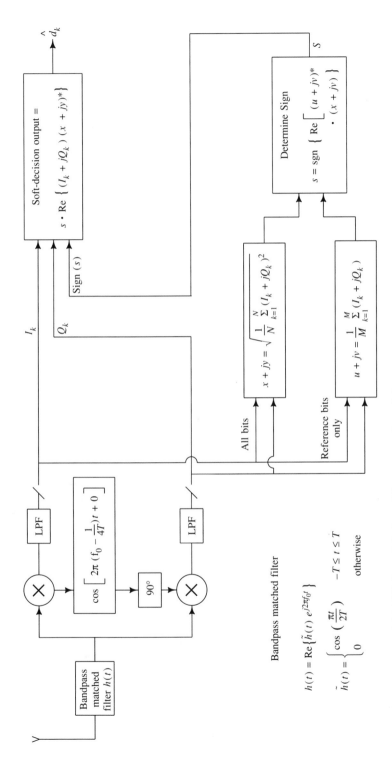

FIGURE 11.16 Serial MSK demodulation with two-pass demodulator.

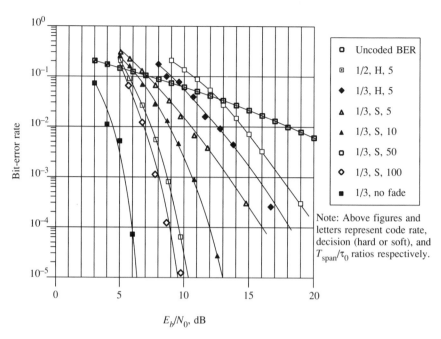

FIGURE 11.17   BER performance of DPSK in Rayleigh fading with convolutional coding* and interleaving. (*3-bit soft decisions.)

volutional coding is preferred over block codes at the present time because soft decision decoding is much easier to implement for convolutional codes than for block codes. Concatenated coding with a convolutional inner code and a block (such as Reed-Solomon) outer code can give attractive performance, at the expense of increased complexity. At the present time, convolutional codes are considered promising for a candidate baseline for SDS sensor data links.

Figure 11.17 illustrates the bit-error-rate performances of convolutional coded DPSK in slow Rayleigh fading, with different coding options ($R = 1/2$, $K = 7$, hard or 3-bit soft decision decoding, and $R = 1/3$, $K = 7$, with hard or 3-bit soft decision decoding). The required E_b/N_0 for different coding options is summarized in Table 11.6. Rate-(1/3) soft decision convolutional code is the preferred choice in the Rayleigh fading channel for it is about 4.4 dB better than rate-(1/2) hard decision. In the current SDS link designs, rate-(1/3) soft decision convolutional code is assumed as the baseline.

In a packet communication system such as the SDS, packet-error rate (PER) probability is a more important measure of performance than bit-error-rate probability. Because of bit-error correlations in the Viterbi decoder outputs, bit errors tend to be grouped in small bursts. The exact relationship

TABLE 11.6 Required E_b/N_0 for a Bit-Error Probability of 10^{-4} in Slow Rayleigh Fading*

Code Rate	Hard or Soft	Interleaver Span T_{span} to τ_0 Ratio	E_b/N_0 (dB)
Uncoded	—	—	34
1/2	Hard	5	20
1/3	Hard	5	18
1/3	Soft	5	15.6
1/3	Soft	10	12
1/3	Soft	50	9.5
1/3	Soft	100	9.0
1/3	Soft	AWGN, no fading	6.0

* DPSK modulation; convolutional coding ($K = 7$), 3-bit soft decision.

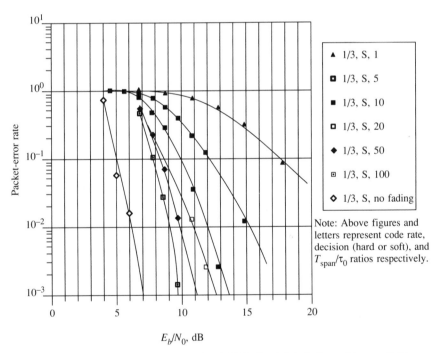

FIGURE 11.18   PER performance of DPSK in Rayleigh fading with rate-(1/3) convolutional coding and packet length 704.

593

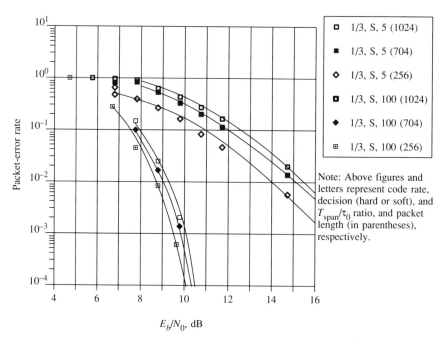

FIGURE 11.19   PER performance of DPSK in Rayleigh fading with convolutional coding, and different packet lengths and interleaver spans.

between packet-error probability and bit-error probability is best determined by computer simulation. Figure 11.18 shows the packet-error probabilities for a 704 bit packet for the various coding options discussed above. Figure 11.19 illustrates the dependence on packet size, for two interleaver spans, at $5\tau_0$ and $100\tau_0$, respectively. The required E_b/N_0 for a packet-error probability of 0.1, with DPSK, rate-(1/3), $K = 7$, 3-bit soft convolutional coding, is summarized in Table 11.7, for three packet sizes.

It is of interest to note that the gain achievable by increasing the interleaver span relative to decorrelation time ratio diminishes after the value exceeds 50. There is an irreducible loss, approximately 3 dB from the AWGN channel, due to Rayleigh fading, that cannot be recovered by increasing the interleaver span alone.

The characteristics of four types of interleavers are summarized in Table 11.8. These are convolutional, block, block-interspersed block, and hybrid (convolutionally interspersed block). The pertinent characteristics of interleavers are memory and delay. The memory requirements shown in Table 11.8 correspond to the case of hard-decision decoding. For soft-decision decoding, the memory required can be adjusted by multiplying the memory requirement shown in Table 11.8 by the factor $1 + 0.5(n - 1)$, where n is the number of bits used in soft-decision decoding. Convolutional or hybrid interleavers are preferred, since their memory requirements and delays are only 1/4 or 1/2 of those required by block or block-interspersed-block interleavers.

Hybrid interleavers (Olsen, 1989) can be considered to be a collection of N_3 convolutional interleavers with dimensions N_1 and N_2. N_1 input bits are written into each of the N_3 convolutional interleavers vertically, one at a time starting from the first one. After the N_3th convolutional interleaver has been written into, the entire cycle is repeated. $N_1 \times N_3$ bits are read into the hybrid interleaver during each of these cycles. Figure 11.20 illustrates the concept of the hybrid interleaver.

Interleaver output bits are read out horizontally from the bank of N_3 convolutional interleavers, starting from the rightmost bit of the first row of

TABLE 11.7 Required E_b/N_0 for a Packet-Error Probability of 0.1 in Slow Rayleigh Fading*

Interleaver span T_{span} to τ_0 Ratio	Packet Size (bits)		
	256	704	1024
	E_b/N_0 (dB)		
5	11	12	12.8
10	9.2	10	10.3
50	8	8.5	8.8
100	7.8	8	8.3
AWGN, no fading	4.8	5	5.2

* DPSK modulation; $R = 1/3$, $K = 7$, 3-bit soft, convolutional.

TABLE 11.8 Interleaver-Type Trade-offs*

Type	Approximate delay requirement	Approximate memory requirement
Convolutional	Span	Span/chip rate
Block	2 × span	4 × span/chip rate
Block-interspersed block	2 × span	4 × span/chip rate
Hybrid (convolutionally interspersed block)	Span	Span/chip rate

Source: Olsen, 1989.
* Either a convolutional or hybrid type is recommended so as to minimize the delay and memory requirements.

A cycle consists of the following:

- Input commutator loads columns of left face ($N_1 \times N_3$ array)
- Output commutator reads rows of right (stepped) face
- Left-face array is shifted right one column

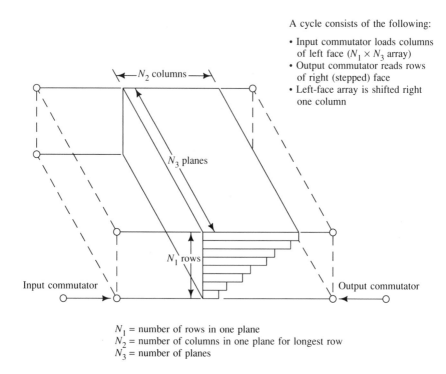

N_1 = number of rows in one plane
N_2 = number of columns in one plane for longest row
N_3 = number of planes

FIGURE 11.20   The hybrid interleaver. (From D. P. Olsen, "Proposed Generic Communication Waveform Standards," The Aerospace Corporation, March 1989).

the first convolutional interleaver, until the rightmost bit of the N_1th row of the N_3th convolutional interleaver. The cycle then repeats itself. During each cycle, $N_1 \times N_3$ bits will be read out.

N_1 is usually chosen to be 32, which relates to the decoder path memory. If N_3 is chosen such that $N_1 \times N_3$ is the length of one packet (or an integer fraction of a packet), then one entire packet will be written into the hybrid interleaver during each of its cycles. The output of the hybrid interleaver is read out in N_3-bit sub-blocks, each sub-block belonging to the same packet, which has been block-interleaved by an $N_1 \times N_3$ block interleaver. These sub-blocks are also convolutionally interspersed among sub-blocks of N_2 different block-interleaved packets.

Thus the hybrid interleaver is very suitable for TDMA applications in which packets intended for different destinations are multiplexed together in a single stream to be transmitted on different beam positions. This technique provides the necessary time interspersing essential to increasing the effective interleaver span of each of the multiplexed packets, without actually inter-leaving bits belonging to different packets. Since N_3 bits belonging to the same packet (also to the same destination) are read out as a sub-block, N_3 is also the size of the sub-block on which basis beam hopping is performed.

Either a convolutional or hybrid interleaver can be used for the SDS sensor data links since these links are node-to-node links, with no particular TDMA requirements. For the links from either SABIR-CV or SSTS to the interceptors, however, the hybrid interleaver is definitely beneficial since TDMA time interspersing must be applied to the data links to different interceptors. This choice is intended to increase the effective interleaving span of each packet to individual interceptors.

From observing the required E_b/N_0 ratios for a packet-error probability of 0.1 for a packet size of 704 bits it becomes clear that with sufficient interleaving and proper coding it is feasible to mitigate most of the degradation caused by flat fading on the SDS medium to high data sensor links. In the AWGN channel, bit errors are random. A bit-error rate of 1.5×10^{-4} is thus required to obtain a packet error rate of 0.1 for the packet size of 704 bits. For uncoded DPSK an E_b/N_0 of 9.1 dB is required. The corresponding E_b/N_0 required for the same packet-error rate in slow Rayleigh fading, with rate-(1/3), $K = 7$, soft-decision convolutional coding, and with $10\tau_0$ interleaving, is only 10 dB. This value decreases to 8 dB with $100\tau_0$ interleaving, which is only 3 dB from the performance of the same modulation and coding arrangement in AWGN.

In the above analyses, only flat Rayleigh fading has been considered. Rayleigh fading corresponds to multipath environment for which the received signal is completely random; there is no specular component at all. As scintillation intensity diminishes, a specular component will appear and the probability distribution makes the transition from a Rayleigh distribution to a more complicated one. A Rician distribution is one example of random fading with a specular component, but it does not accurately describe the fading distribution in a non-Rayleigh environment. For convenience, we use it here to illustrate relative performance behavior.

Tradition is such that $\tau_{0_{ray}}$ is considered to be the nominal longest value of τ_0 for which Rayleigh fading must be assumed. In a more formal sense, $\tau_{0_{ray}}$ corresponds to a Rician distribution for which the ratio β of specular to random components is approximately 10 dB. The parameter S_4, also commonly used, is related to β as shown in Fig. 11.21. (Bogusch, 1974, p. 39). S_4 equals 1.0 for a Rayleigh distribution, and 0.5 when β is 10 dB.

The E_b/N_0 ratio (r) in an example of Rician fading with a specular-to-random ratio of β would have the following probability density:

$$p(r) = \frac{1 + \beta}{r_0} \exp\left[-\left(\frac{r(1 + \beta)}{r_0} + \beta\right)\right] I\left[2\sqrt{\frac{r\beta(1 + \beta)}{r_0}}\right] \qquad (11.10)$$

Averaging the conditional probability of error of DPSK over the above probability density of r leads to

$$P_b = \frac{1}{2}\left(\frac{1 + \beta}{1 + \beta + r_0}\right)\exp\left(-\frac{\beta r_0}{1 + \beta + r_0}\right) \qquad (11.11)$$

Figure 11.22 illustrates this probability of bit error of DPSK in slow Rician fading.

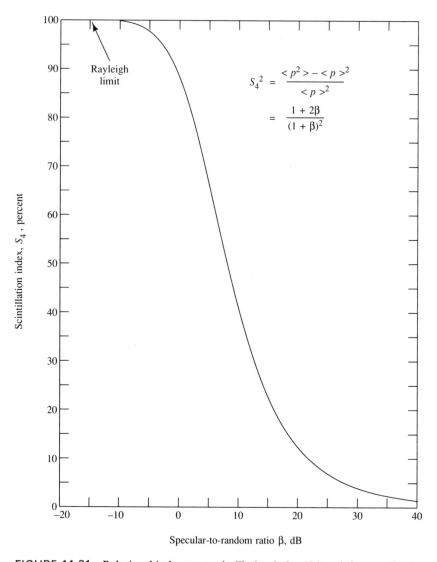

FIGURE 11.21   Relationship between scintillation index (S_4) and the specular-to-random ratio β. (From R. L. Bogusch, "Ionospheric Scintillation Effects on Satellite Communications," Report MRC-N-139, Mission Research Corp., June 1974, p. 39.)

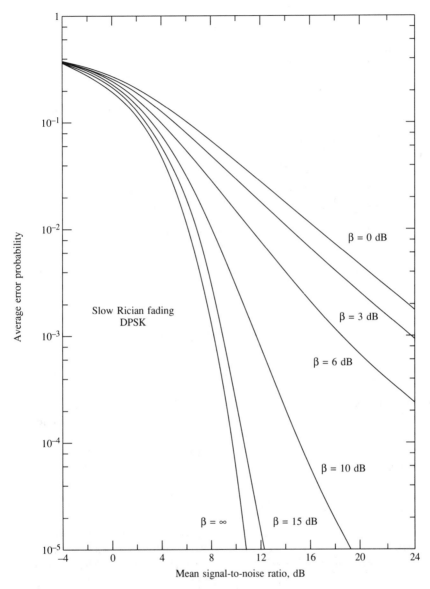

FIGURE 11.22   DPSK error probability, slow fading limit. (From R. L. Bogusch, "Ionospheric Scintillation Effects on Satellite Communications," Report MRC-N-139, Mission Research Corp., June 1974, p. 76.)

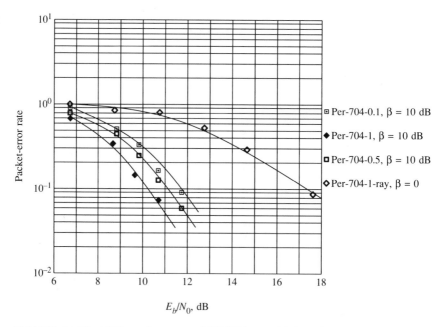

FIGURE 11.23   PER performance of DPSK in Rayleigh and Rician fading with R-$1/3$ soft-decision coding and interleaving.

With a β of $+10$ dB the packet-error-rate performance is also significantly improved. Figure 11.23 shows the error probability of a 704-bit packet, with $R = 1/3$, $K = 7$ convolutional coding, and with $T_{span}/\tau_0 = 0.1$, 0.5, and 1.0, respectively, for $\beta = 10$ dB and for $T_{span}/\tau_0 = 1$ for $\beta = 0$. It is clear that for the same interleaver span, PER performance is much better when a specular component is present.

Low-Data-Rate Links

The SABIR network has a much lower data-rate requirement than the sensor networks. Current estimates for the links from carrier vehicles to interceptors are a few kbps, and in the low tens of kbps for the links from SSTS to CVs or IVs.

11.4.6 Modulation Trade-offs in Fast Scintillation

The transmission symbol durations for the medium- to high-data-rate cases described earlier were short relative to the duration of the scintillation fades, and fading was referred to as "slow fading." For the very low data rates, the fade durations can range from short to long relative to the symbol durations, and is therefore considered to range from "rapid" to "slow." For example, the symbol time for a candidate 8-ary FSK, rate-(1/3) encoded waveform is

333.3 μs for a 3-kbps link, and 50 μs for a 20-kbps link. The corresponding symbol times for rate-(1/3) encoded DPSK are 111.1 μs and 17 μs, respectively. These symbol times are either longer, or have about the same duration as, the minimum decorrelation times of the multiple-burst scenario, even at 60 GHz (see Table 11.4). The τ_0/T ratios for these examples are summarized in Table 11.9.

It follows that both fast and slow scintillation must be considered in selecting the modulation and coding parameters in the low-data-rate waveform.

In a fast fading environment it is known that noncoherent FSK performs better than DPSK or BPSK (Bogusch, 1974). Noncoherent detection of FSK depends on the envelope detector outputs of the tone detectors matched to the modulation frequencies and is not as sensitive to random phase modulations as DPSK or BPSK. In fact, since DPSK requires the carrier phase to be relatively constant only over two symbol times, it is also more robust than BPSK in such an environment. The carrier recovery loop required in BPSK detection usually has a time constant that is 1000 to 2000 times longer than the bit time. Its tracking performance is significantly affected by fast scintillation.

Figure 11.24 compares the required E_b/N_0 (dB) for a channel bit-error rate of 0.03, for 8-ary FSK, BFSK, 4-ary FSK, 16-ary FSK, DPSK, and BPSK modulations, as functions of the ratio τ_0/T (Bogusch, 1976). It shows that BPSK will not function for a τ_0/T ratio less than 14, even with a relatively wide carrier loop bandwidth of $1/30 \times$ symbol rate. DPSK stops functioning when τ_0/T is less than 5.6. With orthogonal tone spacing, BFSK and 8-ary FSK continue to function until the τ_0/T ratio is below 1.3 and 3.5, respectively. With tone spacing at 10 times the symbol rate, the M-ary FSK performance in fast fading can be further improved. For example, with 10 times orthogonal spacing, 8-ary FSK continues to function until $\tau_0/T = 0.22$. Comparing the range of $\tau_{0_{min}}/T$ in Table 11.9 with the above results, it is clear that noncoherent FSK should be used in the SDS low-data-rate links. In the current SDS low-data-rate links, 8-ary FSK is considered to be a promising candidate.

M-ary FSK performance is degraded in fast fading by two related effects due to fast phase variations of the carrier. Fast phase fading is similar to carrier phase noise, which creates a spectral broadening effect on the received signal.

TABLE 11.9 τ_0/τ Ratios for Low-Data-Rate Links and Short τ_0

Data Rate (kbps)	Coding Rate	Modulation	τ_0	
			14 μs	71 μs
3	1/3	8-FSK	0.042	0.213
3	1/3	DPSK	0.126	0.639
20	1/3	8-FSK	0.28	1.42
20	1/3	DPSK	0.84	4.26

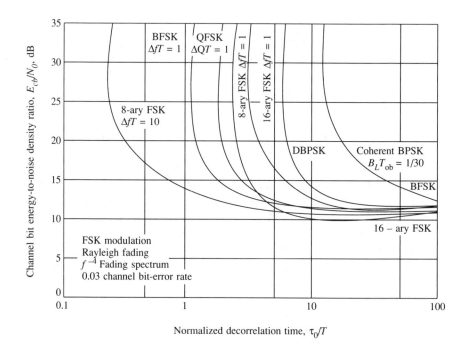

FIGURE 11.24 Comparison of FSK modems in fast nonselective Rayleigh fading. (From R. L. Bogusch, et al., "Effects of Propagation Disturbances on a Specific Satellite Communication Link, Vol. I," Report AFWL-TR-25, Mission Research Corporation, October 1976.)

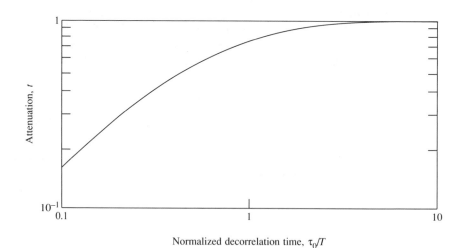

FIGURE 11.25 Attenuation vs. normalized decorrelation time. (From N. P. Shein, "An Approximate Formula for MFSK Symbol Error Probability in Fast Rayleigh Fading," *MILCOM Conf. Rec.*, September 1988. © 1988 IEEE.)

Consequently, the signal tone detector output will be attenuated. Moreover, crosstalk will be created in the other tone detectors orthogonal in frequency to the received signal.

Figures 11.25 and 11.26 illustrate the attenuation and crosstalk effects on 8-ary FSK due to fast fading. Figure 11.25 illustrates the performance of 8-ary FSK as functions of symbol SNR and the τ_0/T ratio (Shein, 1988). Orthogonal tone spacing is assumed here. If the required symbol error rate is 0.01 (an uncoded bit error rate of 0.005), fast fading is no longer very significant if τ_0/T is above 8. Irreducible error rates do exist; their levels depend on τ_0/T. The fast fading effect will be significant until τ_0/T exceeds 50 if a 10^{-3} symbol error rate is required. In fast fading, 8-ary FSK performance can be improved by two easily implementable signaling techniques: (1) chipping, or multiple-chip (noncoherent), combining; and (2) increased tone spacing.

As observed from Fig. 11.26, the crosstalk is the largest at the tone detector at one orthogonal spacing from the signal (the $K = 2$ case), with a maximum of 20 percent crosstalk at short τ_0's. The tone detector at twice the orthogonal spacing has a maximum crosstalk of only 10 percent. Thus by increasing tone spacing to twice orthogonal spacing, a gain of approximately 3 dB can be obtained over the performance with orthogonal spacing.

N-chip combining reduces the FSK symbol time by a factor of N. For example, consider a 20-kbps link with 8-ary FSK, rate-(1/3) encoding. At a τ_0 of 71 μs, the τ_0/T ratio is 1.42, which is below the operable τ_0/T of 3.5 for 8-ary FSK with orthogonal tone spacing. With 4-chip combining, the effective τ_0/T ratio is 5.6. At this τ_0/T ratio, the effect of fast fading is mostly mitigated (see Figs. 11.24 and 11.27). Signaling with twice-orthogonal spacing further improves 8-ary FSK performance in fast fading.

The currently favored waveform for the SDS low-data-rate links assumes 8-ary FSK with 4-chip combining and twice orthogonal tone spacing. With this modulation format the effect of fast scintillation can be mostly mitigated, resulting in the performance close to that achievable in the slow fading limit.

M-ary FSK Performance in Slow Rayleigh Fading

With sufficient chipping the MFSK symbol duration can be made short relative to the fade durations. In this case the performance of MFSK in scintillation can be modeled as slow fading.

The performance of MFSK in slow Rician fading has been analyzed by Lindsey (1964). The mean E_b/N_0 in MFSK is related to the mean symbol SNR, r_0, as follows:

$$r_0 = \left(\frac{E_b}{N_0}\right)\frac{\log_2(M)}{N} \tag{11.12}$$

In addition, r_0 is related to the symbol SNR due to the specular component only (r_S), and the symbol SNR due to the random component only (r_R), as follows:

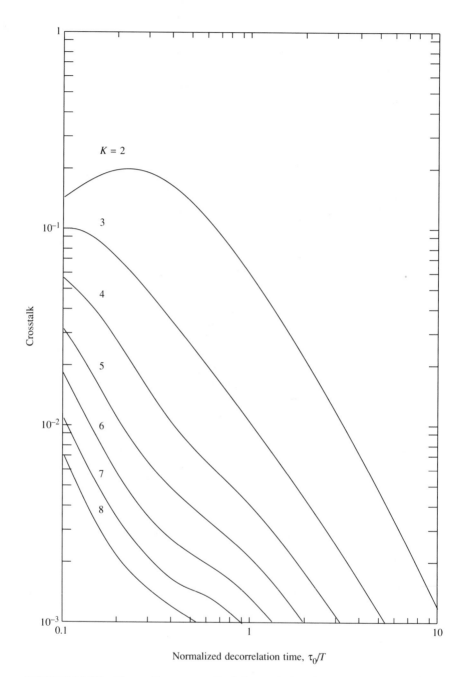

Normalized decorrelation time, τ_0/T

FIGURE 11.26   Crosstalk vs. normalized decorrelation time. (From N. P. Shein, "An Approximate Formula for MFSK Symbol Error Probability in Fast Rayleigh Fading," *MILCOM Conf. Rec.*, September 1988. © 1988 IEEE.)

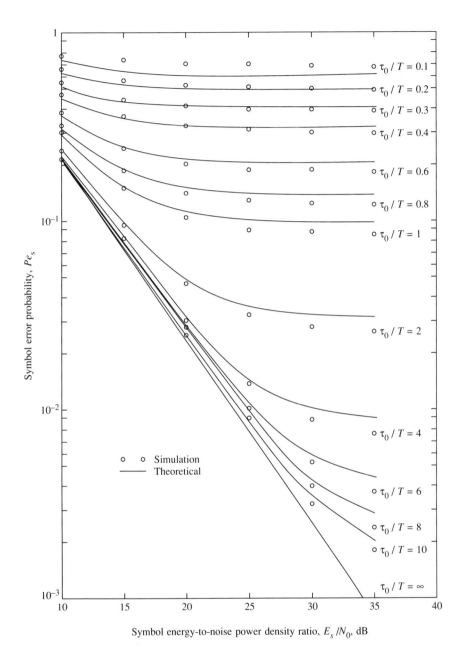

FIGURE 11.27 Symbol error probability in fast fading for $M = 8$ (From N. P. Shein, "An Approximate Formula for MFSK Symbol Error Probability in Fast Rayleigh Fading," *MILCOM Conf. Rec.*, September 1988. © 1988 IEEE.)

$$r_0 = r_S + r_R = r_R(1 + \beta) \tag{11.13}$$

The probability of symbol error, in slow Rician fading, for MFSK with N-chip combining is given by

$$
P_e = 1 - \left[\frac{1}{N\beta\left(\dfrac{r_0}{r_0 + 1 + \beta}\right)} \right]^{(N-1)/2}
$$
$$
\cdot \exp\left(-N\beta \frac{r_0}{r_0 + 1 + \beta} \right) \int_0^\infty x^{(N-1)/2} e^{-x} I_{N-1}\left(\sqrt{4x \frac{N\beta r_0}{r_0 + 1 + \beta}} \right)
$$
$$
\cdot \left\{ 1 - \exp\left[-x\left(1 + \frac{r_0}{1+\beta}\right) \right] \sum_{k=0}^{N-1} \frac{\left(1 + \dfrac{r_0}{1+\beta}x\right)^k}{k!} \right\}^{M-1} dx \tag{11.14}
$$

When there is no specular component (i.e., when $\beta = 0$), Eq. (11.14) is reduced to

$$
P_e = 1 - \int_0^\infty \frac{x^{N-1} e^{-x}}{(N-1)!} \left\{ 1 - \exp[-x(1+r_0)] \sum_{k=0}^{N-1} \frac{[(1+r_0)x]^k}{k!} \right\}^{M-1} dx \tag{11.15}
$$

which is the result for slow Rayleigh fading.

When there is no chip combining (that is, $N = 1$), the above Rician result becomes

$$
P_e = \sum_{k=1}^{M-1} (-1)^{k+1} \binom{M-1}{k} \frac{1+\beta}{kr_0 + (1+\beta)(1+k)}
$$
$$
\cdot \exp\left(-\beta \frac{kr_0}{kr_0 + (1+k)(1+\beta)} \right) \tag{11.16}
$$

The Rayleigh result for $N = 1$ can be obtained from Eq. (11.16) by letting $\beta = 0$.

The probability of bit error is related to the probability of symbol error, for MFSK, through the following relationship:

$$
P_b = \frac{P_e \times (M/2)}{M - 1} \tag{11.17}
$$

Figure 11.28 shows the performance of an M-ary FSK without chip combining (that is, $N = 1$) in both the Rayleigh fading channel and the AWGN-only channel, and for $M = 2, 4, 8,$ and 16. It illustrates that MFSK, like DPSK, also suffers significantly in slow Rayleigh fading. The exponential dependence of symbol error rate on SNR in the AWGN channel becomes a linear dependence.

Figure 11.29 shows the symbol error rate of 8-ary FSK with N-chip combining in slow Rayleigh fading, for $N = 1, 2, 4,$ and 8. With 4- or 8-chip combining, the fading performance is significantly improved.

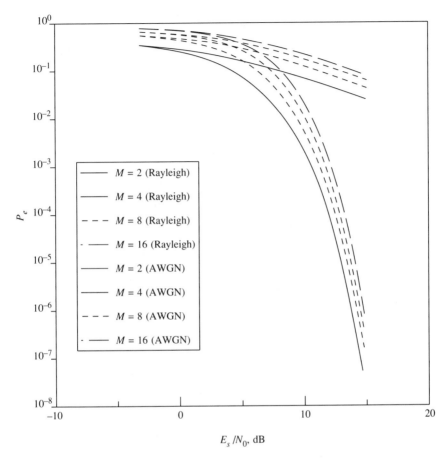

FIGURE 11.28   Theoretical MFSK performance in AWGN and slow Rayleigh fading channels.

Figure 11.30 shows the bit error probability for 8-ary FSK with N-chip combining ($N = 1, 2, 4, 8, 16, 32,$ and 64) at several values of E_b/N_0. When N is too small, there is not sufficient chip diversity and bit-error probability suffers. As N increases, the noncoherent combining loss becomes large, which also causes more bit errors. For low E_b/N_0's (for example, at 5 dB, which is typically the case for coded links) the uncoded bit-error rate is relatively insensitive to N. At E_b/N_0 values of 11, 13, and 15 dB, the optimal combining levels are shown to be $N = 8, 16,$ and 32, respectively.

Coding and Interleaving for SDS 8-FSK Links

In addition to chip combining, coding is also applied to give further diversity improvement on these low-data-rate links to mitigate the effects of slow fading.

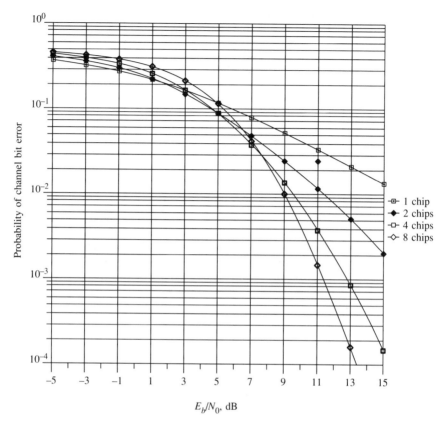

FIGURE 11.29   Typical 8-ary FSK performance in Rayleigh fading.

Like the DPSK links, the $R = 1/3$, $K = 7$ convolutional code with 3-bit soft decision decoding is the current preferred coding candidate for the low-data-rate 8-ary FSK links.

The links from a CV or from an SSTS to the interceptors are one-to-many TDMA links, which are somewhat different from the one-to-one sensor links described earlier. The TDMA packets (approximately 1300 bits in length) are intended for different interceptors in various antenna beam positions. Each interceptor will not be able to receive all packets that are transmitted in a TDMA sequence. Thus coding and interleaving must be applied to individual packets (or individual interceptors) separately.

Block coding on each packet can satisfy this TDMA requirement. On the other hand, either tail-biting (Ma and Wolf, 1986), or direct truncation, or zero tail (i.e., flushing) can be applied to convert convolutional codes to block codes. The disadvantage of direct truncation is that there is little, if any, error protection afforded to the last information digits passed through the decoder. Zero tail requires appending a number of zeros to the input bits, equal to the code's constraint length. This practice introduces a rate (or efficiency) loss.

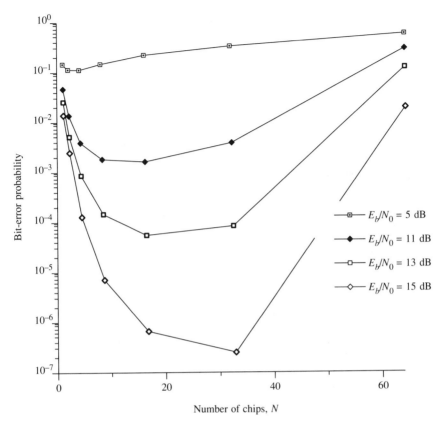

Bit-error probability

Number of chips, N

$E_b/N_0 = 5$ dB
$E_b/N_0 = 11$ dB
$E_b/N_0 = 13$ dB
$E_b/N_0 = 15$ dB

FIGURE 11.30   Theoretical 8-ary FSK performance with N-chip combining (ideal interleaver) in a slow Rayleigh fading channel.

Tail-biting does not have these disadvantages, but it introduces a slight increase in decoder complexity. The current preferred choice is tail-biting.

For this type of TDMA applications the hybrid interleaver (Olsen, 1989) discussed in the preceding subsection is a convenient means for providing time interspersing between separate packets intended for different destinations. The interleaver output bits are separated into sub-blocks of packets and are convolutionally interleaved. Figure 11.31 shows a block diagram of the link transmitting and receiving functions.

Two levels of interleaving provides the time diversity required to mitigate slow fading: bit interleaving and chip interleaving. The first hybrid interleaver performs interleaving on the encoded bits from the convolutional encoder using tail-biting. The interleaved coded symbols are grouped into 8-ary (3-bit) symbols, symbol repeated, and input to the second hybrid interleaver for chip interleaving. Choosing the parameter N_3 to be a multiple of 3 (e.g., 48) in the second hybrid interleaver allows an integer number of 8-ary chips to be (block) interleaved during each input cycle of the hybrid interleaver. This also allows

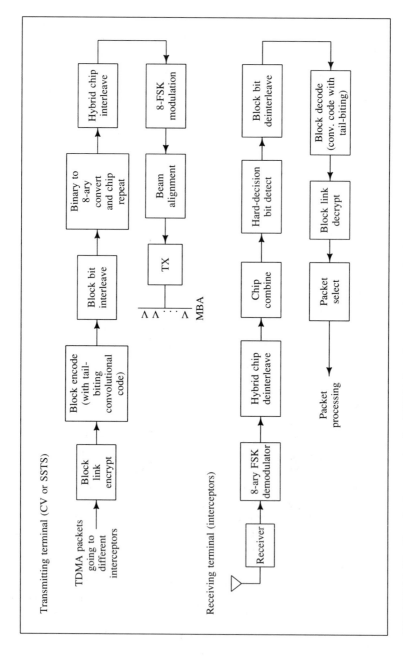

FIGURE 11.31 Terminal processing for TDMA low-data-rate 8-ary FSK links.

time interleaving between sub-blocks of chips from packets, with each sub-block intended for one beam position. Each block will be $N_3/3$ chips in duration.

Performance of this waveform depends on the exact spans of the bit and chip interleavers. Figures 11.32 and 11.33 illustrate typical performances for different T_{span}/τ_0 ratios and for different E_b/N_0, for a packet size of 704 bits. Similar to DPSK, the effects of slowly varying scintillation can be mostly mitigated by interleaving (bit and chip), time interspersing among packets, and by error-correction coding on these low-data-rate FSK links.

With 4-chip combining and twice-orthogonal tone spacing to mitigate the rapid fading effects, and with interleaving and coding to mitigate slow fading, this waveform is considered as the current preferred candidate for the links from SSTS and CV to the interceptors, as well as the links from SSTS to CV.

In the interceptor case the specific packets intended for the receiving interceptor will be selected from the stream of TDMA packets by the inter-

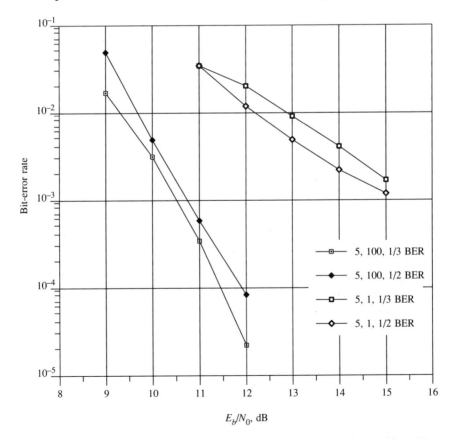

FIGURE 11.32   BER for 8-ary FSK in slow Rayleigh fading (τ_0/T_{span}) with 4-chip combining, $5\tau_0$ chip interleaving, 1 and $100\tau_0$ bit interleaving, and coding rates of 1/2 and 1/3.

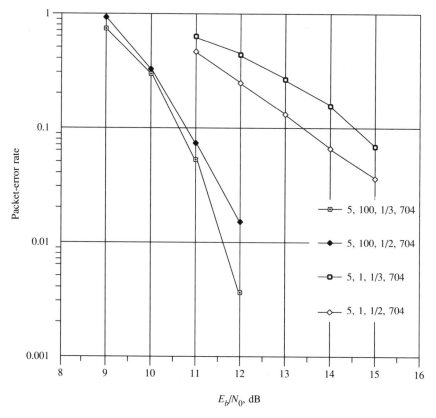

FIGURE 11.33 Performance of 704-bit PER of 8-ary FSK in slow Rayleigh fading ($\tau_0/T_{span} = 50$) with 4-chip combining, $5\tau_0$ chip interleaving, 1 and $100\tau_0$ bit interleaving, and code rates of 1/2 and 1/3.

ceptor itself by checking the packet contents. Because the signals transmitted to the intercept vehicles are beam-switched, the receivers do not necessarily receive a continuous signal. Accordingly, in order to maintain chip synchronization, they must rely on an on-board clock that is kept synchronized by patterns embedded in the transmitted signal at appropriate positions for each interceptor.

In the SSTS-to-CV link all packets are intended for the same receiver (either the CV or the SSTS). These are continuous and one-to-one links. However, the same waveform described here can be applied to either application, with identical performances in nuclear scintillation.

11.5 LINK DESIGN

The SDS link designer is faced with a multiplicity of link conditions that must be accounted for in the waveform design. It is necessary to design for jamming, nuclear-enhanced scintillation, or both, and for a variety of jammer types and fading intensities. These path disturbances can be partially mitigated with a combination of spread spectrum techniques such as frequency hopping, and with various coding and interleaving approaches.

We identified several nuclear-enhanced scintillation effects that must be considered. For each carrier frequency of interest they include:

1. Rayleigh fading

2. Antenna temperature

3. Path absorption

4. Antenna angle-of-arrival losses

5. Correlation bandwidth relative to signal bandwidth

The preceding sections addressed the first of these for both high and low data rates, and for rapid to mild scintillation intensities. Uncoded signal structures operating in a Rayleigh fading environment perform extremely poorly relative to normal AWGN channels. In some cases a few tens of dB more E_b/N_0 is required to obtain the same BER or PER performance. Interleaving/deinterleaving and coding produces striking improvements in required E_b/N_0 performance. Unfortunately, these techniques introduce time delays and/or require a large amount of on-board memory and processing and may not be useful for some links.

What are the fallback options for the links for which total time delay must be minimized? The options include, for instance, designing for jamming but not for scintillation, since longer interleavers required for the scintillation would introduce relatively large time delays for each hop. This approach provides a measure of error correction in jamming that is also useful in a scintillating environment. When scintillation durations, τ_0, are larger, the performance will degrade as described in the following paragraphs, and this degradation must be considered when designing the system.

Link Performance with Variation in Correlation Times, τ_0

Coding with interleaving is a major contributor to improvements in link performance under Rayleigh fading conditions. Unfortunately, coding and interleavers are hardware-specific, so a specific interleaver span must be selected for each SDS link. Scintillation intensity, however, will range from severe to very mild. Once interleaver span is fixed, changing scintillation conditions will cause the system to operate through a large T_{span}/τ_0 range. The link design must be balanced for the complete range of conditions.

As scintillation intensity varies, the same interleaver span can correspond to interleaving depths of $100\tau_0$ for relatively rapid fading where BER/PER performance is relatively good, to $10\tau_0$ for a relatively moderate fading rate where BER/PER is poorer, and to τ_0 for a slow fading rate relative to the interleaver depth where performance is quite poor. (Fast and slow fading limits are terms that are used to characterize link performance in the presence of

Rayleigh fading. These are relative terms, and may be used in two contexts. We have considered fading to be slow if propagation parameters remain reasonably unchanged over the duration of a symbol interval. In the context of interleaver depth, fading is considered relatively rapid if many fades will occur in a time interval corresponding to the interleaver span. Thus, fading may be slow relative to a symbol interval and also rapid relative to the interleaver span. Here fading rates are considered in terms of interleaver spans rather than bit intervals.) In these terms, best performance is achieved for rapid fading (providing certain other requirements relating to symbols/correlation time are still met). Figures 11.17 and 11.18 and Table 11.7 provide relative E_b/N_0 performance for these conditions.

For the multiburst case, antenna temperature can be as high as 4000 K or more (at 60 GHz) when τ_0 is 100 μs. When τ_0 increases to 1 ms, the antenna temperature decreases to 50 K. This finding suggests that a range of E_b/N_0's be investigated. The very high antenna temperature occurs when scintillation is very intense. The very low antenna temperature applies to the longer values of τ_0, for which scintillation intensity is less severe. Similarly, path absorption and antenna angle-of-arrival losses vary with total electron content, and hence correlation time τ_0. Fortunately, when these latter effects are most pronounced, coding with interleaving is most effective. Similarly, as correlation time increases so that interleaving becomes less effective, the other path effects also diminish and their reduction tends to form offsetting effects.

Consider the combined effect: The very high antenna temperature (4000 K) can degrade the performance of a system with a 1500-K system temperature by 5.6 dB. However, this degradation is for very short values of τ_0. When the correlation time increases, more E_b/N_0 is required because the interleaver is less effective, but the antenna temperature drops. The data in Table 11.10 provides an example which compares these cases for 60 GHz, and for a candidate interleaving depth of 10 ms.

TABLE 11.10 Comparative Link Performance for Varying Link Conditions*

| Correlation Time τ_0 (ms): | 0.10 | 1.0 | 2 | 10 | 20 | 100 |
Interleaving depth:	$100\tau_0$	$10\tau_0$	$5\tau_0$	τ_0	$0.5\tau_0$	$0.1\tau_0$
Required E_b/N_0 (dB) with 0.1 PER	8.3	10.3	12.8	18.3	11.2	10.3
Antenna temperature effect (dB)	5.6	0.2	—	—	—	—
Path absorption	1.1	0.3	—	—	—	—
Pointing loss	2.0†	1.0	—	—	—	—
Cumulative effect	16.0	11.8	12.8	18.3	11.2	10.3

* Interleaver span = 10 ms, $\tau_{0_{ray}}$ = 20 ms.
† From 0.1 dB to 1 to 2 dB, depending upon antenna diameters and RF values.

The cumulative effect shown on the bottom line of Table 11.10 is indicative of how required EIRP varies as a function of τ_0. It is high for very small τ_0, decreases for longer values of τ_0, and then increases again for even longer values of τ_0 as coding gain with interleaving becomes less effective. Ultimately, as scintillation intensity decreases, the multipath effect causing scintillation diminishes and the received signal changes from a Rayleigh distribution to one having a specular component. For longer values of τ_0, performance improves markedly. The two worst conditions occur for very short τ_0 and for a value of τ_0 corresponding to the time at which the Rayleigh/Rician transition occurs.

Clearly, the combined effect of changes in coding gain with interleaving, sky temperature, path absorption, and antenna angle-of-arrival losses tend to offset each other as scintillation intensity becomes less severe (as τ_0 increases).

If possible, interleaver spans should be designed with a view to balancing the link design so that the EIRP required for very intense losses (short τ_0) is no greater than for the other extreme (long τ_0). However, if the ideal interleaver span introduces too much path delay for packets that must pass through many hops, the design interleaver span may be shortened and the performance difference made up with additional EIRP.

The overall design depends upon specific design conditions and must ultimately be based upon the nuclear-burst scenarios and path geometries as they drive propagation conditions for the specific frequency of interest.

REFERENCES

Ananda, A., and Srinivasan, B., "An Extensive Bibliography on Computer Networks," *Comput. Communi. Rev.*, Special Interest Group on Data Communications, January–April 1984.

Bogusch, R. L., "Ionospheric Scintillation Effects on Satellite Communications," Report MRC-N-139, Mission Research Corporation, June 1974.

Bogusch, R. L., et al., "Effects of Propagation Disturbances on a Specific Satellite Communication Link, Vol. I," Report AFWL-TR-25, Mission Research Corporation, October 1976.

Cerf, V., and Cain, E., "The DOD Internet Architectural Model," *Comput. Networks*, October 1983.

Dana, R. A., "Propagation of RF Signals Through Structured Ionization, Theory and Antenna Aperture Effect Applications," Report DNA-TR-86-158, prepared for the Defense Nuclear Agency, Mission Research Corporation, p. 64, May 1986.

Day, J., and Zimmerman, H., "The OSI Reference Model," *Proc. IEEE*, vol. 71, December 1983.

Knepp, D. L., "Aperture Antenna Effects After Propagation Through Strongly Disturbed Media," *IEEE Trans. Antennas Propag.*, vol. AP-33, October 1985.

Lindsey, W. C., "Error Probabilities for Rician Fading Multichannel Reception of Binary and N-ary Signals", *IEEE Trans. Inform. Theory*, vol. IT-10, October 1964.

Ma, H. H., and Wolf, J. K., "On Tail Biting Convolutional Codes," *IEEE Trans. Commun. Technol.*, vol. COM-34, February 1986.

Martin Marietta Corporation, "NWE Enhanced IR Backgrounds and Electromagnetic Propagation," Special Studies, Task 7 Survivability Final Report, September 1988.

Middlestead, R. W., et al., "Satellite Crosslink Communications Vulnerability in a Nuclear Environment", *IEEE J. Selected Areas Commun.*, vol. SAC-5, February 1987.

Olsen, D. P., "Proposed Generic Communication Waveform Standards," Viewgraph Presentation, Aerospace Corporation, March 1989.

Olsen, D. P., King, M. A., and Yamada, A. H., "The Hybrid Interleaver," Internal Memo, Aerospace Corporation, July 1988.

Rauch-Hindrin, W., "Communications Standards: OSI Is Not a Paper Tiger," *Syst. Software*, March 1985.

Shein, N. P., "An Approximate Formula for MFSK Symbol Error Probability in Fast Rayleigh Fading," *MILCOM Conf. Rec.*, September 1988.

Siegel, R. S., "Arrival Angle Loss due to Nuclear Induced Scintillation," TRW Interoffice Correspondence to S. B. Franklin, July 22, 1987.

Tannenbaum, A., *Computer Networks*, Englewood Cliffs, NJ: Prentice-Hall, 1981.

Wittwer, L. A., "Nuclear Environment Issues for Exotic Multi-Bursts," Viewgraph Presentation, Defense Nuclear Agency, May 1987.

Wittwer, L. A., "High Altitude Nuclear Effects," Viewgraph Presentation, Defense Nuclear Agency, October 1988.

Index